AF412915

The Wiley Polymer Networks Group Review Series

Volume Two

The Wiley Polymer Networks Group Review Series

Volume Two

Synthetic versus Biological Networks

Edited by

B. T. Stokke and A. Elgsaeter
The Norwegian University of Science and Technology,
Trondheim, Norway

JOHN WILEY & SONS, LTD
Chichester • New York • Weinheim • Brisbane • Toronto • Singapore

Other Wiley Editorial Offices

John Wiley & Sons, Inc., 605 Third Avenue,
New York, NY 10158-0012, USA

WILEY-VCH Verlag GmbH, Pappelallee 3,
D-69469 Weinheim, Germany

Jacaranda Wiley Ltd, 33 Park Road, Milton,
Queensland 4064, Australia

John Wiley & Sons (Asia) Pte Ltd, Clementi Loop #02-01,
Jin Xing Distripark, Singapore 129809

John Wiley & Sons (Canada) Ltd, 22 Worcester Road,
Rexdale, Ontario M9W 1L1, Canada

Library of Congress Cataloging-in-Publication Data

Synthetic versus biological networks / edited by B. T. Stokke and A.
 Elgsaeter.
        p.   cm. — (The Wiley Polymer Networks Group review series ;
 v. 2)
    Includes bibliographical references and index.
    ISBN 0-471-98713-1 (alk. paper)
    1. Polymer networks.   I. Stokke, B. T.   II. Elgsaeter, A.
 III. Series.
 QD382.P67S97   1999
 547'.7 — dc21   99-26510
                                                              CIP

British Library Cataloguing in Publication Data

A catalogue record for this book is available from the British Library

ISBN 0 471 98713 1

Typeset in 10/12pt Times Roman by Laser Words, Madras, India
Printed and bound in Great Britain by Antony Rowe, Chippenham, Wilts
This book is printed on acid-free paper responsibly manufactured from sustainable foresty,
in which at least two trees are planted for each one used for paper production

DEDICATION

REIMUND STADLER

Completely unexpectedly, after falling seriously ill for a few weeks, Professor Reimund Erich Stadler died on June 14, 1998 of stomach cancer. Reimund Stadler was born in 1956 in a small town on the foothills of the Black Forest near the Swiss border, towards Schaffhausen. The remoteness of this area and its scenic environment undoubtedly formed his character and independence of thinking.

Reimund Stadler received his training in chemistry at the University of Freiburg. His outstanding intelligence soon became evident and was acknowledged with a grant from the Studienstiftung des Deutschen Volkes until the end of his PhD work.

Reimund's scientific contributions started with his PhD work, when he studied the viscoelastic behaviour and stretch-induced crystallization of thermoplastic elastomers. An attention provoking milestone was set by his work on the thermoreversible gelation of functionalized polybutadienes, in which cross-linking is brought about via hydrogen bonding between functional groups. At that time, he made significant contributions to the understanding of thermoreversible networks and to the Polymer Networks Group. His work was rewarded by a prize from Gerhard Hess Program, a scheme for the support of highly qualified young scientists. The award included a grant for a few years that enabled him to install equipment for further research.

When he was appointed to the Chair in Organic Chemistry at the University of Mainz, new possibilities arose that gave a completely new direction to his work. He developed further the techniques of living anionic polymerization for the controlled synthesis of block copolymers, and succeeded in the preparation of triblock copolymers of the ABC type, with chemically different blocks A, B and C. On the basis of elementary, but simple, considerations of incompatibility between these blocks, he was able to predict new morphologies that he also realized experimentally.

In 1996, Reimund Stadler moved to the University of Bayreuth, where he was appointed to the Chair in Organic Chemistry. With exceptional energy, he devoted his time to initiating new interdisciplinary cooperation under the programme of the *Sonderforschungsbereich 481* (special research area) of the Deutsche Forschungsgemeinschaft, with the title *Complex macromolecular and hybrid systems under the influence of internal and external fields*. The programme had started when he suddenly became ill and his life ended.

A discussion with Reimund was always fruitful and stimulating. Sometimes his thinking was like a never-stopping motor. He was of cheerful character, was open with everybody and willing to share good ideas. This enabled him to make many friends from all the countries with which he came into contact. One country has to be mentioned particularly, namely Brazil. He stayed twice for long periods at the University of Porte Alegre, where he was active, mainly in education and in organizing improved cooperation between universities in Germany and those in Brazil and Venezuela. He obviously loved the people of Brazil and Venezuela, whose attitudes had a great influence on him and, in the last four years, resulted in a new quality to his private life.

We all, friends and colleagues, mourn for one of our most inspired scientists. Reimund has taught a number of excellent co-workers who eagerly and respectfully continue the lines of his uncompleted work.

Freiburg, March 1999   Walther Burchard

Contents

Part 1: Network Formation

Part 2: Network Characterization

Part 3: Polymer Networks and Precursor Architectures

Part 4: Biopolymer Networks and Gels and their Models

Part 5: Biomedical Applications of Polymer Networks

Part 6: Polymer Networks in Restricted Geometries

List of Contributors

T. Aberle — Institute of Macromolecular Chemistry, University of Freiburg, D-79104 Freiburg, Germany

B. Achenbach — Universität-Gesamthochschule Essen, Institut für Umweltanalytik, Universitätsstrasse 5-7, D-45141 Essen, Germany

E. Adamczak — Polymer Research Group, Department of Dental Biomaterials Science, Karolinska Institute, Box 4064 S-141 04 Huddine (Stockholm), Sweden

J. Aerts — Akzo Nobel Central Research, Physical Chemistry and Mathematics Department, PO Box 9300, NL-6800 SB Arnhem, The Netherlands

V. Alexeev — Petersburg Nuclear Physics Institute, Gatchinal 188350, Russia

Y. Amaike — Narayama Laboratory, Nitta Industries Corporation, 6-5-6 Sakyo, Nara 631-0801, Japan

K. Anseth — Department of Chemical Engineering, University of Colorado, Boulder, Colorado 80309-0424, USA

G. Arcovito — Istituto di Fisica, Facoltà di Medicina e Chirurgia, Università Cattolica S.C., L.go F. Vito n.1-00168 Roma, Italy

O. Aune — SINTEF Applied Chemistry, N-7034 Trondheim, Norway

M.A.V. Axelos — INRA-Laboratoire de Physico-Chimie des Macromolécules, Rue de la Géraudière, BP 71627, 44316 Nantes cedex 03, France

F.A. Bassi — Istituto di Fisica, Facoltà di Medicina e Chirurgia, Università Cattolica S.C., L.go F. Vito n.1-00168 Roma, Italy

J. Bauer	Fraunhofer Institute of Reliability and Microintegration, Branch Lab Polymeric Materials and Composites, Kantstrasse 55, D-14513 Teltow, Germany
M. Bauer	Fraunhofer Institute of Reliability and Microintegration, Branch Lab Polymeric Materials and Composites, Kantstrasse 55, D-14513 Teltow, Germany
A. Berge	Department of Industrial Chemistry, Norwegian University of Science and Technology, NTNU, N-7034 Trondheim, Norway
S. Bernocco	Gruppo di Biostrutture, Istituto Nazionale per la Ricerca sul Cancro (IST), Centro Biotechnologie Avanzate (CBA), L.go R. Benzi 10, 16132 Genova, Italy
C. Bica	Laboratorio de Instrumentacão e Dinâmica Molecular, Instituto de Quimica, Universidade Federal do Rio Grande do Sul, Porto Alegre, Rio Grande do Sul, Brazil
L. Bokobza	ESPCI, Lab. de Physico-Chimie Structurale et Macromoléculaire, 10 rue Vauquelin, 75231 Paris Cedex 5, France
C.N. Bowman	Department of Chemical Engineering, University of Colorado, Boulder, Colorado 80309-0424, USA
S. Bryant	Department of Chemical Engineering, University of Colorado, Boulder, Colorado 80309, USA
T. Budtova	Inst. of Macromolecular Compounds, Bolshoi prosp. 31, St Petersburg 199004, Russia
W. Burchard	Institute of Macromolecular Chemistry, University of Freiburg, D-79104 Freiburg, Germany
A. Buyanov	Inst. of Macromolecular Compounds, Bolshoi prosp. 31, St Petersburg 199004, Russia
I. Capron	Laboratoire de Physique et Mécanique des Milieux Hétérogènes, CNRS UMR 7636 ESPCI, 10 Rue Vauquelin 75231 Paris Cedex 5, France
R. Casado	Polymer Science and Technology Group, Manchester Materials Science Centre, University of Manchester & UMIST, Grosvenor Street, Manchester, M1 7HS, UK
C. Chevillard	INRA-Laboratoire de Physico-Chimie des Macromolécules, Rue da la Géraudière, BP 71627 44316 Nantes cedex 03, France
D. Christova	Bulgarian Academy of Sciences, Inst. of Polymers, 113 Sofia, Bulgaria

S. Costeux	Laboratoire de Physique et Mécanique des Milieux Hétérogènes, CNRS UMR 7636 ESPCI, 10 Rue Vauquelin 75231 Paris Cedex 5, France
M. De Spirito	Istituto di Fisica, Facoltà di Medicina e Chirurgia, Università Cattolica S.C., L.go F. Vito n.1-00168 Roma, Italy
D. Dixon	British Aerospace, Sowerby Research Centre, PO Box 5, Filton, Bristol BS12 7QW, UK
M. Djabourov	Laboratoire de Physique et Mécanique des Milieux Hétérogènes, CNRS UMR 7636 ESPCI, 10 Rue Vauquelin 75231 Paris Cedex 5, France
F.E. Du Prez	University of Ghent, Polymer Chemistry Division, Krijgslaan 281 S4, B-9000 Ghent, Belgium
E.A. Dzhavadyan	Inst. of Problems of Chemical Physics, Russian Academy of Sciences, Chernogolovka, 142432 Moscow Region, Russia
J. Elisseeff	Harvard–MIT Division of Health Sciences and Technology and MIT Department of Chemical Engineering, Cambridge, Massachuseetts 02139, USA
J. Elliot	Department of Chemical Engineering, University of Colorado, Boulder, Colorado 80309-0424, USA
G. Erdödi	Material Science of Polymers Group, Institute of Chemistry, Chemical Research Center, Hungarian Academy of Sciences, H-1525 Budapest, Pusztaszeri u. 59-67, PO Box 17, Hungary
G. Evmenenko	Petersburg Nuclear Physics Institute, Gatchinal 188350, Russia
F. Ferri	Istituto di Science Mathematiche, Fisiche e Chimiche, Università degli Studi di Milano a Como, via Lucini, 3-22100 Como, Italy
S. Frenkel	Inst. of Macromolecular Compounds, Bolshoi prosp. 31, St Petersburg 199004, Russia
E. Frey	Physik-Department der Technischen Universität München, Institut für Theoretische Physik, James-Franck-Straße, D-85747 Garching, Germany
H. Galina	Department of Industrial Materials Chemistry, Faculty of Chemistry, Rzeszów University of Technology, 35-959 Rzeszów, Poland
E. Geissler	Laboratoire de Spectrométrie Physique, CNRS UMR 5588, Université J. Fourier de Grenoble, BP 87, 38402 St Martin d'Hères Cedes, France

P.M. Gilsenan   Biopolymers Group, Division of Life Science, Kings
                    College London, Franklin-Wilkins Building,
                    150 Stamford Street, London SE1 8WA, UK

E.J. Goethals   University of Ghent, Polymer Chemistry Division,
                    Krijgslaan 281 S4, B-9000 Ghent, Belgium

M. Gottlieb   Department of Chemical Engineering, Ben-Gurion
                    University of the Negev, Beer-Sheva 84105, Israel

G. Groszek   Faculty of Chemistry, Rzeszów University of
                    Technology, 35-959 Rzeszów, Poland

S. Harris   British Aerospace, Sowerby Research Centre, PO Box 5,
                    Filton, Bristol BS12 7QW, United Kingdom

T. Hashimoto   Hashimoto Polymer Phasing Project, ERATO, JST, 15
                    Morimoto-cho, Shimogamo, Sakyo-ku, Kyoto 606-0805,
                    Japan, and Department of Polymer Chemistry, Graduate
                    School of Engineering, Kyoto University, Kyoto
                    606-8501, Japan

T. Hirai   Faculty of Textile Science and Technology, Shinshu
                    University, Tokida 3-15-1, Ueda 386-8567, Japan

Y. Hirokawa   Hashimoto Polymer Phasing Project, ERATO, JST, 15
                    Morimoto-cho, Shimogamo, Sakyo-ku, Kyoto 606-0805,
                    Japan

F. Horkay   General Electric Corporate Research and Development,
                    PO Box 8, Schenectady, N.Y. 12301 USA

M. Husmann   Universität-Gesamthochschule Essen, Institut für
                    Umweltanalytik, Universitätsstrasse 5-7, D-45141 Essen,
                    Germany

F. Ikkai   Department of Polymer Science and Engineering, Kyoto
                    Institute of Technology, Matsugasaki, Sakyo-ku, Kyoto
                    606-8585, Japan

C.E. Ioan   'P. Poni' Institute of Macromolecular Chemistry, 6600
                    Jassi, Romania

V.I. Irzhak   Inst. of Problems of Chemical Physics, Russian
                    Academy of Sciences, Chernogolovka, 142432 Moscow
                    Region, Russia

M.H. Islam   Faculty of Textile Science and Technology, Shinshu
                    University, Tokida 3-15-1, Ueda-Shi 386-8567, Japan

B. Iván   Material Science of Polymers Group, Institute of
                    Chemistry, Chemical Research Center of the Hungarian
                    Academy of Sciences, H-1525 Budapest, Pusztaszeri u
                    59-67, PO Box 17, Hungary

A. Janecska	Material Science of Polymers Group, Institute of Chemistry, Chemical Research Center of the Hungarian Academy of Sciences, H-1525 Budapest, Pusztaszeri u 57-67, PO Box 17, Hungary
M. Johansson	Department of Polymer Technology, Royal Institute of Technology, S-100 44 Stockholm, Sweden
N. Jun	Polymer Research Group, Department of Dental Biomaterials Science, Karolinska Institute, Box 4064 S-141 04 Huddine (Stockholm), Sweden
K. Kajiwara	Faculty of Engineering and Design, Kyoto Institute of Technology, Kyoto, Sakyo-ku, Matsugasaki 606-8585, Japan
M. Kané	Laboratoire de Physique et Mécanique des Milieux Hétérogènes, CNRS UMR 7636 ESPCI, 10 Rue Vauquelin 75231 Paris Cedex 5, France
E. Kato	Marine Engineering, Kobe University of Mercantile Marine, 1-1, 5, Fukae-Minami, Higashinada, Kobe 658, Japan
A.R. Khokhlov	Physics Department, Moscow State University, Moscow 117234, Russia
L. Kilaas	SINTEF Applied Chemistry, N-7034 Trondheim, Norway
C. Kloster	Laboratoire de Spectrométrie Physique, UMR 5588, B.P. 87, 38402 St Martin d'Hères Cedex, France, Laboratorio de Instrumentacão e Dinâmica molecular, Instituto de Quimica, Universidade Federal do Rio Grande do Sul, Porto Alegre, Rio Grande do Sul, Brazil
S. Kohjiya	Institute for Chemical Research, Kyoto University, Uji, Kyoto-fu 611, Japan
K. Kroy	Physik-Department der Technischen Universität München, Institut für Theoretische Physik, James-Franck-Straße, D-85747 Garching, Germany
R. Langer	Harvard–MIT Division of Health Sciences and Technology and MIT Department of Chemical Engineering, Cambridge, Massachusetts 02139, USA
C. Lartigue	Laboratoire de Spectrométrie Physique, UMR 5588, B.P. 87, 38402 St Martin d'Hères Cedex, France
J.B. Lechowicz	Department of Industrial Materials Chemistry, Faculty of Chemistry, Rzeszów University of Technology, 35-959 Rzeszów, Poland

J. Lefebvre	INRA-Laboratoire de Physique-Chimie des Macromolécules, Rue de la Géraudière, BP 71627 44316 Nantes cedex 03, France
L.-Å. Lindén	Department of Dental Materials Science, Umeå University, S-901 97 Umeå, Sweden
P.A. Lovell	Polymer Science and Technology Group, Manchester Materials Science Centre, University of Manchester & UMIST, Grosvenor Street, Manchester, M1 7HS, UK
G.B. McKenna	Polymers Division, National Institute of Standards and Technology, Gaithersburg, MD 20899, USA
C. Macron	ESPCI, Lab. de Physico-Chimie Structurale et Macromoléculaire, 10 rue Vauquelin, 75231 Paris Cedex 5, France
P. Martens	Department of Chemical Engineering, University of Colorado, Boulder, Colorado 80309, USA
K. Matsuzaka	Hashimoto Polymer Phasing Project, ERATO, JST, 15 Morimoto-cho, Shimogamo, Sakyo-ku, Kyoto 606-0805, Japan
M. Mimura	Faculty of Engineering and Design, Kyoto Institute of Technology, Kyoto, Sakyo-ku, Matsugasaki 606-8585, Japan
H. Morikawa	Faculty of Textile Science and Technology, Shinshu University, Tokida 3-15-1, Ueda-Shi 386-8567, Japan
T. Murakami	Yuge National College of Maritime Technology, Ehime 794-25, Japan
T. Norisuye	Department of Polymer Science and Engineering, Kyoto Institute of Technology, Matsugasaki, Sakyo-ku, Kyoto 606-8585, Japan
T. Okamoto	Hashimoto Polymer Phasing Project, ERATO, JST, 15 Morimoto-cho, Shimogamo, Sakyo-ku, Kyoto 606-0805, Japan
E. Paganini	CISE, PO Box 12081, 20134 Milano, Italy
O.E. Philippova	Physics Department, Moscow State University, Moscow 117234, Russia
J.F. Rabek	Polymer Research Group, Department of Dental Biomaterials Science, Karolinska Institute, Box 4064 S-141 04 Huddine (Stockholm), Sweden
M. Randolph	Department of Surgery, Massachusetts General Hospital, Boston, Massachusetts 02114, USA

H. Rehage	Universität-Gesamthochschule Essen, Institut für Umweltanalytik, Universitätsstrasse 5-7, D-45141 Essen, Germany
D. Renard	INRA-Laboratoire de Physico-Chimie des Macromolécules, Rue de la Géraudière, BP 71627 44316 Nantes cedex 03, France
M. Rocco	Gruppo di Biostrutture, Istituto Nazionale per la Ricerca sul Cancro (IST), Centro Biotechnologie Avanzate (CBA), L.go R. Benzi 10, 16132 Genova, Italy
C. Rochas	Laboratoire de Spectrométrie Physique, UMR 5588, B.P. 87, 38402 St Martin d'Héres Cedex, France
S.B. Ross-Murphy	Biopolymers Group, Division of Life Science, Kings College London, Franklin-Wilkins Building, 150 Stamford Street, London SE1 8WA, UK
B.A. Rozenberg	Inst. of Problems of Chemical Physics, Russian Academy of Sciences, Chernogolovka, 142432 Moscow Region, Russia
D. Samios	Laboratorio de Instrumentacão e Dinâmica molecular, Instituto de Quimica, Universidade Federal do Rio Grande do Sul, Porto Alegre, Rio Grande do Sul, Brazil
R. Schmid	SINTEF Applied Chemistry, N-7034 Trondheim, Norway
M. Shibayama	Department of Polymer Science and Engineering, Kyoto Institute of Technology, Matsugasaki, Sakyo-ku, Kyoto 606-8585, Japan
G.M. Sigalov	Inst. of Problems of Chemical Physics, Russian Academy of Sciences, Chernogolovka, 142432 Moscow Region, Russia
M. Stading	Chalmers University of Technology/SIK, The Swedish Institute for Food and Biotechnology, PO Box 5401, S-402 29 Göteborg, Sweden
J.L. Stanford	Polymer Science and Technology Group, Manchester Materials Science Centre, University of Manchester & UMIST, Grosvenor Street, Manchester, M1 7HS, UK
P. Stenstad	SINTEF Applied Chemistry, N-7034 Trondheim, Norway
R.F.T. Stepto	Polymer Science and Technology Group, Manchester Materials Science Centre, University of Manchester & UMIST, Grosvenor Street, Manchester, M1 7HS, UK
M. Suzuki	Narayama Laboratory, Nitta Industries Corporation, 6-5-6 Sakyo, Nara 631-0801, Japan

N. Takahashi	Faculty of Textile Science and Technology, Shinshu University, Tokida 3-15-1, Ueda 386-8567, Japan
M. Takeda	Department of Polymer Science and Engineering, Kyoto Institute of Technology, Matsugasaki, Sakyo-ku, Kyoto 606-8585, Japan
D.J.R. Taylor	Polymer Science and Technology Group, Manchester Materials Science Centre, University of Manchester & UMIST, Grosvenor Street, Manchester, M1 7HS, UK
T. Ueda	Narayama Laboratory, Nitta Industries Corporation, 6-5-6 Sakyo, Nara 631-0801, Japan
J. Uhlig	Fraunhofer Institute of Reliability and Microintegration, Branch Lab Polymeric Materials and Composites, Kantstrasse 55, D-14513 Teltow, Germany
H. Urakawa	Faculty of Engineering and Design, Kyoto Institute of Technology, Kyoto, Sakyo-ku, Matsugasaki 606-8585, Japan
K. Urayama	Institute for Chemical Research, Kyoto University, Uji, Kyoto-fu 611, Japan
N. Watanabe	Faculty of Textile Science and Technology, Shinshu University, Tokida 3-15-1, Ueda 386-8567, Japan
M. Watanabe	Faculty of Textile Science and Technology, Shinshu University, Tokida 3-15-1, Ueda 386-8567, Japan
J. Wilhelm	Physik-Department der Technischen Universität München, Institut für Theoretische Physik, James-Franck-Straße, D-85747 Garching, Germany
A. Wrzyszczynski	Faculty of Chemical Technology and Engineering, University of Technology and Engineering, Seminaryjna 3 85-326 Bydgoszcz, Poland
K. Yokoyama	Institute for Chemical Research, Kyoto University, Uji, Kyoto-fu 611, Japan
R.J. Young	Polymer Science and Technology Group, Manchester Materials Science Centre, University of Manchester & UMIST, Grosvenor Street, Manchester, M1 7HS, UK
Y. Yuguchi	Faculty of Engineering and Design, Kyoto Institute of Technology, Kyoto, Sakyo-ku, Matsugasaki 606-8585, Japan
K.B. Zeldovich	Physics Department, Moscow State University, Moscow 117234, Russia
Z.-R. Zhang	Department of Chemical Engineering, Ben-Gurion University of the Negev, Beer-Sheva 84105, Israel

Series Preface

We are very pleased to introduce Volume Two of *The Wiley Polymer Networks Group Review Series*. Volume One, published in May 1997, was edited by Klaas te Nijenhuis and Willem Mijs and was entitled Chemical and Physical Networks—Formation and Control of Properties. It was based on papers presented at the 13th Polymer Networks Group International Conference, Polymer Networks 96, held in Doorn, The Netherlands. The present volume continues the series, comprising papers presented at Polymer Networks 98, the 14th Polymer Networks Group International Conference. The conference was held at the Norwegian University of Science and Technology (NTNU) in Trondheim, and organized, under the Chairmanship of Bjørn Torger Stokke, by representatives of the NTNU, the SINTEF Group and the Norwegian Pulp and Paper Research Institute.

The main theme of Polymer Networks 98 was Synthetic versus Biological Networks. Thus the emphases of Volume Two of *The Wiley Polymer Networks Group Review Series* are distinct from those of Volume One. In fact, the themes of Polymer Networks Group International Conferences are not fixed, but are at the behest of the Conference Chairmen and their Programme Committees. Thus, basing *The Wiley Polymer Networks Group Review Series* on papers presented at the conferences, ensures that the Series will present refereed articles at the forefront of research into polymer netwoks. We believe that, like Volume One, Volume Two contains many excellent and authoritative articles and makes a significant contribution to the literature on polymer networks.

It is planned that Volume Three will present articles based on the 15th Polymer Networks Group International Conference, Polymer Networks 2000, to be held in Kraków, Poland in 2000 under the Chairmanship of Henryk Galina. The theme of that conference will be Polymer Networks: Formation–Structure–Properties

Note: Any reader interested in learning more of the activities of the Polymer Networks Group is invited to contact Klaas te Nijenhuis, Secretary, Polymer Networks Group, Delft University of Technology, Faculty of Applied Sciences, Department of Polymer Science and Technology, Julianalaan 136, 2628 BL Delft, The Netherlands. Facsimile: +31 (0)15 2787415; e-mail k.tenijenhuis@stm.tudelft.nl.

and the topics will include Network Formation, the Topology, Structure and Heterogeneity of Networks, Static and Dynamic Properties of Gelling Systems and Networks, and the Phase Behaviour of Multiphase Networks.

We are confident that Volume Two will uphold the high standard set by Volume One and that Volume Three will continue the tradition into the future.

Walther Burchard	Robert Stepto
Freiburg	Manchester
Chairman	Managing Series Editor
Polymer Networks Group	*The Wiley Polymer*
	Networks Group
	Review Series

January, 1999

Preface

Polymer Networks — Synthetic versus Biological

Polymer Networks 98, the 14th international Polymer Networks Group Conference was arranged in Trondheim, Norway from 28 June to 3 July 1998. It was organized by representatives from the Norwegian University of Science and Technology (NTNU), the SINTEF Group, and the Norwegian Pulp and Paper Research Institute. The program highlighted the different ideas emerging from investigations into synthetic polymer networks as opposed to and in comparison with polymer networks of biological origins. There were 17 invited lectures, 36 contributed oral presentations and 75 posters covering diverse topics within the themes highlighted at the conference. The present volume contains contributions selected from the papers presented at the conference. These are divided into six sections: Network Formation, Network Characterization, Polymer Networks and Precursor Architectures, Biopolymer Networks and Gels, Biomedical Applications of Polymer Networks, and Polymer Networks in Restricted Geometries.

Ongoing continuing developments of insights into the chemistry and physics of formation, description, determination and application of polymer networks were well presented at the conference. Also, a general trend of going towards a more molecularly based science in the polymer networks field was evident in numerous contributions at the conference. This was highlighted in the session detailing the molecular architecture of the chemical units comprising polymer networks. It was reported that the presently conventional use of hydrophobically modified polymers to construct bulk polymer networks sensitive to the possible states of hydrophobe association now can be extended to crosslink surfactants directly. This highlights that functional monomers can be used not only to modify bulk properties of networks, but also to control the macroscopic shape of the networks being formed. The finding that nature provides us with similar rheological properties based on various ways of organizing the polymer network constituents, the influence of external fields on the build-up or degradation of networks, and the application of network descriptions of biological structures *in vivo* exemplifies the impact of the polymer network field within biology. Biomedical applications of polymer networks, whether synthetic or biological, that impose the additional constraint of compatability with the organism, illustrate one direction in which this multidisciplinary field is expanding.

The editors are indebted to the authors for reporting, within the scope of the Polymer Networks 98 conference, many new and exciting results. We hope that the present volume will serve as a continuing stimulus within this multidisciplinary research field to a broad range of researchers employed both in industry and academia.

B.T. Stokke and A. Elgsaeter
The Norwegian University of Science and Technology

PART 1
Network Formation

1

In Situ Studies on Gelation by Dynamic and Time-resolved Light Scattering Technique

MITSUHIRO SHIBAYAMA, TOMOHISA NORISUYE and MASANAO TAKEDA

Department of Polymer Science and Engineering, Kyoto Institute of Technology, Matsugasaki, Sakyo-ku, Kyoto 606-8585, Japan

ABSTRACT

A non-destructive method for determination of the gelation threshold has been proposed, which incorporates time-resolved light scattering and dynamic light scattering. The

---

Wiley Polymer Networks Group Review Series Vol. 2. Edited by B.T. Stokke and A. Elgsaeter
© 1999 John Wiley & Sons Ltd

gelation threshold was characterized by (1) an pulse-like rise in the scattered intensity during the course of polymerization/gelation, (2) a power-law behavior of the time intensity correlation function (ICF); $g^{(2)}(\tau) - 1$, where τ is the decay time, (3) appearance of a long tail in the characteristic decay time distribution function, and (4) reduction of the initial amplitude of ICF. N-isopropylacrylamide (NIPA) aqueous solutions with different monomer concentrations, C, were chosen as a gelling system to be studied. First of all, the scattered intensity at the scattering angle of $60°$ was monitored as a function of polymerization time, t, for NIPA monomer solutions with various Cs. A characteristic pulse-like intensity rise was observed around $t = t_{th} \approx 20$ min, irrespective of $C(> C_{th} \approx 100$ mM), where t_{th} and C_{th} were confirmed to be the time at the gelation threshold and the monomer concentration leading to the gelation threshold, respectively. This pulse-like intensity rise was followed by relatively low scattering for $t \gg t_{th}$. The ICF for the reaction-completed cross-linked poly-NIPA with $C \approx C_{th}$ exclusively showed the characteristic phenomena (2)–(4) described above. Furthermore, an appearance of speckle patterns was observed for $C \geqslant C_{th}$, which also indicates formation of infinite clusters, i.e. gels. On the other hand, the so-called gel mode scattering, where the ICF is well described with a single exponential function of τ, was recovered for $C \gg C_{th}$.

INTRODUCTION

Gelation, i.e., sol–gel transition, of polymers has been an attractive subject for more than a half century. Classical theories of gelation were developed on the basis of the Bethe lattice in the 1940s [1,2]. Then, de Gennes [3] and Stauffer [4] treated a gelation problem with the percolation theory in the 1970s. These theories of sol–gel transition and the efforts of a large number of experimental studies have led to a relatively clear understanding of the nature of cross-linked clusters at the gelation threshold, as recently reviewed by Adam and Lairez.[5] Scattering techniques, such as small-angle X-ray (SAXS), small-angle neutron scattering (SANS), and light scattering (LS), are particularly useful to characterize a gelling system and to examine the validity of percolation theory [6–9]. Rheological measurements have also been employed to study a gelation mechanism [10–12]. However, novel methods which allow in situ and non-destructive determination of the gelation point have been pursued in industry. This is due to the fact that the gelation or vulcanization is of particular importance in controlling the physical properties and chemistry of commercial products, such as cosmetic and foods. LS seems to be one of the most suitable means for such a demand, provided the gelling system is transparent. We demonstrate, in this paper, four methods which allow one to determine the gelation point. Poly(N-isopropylacrylamide) (PNIPA) in water with and without cross-links are used. These methods include detection of (1) a pulse-like rise in the scattered intensity during the course of polymerization/gelation, (2) a power law behavior of the time intensity correlation function ICF, (3) broadening of the characteristic decay time distribution function, $P(\Gamma^{-1})$, and (4) reduction of the initial amplitude of ICF. Since these methods originate from the characteristic physical nature of gels, i.e. nonergodicity and divergence of connectivity, each of these methods can be used independently. A comparison between two PNIPA systems with and without cross-link clearly showed the effect of cross-linking on the polymer chain dynamics.

Another objective of this paper was to examine the validity of single-exponential-fit analysis for the intensity–time correlation function (ICF), $g^{(2)}(\tau)$, of gels, which has been frequently used after the report by Tanaka *et al.* [13]. They showed that $g^{(2)}(\tau)$ for gels can be expressed as a single exponential function of the decay time, τ, even though the chain length between neighboring cross-links is widely distributed. This is due to the cooperative nature of polymer networks in a solvent. However, it has still been an open question whether such single-exponential behavior in $g^{(2)}(\tau)$ is universal for any kind of gels or when it is attained during a gelation process.

In this paper, therefore, we demonstrate the feasibility of LS to investigate a gelation process as well as the validity of a single exponential analysis for $g^{(2)}(\tau)$ and discuss gelation mechanism on the basis of two types of experimental results; (1) a time-resolved light scattering study (TRLS)[14] and (2) dynamic light scattering analysis for a series of PNIPA gels prepared with different monomer concentrations [15].

EXPERIMENTAL SECTION

Poly (*N*-isopropylacrylamide) (PNIPA) gels and corresponding polymer solutions were prepared by redox polymerization. The *N*-isopropylacrylamide monomer concentration, C, was varied from 48 to 1000 mM, while the N, N'-methylenebisacrylamide (BIS; cross-linker) concentration was fixed to be 8.62 mM (for the cross-linked PNIPA) and 0 mM (for linear PNIPA). After adding 1.75 mM of ammonium persulfate (APS; initiator), the monomer solution was degassed for 15 min with a vacuum pump and cooled in a refrigerator for another 15 min in order to decelerate the polymerization reaction. Then, 8 mM of N, N, N', N'-tetramethylethylenediamine (TMEDA; accelerator) was added to the system. In the case of time-resolved light scattering (TRLS), polymerization was initiated in a 10 mm-diameter test tube at 25 °C, and the scattered intensity was measured every 500 ms with a DLS-7000, Otsuka Electric Co., Japan in order to detect a rapid change in the scattered intensity with respect to the polymerization time. On the other hand, DLS/SLS-5000, ALV, Langen, Germany, was employed for dynamic LS (DLS) of post-polymerized PNIPAs at 20 °C. In order to obtain ensemble averages, the intensity correlation function, $g^{(2)}(\tau)$, was obtained at 100 different sample positions. All of the DLS measurements were carried out at 20 °C.

RESULTS AND DISCUSSION

TIME-RESOLVED LS

Figure 1.1 shows time evolution of the scattered intensity obtained at the scattering angle of 60° during the polymerization process of cross-linked PNIPA (left) and linear PNIPA (right) with (a and e) $C = 88$, (b and f) 138, (c and g) 330, and (d and h) 690 mM. The intensity is quite low at $t < 20$ min, and then increases stepwise at the threshold polymerization time, $t_{th} \approx 20$ min. For $C \leqslant 88$ mM, the intensity reaches a plateau value and remains fairly constant. In this case, no gelation took place, indicating that the concentration was too low to form an infinite cluster. However, for $C \geqslant 330$ mM, the intensity increased abruptly, followed by a gradual decrease with strong fluctuations. Then the intensity leveled off with less fluctuations. Therefore, we classify the intensity variation with time in four stages characterized by (I) the induction stage ($t \leqslant 20$ min), (II) gelation threshold stage ($t \approx 20$ min), (III) the strong fluctuation stage ($40 \leqslant t < 60$ min), and (IV) the plateau stage with $I = I_{plateau}$ (80 min $\leqslant t$). A similar phenomenon is also observed in the polymerization process of linear PNIPAs as shown in the right side of Figure 1.1. However, there are at least two characteristic features in the time evolution of the scattered intensity for the linear PNIPAs. One is that the peak and plateau intensities for linear PNIPAs are significantly lower than those for cross-linked PNIPAs. The other is that the fluctuations in the plateau region are rather small compared to those of cross-linked PNIPAs. The intensity fluctuations observed in cross-linked PNIPAs are now know to be the result of nonergodicity of gels [16]. It should be noted here that the time evolution of the scattered intensity described above was insensitive to the scattering angle, e.g. 45°, 75°, and 90°, and the intensity rise was observed at the same time [14].

Figure 1.2 shows $I_{plateau}$ of cross-linked (closed circles) and linear PNIPAs (open circles) as a function of C. The scattering behavior of cross-linked PNIPA can be classified into four regimes; (C_I) the finite cluster regime ($C \leqslant 88$ mM), (C_{II}) the gelation threshold regime ($88 < C < 200$ mM), (C_{III}) the gel regime ($200 \leqslant C \leqslant 801$ mM), and (C_{IV}) the phase-separated gel regime (801 mM $< C$). In regime C_I the intensity increases stepwise with time (see Figure 1.1(a)). This tendency is also observed in linear PNIPAs (see Figure 1.1(e)). Regime C_{II} is characterized by the appearance of the abrupt increase in the scattered intensity as shown in Figures 1.1(c), (d) and (g), (h). In this regime, $I_{plateau}$ is a strong function of C. As shown in the figure, $I_{plateau}$ for gels is highest around $C = 171$ mM and decreases with C. This concentration may correspond to the chain overlap concentration, C^*. In regime C_{III}, a decrease in $I_{plateau}$ with C is observed. This means that large finite polymer chain clusters giving strong scattering do not exist due to formation of infinite clusters and a relatively weak scattering resulting from concentration fluctuations in a gel network is observed. Furthermore, $I_{plateau}$ is much stronger for the gels than for the polymer solutions. This is due to the presence of frozen inhomogeneity in gels introduced by cross-link formation [17]. In this regime, the so-called gel mode may be attained and a single-exponential ICF fit becomes adequate, from which the collective diffusion coefficient is evaluated [13]. Rigorously speaking, however, a single-exponential ICF fit was successful only in the case of $C = 690$ mM in

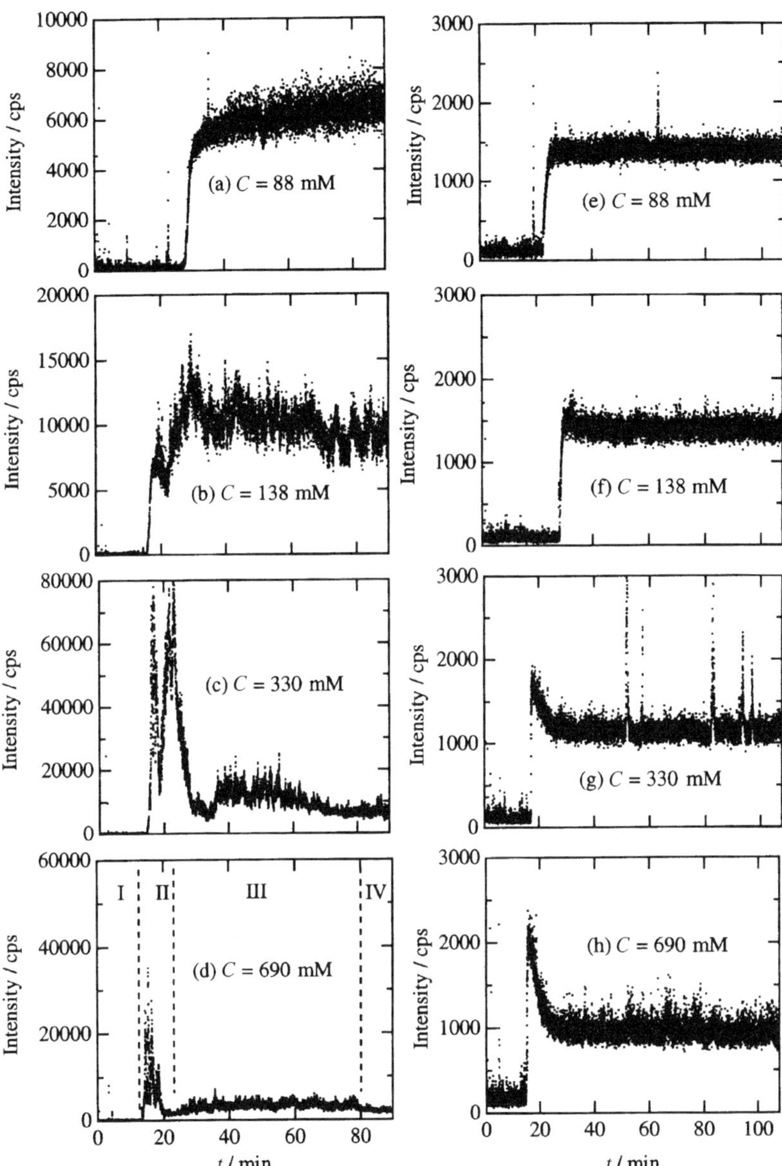

Figure 1.1 Time evolution of the scattered intensity during the polymerization process; cross-linked PNIPA (left) and linear PNIPA (right). In (d) the gelation process is divided into four stages: (I) induction stage, (II) gelation stage, (III) strong fluctuation stage, and (IV) plateau stage

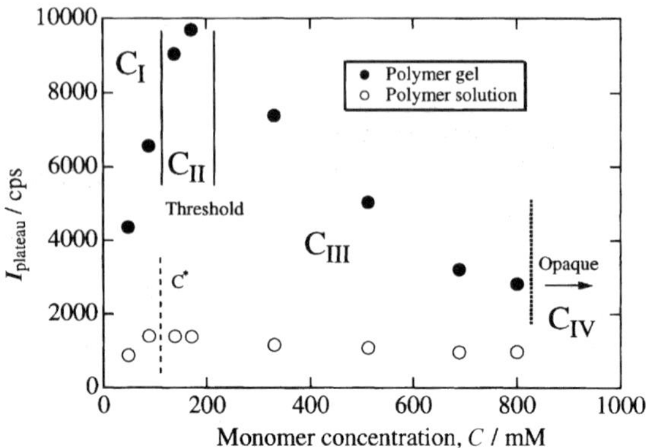

Figure 1.2 Monomer concentration, C, dependence of the plateau intensity, I_{plateau}

this study. In regime C_{IV}, the monomer concentration is too high as a uniform gel, resulting in a phase separation. This was confirmed by the fact that the gel became opaque. Since such a phase separation was not observed in linear PNIPAs even for $C > 801$ mM, this clearly indicates the equivalence of cross-link formation to an increase of attractive interaction as pointed out by de Gennes [18].

ESTIMATION OF C* AND MOLECULAR WEIGHT AT C*

The chain overlap concentration, C^*, can be obtained from the intrinsic viscosity, $[\eta]$, via

$$C^* = \frac{3 \cdot 6^{3/2} \Phi}{4\pi N_A [\eta]} \tag{1.1}$$

where N_A is Avogadro's number and Φ is a universal constant. The value of Φ is known to be 2.1×10^{23} for polydisperse dilute polymer solutions [19]. In our previous paper, the intrinsic viscosity, $[\eta]$, for a linear PNIPA aqueous solution with $C = 88$ mM, was estimated to be 118.1 cm^{3}g^{-1} at $20\,^\circ$C, which readily leads to $C^* \approx 91.6$ mM [20]. Therefore, the lowest concentration of NIPA monomers to form an infinite cluster is estimated to be around 110 mM for PNIPA aqueous solutions prepared by redox polymerization at $25\,^\circ$C. The viscosity-average molecular weight, M_v, for the PNIPA solution at $C = 88$ mM was estimated to be 6.0×10^5 by using the Mark–Houwink–Sakurada equation,

$$[\eta] = KM^a \tag{1.2}$$

where the values of K and a were chosen to be the reported values by Kubota *et al.*, i.e., $K = 0.11$ and $a = 0.51$ at $20\,^\circ$C [21].

DYNAMIC LIGHT SCATTERING ON POST-POLYMERIZED PNIPA IN WATER

Since the gelation of NIPA took place in a very limited time range around $t \approx 20$ min, a time-resolved dynamic light scattering experiment could not be conducted. Thus, the gelation was controlled by varying the monomer concentration, C. Then, the dynamics of post-polymerized PNIPAs with and without cross-links in aqueous media was investigated with a decay time spectrum $P(\Gamma^{-1})$ obtained from $g^{(2)}(\tau)$ with a inverse Laplace transform (i.e., CONTIN; a constrained regularization program). Here, Γ^{-1} is the characteristic decay time. In addition, the validity of a single exponential analysis of ICF for gels was examined.

ICF

Figure 1.3 shows double logarithmic plots of the intensity–time correlation function, $\log[g^{(2)}(\tau) - 1]$, for cross-linked PNIPA in a solvent having various Cs. For $C \leqslant 75$ mM, $g^{(2)}(\tau)$ has a characteristic decay at $\tau \approx 10^0$ ms. For $C = 88$ and 100 mM, $g^{(2)}(\tau)$ has a long tail at the larger value of τ. This corresponds to the gelation threshold at which the connectivity correlation diverges. According to Martin *et al.* [9], the scattering field correlation function, $g^{(1)}(\tau) \equiv [g^{(2)}(\tau) - 1]^{1/2}$, at the gelation threshold is given by a sum of two contributions, i.e., diffusive mode and power-law behavior, as follows,

$$g^{(1)}(\tau) = A \exp[-\Gamma\tau] + (1 - A)\left(1 + \frac{\tau}{\tau^*}\right)^{-(1-D_p)/2} \tag{1.3}$$

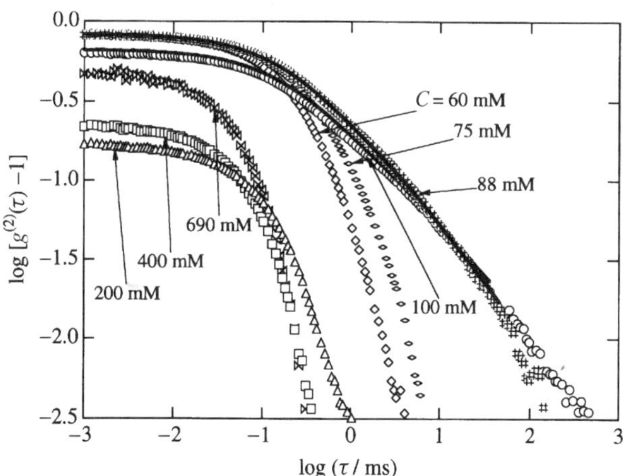

Figure 1.3 Double logarithmic plot of the intensity correlation function, $g^{(2)}(\tau) - 1$, for cross-linked PNIPA prepared from different monomer concentration, C

where τ^* is a characteristic decay time above which a power-law behavior appears and D_p is the fractal dimension of the detected photons scattered from the gel. A $(0 \leqslant A \leqslant 1)$ is the relative strength of the diffusive mode. The solid lines in Figure 1.3 were obtained by fitting ICF with equation (1.3) to the case of $C = 88$ and 100 mM. This fit gives $D_p = 0.39(C = 88$ mM$)$ and $0.42(C = 100$ mM$)$. These values are significantly smaller than the values reported in the literature, e.g., $D_p = 0.73$ for silica gels [9]. However, as Muthukumar discussed [12], D_p depends on the degree of branching. In our case, the degree of branching seems to be much smaller than the case of silica gels. D_p is the same as the exponent n for the frequency dependence of the storage G' and loss moduli G'', i.e., $D_p = n$, where $G' \sim G'' \sim \omega^n$ and ω is the angular frequency. Percolation theory gives $n = 2/3$, and hence $D_p = 2/3$. This is the case of highly branched clusters. On the other hand, Winter obtained $n = 1/2$ for clusters with relatively long chains between cross-links [11], which gives $D_p = 1/2$. Our system seems to be classified into the latter case.

For $C \geqslant 200$ mM, $g^{(2)}(\tau)$ recovers a similar behavior to the case of $C \leqslant$ 75 mM. Note that there are two characteristic features in $g^{(2)}(\tau)$ for $C \geqslant 200$ mM; (i) the characteristic decay time being much smaller, i.e., $\tau \approx 10^{-1}$ ms, and (ii) the value of $g^{(2)}(\tau \to 0)$ being much lower than unity. The latter indicates that the system becomes nonergodic and $g^{(2)}(\tau \to 0) - 1(= \sigma_I^2)$ depends on the sample position [16]. The behavior of $g^{(2)}(\tau) - 1$ for $C > 200$ mM is also very different from that obtained by Martin *et al.* [9], who obtained a power-law functional form even for a well developed gel. Therefore, it is clear that such a power law function analysis is not relevant for well-developed gels studied in this work., e.g., a chemically cross-linked gel in a reactor batch prepared by radical polymerization from a vinyl monomer solution. This is due to larger mobility of long PNIPA chains between cross-links than highly branched silica gel. In order to investigate the change in $g^{(2)}(\tau)$ with C more quantitatively, we examine the characteristic decay time distribution function, $P(\Gamma^{-1})$, as a function of C.

Characteristic Decay Time Distribution Analysis

The characteristic behavior in ICF can be discussed by using the characteristic decay time distribution function, $P(\Gamma^{-1})$, which is obtained by Laplace transform of $g^{(2)}(\tau)$, as follows,

$$g^{(2)}(\tau) - 1 = \left[\int_0^\infty G(\Gamma) \exp(-\Gamma\tau) \, d\Gamma \right]^2 \tag{1.4}$$

where Γ is the characteristic decay rate (an inverse of the characteristic decay time, Γ^{-1}). We define the characteristic decay time distribution function being $P(\Gamma^{-1}) \equiv G(\Gamma)$. Not only the characteristic behavior at the gelation threshold but also the validity of a single exponential analysis can be examined by evaluating a decay time spectrum $P(\Gamma^{-1})$, as will be discussed in the following.

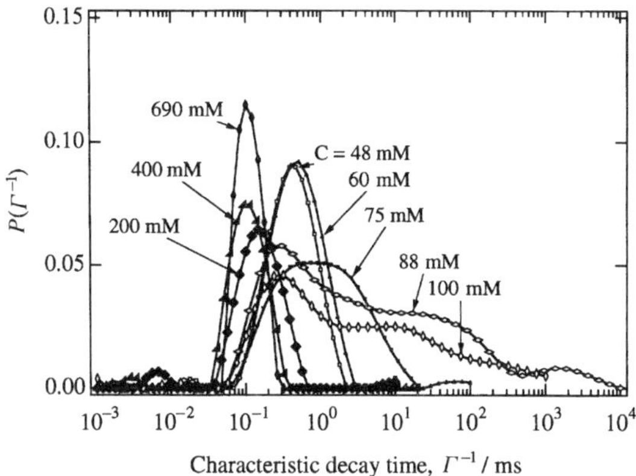

Figure 1.4 Characteristic decay time distribution function $P(\Gamma^{-1})$ for cross-linked PNIPA prepared from different Cs

Figure 1.4 summarizes the C dependence of $P(\Gamma^{-1})$'s obtained for NIPA gels after completion of gelation. This figure clearly shows the variation of $P(\Gamma^{-1})$. $P(\Gamma^{-1})$ is unimodal for $C \leqslant 60$ mM, but becomes broader for $C \geqslant 75$ mM. Note that the maximum position moves toward a larger value of Γ^{-1} with C. A very broad distribution is observed at the concentration regime between 88 and 100 mM where the gelation threshold is located. For $C \geqslant 200$ mM, a unimodal distribution is recovered, which corresponds to the so-called gel mode scattering, i.e., a single exponential behavior in $g^{(2)}(\tau)$. In the case of linear PNIPA, such a broadening in $P(\Gamma^{-1})$ was not observed. These results were interpreted as follows: In the case of cross-linked PNIPA, segregation between monomers belonging to different clusters may be stronger than in the case of linear PNIPA due to the presence of cross-links. The cross-links also prevent mixing of different clusters against thermal fluctuations. As a result, a wide distribution of polymer chain clusters is observed. For $C \geqslant 200$ mM, however, a unimodal distribution of $P(\Gamma^{-1})$ is recovered, indicating a formation of an infinite network. Therefore, a gelation threshold can be determined by the characteristic decay time spectral analysis.

Initial Amplitude

Figure 1.5 shows C dependence of the ensemble average initial amplitude of ICF, $\langle \sigma_I^2 \rangle_E$, for cross-linked PNIPA (solid circles) where $\langle \sigma_I^2 \rangle_E$ was simply obtained by taking an average of $\langle g^{(2)}(\tau = 0) - 1 \rangle_T$ for 100 data points. Hence, it is not the same as the average estimated from $\langle g^{(2)}(\tau = 0) - 1 \rangle_E$, where $\langle .. \rangle_T$ and $\langle .. \rangle_E$ are the time and ensemble averages, respectively. As shown here, $\langle \sigma_I^2 \rangle_E$ for linear PNIPAs does not depend on C, while that for cross-linked PNIPAs drops at $C \approx C_{th}$ (indicated with an arrow). This is ascribed to the fact that a nonergodic nature appears when an infinite cluster, i.e. a gel, is formed. Therefore, by examining $\langle \sigma_I^2 \rangle_E$ as a function of C, one can determine

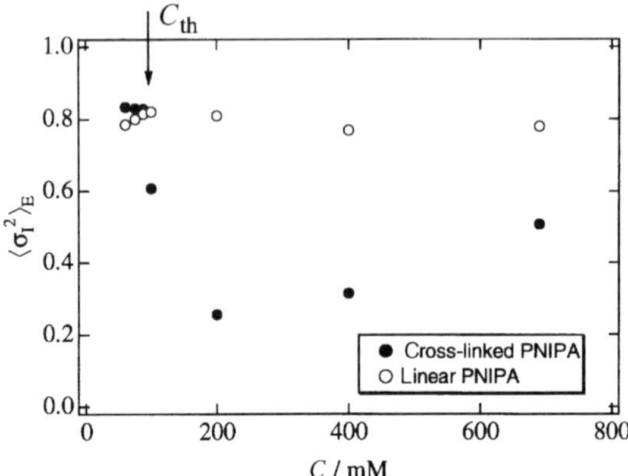

Figure 1.5 C dependence of the initial amplitude of $g^{(2)}(\tau) - 1$ for cross-linked PNIPAs. C_{th} indicates the NIPA monomer concentration which leads to gelation threshold

the onset of gelation. This method can apply to a gelling system in which polymerization/gelation reaction goes on. Recently, we have succeeded in determining the gelation threshold for polycondensation of tetramethoxysilane [22]. Another interesting feature in Figure 1.5 is that $\langle \sigma_I^2 \rangle_E$ for cross-linked PNIPAs increases with C for $C \geqslant 200$ mM. This is simply accounted for as follows: the cross-linked PNIPA with $C = 200$ mM has the highest ratio of C_{BIS}/C for $C > C^*$, since we fixed the concentration of the cross- linker (BIS), C_{BIS}. In our previous paper, we reported that the ensemble average scattered intensity, $\langle I \rangle_E$, increase with C_{BIS}, while the time average scattered intensity of the fluctuating component, $\langle I_F \rangle_T$, is fairly invariant with C_{BIS} [17]. Since $\langle \sigma_I^2 \rangle_E$ is related to the ratio of $\langle I_F \rangle_T / \langle I \rangle_E$, the increase of $\langle \sigma_I^2 \rangle_E$ with $C(>200$ mM) is quite reasonable.

CONCLUSION

We have shown in this work that time-resolved dynamic light scattering and dynamic light scattering allow one to determine the gelation threshold. Four types of characteristic symptoms indicating gelation point were demonstrated in the case of poly(N-isopropylacrylamide) aqueous systems. These are (1) pulse-like increase in the scattered intensity during the polymerization process, (2) appearance of a long tail (or a power-law type behavior) in the intensity correlation function, $g^{(2)}(\tau)$, (3) broadening of the characteristic decay time distribution function, $P(\Gamma^{-1})$, and (4) deviation of the initial amplitude of $g^{(2)}(\tau) - 1$ from unity. All of those phenomena are ascribed to divergence of the connectivity correlation at the gelation threshold and the appearance of a nonergodic nature above the gelation threshold.

ACKNOWLEDGMENT

This work is partially supported by the Ministry of Education, Science, Sports and Culture, Japan (Grant-in-Aid, 09450362 and 10875199 to M.S.).

REFERENCES

1. P.J. Flory, *J. Am. Chem. Soc.*, **63**, 3038 (1941).
2. W.H. Stockmayer, *J. Chem. Phys.*, **11**, 45 (1943).
3. P.G. de Gennes, *J. Physique Lett.*, **37**, L1 (1976).
4. D Stauffer, *J. Chem, Soc, Farad. Trans.*, **II 72**, 1354 (1976).
5. M. Adam and D. Lairez, in *The Physical Properties of Polymeric Gels*, (J.P. Cohen Addad, ed.) John Wiley & Sons, New York, 1996, pp. 87.
6. J.E. Martin and A.J. Hurd, *J. Appl. Cryst.*, **20**, 61 (1987).
7. J.E. Martin, J. Wilcoxon and D. Adolf, *Phys. Rev. A.*, **36**, 1803 (1987).
8. J.E. Martin and J. Wilcoxon, *Phys. Rev. Lett.*, **61**, 373 (1988).
9. J.E. Martin, J. Wilcoxon and J. Odinek, *Phys. Rev. A.*, **43**, 858 (1991).
10. H.H. Winter, P. Morganelli and F. Chambon, *Macromolecules*, **21**, 532 (1988).
11. H.H. Winter and F. Chambon, *J. Rheology*, **30**, 367 (1986).
12. M. Muthukumar, *Macromolecules*, **22**, 4656 (1989).
13. T. Tanaka, L.O. Hocker and G.B. Benedek, *J. Chem. Phys.*, **59**, 5151 (1973).
14. T. Norisuye, M. Shibayama and S. Nomura, *Polymer*, **39**, 2769 (1998).
15. T. Norisuye, M. Takeda and M. Shibayama, *Macromolecules*, **31**, 5316 (1998).
16. P.N. Pusey and W. van Megen, *Physica A*, **157**, 705 (1989).
17. M. Shibayama, T. Norisuye and S. Nomura, *Macromolecules*, **29**, 8746 (1996).
18. P.G. de Gennes, *Scaling Concepts in Polymer Physics*. Cornell University, Ithaca, 1979.
19. H. Fujita, *Polymer solution*, Elsevier, Amsterdam, 1990.
20. M. Shibayama, Y. Isaka and Y. Shiwa, *Macromolecules*, in press.
21. K. Kubota, S. Fujishige and I. Ando, *Polymer J.*, **22**, 15 (1990).
22. T. Norisuye, M. Shibayama, R. Tamaki and Y. Chujo, *Polym. Prep., Jpn.*, **47**, 904 (1998); *Macromolecules*, **32**, 1528 (1999).

2

Modelling of Network Polymerization with Intramolecular Cyclization

HENRYK GALINA and JAROMIR B. LECHOWICZ
Department of Industrial and Materials Chemistry, Faculty of Chemistry,
Rzeszów University of Technology, 35-959 Rzeszów, Poland

ABSTRACT

A Smoluchowski-like coagulation equation which describes the time evolution of size distribution in a model homopolymerization of a three-functional monomer has been rederived. The functional groups were allowed to react with substitution effect and the rates of cycle closing reactions were premultiplied by a cyclization parameter. The gelation was found shifted towards high conversion, but only when the cyclization parameter was sufficiently large. The critical exponent at which the weight average polymerization degree diverged had the classical value.

INTRODUCTION

In the classical Flory–Stockmayer theory of network formation [1,2] the possibility of intramolecular reactions was disregarded although both authors were

Wiley Polymer Networks Group Review Series Vol. 2. Edited by B.T. Stokke and A. Elgsaeter
© 1999 John Wiley & Sons Ltd

aware of the effect of cycle formation on the size distribution of polymer species and the position of gel point, the existence of which they predicted and defined. This problem was on the agenda already well over fifty years ago when the classical theory was formulated [3].

For the polymerization of bi-functional monomers there exist mean-field theories of cyclization in systems both in equilibrium [4,5] or rate controlled ones [6,7]. The extent of cyclization in the early stages of a polymerization depends on the type of reaction. It is small in a polycondensation or polyaddition, and quite substantial in a radical vinyl–divinyl copolymerization [8]. The state of the art in the mean-field theory describing cyclization during network formation can be found in a review by Ross-Murphy and Stepto [9]. The methods of predicting size distributions of species or related properties of polymerizing systems described in this review were the statistical ones based on the spanning-tree approximation introduced by Gordon and coauthors [10] or the rate theory of Ahmad and Stepto [11]. A kinetics analysis of the problem of cyclization in the polymerization of multifunctional monomers based on the Smoluchowski coagulation equation was presented by Lu and Bak [12]. Following a similar idea, one of the present authors derived a Smoluchowski-like equation to model the kinetics of polymerization of multifunctional monomers reacting with substitution effect, where intramolecular reactions were taken into account [13]. A simplified version of this equation along with its derivation is presented in this paper. Since, however, we were unable to solve this equation, we present results of Monte Carlo calculations based on the same model. These results are in fact the numerical solution of the equation.

THE MODEL

We consider a purely graph-theoretical static model [14] of a homopolymerization of a three-functional monomer. The structure of each molecule is coded with three parameters:

- the number of units of substitution degree 1 (pendant units): i,
- the number of units with two reacted functional group (linear chain units): j, and
- the number of cycles in a molecule (its cycle rank): k.

A molecule coded as the $\{i, j, k\}$-mer is not a unique one. The same symbol describes all structural isomers that share the same code values. From the point of view of reactivity, though, the rate at which all $\{i, j, k\}$-mers react should be the same. The unreacted monomer molecules are counted separately.

The concentration of an $\{i, j, k\}$-mer, $c_{i,j,k}$, is expressed in terms of the number of these molecules divided by the total number of units in the system. The concentration of monomer is denoted as X_0.

The set of three parameters describing a molecule is sufficient to determine its size. The number of units, n, it contains is

$$n = i + j + m \tag{2.1}$$

where m is the number of units having all groups reacted. This number is irrelevant from the point of view of the molecule's reactivity. The number of links between units in an $\{i, j, k\}$-mer is

$$n + k - 1 = \tfrac{1}{2}(i + 2j + 3m) \tag{2.2}$$

Thus,

$$n = 2i + j + 2k - 2 \tag{2.3}$$

Clearly, similar relations can be written for molecules obtained in other types of polymerization [15–17]

We assume that all functional groups on a monomer molecule have the same reactivity. This assumption does not limit the generality of conclusions, since it may easily be relaxed at the expense of heavier algebra. The groups may react with the first shell substitution effect [18]. After one group on a monomer molecule has reacted, the remaining two again have the same reactivity, but this reactivity may be different than that of the groups of the original monomer. Similarly, the reactivity of the functional group on a unit with two groups already reacted may differ from that on the pendant unit (which may, or may not, also be different from that of monomer).

In order to reduce the number of parameters describing the substitution effect we take the rate constant of the reaction between i and j substituted units ($i, j = 0, 1, 2$) to be $K_{i,j} = (3 - i)(3 - j)k_i k_j$. It has a product form with independent contributions from the reacting units. This form of the rate constant is equivalent to the assumption that the activation energy of the reaction between units has additive terms due to the substitution degree [18].

Next, we define the relative rates at which the whole units (not just the groups) react. They are $a = 2k_1/3k_0$ and $b = k_2/3k_0$. Now, if all reactions are second order, multiplying time by $3k_0$ ($\tau = 3k_0 t$, where t is the actual time) we find that the reaction rate between two monomers is 1.

The elementary intermolecular reactions in the homopolymerization of a three-functional monomer are shown in Figure 2.1.

Let the size distribution of molecules in the system be described by the counting polynomial function

$$H(\tau, x, y, z) = X_0(\tau) + \sum_{i=2}^{\infty} \sum_{j=0}^{\infty} \sum_{k=0}^{\infty} c_{i,j,k}(\tau)[ax]^i [by]^j z^k \tag{2.4}$$

where x, y and z are the dummy variables of no physical meaning. When the cycle closing reactions are disregarded, the single partial differential equation

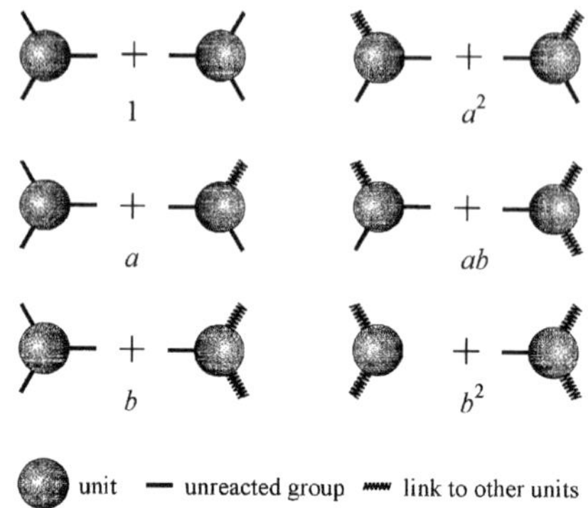

Figure 2.1 The elementary reactions in the homopolymerization of a three-functional monomer reacting with substitution effect. The relative rates of the reaction between functional groups shown are indicated

describing the time evolution of size distribution in system is [15,19]

$$\frac{\partial H}{\partial \tau} = \frac{1}{2}\left[(ax)X_0 + (by)\frac{\partial H}{\partial x} + \frac{\partial H}{\partial y}\right]^2 - (X_0 + X_1 + X_2)\left[X_0 + (ax)\frac{\partial H}{\partial x} + (by)\frac{\partial H}{\partial y}\right] \tag{2.5}$$

where

$$X_1 = \left.\frac{\partial H}{\partial x}\right|_{x=1/a,\,y=1/b,\,z=1} \qquad X_2 = \left.\frac{\partial H}{\partial y}\right|_{x=1/a,\,y=1/b,\,z=1} \tag{2.6}$$

This equation was used to demonstrate the role and extent of time correlations that may modify the size distribution in kinetically controlled polymerization reactions [15].

The kinetics analysis can be extended to include formation of intramolecular links. We now present a derivation of the Smoluchowski-like equation simplified compared to its original version [13].

To the set of elementary reactions presented in Figure 2.1 one has to add three more reactions leading to cycle formation, as shown in Figure 2.2.

The relative rates of these reactions are premultiplied by three functions λ_{11}, λ_{12}, and λ_{22} called the cyclization parameters. They modify the rates at which the reactions between units of the respective substitution degree take place. They are proportional to the probability that two units belonging to the same molecule will meet in the same volume element. The cyclization parameters are expected to depend on many factors: the chain flexibility, total concentration of units (dilution of the system), topology of molecules, etc. To a first approximation, one may consider them to be independent of the distribution, or just constant.

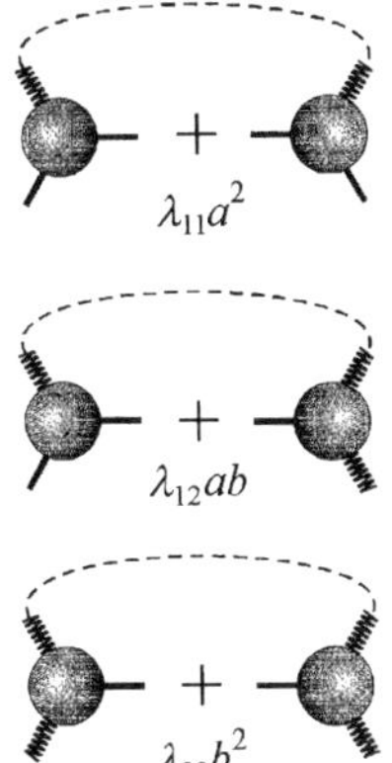

Figure 2.2 The additional elementary reactions leading to intramolecular cyclization and the relative rates thereof

With this approximation, intramolecular reaction between units of substitution degree 1 in an $\{i, j, k\}$-mer takes place at the relative rate

$$-\lambda_{11}a^2\tfrac{1}{2}i(i-1)c_{i,j,k} \tag{2.7}$$

since the number of ways the monosubstituted units may react with each other is $\binom{i}{2} = \tfrac{1}{2}i(i-1)$. The reaction produces an $\{i-2, j+2, k+1\}$-mer. On the other hand, the $\{i, j, k\}$-mer is obtained from an $\{i+2, j-2, k-1\}$-mer at the relative rate

$$\lambda_{11}a^2\tfrac{1}{2}(i+2)(i+1)c_{i+2,j-2,k-1} \tag{2.8}$$

The Smoluchowski-like equation is formed by multiplying the rates of appropriate reactions by their corresponding $[ax]^i[by]^jz^k$ followed by summing up all terms. It is not difficult to verify that by doing this one obtains from equation (2.7)

$$-\frac{1}{2}\lambda_{11}\frac{\partial^2 H}{\partial x^2}[ax]^2 \tag{2.9}$$

and from equation (2.8)

$$\frac{1}{2}\lambda_{11}\frac{\partial^2 H}{\partial x^2}[by]^2z \tag{2.10}$$

Similar reasoning leads to the expressions for the relative rates at which the size distribution is modified by intramolecular reactions between mono- and disubstituted units in an $\{i, j, k\}$-mer

$$-\lambda_{12}\frac{\partial^2 H}{\partial x\partial y}[ax][by] \tag{2.11}$$

or by formation of $\{i, j, k\}$ from $\{i+1, j, k-1\}$

$$\lambda_{12}\frac{\partial^2 H}{\partial x\partial y}[by]z \tag{2.12}$$

By intramolecular reaction involving two bisubstituted units, an $\{i, j, k\}$-mer is formed from an $\{i, j + 2, k - 1\}$-mer. The respective terms of the Smoluchowski-like coagulation equation are

$$-\frac{1}{2}\lambda_{22}\frac{\partial^2 H}{\partial y^2}[by]^2 \tag{2.13}$$

for disappearance and

$$\frac{1}{2}\lambda_{22}\frac{\partial^2 H}{\partial y^2}z \tag{2.14}$$

for formation of an $\{i, j, k\}$-mer.

Summation of all terms leads to a simpler equation than that proposed previously [13]:

$$\frac{\partial H}{\partial \tau} = \frac{1}{2}\left[(ax)X_0 + (by)\frac{\partial H}{\partial x} + \frac{\partial H}{\partial y}\right]^2$$

$$- (X_0 + X_1 + X_2)\left[X_0 + (ax)\frac{\partial H}{\partial x} + (by)\frac{\partial H}{\partial y}\right]$$

$$+ \frac{1}{2}\left[\lambda_{11}([ax]^2 - [by]^2z)\frac{\partial^2 H}{\partial x^2} + 2\lambda_{12}([ax] - z)[by]\right.$$

$$\left.\times\frac{\partial^2 H}{\partial x\partial y} + \lambda_{22}([by]^2 - z)\frac{\partial^2 H}{\partial y^2}\right] \tag{2.15}$$

MONTE CARLO SIMULATIONS AND RESULTS

The algorithm used in Monte Carlo calculations of the model described by equation (2.15) was similar to that described elsewhere [20]. The reaction system consisted of 10^5 3-functional units. Just one value of the cyclization parameter was used in each run $\lambda = \lambda_{11} = \lambda_{12} = \lambda_{22}$. To verify the algorithm, the critical conversion at gelation were calculated for various magnitudes of the substitution effect and no cyclization. The results were compared with those obtained by direct application of equation (2.5) (cf. [15]). The gel point was located as the conversion degree where the weight average polymerization degree of the sol fraction was maximal. In Monte Carlo calculations the gel molecule is the one which is the biggest and all that remains is considered the sol fraction. The difference between gel conversions calculated by the two methods never exceeded ca. 3% which indicated that the size of the system was sufficiently large to suppress possible random fluctuations.

In the next series of calculations, the values of cyclization parameter λ was changed for various sets of rate constants $k_1(= 3a/2)$ and $k_2(= 3b)$. The results of this series of calculations are shown in Figure 2.3. One can see in the plot in

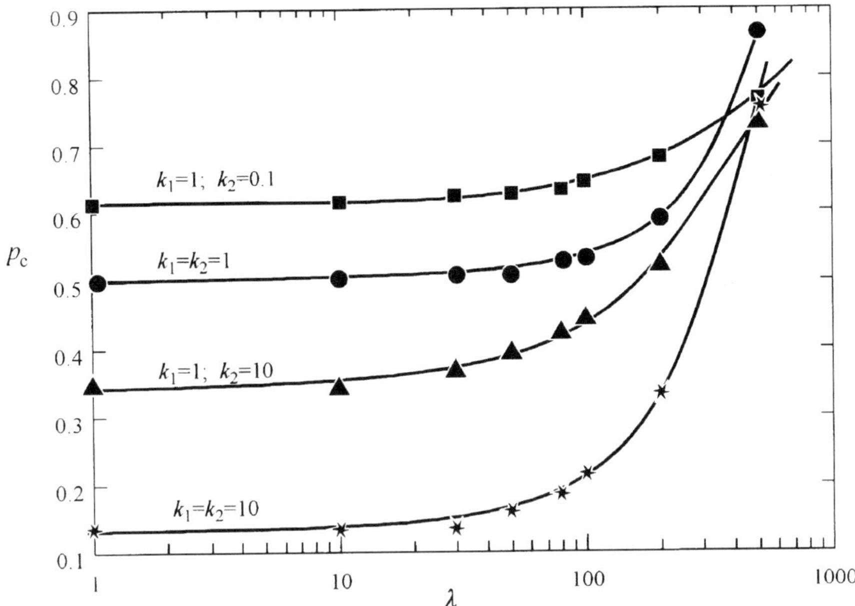

Figure 2.3 Critical conversion at gelation vs. cyclization parameter for several sets of kinetics constants. The results of Monte Carlo calculations for systems consisting of $N = 10^5$ units

Figure 2.3 that the conversion at gelation increases as the cyclization parameter becomes larger. The increase in the critical conversion is obviously related to the wastage of functional groups engaged in intramolecular links.

It is interesting, however, that initially, i.e., for smaller λ, the shift in the gel point conversion towards higher values is very small. One may conclude that for a value of cyclization parameter much smaller than the size of the whole system, the rates of intermolecular reactions prevail as the system reacts according to the mass action law, i.e., according to the mean field approximation. We observe in Figure 2.3, however, a rapid growth of the critical conversion when λ exceeds about 100. This value is still relatively small compared to the system size and may not substantially change the relative rates of intramolecular to intermolecular reactions. The key factor seems to be the topologies of the molecules formed in the system.

Some insight into the topology of species can be gained by examining the relation between critical conversion degree at the gel point and the content of cycles per unit (cycle rank per unit). The relations for several sets of kinetics parameters are shown in Figure 2.4. All relations are linear, but not parallel. This means that both rate and cyclization parameters affect the content of cycles at the gel point.

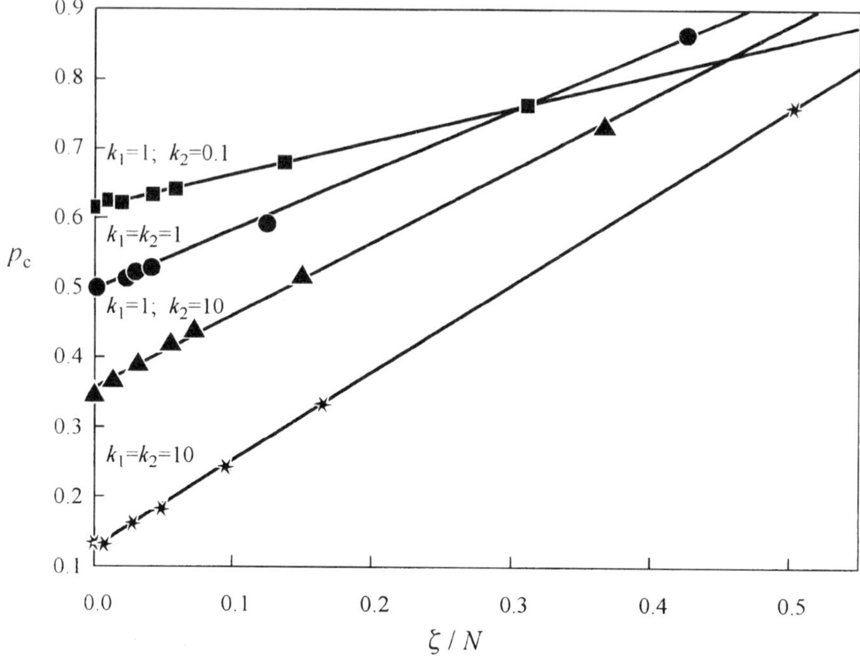

Figure 2.4　Critical conversion at gelation vs. cycle rank (number of independent cycles) per unit for several sets of kinetics constants

Figure 2.5. on the other hand, shows how the weight-average polymerization degree of the sol fraction changes with conversion for different values of cyclization parameter. The rate constants are in all cases the same (functional groups reacting at random, without substitution effect). In the system without cyclization ($\lambda = 0$), the average size of molecules tend to zero soon after the gel point, whereas for a system with cyclization, the average size of sol molecules oscillates around a finite value up to the end of a simulation. This limiting value is not necessarily very small. Clearly, microgel structures are formed which cannot be absorbed by the gel molecule since all functional groups in the microgel are already reacted. It should be pointed out that in other Monte Carlo studies, when the cyclization mechanism was modelled in a different way [20], the formation of microgels was not observed.

The microgel structure of the sol fraction in a homopolymerization of a three-functional monomer as modelled by eq. (15) with a single cyclization parameter of constant value can even more clearly be seen in Figures 2.6 through 2.8, where the concentration of the smallest components in the systems (monomer through pentamers) are plotted against conversion degree.

When no cyclization is allowed (Figure 2.6, $\lambda = 0$), the concentrations of all low-molecular components drop down to zero at a conversion of 0.67, since

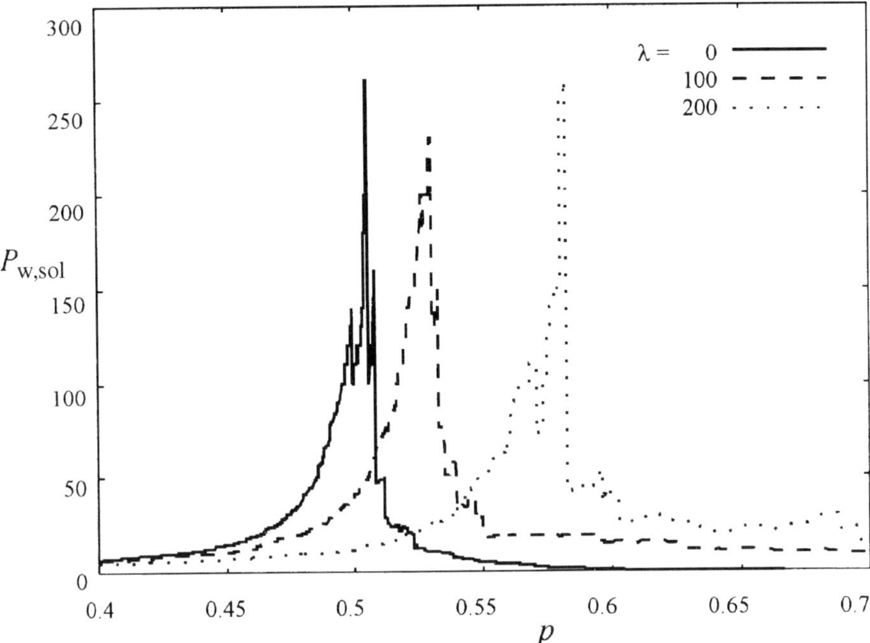

Figure 2.5 The changes of weight-average degree of polymerization of molecules in the sol fraction in the system with equal reactivity of all functional groups and different cyclization parameters. Monte Carlo calculations for systems consisting of $N = 10^5$ units

then just one molecule remains, the gel molecule. With $\lambda = 500$ and the random reaction of functional groups ($k_1 = k_2 = 1$) one observes a large proportion of small molecules, particularly with 2, 4 and other even numbers of units present in the sol. (The concentration of 4-mers in Figure 2.7 is higher that that of 3-mers at the end of a simulation). The presence of 'acetylenic' dimers becomes extremely high in the system with the positive substitution effect ($k_1 = k_2 = 10$, Figure 2.8). Then, the second and third functional group react faster than the first and, as soon as a dimer appears its functional groups react among themselves at a high rate. In Figure 2.8, one can see that, at the end of reaction, as many as over 30% of units form dimers.

The forced cyclization in the system modelled by introducing a very high rate of cycle closing reactions should in real systems correspond to a high dilution of monomers or a reaction carried out in a highly viscous medium, where the diffusivity of reactants is highly suppressed. Nevertheless, the model defined by equation (15) with a constant cyclization parameter seems more artificial than that with an in situ determined cyclization rate studied previously [20,21]. However, it seems somewhat better than the model of Lu and Bak [12] based on the original

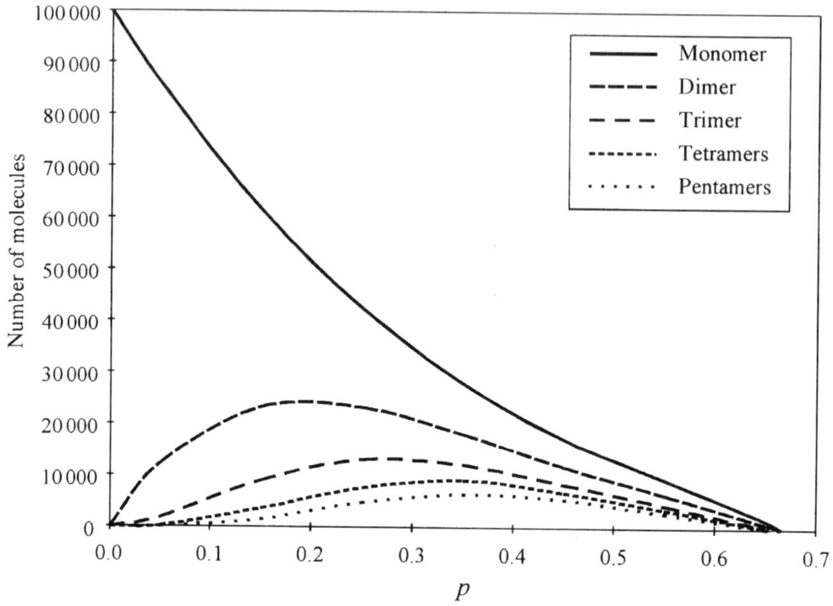

Figure 2.6 The number of small molecules, monomers through pentamers, in the polymerization where $N = 10^5$ three-functional units react randomly without cyclization allowed (cyclization parameter $\lambda = 0$)

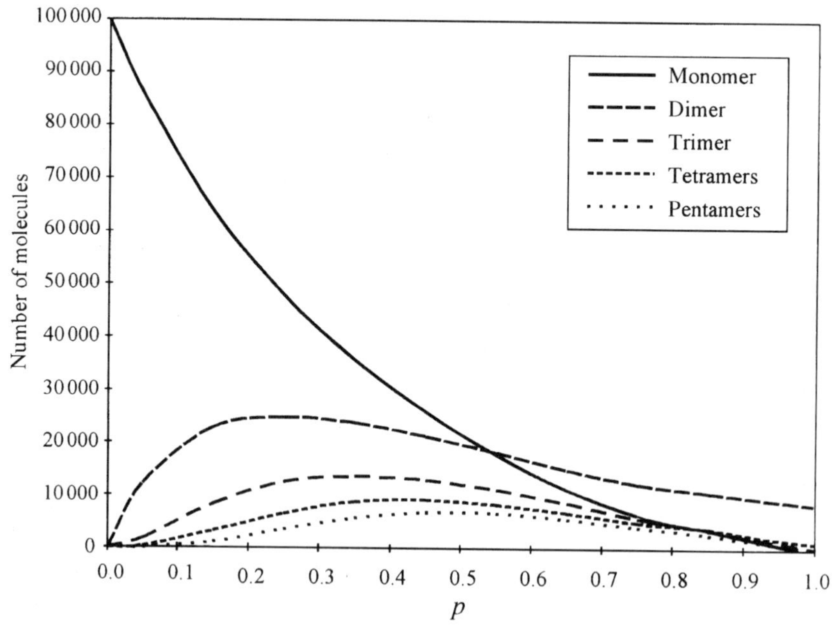

Figure 2.7 As in Figure 2.6, but with the cyclization parameter $\lambda = 500$

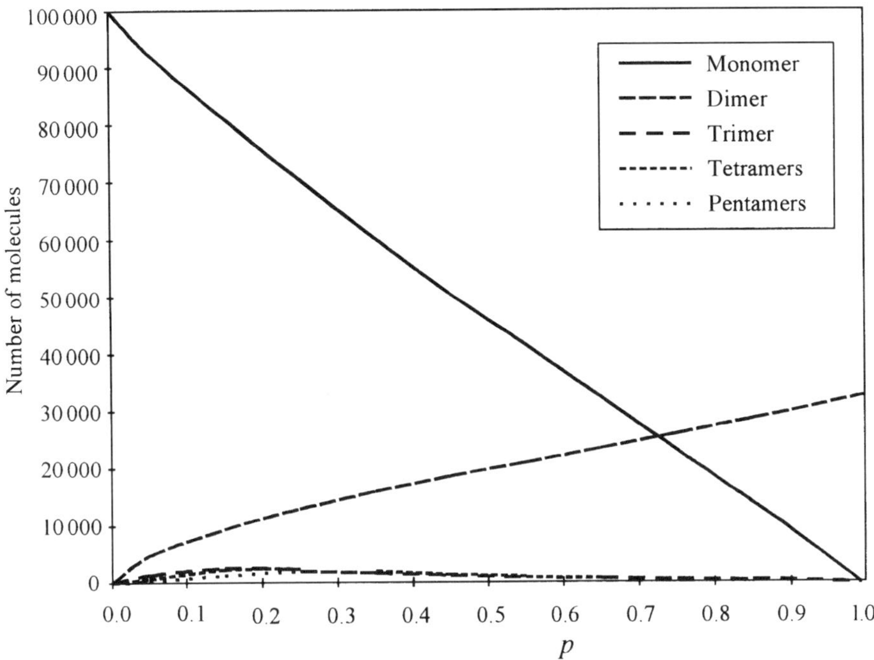

Figure 2.8 As in Figure 2.7, but with the rate parameters $k_1 = k_2 = 10$, indicating that the functional groups on a unit already connected to a molecule react 10 times faster than those on a monomer molecule

Smoluchowski equation, where it was the monomer which underwent cyclization reactions to the highest extent.

In all computer experiments performed by us the critical exponent γ in the relationship $P_w = |p_c - p|^\gamma$ assumed the classical value of -1, even for large values of the cyclization parameter.

ACKNOWLEDGMENTS

The financial support of this work from the Polish Committee of Scientific Research, grant no. 3T09A 038 011 is gratefully acknowledged.

REFERENCES

1. (a) P.J. Flory, *J. Am. Chem. Soc.*, **63**, 3083–90, 3091–6, 3096–100 (1941); (b) P.J. Flory, *Chem. Revs.*, **39**, 137–97 (1946); (c) P.J. Flory, *Principles of Polymer Chemistry*, Cornell Univ. Press, Ithaca 1953.
2. W.H. Stockmayer, *J. Chem. Phys.*, **11**, 45–55 (1943); **12**, 125–31 (1944).
3. W.H. Stockmayer and L.L. Weil, *Polym. Int.*, **44**, 221–4 (1997).

4. H. Jacobson and W.H. Stockmayer, *J. Chem. Phys.*, **18**, 1600–7 (1950).
5. G. Ercolani, L. Mandolini, P. Mencarelli and S. Roelens, *J. Am. Chem. Soc.*, **115**, 3901–8 (1993).
6. T.F. Irzhak, N.I. Peregudov, V.I. Irzhak and B.A. Rozenberg, *Vysokomol. Soedin. B*, **35**, 905; 1545–9 (1993).
7. A.D. Cort, G. Ercolani, A.L. Iamicelli, L. Mandolini and P. Mencarelli, *J. Am. Chem. Soc.*, **116**, 7081–7 (1994).
8. K. Dušek, *Br. Polym. J.*, **17**, 185–9 (1985).
9. S.B. Ross-Murphy and R.F.T. Stepto, in *Cyclization, Gelation and Network Formation* (J.A. Semlyen ed.), Elsevier, London, 1986.
10. M. Gordon and G.R. Scantlebury, *Proc. Roy. Soc. (London) A*, **292**, 380 (1966); *J. Polym. Sci.*, **16**, 3933–42 (1968); K. Dušek, M. Gordon and S.B. Ross-Murphy, *Macromolecules*, **11**, 236 (1978).
11. Z. Ahmad and R.F.T. Stepto, *Colloid Polym. Sci.*, **258**, 663–74 (1980).
12. B. Lu and T.A. Bak, in *Biological and Synthetic Polymer Networks* (O. Kramer, ed.), Elsevier, Amsterdam, 1988.
13. H. Galina, *Makromol. Chem., Macromol. Symp.*, **40**, 42–45 (1990); H. Galina and J. Lechowicz, *Adv. Polym. Sci.*, **137**, 135–172 (1998).
14. We consider the static properties which include the size distribution, gel point, and other topology-related properties as opposed to dynamic properties of polymerizing systems such as those studied recently by M. Adam, D. Lairez, M. Karpasas and M. Gotlieb, *Macromolecules*, **30**, 5920–9 (1997).
15. H. Galina and A. Szustalewicz, *Macromolecules*, **22**, 3124–9 (1989).
16. H. Galina and A. Szustalewicz, *Macromolecules*, **23**, 3833–8 (1990).
17. K. C. Cheng and W.Y. Chiu, *Macromolecules*, **26**, 4658–64; 4665–9 (1993).
18. M. Gordon, G.R. Scantlebury, *Trans. Faraday Soc.*, **60**, 604–21 (1964).
19. H. Galina, *Europhys. Lett.*, **3**, 1155–9 (1987).
20. H. Galina and J. Lechowicz, *Comput. Polym. Sci.*, **5**, 197–201 (1995).
21. H. Galina and J. Lechowicz, *Progr. Colloid Polym. Sci.*, **102**, 1–3 (1996).

3

Primary Cyclization Reactions in Crosslinked Polymers

JEANNINE E. ELLIOTT and CHRISTOPHER N. BOWMAN
Department of Chemical Engineering, University of Colorado at
Boulder, Boulder, CO 80309-0424, USA

ABSTRACT

The kinetics of cyclization reactions which occur during free radical photopolymerizations of multifunctional vinyl monomers are difficult to characterize because of the varying pendant double bond reactivities and inhomogeneities in the system. To date, no comprehensive model exists to predict the network formation in highly crosslinked polymers which is able to predict primary cyclization rates (i.e. where a pendant double bond loops back intramolecularly with the propagating radical that formed it). A numerical approach is presented which solves the differential kinetic equations and can be used to investigate the factors influencing cyclization versus crosslinking reactions. The effects of monomer size and monomer concentration on primary cyclization were examined.

Wiley Polymer Networks Group Review Series Vol. 2. Edited by B.T. Stokke and A. Elgsaeter
© 1999 John Wiley & Sons Ltd

The model successfully predicted that decreasing monomer size from 100 Å to 10 Å increases the fraction of reacting pendant double bonds that cycle by about two and a half times at 60% double bond conversion. When the monomer concentration is diluted by adding solvent, the fraction of primary cycles increases significantly as compared to a bulk polymerization.

INTRODUCTION

Because of the wide array of industrial and scientific applications for crosslinked polymers formed by photopolymerization, understanding the polymerization kinetics and the resulting material properties is essential. Applications for crosslinked photopolymers include photolithography and microelectronics, protective and decorative coatings, dental restorative materials, aspherical lenses, laser videodisks, and optical fiber coatings [1–4]. These photopolymerizations have the advantage of rapid processing time, spatial control of the polymerization, and ambient processing temperature.

In general, the polymerization of multifunctional monomers results in pendant double bonds on the growing polymer chains. These pendant double bonds can react with propagating radicals to form primary cycles, secondary cycles, or crosslinks (Figure 3.1). Primary cyclization, i.e., when a pendant double bond reacts with a radical on the same chain, causes microgels and heterogeneity to occur in the polymer. The effects of heterogeneity are often pronounced including a significant delay in the gel point conversion, a dramatic change in the kinetics of polymerization (since cycles hinder mobility less than crosslinks), and a reduction in some properties such as mechanical strength, solvent resistance, and glass transition temperature [5–8].

Previously, models for the structural evolution, the gel point conversion, and sol and gel fractions have been developed for chain polymerizations which form crosslinked polymers [5–17]. Generally, these models include the effects of cyclization on crosslinked polymer structure without providing a methodology for determining cyclization rates. In this work we will present a first principles model for calculating the rates of primary cyclization throughout the polymerization.

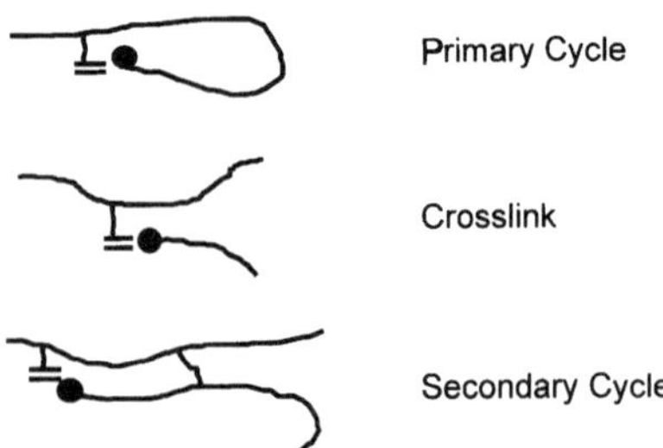

Figure 3.1   Three possible ways pendant double bonds react

The kinetics of cyclization reactions occurring during free radical photopolymerization of multifunctional monomers are difficult to characterize because of the varying pendant reactivity and inhomogeneities of the system. This work will focus on a novel modeling approach, which solves the differential kinetic equations to investigate the factors influencing cyclization versus crosslinking. The model follows the formation time and lifetime of each pendant double bond added to the polymer, starting when the first double bond on a multifunctional monomer is consumed. Calculation the concentration of radicals in the bulk solution and the concentration of radicals on the same propagating chain (local radicals) is used to determine the rate of reaction of each pendant. A population balance approach is taken to account for varying 'birth times' of each pendant vinyl in the reaction as well as the local radical concentration associated with each pendant vinyl. In this manner it is possible to account for the observed experimental behavior.

BACKGROUND

The effect of cyclization of polymerization of crosslinked systems has been observed experimentally including when the actual gel point conversion in a system was significantly greater than that predicted by classical Flory–Stockmayer theory [11,12]. Dusek and Ilavsky [9] indentified that primary cyclization was the dominant reaction of pendant functional groups at low conversions. Other researchers have theorized that the pendant's reactivity then decreases with conversion due to the polymer gel forming around the pendants, shielding them from further reacting [1,10]. Dusek and Ilavsky developed a statistical model for crosslinking during copolymerizations, but they assumed that all the pendant functional groups were equivalent in reactivity [9]. The model was thus limited in the degree of crosslinking for which it was applicable.

A percolation type model which accounts for varying pendant reactivity in crosslinking chain polymerizations was first proposed by Manneville and de Seze [5] and has been utilized and further developed by numerous researchers over the last decade [5–8]. As discussed by Kloosterboer [1], in this model type a cubic lattice structure was assumed to simulate the network structure in space. Each monomer was considered to be only one point on the lattice. Sites were arbitrarily chosen to contain free radicals. At each unit of time a neighbor was chosen and a chemical bond was formed if the pendant double bonds of the neighbor were not yet fully reacted. The model, thus, simulates the radicals randomly moving through the lattice connecting monomer molecules. Most importantly, these types of models were able to include the non-mean field nature of the polymerization reaction.

Thus, these simulations were able to predict successfully the inhomogeneities that existed as microgels formed at low conversions. These microgels form initially because of the high pendant double bond reactivity, which is represented in the lattice model by the greater spatial proximity of pendants to propagating radicals. Thus, more primary cycles were created at low conversion. The simulation also showed that inhomogeneities were diminished with decreasing maximum kinetic chain length and were dependent on the rate of initiation. The percolation model, however, underestimated the shielding of pendant double bonds at high conversion.

Anseth amplified the percolation type simulation creating a new kinetic gelation model [8,15]. In this simulation a face centered cubic lattice was used rather than the simple cubic lattice. Molecules were allowed to occupy more than one site so that the effect of the different sized monomers could be investigated. Further, the initiator concentration was assumed to decay through a first order decay reaction rather than assuming all initiator is present at the beginning of the reaction. Termination by combination of initiator radicals was also included. These assumptions put additional information in the simulation that allowed it to be more accurate. Primary cyclization was allowed to occur in the model, but the relative flexibility or stiffness of monomers in cycling back on themselves was not accounted for.

The Anseth model was able to predict for monomers more accurately the fraction of fully reacted versus conversion when compared to the experimental data from Kloosterboer [1], and the understanding of highly crosslinked polymerization kinetics was furthered. This lattice simulation type model is limited, however, because it uses a fixed lattice structure and there is difficulty in introducing realistic mobility of the reacting species, although it has been the current model of choice for modeling highly crosslinked polymers.

An additional approach to modeling the network evolution of crosslinked polymers is pseudokinetic modeling, which was developed by Tobita and Hamielec [13,14]. In this method all the important elementary reactions in the copolymerization are considered, but the kinetics of these reactions are simplified by averaging the effect of chain size into pseudokinetic rate constants that are functions of time. Thus, the kinetic expressions are reduced to be analogous to expressions for a homopolymerization. Okay *et al.* [16] continued this approach while also including cyclization into the possible kinetic mechanisms, fitting experimental data to the model. His work elucidated the importance of cyclization in the beginning of the reaction, where it was estimated to be 30–60% at zero monomer conversion. Because of the pseudokinetic approach a constant rate of primary and secondary cyclization and crosslinking is assumed throughout the reaction [16,17].

METHOD

Little non-mean field kinetic modeling has been done because of the difficulty in developing the appropriate differential equations, which represent the varying pendant double bond reactivities as a function of conversion. In this work a

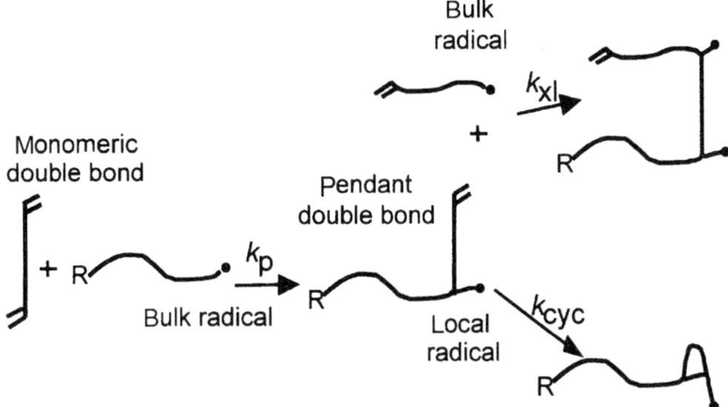

Figure 3.2 Mechanism of monomeric and pendant double bond reaction

kinetic model is proposed that develops and solves the relevant species balances and accounts for the difference in reactivity before monomeric double bonds and pendant double bonds. This approach captures the difference in reactivity of double bonds throughout the entire polymerization.

In Figure 3.2 the different pathways for reactions of double bond are illustrated. As shown, the consumption of a monomeric double bond is a function of monomer concentration, bulk radical concentration, and the propagation kinetic constant (k_p), while the reactivity of a pendant double bond is controlled by two competing the reactions: cyclization and crosslinking. Pendant vinyls reacting with a radical on the same propagating chain (local radical) produce primary cycles while reaction with a bulk radical produces either secondary cycles or crosslinks. Cyclization and crosslinking reactions are controlled by the kinetic constant for cyclization (k_{cyc}) and the local radical concentration, and the kinetic constant for crosslinking (k_{xl}) and bulk radical concentration, respectively. Thus, pendants and monomeric double bonds are given different reactivities based on differing kinetic constants and radical concentrations. The overall propagation rate of the reaction (R_p) is the sum of the rate of monomeric double bond consumption (R_m) and the rate of pendant double bond consumption (R_{pen}) by cyclization and crosslinking:

$$R_p = R_m + R_{pen} \tag{3.1}$$

MONOMERIC DOUBLE BOND REACTIVITY

The rate of monomer consumption is defined for a bimolecular reaction as the product of the kinetic constant and the concentration of each the reactants.

$$R_m = k_p[M][R_b] \tag{3.2}$$

The bulk radicals are produced from decaying initiator. Assuming pseudo-steady state on the bulk radical concentration, the bulk radical concentration ($[R_b]$) can be expressed as the following:

$$[R_\mathrm{b}] = \left(\frac{R_\mathrm{i}}{2k_\mathrm{t}}\right)^{1/2} \tag{3.3}$$

where $[R_\mathrm{i}]$ is the rate of initiation and k_t is the termination kinetic constant. This expression is the standard method of calculating the radical concentration in free radical polymerization. The initiator concentration is assumed to be constant or can be allowed to decay exponentially. In the model results presented in this paper, the rate of initiation is assumed constant.

PENDANT DOUBLE BOND REACTIVITY

For each time step in the model the concentration of pendants created by the consumption of monomer is stored. Pendants born at each time are tracked separately for their reactivity with bulk radicals and with the radical on the same propagating chain (local radical). The rate of consumption of pendants by bulk radicals is calculated by the product of the kinetic constant for crosslinking (k_xl) and concentration of bulk radicals, as defined above. Consumption of pendants by cyclization is dependent on the local radical concentration, which is less straightforward to define as it is correlated directly to pendants born at a specific birth time.

For the pendants of each birth time, a volume is defined which includes both the pendant vinyl and propagating radical on the same kinetic chain that initially formed the pendant. The volume has a radius, which represents the distance between the pendant double bond and the radical. This radius is assumed to be the average distance between double bonds in a monomer (r_o), plus the end-to-end distance ($r_\mathrm{e-e}$) of the polymer that has the same number of repeat units that are between the pendant and propagating radical (Figure 3.3). The local radical concentration ($[R_1]$) for the pendant is thus, the moles of radicals, which is one divided by Avogradro's number (N_A), over the volume that contains the pendant on the same chain:

$$[R_1] = \frac{1}{N_\mathrm{A}(4/3\pi(r_\mathrm{o} + r_\mathrm{e-e})^3)} \tag{3.4}$$

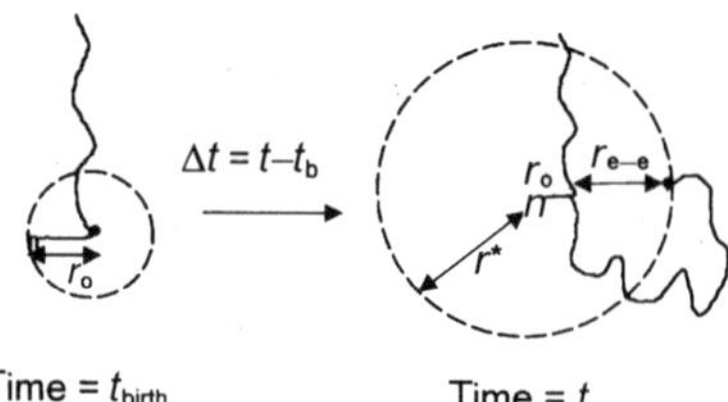

Figure 3.3   Radius containing the local radical for a particular pendant

The value of r_0 can be varied to represent different sized monomers or monomers with different stiffness. The polymer end-to-end distance is assumed to have the following Gaussian form based on statistical arguments for the freely jointed chain model [18].

$$r_{e-e} \approx n^{1/2}l \tag{3.5}$$

The variable l represents how much length each monomer adds to the overall chain length and is estimated as 4 Å in the kinetic model presented. The variable n represents the number of repeat units between the monomeric double bond that reacted to form the pendant and the radical. It is defined as the rate of propagation (R_P) divided by the bulk radical concentration ($[R_b]$) integrated over the time since the pendant was formed:

$$n = \int_{t_{\text{birth}}}^{t} R_p(t')/[R_b]\mathrm{d}t' \tag{3.6}$$

Thus, the expression for the local radical concentration is given below:

$$[R_1] \approx \frac{1}{N_A[4/3\pi(r_0 + n^{1/2}l)^3]} \tag{3.7}$$

A bimolecular reaction is assumed where the rate of consumption of pendants depends directly on the local concentration of radicals. Using the concentration of radicals developed in equation (3.3) and equation (7), the rate of consumption of pendants is the sum of the rate of pendant reaction with bulk radicals and local radicals:

$$R_{\text{pen}|_{t_b}} = \left[k_{xl}\left(\frac{R_i}{2k_t}\right)^{1/2} + \frac{k_{\text{cyc}}}{N_A[4/3\pi(r_0 + n^{1/2}l)^3]} \right] [\text{pen}]|_{tb} \tag{3.8}$$

where k_{xl} is the kinetic constant for crosslinking and k_{cyc} is the kinetic constant for cyclization. The total rate of pendant disappearance is then calculated by summing R_{pen} for pendants born at all times before the current time. Using this kinetic expression and the rate of consumption of monomer, a numerical model has been developed which tracks the formation of pendant double bonds and their subsequent reaction with bulk and local radicals.

INPUT PARAMETERS AND KINETIC CONSTANTS

For the model runs presented in this paper, the values of input parameters were taken from experimental data for DEGDMA [19]. Diffusion controlled kinetics are incorporated into the model. Data presented in this paper, however, does not include the effect of diffusion control in order to isolate the effect of other parameters on the kinetics of cyclization. Thus, the propagation and termination kinetic constants are fixed at the more diffusion controlled values. As the rate of initiation is also held constant, the bulk radical concentration is unchanging throughout the reaction. Thus, the effect of the local radical concentration can be isolated from the effect of the bulk radical on cyclization. When diffusion controlled kinetics are included, the expressions for the kinetic parameters, k_p, k_t, k_{cyc}, and k_{xl}, incorporated are those given by Anseth and Bowman [20]. These expressions include autoacceleration and deceleration effects, reaction diffusion termination, and the transition between when propagation and termination are diffusion controlled and reaction controlled. The kinetic constants for propagation (k_p) and cyclization (k_{cyc}) are assumed to be the same. To account for steric limitations in the reaction of a pendant with a bulk radical, the kinetic constant for crosslinking (k_{xl}) is assumed to be two orders of magnitude lower than the propagation kinetic constant (k_p). This reduction is approximately the value of k_p in a diffusion-controlled system at 80% double bond conversion. It is assumed that viscous limitations reducing k_p at 80% conversion are approximately the equivalent the reactivity limitations on k_{xl} due to steric hindrance.

RESULTS AND DISCUSSION

Because of the first principles approach, this kinetic model is able to investigate factors which control primary cyclization in chain polymerization. The effects of monomer size and solvent concentration on the degree of primary cyclization were considered. Results are presented for a bulk homo-polymerization unless otherwise specified.

The monomer and double bond concentrations decrease exponentially with time (Figure 3.4). The total amount of pendants present begins at zero and

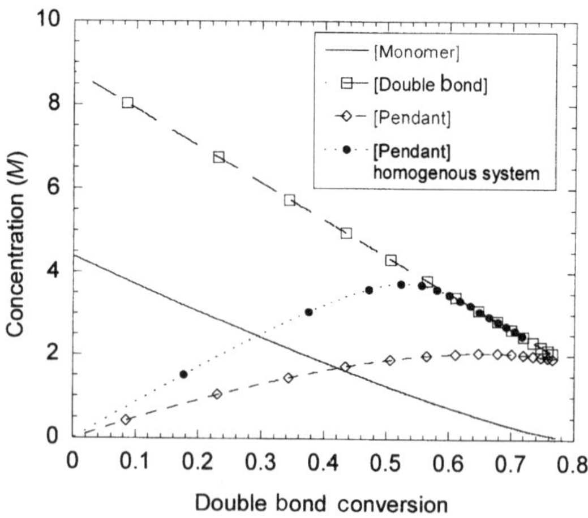

Figure 3.4 Model prediction of double bond, unreacted monomer, and total pendant concentrations (monomer size = 20 Å, light intensity = 1 mW/cm²)

increases as monomer is being consumed rapidly. At 55% double bond conversion, the total amount of pendants begins to decrease as the rate of monomer consumption falls off and more pendants are reacting away than are being formed (Figure 3.4). The pendant concentration of a homogenous system (i.e. a system where local radical reactions are neglected) is also shown in Figure 3.4, illustrating that pendants are reacting away faster due to cyclization than they would be expected to. In this homogenous system k_{xl} is assumed to be reduced because of steric hindrance and k_{cyc} is assumed to be zero.

Because of increased viscosity effects and the increasing shielding of the pendant double bond, it is known that the pendant double bond reactivity varies over time. As seen in Figure 3.5, the fraction of reacting pendants forming primary cycles varies with conversion. The highest fraction of pendants cycling occurs in the beginning of the reaction because the limited propagation of radicals away from the newly created pendants (smaller end-to-end distance). As time goes on, the fraction to cycle deceases because of the decreasing local radical concentrations and the increasing relative amounts of pendants which can react with bulk radicals. The effect of monomer size on cyclization is also shown in Figure 3.5. Decreasing monomer size from 100 Å to 10 Å increases cyclization by about two and a half times at 60% double bond conversion. This trend is consistent with data from the Anseth kinetic-gelation model [8].

The relationship between cyclization and solvent concentration is illustrated in Figure 3.6. Upon the addition of 60% solvent the fraction of primary cycles is about 140% higher than with no solvent at 60% double bond conversion. This trend is consistent with what we would expect as the consumption of monomeric

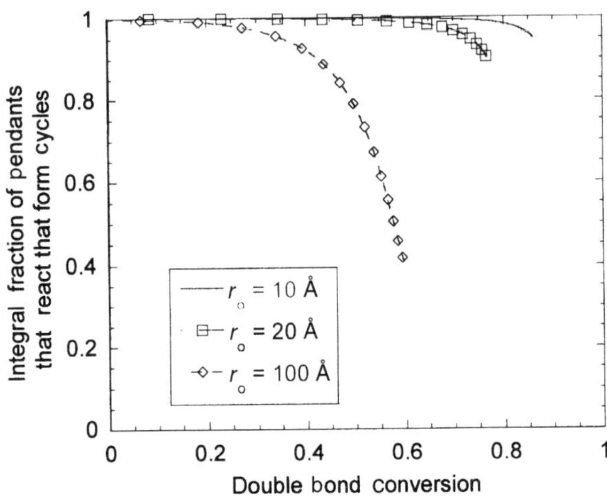

Figure 3.5 Effect of monomer size on the integral fraction of reacting pendant forming primary cycles (monomer size = 20 Å, light intensity = 1 mW/cm^2)

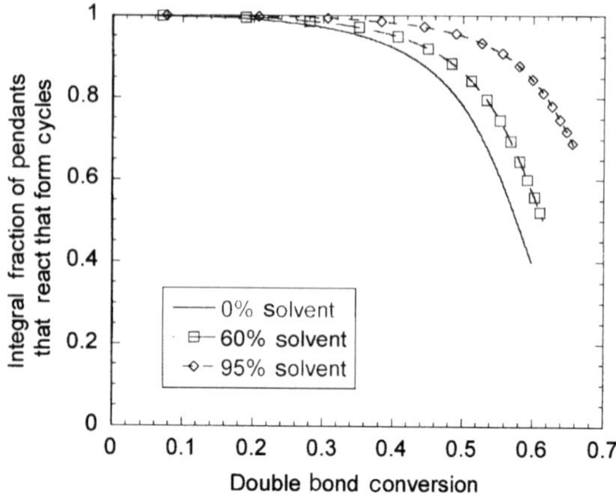

Figure 3.6 Effect of solvent concentration on the integral fraction of reacting pendant forming primary cycles (monomer size = 100 Å, light intensity = 1 mW/cm²)

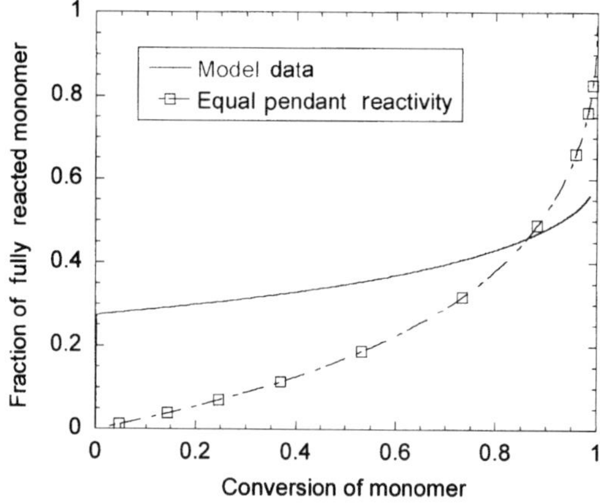

Figure 3.7 Fraction of fully reacted monomer versus monomer conversion (monomer size = 20 Å, light intensity = 1 mW/cm²)

and pendant double bonds are competing reactions. At lower monomer concentrations (60% solvent system), the pendant double bonds are more reactive because the effective local concentration of radicals of the pendant double bonds is maintained while the monomeric double bond concentration is decreased by the addition of solvent. At solvent concentrations of 95%, the fraction of reacting pendants that cycle is nearly 100% until 50% double bond conversion. The effect of solvent on cyclization is seen here even without incorporating changes in free volume because of increased mobility of the polymer in solvent.

Because cyclization is difficult to measure experimentally, very little experimental data is available to test the validity of the model. One study by Kloosterboer about pendant reactivity was used for comparison with the model results [1]. In his work Kloosterboer determined the fraction of fully reacted monomer units as a function of monomer conversion for hexanediol diacrylate. In Figure 3.7 model predictions are compared with mean-field assumption data, where the reactivity of all double bonds is assumed to be equal. Model data is consistent in its trend with the experimental data. The amount of cyclization is approximately 35% until 80% monomer conversion. At the end of the reaction (above 80% monomer conversion) the consumption of pendants increases relative to the consumption of pendants because the pendant concentration is at least an order of magnitude higher than the monomer concentration.

CONCLUSIONS

The kinetic model presented provides insight into the factors effecting cyclization during the photopolymerization of multifunctional monomers. Results demonstrate how pendant reactivity varies with conversion. The integral fraction of reacting pendants that form cycles is highest initially, when the radical on the same chain has undergone limited propagation. The model also shows how cyclization increases with decreasing monomer size and decreasing initial monomer concentration (increasing solvent concentration). Additionally, the model is able to predict experimental data for quantitatively relative rate of consumption of monomeric and pendant double bonds.

In future work, the effect of light intensity, copolymerization, and monomer functionality will be evaluated. The model will also be used to investigate the effect of diffusion controlled kinetics on crosslinked polymerization to provide more insight into primary cyclization in crosslinked polymer systems.

ACKNOWLEDGMENTS

The authors would like to acknowledge Jay Anseth for his work in running the model simulations. This work was funded by the Camille Dreyfus Teacher–Scholar Program, the Presidential Faculty Fellow Program at the National Science Foundation, and a National Science Foundation Graduate Fellowship.

REFERENCES

1. J.G. Kloosterboer, *Adv. Polym. Sci.*, **84**, 1 (1988).
2. R.J.M. Zwiers and G.C.M. Dortant, *Applied Optics*, **24**, 4483 (1985).
3. G.H. Johnson, G.E. Thomas, S.B. Luitjens, J.E. Hudson, P.D. Scholten, D.J. Whilhite, U. Enz, J.G. Kloosterboer, P.E.J. Legierse, D.J. Graversteijn, P. Hansen and H. Heirtmann, Information Storage Materials in *Ullmann's Encyclopedia of Industrial Chemistry*, Vol. A14, VCH Publishers: New York (1990).

4. K.S. Anseth, S.M. Newman and C.N. Bowman, *Adv. Polym. Sci.*, **122**, 177 (1995).
5. Manneville and L. de Seze, in *Numerical Methods in the Study of Critical Phenomena* (J. Della Dora, J. Demonogoet and B. Lacolle, eds), Springer, Berlin, 1981, p. 116.
6. H. Boots and R. Pandey, *Polym. Bull.* **11**, 415 (1984).
7. G. Simon, P. Allen, D. Bennett, D. Williams and E. Williams, *Macromolecules*, **22**, 3555 (1989).
8. K.S. Anseth and C.N. Bowman, *Chem. Eng. Sci.*, **49**, 2207 (1994).
9. K. Dusek and M. Ilavsky, *J. Polym. Sci., Symp.*, **53**, 57 (1975).
10. K. Dusek and J. Spevack., *Polymer*, **21**, 750 (1980).
11. K. Dusek, in *Developments in Polymerization — 3. Network Formation and Cyclization in Polymer Reactions* (R.N. Haward, ed.), Applied Science Publishers: Englewood Cliffs, NJ 1982.
12. H.M.J. Boots, J.G. Kloosterboer, G.M.M. van de Hei and R.B. Pandey, *Brit. Polym. J.*, **17**, 219 (1985).
13. H. Tobita and A.E. Hamielec, *Makromol. Chem. Macromol. Symp.*, **20/21**, 501 (1988).
14. H. Tobita and A.E. Hamielec, *Macromolecules*, **22**, 3098–3105 (1989).
15. K.S. Anseth, Ph.D. Thesis, University of Colorado, 1995.
16. O. Okay, M. Kurz, K. Lutz and W. Funke, *Macromolecules*, **28**, 2728–37 (1995).
17. H.J. Naghash, O. Okay and Y. Yagci, *Polymer*, **38**, 1187–96 (1997).
18. P.J. Flory, in *Principles of Polymer Chemistry*, Cornell University Press, Ithica and London, 1993, pp. 399–408.
19. M.D. Goodner, H.R. Lee and C.N. Bowman, *Ind. Eng. Chem. Res.*, **36**, 1247–52, (1997).
20. K.S. Anseth and C.N. Bowman, *Polymer Reaction Engineering*, **1**, 499–520 (1993).

4

Networks Monte Carlo Simulations for Coatings Research in Industry

J. AERTS

Akzo Nobel Central Research, Arnhem, the Netherlands

ABSTRACT

Thermosets, materials forming polymeric networks upon heating or drying, are the main components of coatings, adhesives and some plastics. Their properties mainly depend on the density of elastic effective chains between crosslinks and the chemistry between the crosslinks. The application of the Eichinger Monte Carlo method for the prediction of the network properties of thermosets is presented here. The Eichinger Monte Carlo method uses information from RIS-calculations to build a three-dimensional reaction box in which the monomers are reacted by using a capture sphere approach. This method

Wiley Polymer Networks Group Review Series Vol. 2. Edited by B.T. Stokke and A. Elgsaeter
© 1999 John Wiley & Sons Ltd

ensures that ring formation is intrinsically taken into account. The method has been tested on a stoichiometric system of three-functional isocyanate with a diol in the presence of extra water. The gel point retardation is predicted as a function of the extra amount of water and the reactivity of water relative to diol. Other properties such as the gel fraction and the shear modulus are calculated and compared to those of an ideal network from Miller–Macosko calculations. When using the Eichinger Monte Carlo approach, a size effect of the reacting molecules has been detected: even when assigned equal reactivities, the reaction of functional groups on small diols is retarded relative to functional groups on large diols. The effect of gradually replacing a large-chain diol by a short-chain diol on the shear modulus is evaluated. Calculations have been performed for two regimes which can be designated as 'reaction controlled' and 'diffusion controlled'. The Eichinger Monte Carlo method is compared with other methods that are used in the coatings industry such as Dry Add. Also possible improvements to the Eichinger method are proposed. Important future developments of computational methods in the coatings industry will be the combination of network calculations with QSPR methods in order to predict properties like T_g, viscosity, hardness, etch- and scratch-resistance, resistance to weathering, etc.. An example is given of a QSPR predicting the chemical contribution to the T_g of a polymeric network.

INTRODUCTION

Thermosetting polymers have been used for more than 100 years in coatings, adhesives and plastics. Despite this long history, the use of computational methods to describe and predict the properties of thermosets is still underdeveloped in industry.

Thermosets typically originate from small molecules or mixtures of small molecules (monomers) or prepolymers (oligomers). These precursors are typically in the fluid state, or in the form of a powder (e.g. powder coatings). When heated, a crosslinking reaction occurs (condensation polymerization), leading to a polymeric network, with very high molecular weights. Thermosets can also be produced by a radical reaction mechanism, usually by mixing the monomers with a radical precursor, with subsequent exposure to heat, or visible or UV-light (e.g. UV-curable coatings). In the case of condensation polymerization, to which we will limit ourselves in this chapter, a polymeric network can only develop if at least one of the monomers has at least three reactive groups, which can react with itself or with other monomers with at least two reactive groups. Classic reactions are epoxy with amine, acid with alcohol (polyesters) or isocyanate with alcohol (polyurethanes). Also the classic reaction of rubber with sulphur can be treated as a condensation polymerization leading to a polymeric network.

In industry, prediction of the physical properties of a polymeric network can lead to considerable cost reductions when a new or improved thermosetting system is developed. Therefore, many industrial research groups have implemented the Miller–Macosko methodology [1,2] based on Flory–Stockmayer theory [3–9], into software packages in order to solve formulation problems or optimize formulations of thermosetting systems. The implementation of this methodology however is essentially limited to ideal systems (no ring formation, equal reactivities, etc.). In order to take ring formation (which leads to retardation of the gel point and reduced mechanical properties) into account the Ahmad–Rolfes–Stepto (A–R–S) theory [10,11] has been developed, giving an excellent description of the change of polymer networks properties as a function of dilution of the starting composition. Dilution typically leads to increased probability of intramolecular reaction, and hence to ring formation. However, also in the A–R–S methodology a treatment of systems containing groups with unequal reactivities, as encountered in many commercial compositions, is not (yet) possible.

Therefore, some industrial research groups have developed Monte Carlo methods enabling the treatment of monomers with chemical groups with unequal reactivities, or where the reactivity of a chemical group in a monomer is dependent on the probability that a neighbouring chemical group has reacted or not. One of these methods has been developed at the ICI laboratories [12] and the software package is now being commercialized by Oxford Materials under the name DryAdd [13].

On the other hand, Eichinger *et al.* developed a Monte Carlo (MC) method for the prediction of the physical properties of rubber networks [14,15]. Recently, the method, as implemented in the MSI suite of molecular and mesoscale polymer modelling packages, has been adapted for thermosetting systems from small molecule monomers [16]. An important difference between DryAdd and the Eichinger method is that the latter uses spatial information of the branching and end points of the monomers in order to predict and take ring formation into account.

We have now tested the Eichinger method for its applicability to thermosets from isocyanates and diols in the presence of water.

OUTLINE OF THE EICHINGER METHOD

The Eichinger method for the calculation of properties of polymer networks uses information from rotational isomeric state theory (RIS) [17,18], for the calculation of end-to-end distance distributions of reactive groups in the starting monomers. In practice, mostly the RIS–Metropolis–Monte Carlo (RMMC) [19] method, as implemented in the MSI software, is used. This RMMC is based on a molecular mechanics calculations, using high quality force fields such as the recently released COMPASS forcefield [16,20]. This information is necessary as the amount of ring formation will be determined by the distances between reactive groups *within* one molecule or particle relative to the distances between reactive groups *between* molecules or particles.

In the first stage of an Eichinger Monte Carlo calculation a large number of monomers is created with reactive group/reactive group distances according to the distribution calculated by RIS or RMMC. These monomers are then placed at random places and with random orientation in a (virtual) closed reaction vessel (a cube) with a volume corresponding to the given density (typically a volume of 10^7Å^3 is used). Essentially, only coordinates of the branching points and reactive groups are generated. This corresponds to what molecular modellers call 'coarse-graining'. Excluded volume of the molecules is not taken into account and no periodic boundary conditions for the reaction cube are used. As the monomers have been defined as polymeric units during building, molecular weight distributions of branches and backbones (as present in many industrial systems) can be taken into account. Reaction probabilities between chemical groups are then defined (relative reactivities) and 'steric penalties', i.e. reduced reactivities of chemical groups on small crosslinker molecules when one of the neighbour chemical groups has already reacted, can be assigned. A typical example where this is necessary is the reaction of epoxy with amine, where the second reaction of the amine group is less probable (lower reactivity).

At the start of the actual calculation a 'capture sphere' with a small radius is being put on each of the reactive groups in the system (Figure 4.1). If this capture sphere overlaps with a capture sphere of another reactive group, it is decided whether reaction takes place based on the assigned reaction probabilities by a Monte Carlo algorithm. If reaction takes place between groups on two different molecules, both the molecules are moved towards each other with a relative amount based on the reciprocal of their molecular weights. If the capture

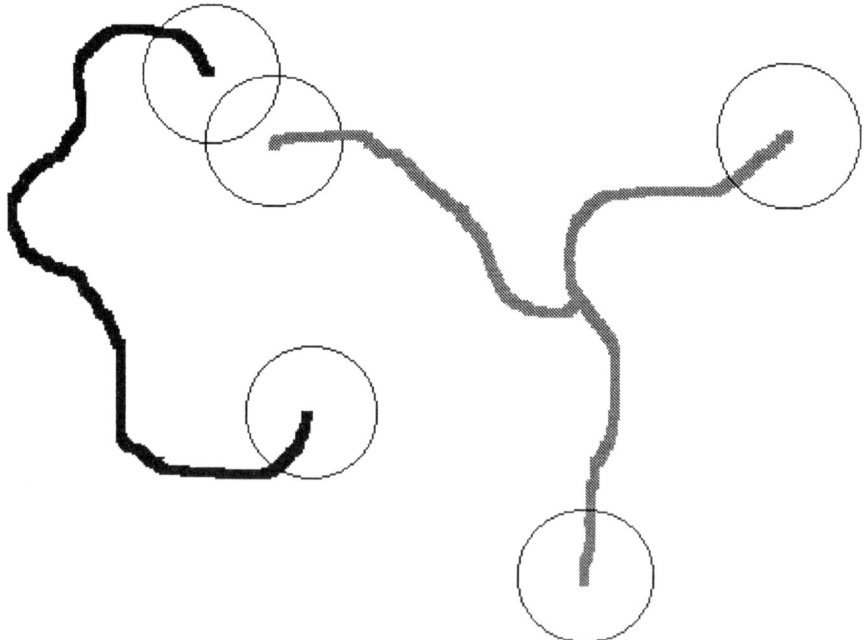

Figure 4.1 The capture radius concept

spheres on groups within the same molecule (or later, particle) overlap and reaction takes place, a bond between the groups is defined (intramolecular reaction with ring formation). The size of the capture sphere is then increased and the process repeated. Just before each increase of capture sphere radius, properties are calculated and (if desired) printed into a log file. The increase of capture sphere together with the movement of the molecules successfully mimics diffusion in the system, as has been shown by a number of calculations of polyethylene glycols with hexamethylene diisocyanate [16]. During the linkage process, a record is kept of the connectivity of all molecules through a spanning-tree algorithm. In the most recent implementation of the software, it is also possible to have an output of the coordinates and bonds between the branches and end points of the gel particle, for visualization or further analysis.

The gel particle is defined as the largest particle in the system. The gel point is calculated from the maximum in the slope of the number averaged molecular weight of the whole system (upper bound) and from the maximum in the number averaged molecular weight of the non-gel particles (i.e. the sol) (lower bound) which is typical for gelation. A gel point is assigned only if the weight fraction of the largest particle is at least 0.5. The difference in gel point as expressed in degree of conversion between lower and upper bound is usually within 0.01–0.02 if a sufficiently large system has been chosen. At the end of each step also an analysis of the gel particle using a graph-theoretical approach is performed.

Note that the calculation is performed as a function of conversion rather than as a function of time, as no information about kinetics is provided to, nor generated by, the method. This also means that there is no temperature applied to the system. The software requests a temperature, but it is only used for the calculation of the mechanical properties of the gelled system (see further).

In order to get good statistics, we usually perform a set of three independend calculations for each system, i.e. using three different starting configurations and each time a new random number seed for the Monte Carlo procedures. Typical CPU times on an SGI workstation with and R8000 processor are then in the order of 20 minutes per configuration (for a system size of 10^7Å^3.

CALCULATIONS

A set of calculations was performed on a system which was at the same time as close as possible to real industrial systems, as well as suitable for analysis by experimental methods (not reported here). The system we choose consists of a mixture of isocyanates (Desmodur LS2025), mainly of 3-functional isocyanate (designated Desmo-f3), 4-functional isocyanate (Desmo-f4) and 5-functional isocyanate (Desmo-f5). The mole fractions are 0.81, 0.14 and 0.05 respectively. The molecular structures are given in Figure 4.2. These isocyanates can react with hydroxy groups from a long-tail flexible diol (propriety information) or

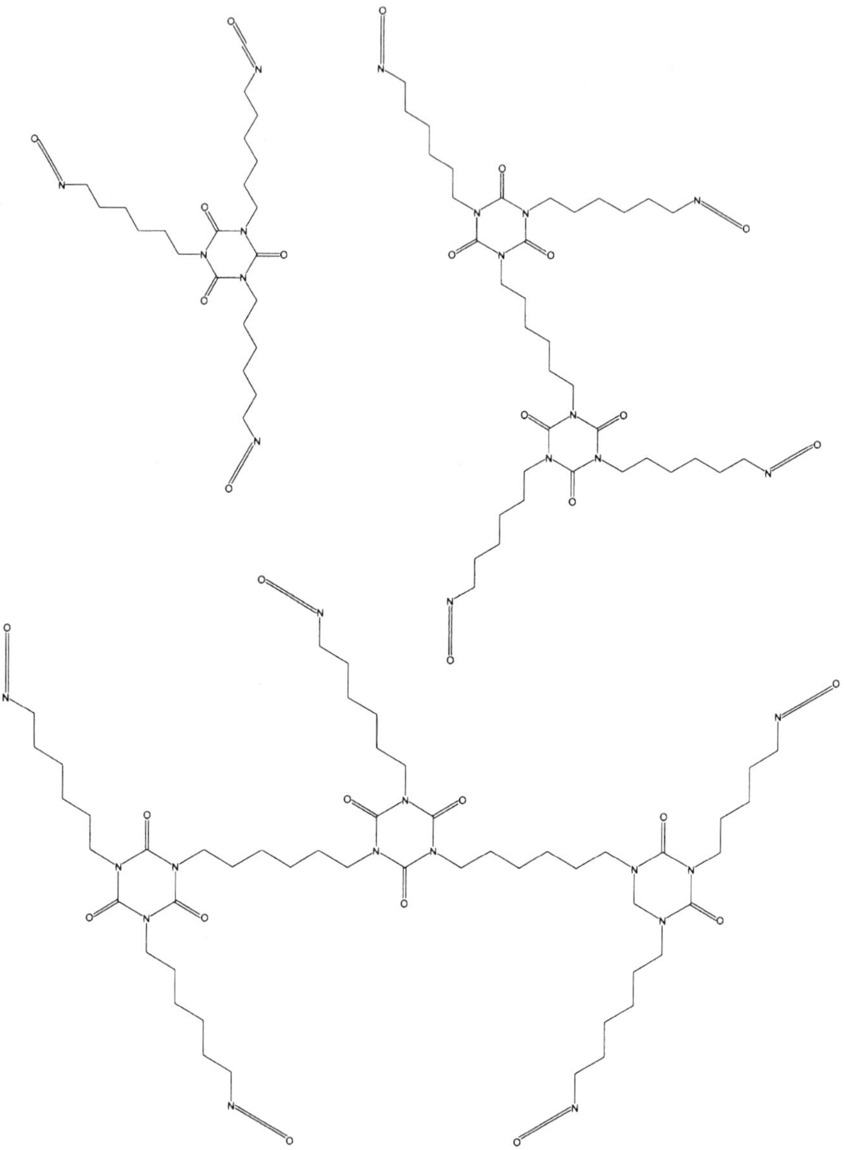

Figure 4.2 Molecular structures of the monomers in Desmodur LS2025

with water. In the former case a urethane group is formed, in the latter case an amine group is formed (with splitting off of carbon dioxide), which can react with a second isocyanate group to form an urea. In the former case a long, flexible chain is created between branches or crosslinks, whereas in the latter case a short, hard urea block is created.

                                        45

$$R-N=C=O + R'-OH \rightarrow R-NH-CO-O-R'$$

$$R-N=C=O + H_2O \rightarrow R-NH2 + CO_2$$

$$R-NH_2 + R-N=C=O \rightarrow R-NH-CO-NH-R$$

Water, OH of diol and amine each have a different reactivity for isocyanate, but in the first instance, in order to test the main features of the software, we choose to assume equal reactivies for amine and for water. Also we first worked with the 3-functional isocyanate (Desmo-f3) only. Group-to-group distance distributions of the reactive groups on the monomers were calculated using RMMC and used as input for the Monte Carlo calculations. The calculated OH–OH (diol) and NCO–NCO (Desmo-f3) distance distributions are shown in Figure 4.3.

The system was stoichiometric in OH (of diol) and isocyanate groups. Extra amounts of water molecules were added to the system with molar ratios water/diol ranging from 0.0 to 1.0 (i.e. the latter system contains as many moles of water as of diol). The effect of water is not only gel retardation due to unequality of the amount of OH groups (from diol or from water) and isocyanate groups, but the reaction product with water leads to a shorter chain than with OH of the diol, leading to increased cyclization.

Figure 4.4 shows the calculated gel points for reactivities of water/amine relative to OH of diol of 0.2, 1.0, 5 and 20. The latter means that water (and also its amine reaction product) is taken as 20 times are reactive as OH of diol. It is seen that the gel points for a relative reactivity of 1.0 are close to the

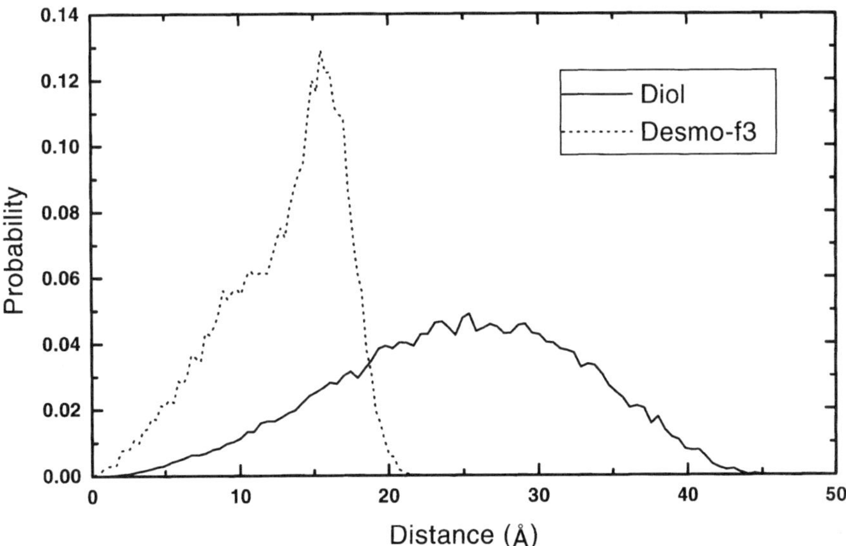

Figure 4.3 Distance distribution between the reactive groups in the 3-functional isocynate and the diol

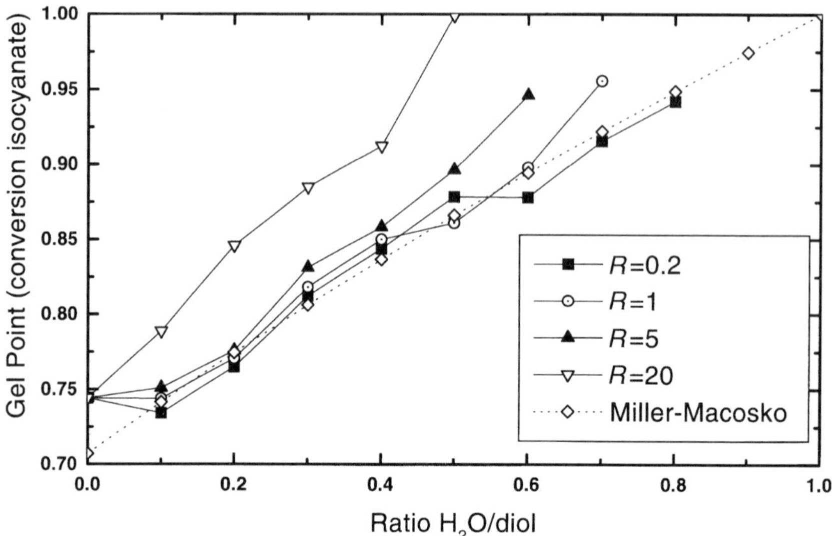

Figure 4.4 Calculated gel points (expressed as conversion of isocyanate) as a function of the amount of water added and relative reactivity of water

Miller–Macosko gel points, giving the impression that cyclization is not important. However, the weight fractions of gel at the end of reaction (Table 4.1) are rather different from the Miller–Macosko ones. Even at low amounts of water the weight fraction of gel is considerably below the Miller–Macosko value, which shows that cyclization is not unimportant, even though the gel points are close to the Miller–Macosko values. At high amounts of water, the weight fraction of gel is above the Miller–Macosko value. However, one must realize that in the Eichinger concept, the definition of a gel (largest particle) is different from that in the Miller–Macosko concept (infinite network). In the Eichinger concept, at high amounts of water, gelation does even not occur, as the weight fraction of the largest particle is below 0.5.

Figure 4.4 also nicely reveals the effect of the (relative) reactivity of water (set equal to that of amine) versus that of OH from the diol. When water is less reactive than diol ($R = 0.2$), no strong effect is seen on the gel point, except that it follows the Miller–Macosko line somewhat longer than for $R = 1$ (if no point is plotted in the graph there is no gel point). If however water is of higher reactivity ($R = 5$ and $R = 20$), gelation occurs at higher isocyanate conversion, or, depending on the amount of water, the system does not gel at all (weight fraction of the largest particle in the system < 0.5). These gel points cannot be calculated with classic Miller–Macosko theory as the latter does not allow unequal reactivities. More important, the Eichinger method (as other MC methods such as DryAdd) allows the calculation of a large set of physical properties of these networks.

Table 4.1 Weight fractions of gel at reaction completion, as function of amount of extra water (relative to diol) for a system with stoichiometric amounts of OH (from diol) and isocyanate

No. moles of water relative to diol	Weight fraction of gel at end of reaction Eichinger model	Weight fraction of gel at end of reaction Miller–Macosko
0	0.989	1.0
0.1	0.939	0.999
0.2	0.894	0.972
0.3	0.834	0.934
0.4	0.756	0.876
0.5	0.710	0.797
0.6	0.621	0.695
0.7	0.526	0.566
0.8	0.441	0.408
0.9	0.402	0.221
1.0	0.287	0.0

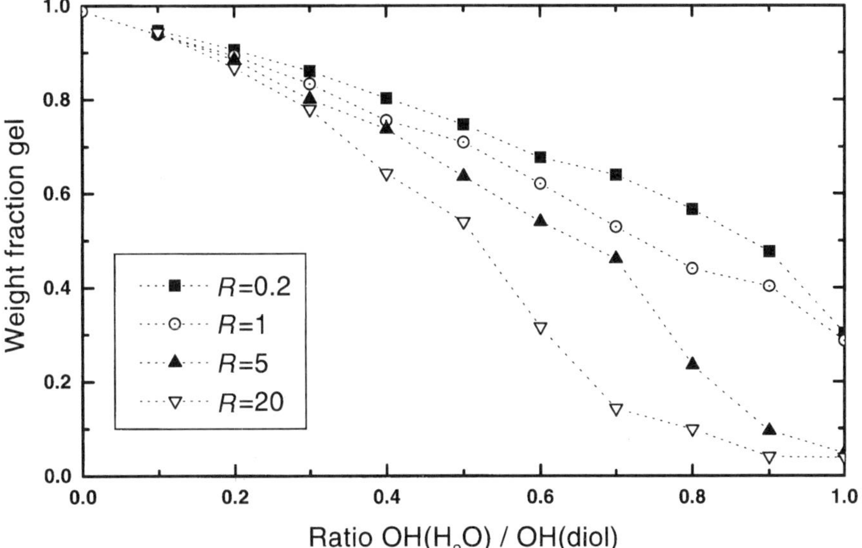

Figure 4.5 Weight fractions of the largest particle (the gel) as a function of the amount of water added and reactivity of water

Figure 4.5 shows the weight fractions of the largest particle (if >0.5 this is the gel) at the end of reaction for different amounts of water and different reactivity ratios.

The Eichinger Monte Carlo method has essentially been developed in order to predict mechanical properties of rubbery networks. For affine deformations, the

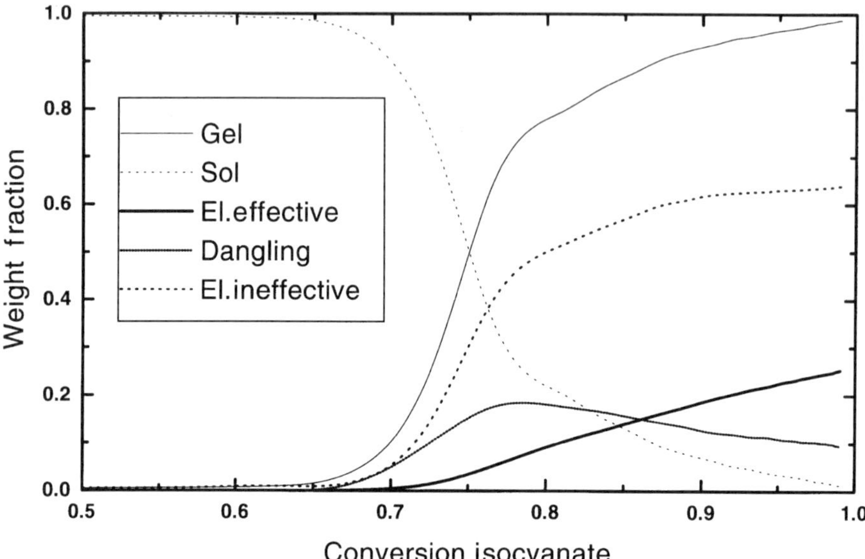

Figure 4.6 Evolution of weight fractions of gel, dangling chains and elastic effective chains for a system without water. The difference between gel fraction and elastic-effective and dangling chains is assigned as 'elastic-ineffective' chains

shear modulus is proportional to the number density of elastic effective chains (and inversely proportional to the temperature), whereas in the phantom chain limit, it is proportional to the cycle rank of the system. In the following we shall concentrate on the fraction of elastic effective chains.

Figure 4.6 shows the evolution of weight fractions of sol (being defined as everything but the largest particle), gel, dangling chains (also named pendant chains), and elastic effective chains for a system without water. The difference between gel fraction and elastic and dangling chains is assigned as 'inelastic' chains. In reality, the molecular weight of effective chains is calculated as the harmonic mean of all shortest chains between two crosslinks when there is more than one such chain between two crosslinks (multiple edges in the graph).

In particular, these multiple chains between crosslinks are responsible for a modulus reduction of at least a factor 2 relative to the ideal network.

Figure 4.7 shows the calculated modulus (at full conversion of isocyanate) as a function of amount of added water and relative reactivity. The modulus of the ideal network is more than twice the highest calculated value. From the figure it is clear that the modulus drastically decreases when even small amounts of water are present in the system.

Figure 4.8 shows the weight fraction of smallest cycles in the largest particle as a function of conversion of isocyanate for a system consisting of stoichiometric

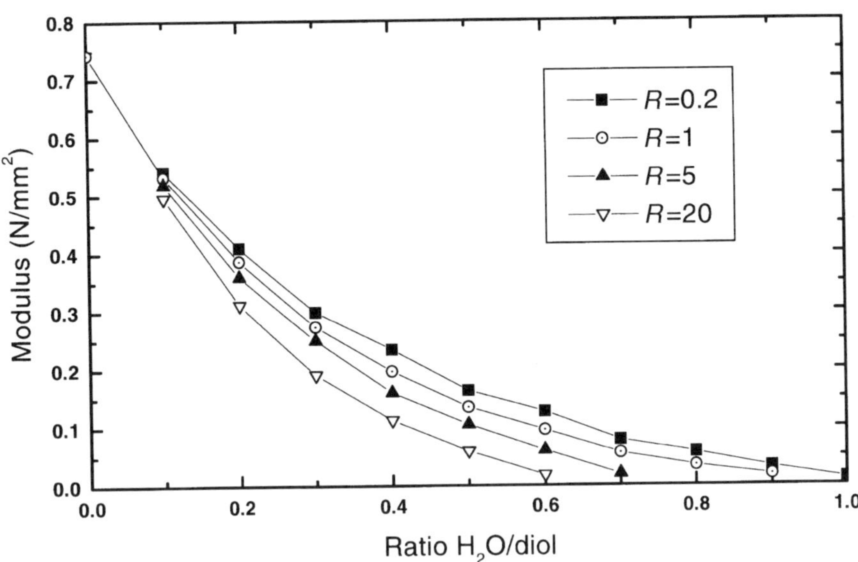

Figure 4.7 Calculated modulus (at full conversion of isocyanate) as a function of the amount of added water and relative reactivity

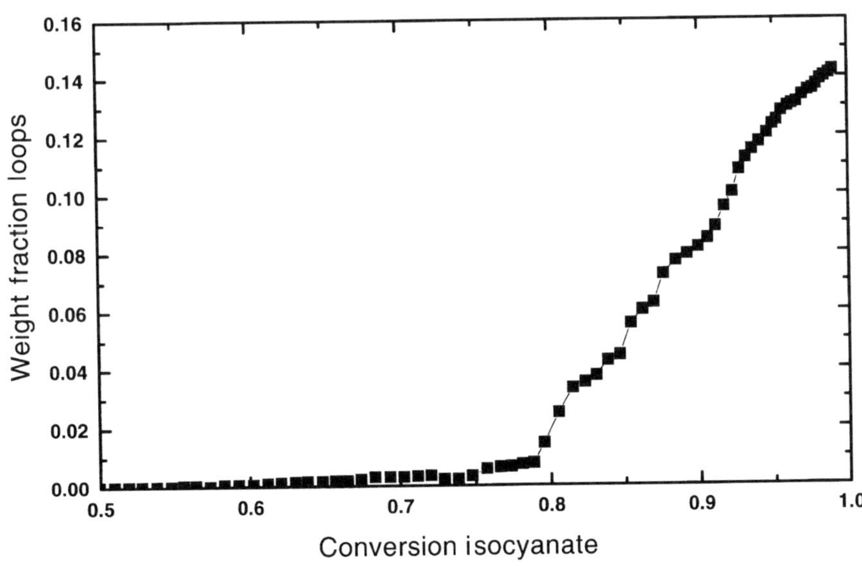

Figure 4.8 Weight fraction of smallest cycles (loops) in the largest particle as a function of conversion of iscyanate for a system consisting of stoichiometric amounts of Desmo-f3 and water

amounts of Desmo-f3 and water. At full conversion there is about 15% of smallest cycles. The gel point is at 83% conversion. The formation of the smallest cycles essentially starts just below the gel point and then rapidly increases more or less linearly.

EFFECT OF THE SIZE OF THE DIOL/WATER

Figure 4.9 shows the conversion of OH of the diol and H_2O as a function of isocyanate conversion where OH and H_2O were given the same reactivity $R = 1.0$. From the figure it is clear that the conversion of water is retarded relative to the OH conversion. The reason for this is that the OH groups of the diol react almost independently, as they are spatially well separated. The two functionalities of water, however (once as OH, the second as amine, see above), are located at the same point in space. As with the capture sphere concept (at least when $R = 1.0$) the OH groups and water molecules essentially react with the nearest isocyanate group, a second reaction of the water molecule (as amine) will be retarded as it will have to react with the second-neighbour isocyanate group, which will be at larger distance.

A set of calculations with ethyleneglycol ($HO-CH_2-CH_2-OH$) replacing the water molecules reveals an analogous situation but with a lower retardation than

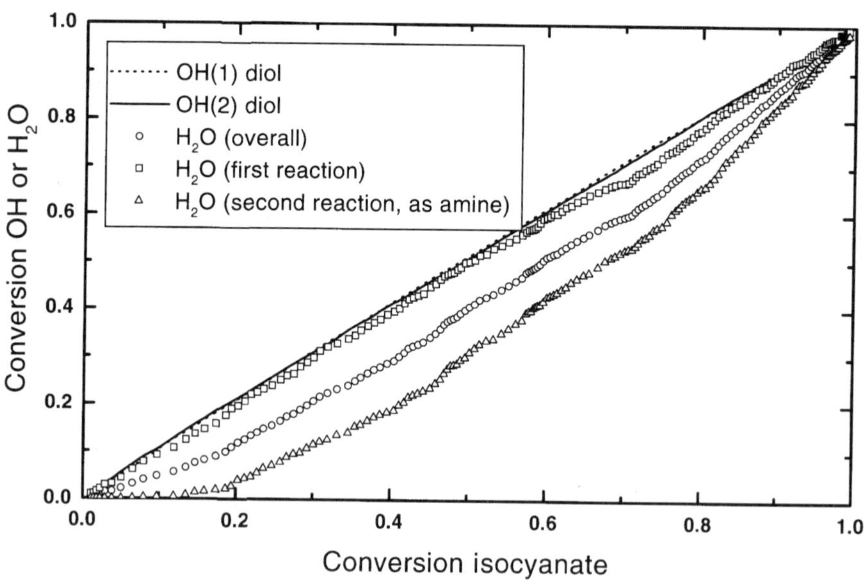

Figure 4.9 Conversion of OH-functionalities of the diol (upper two lines) and H_2O (symbols) as a function of isocyanate conversion for a system where OH and H_2O were given the same reactivity ($R = 1.0$)

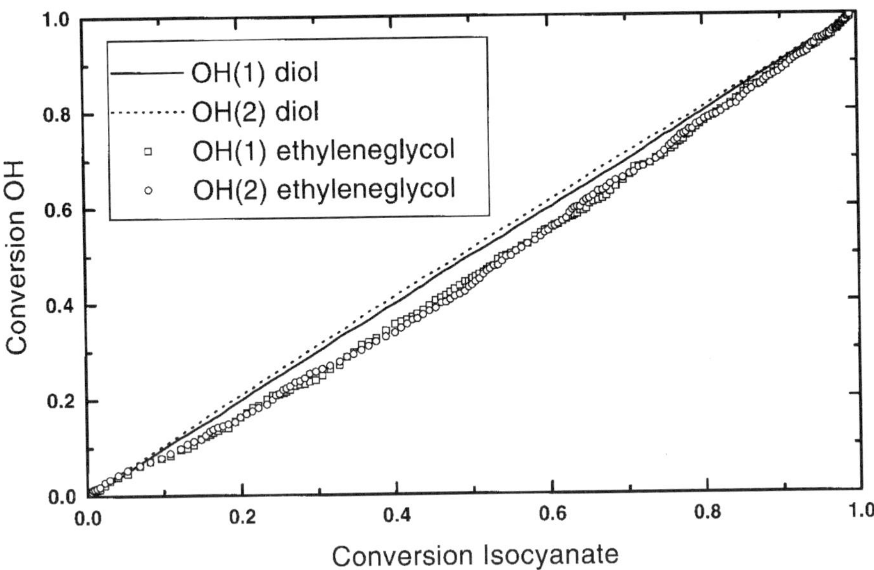

Figure 4.10 Conversions of OH groups of the diol and of ethyleneglycol as a function of the conversion of isocyanate for a system with equal reactivities assigned to OH of diol and of ethyleneglycol

with water (Figure 4.10). The conversion of OH of ethyleneglycol is at all times lower than that of the long-chain diol, but the difference is smaller than in the case of water. Of course in practice it is not easy or even possible to make an experimental distinction between the size effect and the real chemical reactivity, as these are mostly measured as one single property. So, in order to see whether this 'size effect' also occurs in reality, it will be necessary to perform a number of well-designed experiments.

CROSSLINK DENSITY VERSUS RING FORMATION

Imagine that one would like to increase the crosslink density in order to increase the modulus of the network. One way to do so is to replace the long-chain diol by a short-chain diol, e.g. hexamethylenediol ($HO-(CH_2)_6-OH$) or even ethyleneglycol. A simple Miller–Macosko calculation predicts an increase of the crosslink density by a factor of over 2.2 for ethyleneglycol relative to the long-chain diol. However, replacing a long-chain diol by a short-chain diol also enhances ring formation and thus also should decrease the modulus. Which effect is stronger, the increase in crosslink density or the increase in the formation of rings? Eichinger Monte Carlo calculations give the answer.

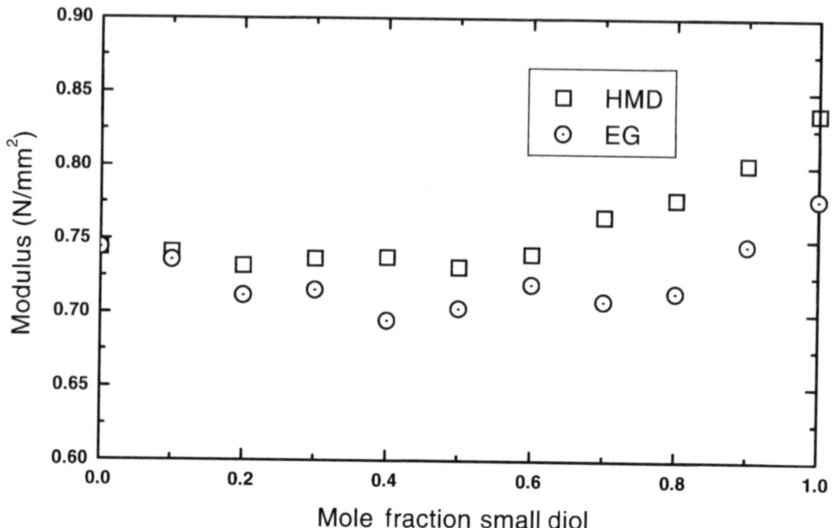

Figure 4.11 Calculated shear modulus (from the number of elastic effetive chains per unit of volume) as a function of replacement of the long-chain diol by ethyleneglycol (EG) or by a hexamethylenediol (HMD)

Figure 4.11 shows a plot of the calculated shear modulus (from the number of elastic effective chain per unit of volume) as a function of the amount of replacement of the long-chain diol by ethyleneglycol or hexamethylenediol. It is seen that the modulus does not increase at all with a factor of 2.2 for ethyleneglycol, as predicted by the Miller–Macosko method, but that the modulus increase is very modest. For small amounts of replacement, one even finds a reduction in the calculated modulus. This means that for small amounts of ethyleneglycol, loop formation is a more important factor, whereas at large amounts of ethyleneglycol, loop formation is counterbalanced by the shortening of the chains between the crosslinks which means more elastic-effective chains per unit of volume. For hexamethylenediol, the results are even more surprising: as for ethyleneglycol, a modulus reduction is found at small amounts of replacement, but at larger amounts a higher modulus is found than when using ethyleneglycol. This means that hexamethylenediol is more effective to increase the modulus as for ethylene-glycol the effect of ring formation is very dominant. This is in opposition to what one should expect from Miller–Macosko calculations, i.e. the hexamethylenediol increases the modulus less than ethyleneglycol.

DIFFUSION VERSUS REACTION CONTROLLED

Using the capture sphere concept in combination with high reaction probabilities essentially means that every isocyanate group will react with the

Table 4.2 Differences in calculated properties for 'diffusion-controlled' ($R_1 = R_2 = 1.0$) versus "reaction controlled" ($R_1 = R_2 = 0.001$) simulated kinetics

	$R_1 = R_2 = 1.0$ (diffusion-controlled)	$R_1 = R_2 = 0.001$ (reaction-controlled)
Gel point (conv.isocyanate)	0.73	0.79
Conv.OH at end-of-reaction	0.91	0.91
Conv. H_2O at end-of-reaction	0.88	0.81
Weight fraction gel	0.94	0.98
Weight fraction el. effective chains	0.32	0.39
Weight fraction dangling chains	0.26	0.17
Modulus (N/mm^2)	0.54	0.66
Max.capture radius (Å)	47	104

nearest-neighbour OH group of diol or water molecule. One may call this the diffusion-controlled situation. The reaction-controlled situation can be modelled by taking all the reaction probabilities very low, e.g. at 0.001. In this case, when the capture sphere radius is increased, and two groups satisfy the distance criterion, the probability of reaction is very low, and the capture sphere will have to be increased many times again, so that also groups at large distances will have the capability to react. This means that an isocyanate group can react with an OH group or water molecule which is far away, even at low degree of conversion. This is also seen from the output of the program as much larger capture radii are needed for reaction completion than in the diffusion-controlled case (typically 100 Å versus 40 Å). Table 4.2 shows the most important differences for a system of stoichiometric amounts of isocyanate and diol with additionally 0.1 moles of water (relative to diol). We find a shift in gel point from 0.73 to 0.79 when going from $R = 1.0$ to $R = 0.001$ for all possible reactions. The gel fraction at the end of the reaction increases from 0.94 to 0.98 and the elastic effective fraction from 0.32 to 0.39, whereas the dangling chain fraction decreases from 0.26 to 0.17. This means that in the 'reaction-controlled' case, a more perfect network is formed with better mechanical properties. Although not fully understood yet, the effect of reaction-controlled versus diffusion controlled seems to be non-negligible.

ADVANTAGES AND DISADVANTAGES OF NETWORK MONTE CARLO METHODS

Network Monte Carlo methods have a good number of advantages over analytical methods. They can easily be tailored towards special situations and are very versatile. All kinds of reactions with different reactivities can be treated. 'Electronic' and 'steric' penalties can be assigned to chemical groups when another nearby group in the same monomer has already reacted. An example of 'electronic' effect is e.g. toluene–diisocyanate, where the reactivity of one isocyanate group is changed when the other has already reacted. Another advantage is that ring formation is easily taken into account. Also in the Eichinger MC method as implemented in the MSI software, molecular weight distributions of prepolymers can easily be taken into account.

Disadvantages of the MC methods are surely the computing times. For a sufficiently large system and in order to avoid size effects ($>10\,000$ monomers) the CPU time for a single Eichinger MC calculation amounts to 20 minutes on an SGI workstation with an R8000 processor, whereas the Miller–Macosko method only takes a fraction of a second. This disadvantage is much less for the DryAdd software, which only takes a few seconds or minutes on a modern PC. However, Dry Add does not use spatial information (i.e no coordinates in space are assigned), and instead of using RIS-information for group-to-group distance distributions, the user has to give the probability of intramolecular (ring-formation) versus intermolecular reaction as input. The later has to be calculated using molecular modelling calculations in combination with A–R–S theory, or derived from experimental data. This ratio is then used during the whole reaction. Not using spatial information also means that no distinction can be made between diffusion- and reaction-controlled systems. Also molecular weight distributions of prepolymers cannot be taken into account as elegantly as in the Eichinger MC code.

Dry Add, however, has another important advantage: it is possible to 'add' monomers slowly to the reaction medium, or at a specific time during the reaction (e.g. at a certain conversion of one of the chemical groups). This enables e.g. to make a better model for curing under humid circumstances, as it is believed that in that case the concentration of free (unreacted) water in the reaction medium is more or less constant during the reaction, or to simulate curing systems for which a component is added after some pre-curing time.

One can state that for calculations on very simple systems (no serious ring formation expected, equal reactivities), Miller–Macosko software can very quickly give information whether a formulation can or cannot gel, and give an upper limit of the crosslink density and mechanical properties. The A–R–S method is very suitable for predicting the effect of dilution, but the ring-forming parameter has to be known or calculated in advance. When unequal reactivities are encountered, DryAdd is a very good choice if at least one has a good idea of the probability of intramolecular versus intermolecular reaction. It is very user-friendly and ideally suited for the formulation chemist. The relative probability of intramolecular versus intermolecular reaction has to be derived from experimental data or calculated by a computational chemist. If one does not a priori know the relative probability of intramolecular reaction, the best choice is the Eichinger MC method, where RIS calculations are used to calculate group-to-group distance distributions, which in the MC calculation itself is then used for determining whether intramolecular reaction occurs or not.

FUTURE DEVELOPMENTS

Many improvements are still possible for the MSI implementation of the Eichinger MC method. Better analysis methods can be provided for analysing the sol fraction. This is especially important for the development of resins as the classic methods often predict gelation, whereas this is just avoided in practice. Also the possibility of addition (or extraction) of (extra) molecules during or at certain stages of the reaction (as possible in the DryAdd software) is desirable, as this is common practice when manufacturing resins. For the algorithm this would mean that the total volume can change during the reaction. An important future direction might be the use of the generated RIS-information for backmapping the atomistic information onto the network topology. To be useful, however, faster and larger computers will be necessary as the resulting (set of) giant molecule(s) will have to be further treated with molecular mechanics and molecular dynamics.

In order to predict physicochemical properties such as the glass transition temperature (T_g) of the network, very probably network calculations will have to be combined with quantitative structure property relationship (QSPR) methodology. As an example we recently used the Stutz methodology [21], where the crosslink and branching density is calculated using a addition to Miller–Macosko methodology, for the prediction of T_gs of polymeric networks from epoxies with amines. The T_g^∞ parameters (which corresponds to the T_g of the equivalent infinite uncrosslinked polymer) were calculated using a QSPR. The T_g^∞ parameter essentially defines the contribution of the chemistry between the crosslinks (and branches) to the T_g of the network. In the development of our QSPR we could successfully apply the QSPR for T_g of Bicerano [22] for linear polymers by recalculating the coefficients [23]. The resulting QSPR shows calculated T_g^∞ values which are mostly within $10\,^\circ$C of the experimental ones, with a standard deviation of only $8\,^\circ$C (Figure 4.12). This allowed us to predict the T_g of thermosets resulting from many more combinations of epoxies and amines. Also other properties, such as viscosity, hardness, resistance towards UV and weathering, etch and scratch resistance, etc., will very probably in future be calculated or predicted with a combination of network calculations and QSPR methods.

CONCLUSIONS

The application of the Eichinger Monte Carlo method for the prediction of properties of thermosets has been presented. Evaluation versus experimental work is underway. The method has been shown to be a major improvement over methods that are currently used in the coatings industry and which do not take the effect of ring formation into account. The implementation of the method can still be improved and made more applicable to simulate practices that are common in resins and coatings manufacturing. The combination of Eichinger Monte Carlo methods with QSPR methods will in future enable to predict properties that are

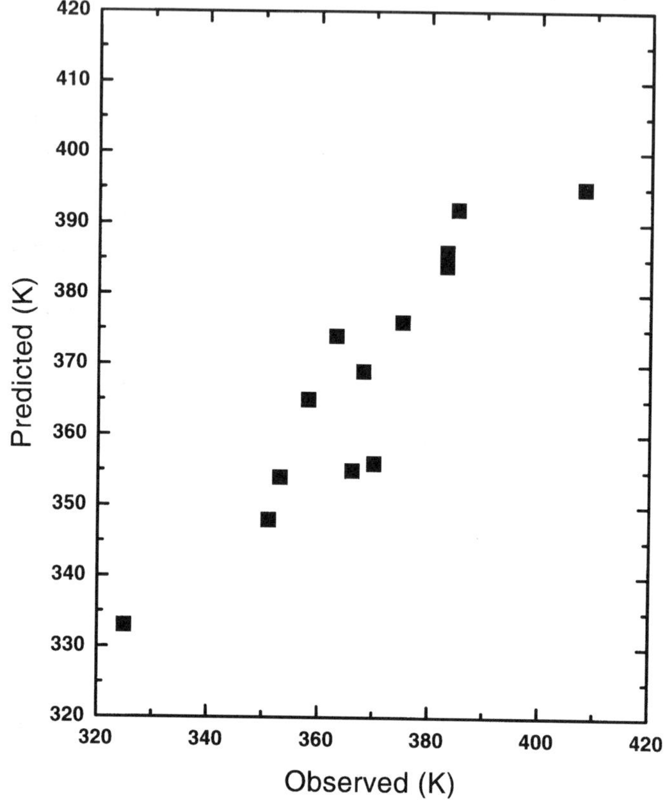

Figure 4.12 QSPR for T_g^∞ of Stutz *et al.* based on the QSPR of Bicerano for T_g^∞ of linear polymers with recalculation of the coefficients

important to the end-user of the coating, such as resistance to exposure to the environment.

REFERENCES

1. C.W. Macosko and D.R. Miller, *Macromolecules*, **9**, 199 (1976).
2. D.R. Miller and C.W. Macosko, *Macromolecules*, **9**, 206 (1976) .
3. P.J. Flory, *J. Am. Chem. Soc.*, **63**, 3083 (1941).
4. P.J. Flory, *J.Am.Chem.Soc.*, **63**, 3097 (1941).
5. P.J. Flory, *Principles of Polymer Chemistry*, Cornell University Press,Ithaca, N.Y., 1953.
6. W.H. Stockmayer, *J.Chem.Phys.*, **11**, 45 (1943).
7. W.H. Stockmayer, *J.Chem.Phys.*, **12**, 125 (1944).
8. W.H.Stockmayer, *J.Polym.Sci.*, **9**, 69 (1952).
9. W.H. Stockmayer, *J.Polym.Sci.*, **11**, 424 (1953).

10. Z. Ahmad and R.F.T. Stepto, *Colloid Polym.Sci.*, **258**, 663 (1980).
11. H. Rolfes and R.F.T. Stepto, *Makromol.Chem.*, Macromol.Symp., **76**, 1 (1993).
12. M. Claybourn and M. Reading, *J.Appl.Polym.Sci.*, **44**, 565 (1992).
13. Dry Add, Oxford Materials Ltd, Worley Court, Bolesworth Road, Tattenhall, Cheshire CH39HL, United Kingdom.
14. Y-K. Leung and B.E. Eichinger, *J.Chem.Phys.*, **80**, 3877 (1984).
15. Y-K. Leung and B.E. Eichinger, *J.Chem.Phys.*, **80**, 3885 (1984).
16. D. Rigby, H. Sun and B.E. Eichinger, Polym.Int., **44**, 311 (1997).
17. P.J. Flory, *Statistical Mechanics of Chain Molecules*, Interscience, New York, 1969.
18. P.J. Flory, *Macromolecules*, **7**, 381 (1974).
19. D. Honeycutt, *J.Theor.Comp.Polym.Sci.*, **8**, 1 (1998).
20. H. Sun and D. Rigby, *Spectrochimica Acta A*, **53**, 1301 (1997).
21. H. Stutz, K.H. Illers and J. Mertes, *J.Polym.Sci.B, Polym.Phys.*, **28**, 1483 (1990).
22. J. Bicerano, *Perdiction of Polymer Properties*, Marcel Dekker Inc., New York, (1990).
23. J. Aerts, to be published.

5

Formation of Diacetylene-containing Networks for Use in Deformation Micromechanics Studies on Polymers

J.L. STANFORD, R. CASADO, P.A. LOVELL and R.J. YOUNG

Polymer Science and Technology Group, Manchester Materials Science Centre, University of Manchester and UMIST, Grosvenor Street, Manchester, M1 7HS, United Kingdom

Wiley Polymer Networks Group Review Series Vol. 2. Edited by B.T. Stokke and A. Elgsaeter
© 1999 John Wiley & Sons Ltd

ABSTRACT

Diacetylene diols of structure $HO-(CH_2)_n-C{\equiv}C-C{\equiv}C-(CH_2)_n-OH$, where $n = 1$, 2, 3, 4 and 9, have been used to prepare diacetylene-containing polyurethanes (PUs) and polyesters (PEs). The PUs from the diol with $n = 1$ were formed via phase separation during copolymerization with MDI and a polyether diol (PPG400). A range of diacetylene-containing PEs from the diols was prepared by polycondensation with terephthaloyl chloride, and the PE from the diol with $n = 3$, PE ($n = 3$), was blended with isotactic polypropylene (PP). Cross-polymerization of the diacetylene-containing phases in copolymers and blends was effected by heat and ^{60}Co γ-radiation, respectively. The results from characterization of the copolymers, the PEs and the PE($n = 3$)/PP blends by DSC and Raman spectroscopy are presented. Simultaneous tensile testing and resonance Raman spectroscopy were used to determine the shift in wavenumber of the $C{\equiv}C$ stretching band at ~2100 cm^{-1} of the polydiacetylene crosslinks. The shifts were used to determine the local stress in the crosslinked phases independent of the overall applied stress, thereby facilitating determination of the efficiency of stress transfer in the multiphase materials.

INTRODUCTION

Polydiacetylenes are network materials formed by solid-state topochemical polymerisation of conjugated diacetylene units, present either in macromonomer crystals [1] or in the crystalline phases of semi-crystalline polymers [2–4]. Such network materials display very useful optomechanical properties [5–9] that enable detailed studies on deformation micromechanics of materials to be made directly at the molecular level. In particular, the stress dependence of the wavenumbers of the polydiacetylene $C=C$ and $C{\equiv}C$ stretching bands in Raman spectra can be used to determine directly the stress transfer between phases during the deformation of multiphase polymers [5,9]. This paper describes the formation of a range of diacetylene-containing polymers, either as linear polyurethane hard segments via *in situ* phase separation during copolymerization or as isolated linear polyesters for use as one component in polymer blends. Comparative results of the deformation micromechanics of the cross-polymerized copolymers and blends, determined from simultaneous tensile testing and Raman spectroscopy, are also presented.

EXPERIMENTAL

PREPARATION, MOULDING AND CROSS-POLYMERIZATION OF COPOLYMERS AND BLENDS

A series of diacetylene diols of general structure $HO-(CH_2)_n-C{\equiv}C-C{\equiv}C-(CH_2)_n-OH$ were prepared by oxidative coupling of the corresponding acetylenic alcohol ($HO-(CH_2)_n-C{\equiv}CH$) [10]. The acetylenic alcohol with $n = 1$, 2, 3 and 4 were commercially available. The acetylenic alcohol with $n = 9$ was prepared in two stages from $HO-(CH_2)_9-CH{=}CH_2$, by bromination and then elimination of HBr using potassium hydroxide.

The formation and cross-polymerisation of diacetylene-containing polyurethanes and polyesters are shown in Figure 5.1. Segmented copolyurethanes were prepared from 4,4 '-methylenediphenylene diisocyanate, MDI, polypropylene glycol, PPG400 (Mn $\sim$433 g mol^{-1}) and the diacetylene diol with $n = 1$ (namely, 2,4-hexadiyne-1,6-diol, HDD). Linear copolymers were produced via a one-stage, bulk polymerization process using stoichiometric equivalent amounts of isocyanate and total hydroxyl groups. The required amounts of PPG400 and HDD were blended at 80 °C under nitrogen in a sealed, flanged reaction vessel. The stochiometric amount of molten MDI was added and after complete mixing, the flask was quickly cooled to $\sim$20 °C to minimize the initial temperature rise due to the reaction exotherm. The mixture was then allowed to react with stirring at 60 °C for 2 h, cast into a pre-heated picture-frame mould and placed in an oven at 60 °C for a further 2 h to complete copolyurethane formation. Cross-polymerization was effected by thermal treatment (40 h at 100 °C) to produce highly crosslinked and deep coloured copolyurethane networks.

Diacetylene-containing polyesters were prepared from each diacetylene diol by reaction in solution with terephthaloyl chloride in 1:1 stoichiometry for 5 h at 0 °C. Each polyester was obtained in the form of a white powder which was washed first with 0.1 M HCl and then with deionised water before drying. The nomenclature for the polyesters is exemplified by PE($n = 3$), which defines the polyester prepared from HO–(CH$_2$)$_3$–C≡C–C≡C–(CH$_2$)$_3$–OH. Blends of PE($n = 3$) with isotactic polypropylene (Himont S30S), PP, were prepared, moulded and cross-polymerized as follows. In each case the polyester was first blended in the appropriate proportion with the PP by dissolution in boiling xylene, followed by co-precipitation into a large excess methanol which was at room temperature. After drying, the blend was compression-moulded directly into dumbbell specimens (gauge section 28 × 6 × 3 mm) in a pre-heated mould at 180 °C for a period of about 3 minutes before cooling to room temperature. The diacetylene-containing polyester phases were then cross-polymerized by exposure to ^{60}Co γ-radiation to a dose of 40 Mrad, causing the specimens to attain the deep green/purple colour characteristic of polydiacetylenes. Finally, the specimen surfaces were polished using successively finer grades of metallographic grinding paper.

CHARACTERIZATION METHODS

Differential scanning calorimetry, DSC, was performed on a DuPont 2000 Thermal Analyst instrument with a DuPont 910 cell base equipped with a DSC cell. Powdered samples ($\sim$10 mg) were analysed over the range -100 to 300 °C for copolymers and -50 to 400 °C for blends at a heating rate of 20 °C min^{-1} under nitrogen flowing at 20 ml min^{-1}. Calibration was carried out using indium. Resonance Raman spectra of cross-polymerized polyester powders were recorded using a Renishaw 1000 Raman spectrometer fitted with

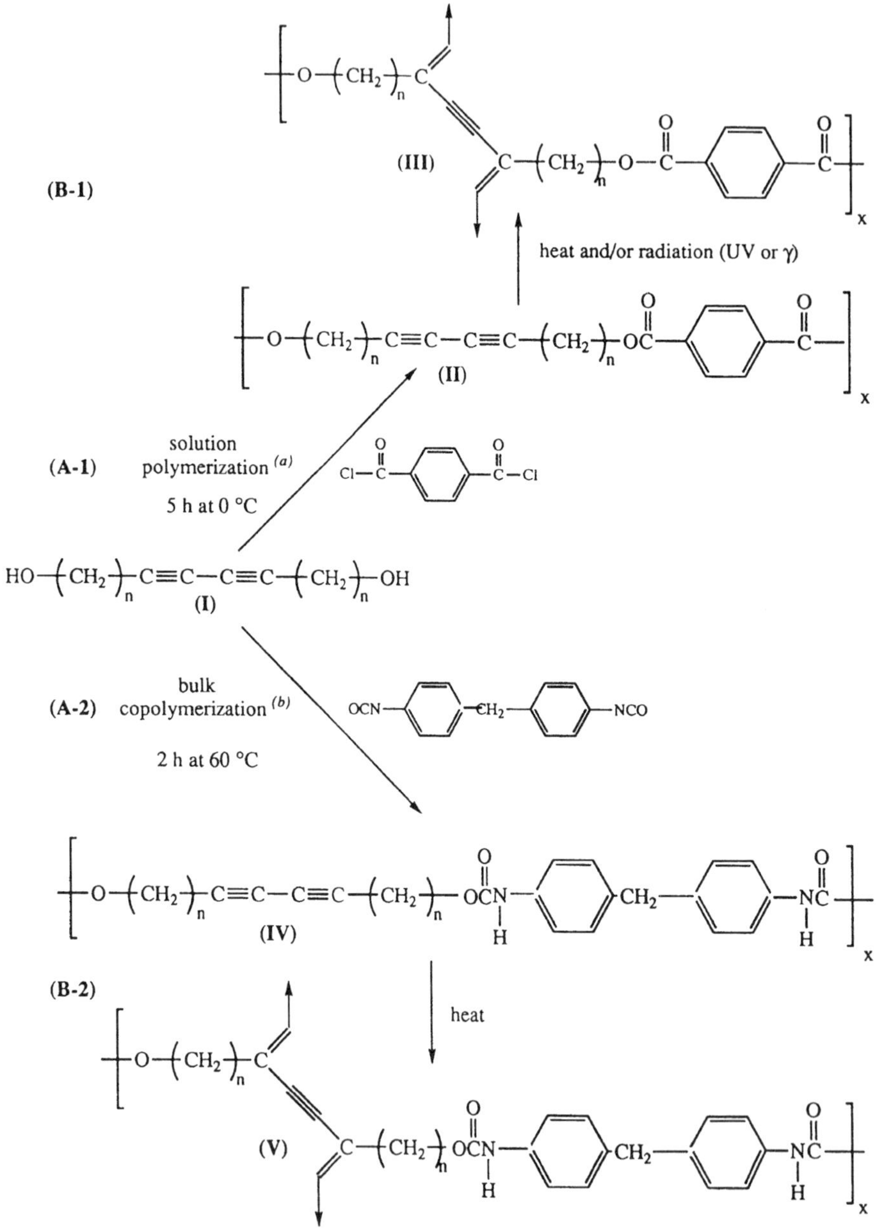

(B-1)

(III)

heat and/or radiation (UV or γ)

(II)

(A-1)　solution polymerization [a]

5 h at 0 °C

(I)

(A-2)　bulk copolymerization [b]

2 h at 60 °C

(IV)

(B-2)

heat

(V)

an Olympus optical microscope, a charge-coupled device detector, and a 25 mW He–Ne laser ($\lambda = 632.8$ nm) which was focused via a $\times 50$ objective lens to give a spot of diameter of ~ 2 µm with a power of ~ 1.0 mW at the sample. The scattered light was collected using a $180°$ back-scattering geometry and spectra recorded over wavenumber windows of 40 and 600 cm^{-1}, using scanning times of 10 and 1 s, respectively, for copolymers and blends.

SIMULTANEOUS TENSILE TESTING AND RAMAN SPECTROSCOPY

Cross-polymerized dumbbell specimens were subjected to simultaneous tensile testing and Raman spectroscopy measurements at $20 \pm 3\,°C$. The specimens were deformed to failure in a stepwise manner at nominal cross-head displacement rates of 1.0 and 0.4 mm min^{-1}, respectively, for copolymers and blends, using strain increments of about 0.04% on a Polymer Laboratories Minimat miniature mechanical tester fitted with a 1 kN load cell. The overall strain on a specimen was measured using a resistance strain gauge (accuracy $\pm 0.0014\%$ strain) bonded to the specimen. Raman spectra containing the $C{\equiv}C$ stretching band were recorded immediately after each increment in deformation using the apparatus and procedures described above. At least three specimens of each material were tested and mean values of tensile properties determined.

RESULTS AND DISCUSSION

PREPARATION AND PROPERTIES OF THE DIACETYLENE-CONTAINING POLYESTERS AND BLENDS

The diacetylene-containing polyesters prepared according to the reaction Scheme 1A (Figure 5.1) were obtained as semi-crystalline white powders in yields of 86–94%. They were analysed by DSC to determine their melting points (T_m) and the temperature (T_{exo}) associated with thermal cross-polymerization of the diacetylene units (see Figure 5.2). The results show that $PE(n = 1)$

Figure 5.1 Preparation of diacetylene-containing polymers from diacetylene diols **(I)**. **(A-1)**: Solution polymerization[a] using as solvents, N-methylpyrrolidone ($n = 1$, 2) and chloroform ($n = 3$, 4, 9), and pyridine as base, to give diacetylene-containing polyesters **(II)**. **(A-2)**: Bulk copolymerisation[b] using as co-reactant polypropylene glycol (PPG400, $M_n \sim 433$ g mol^{-1}), to give diacetylene-containing polyurethanes **(IV)** with $n = 1$. **(B-1)**, **(B-2)**: Solid-state topochemical cross-polymerization of the diacetylene units in crystalline regions of the polyesters and polyurethanes is induced by heat and/or radiation to produce highly oriented polydiacetylene crosslinks. The vertical arrows in the cross-polymerized networks **(III)** and **(V)** represent continuing polydiacetylene chains

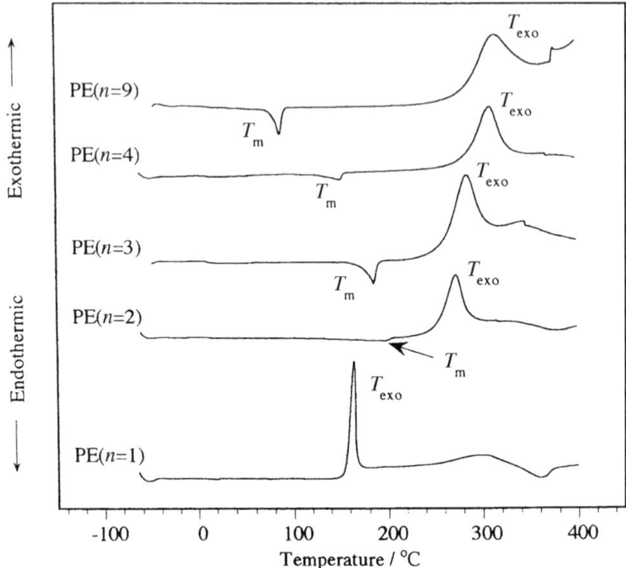

Figure 5.2 DSC curves of the as-prepared diacetylene-containing polyesters showing melting endotherms (T_m) and cross-polymerization exotherms (T_{exo}). PE($n = 1$) shows only T_{exo} because T_m for this polyester is above T_{exo}

did not melt, but underwent solid-state topochemical cross-polymerization (see Scheme B-1 in Figure 5.1) with $T_{exo} = 167\,°C$, in accord with previous observations [8,11]. Each of the other polyesters showed both a melting endotherm and a cross-polymerization exotherm, with T_m decreasing and T_{exo} increasing as the number of methylenes in the diacetylene unit increased. Thus, the objective of increasing the processing window (i.e., $T_m \to T_{exo}$) was achieved, enabling blending with polypropylene to be achieved without (thermal) cross-polymerization.

A series of blends of PP with PE($n = 3$) was successfully prepared and compression-moulded, without thermal cross-polymerization of the polyesters, using the procedures described in the experimental section. These materials, PB-1→PB-4, comprised 20, 40, 60 and 80 wt-% polyester and were a slightly off-white colour after moulding. The blends were analysed by DSC which showed the characteristic melting endotherms of the polyester and the PP ($T_m = 167\,°C$), and the cross-polymerization exotherm of the polyester. The magnitude of these transitions changed in accordance with the blend composition as shown by the comparative DSC curves in Figure 5.3. Exposure of the moulded tensile specimens to ^{60}Co γ-radiation led to solid-state topochemical cross-polymerization of the diacetylene units in the crystalline regions of the polyester phases.

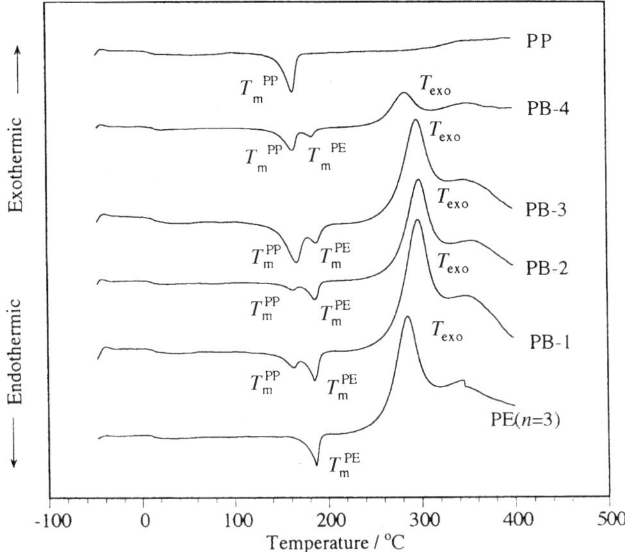

Figure 5.3 DSC curves of the as-prepared polyester, $PE(n = 3)$, isotactic polypropylene, PP, and polyester/polypropylene blends, PB-1 to PB-4, containing different percentage weight fractions (80–20%) of $PE(n = 3)$. Indicated on the DSC curves are melting endotherms T_m^{PE} and T_m^{PP} for the $PE(n = 3)$ and PP components, respectively, and cross-polymerization exotherms (T_{exo})

PREPARATION AND PROPERTIES OF THE DIACETYLENE-CONTAINING COPOLYURETHANES

The formation of diacetylene-containing polyurethanes, as shown by the reaction Scheme A-2 (Figure 5.1), occurs *in situ* during bulk copolymerization with the polyether diol PPG400. During copolymerization, phase separation results due to the thermodynamic incompatibility between the aromatic diacetylene-containing hard segments and the polyether-based soft segments. The development of hydrogen bonding and potential crystallinity of the diacetylene-containing hard segments further enhance the driving force for phase separation. The linear copolyurethanes thus form an essentially two-phase morphology comprising rigid, highly hydrogen-bonded hard segment domains (with a distribution of sizes) dispersed in a ductile, poly(ether–urethane) matrix. Spherulitic morphologies were observed [12] by transmission electron microscopy for the copolyurethanes with the highest hard segment contents ($\sim$35 wt-%), and spherulites up to 1 μm in size were typical.

Linear copolyurethanes were cross-polymerized by thermal treatment and optimum conditions producing maximum cross-polymerization but with minimum thermal degradation needed to be established [12]. Analysis by DSC showed no melting behaviour, despite the apparent hard-segment crystalline structure. This is probably due to the low hard segment contents and the paracrystalline nature of the domains, which provide relatively small energy changes that would be masked completely by the much more intense and extensive cross-polymerization exotherm. Complete cross-polymerization was not possible without causing significant disruption of the hard segment domains and some degradation. Conditions of 100 °C and 40 h for thermal treatment were therefore shown to be optimum [12].

The effects of copolymer composition on thermal behaviour are shown by the DSC curves in Figure 5.4 which show the expected trends in transition behaviour with increasing hard segment content. The intensity of the glass transition decreases and the value of T_g increases steadily as hard segment content increases, indicating increased phase-mixing between soft-and hard-segment phases. The intensity of the cross-polymerization exotherm increases markedly and gradually shifts to lower temperatures, reflecting not only the increase in diacetylene content but also a more efficient packing of the diacetylene units within the hard-segment domains. Thermal treatment at 100 °C for 40 hours shifts both T_g and T_{exo} to higher temperatures which also indicate increased phase-mixing with some disruption in the packing of diacetylene units within hard-segment domains.

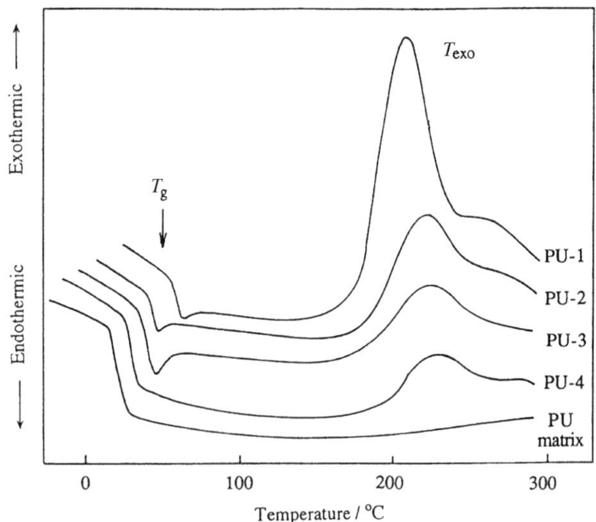

Figure 5.4 DSC curves of the thermally cross-polymerised copolyurethanes, PU-1 to PU-4 containing different percentage weight fractions (34–12%) of diacetylene-containing polyurethane hard segments. Indicated on the DSC curves are glass transition temperatures T_g and cross-polymerization exotherms (T_{exo}). The DSC curve for the polyurethane matrix phase (bottom curve) shows only T_g. No melting endotherms were observed for the copolymers as T_m for the hard segments is above T_{exo}

Table 5.1 Thermal behaviour (DSC) and deformation micromechanical properties (Raman spectroscopy) of cross-polymerized diacetylene-containing polymer blends and copolymer networks

Material	Composition (w:w/%)	$^d T_g$ (°C)	$^d T_m$ (°C)	$^g T_{exo}$ (°C)	$^h E$ (GPa)	$^i d(\Delta v)/d\varepsilon$ (cm^{-1}/1%)	$^j d(\Delta v)/d\sigma$ (cm^{-1}/MPa)
Polymer blends[a]							
PE($n = 3$)	[c]100:0	61, —	—, 159	314	3.15	−3.5	−0.11
PB-1	80:20	58, 16	150, 179	313	2.91	−3.4	−0.12
PB-2	60:40	58, 16	150, 170	315	2.56	−2.8	−0.11
PB-3	40:60	58, 16	150, f	312	2.18	−2.4	−0.11
PB-4	20:80	58, 16	153, f	315	1.78	−2.0	−0.11
Copolymer networks[b]							
PU-1	[c]34:66	[e]51	f	209	1.66	−6.23	−0.37
PU-2	27:73	42	f	218	1.54	−3.81	−0.25
PU-3	19:81	34	f	226	0.93	k	k
PU-4	12:88	26	f	240	0.12	k	k

[a]polyester ($n = 3$)/polypropylene: [b]polyurethane ($n = 1$)/polyether. [c]first figure refers to the diacetylene-containing phase; i.e. polyester in blends and polyurethane hard segments in copolymers. [d]values, respectively, for polyester and polypropylene phase in blends; (isolated polypropylene, $T_m = 157\,°C$). [e]single value only for copolymer; (isolated polyether soft-segment matrix, $T_g = 17\,°C$). [f]melting not observed; ($T_m > 200\,°C$ for polyurethane hard segments). [g]cross-polymerization exotherm (as-prepared materials). [h]tensile modulus. [i], [j]Raman (C≡C) shift factor [i]per unit overall strain, [j]per unit overall stress. [k]very weak Raman signals.

The composition and thermal behaviour (obtained from DSC analysis) of the various cross-polymerized polymer blends and copolymer networks are summarized in Table 5.1

DEFORMATION MICROMECHANICS STUDIES USING RAMAN SPECTROSCOPY

Very well-defined resonance Raman spectra were obtained for all of the blends but for only copolymers containing >20 wt-% of diacetylene-containing hard segments (see Figure 5.5). Both sets of materials show the strong C=C and C≡C stretching bands at ∼1480 and 2090 cm^{-1}, respectively, characteristic of polydiacetlylenes and, as stated, the C≡C stretching band is particularly sensitive to deformation and shifts to lower wavenumbers by up to -20 cm^{-1}, per 1% increase in tensile strain [13].

Deformation micromechanics data for blends and copolymers were obtained using dumbbell specimens subjected to tensile testing with simultaneous resonance Raman spectroscopy. For each material, the shift, $\Delta\nu$, in position of the C≡C stretching band was monitored as a function of increasing tensile deformation and, as an example, a typical set of data is shown in Figure 5.6 for the polymer blends. The dependence of the C≡C Raman peak position on either tensile strain (d($\Delta\nu$)/dε) or stress (d($\Delta\nu$)/dσ) was determined for each material from the initial slopes of plots such as those shown in Figure 5.6. Values of the Raman shift factor and the tensile modulus for each material are given in the final three columns of Table 5.1. Significantly, in contrast to the copolymers, good linear relationships between the Raman wavenumber shift and applied stress were obtained for all of the blends, essentially independent of blend composition as shown by the constant value of d($\Delta\nu$)/dσ. For materials containing similar proportions of diacetylene-containing phases, the copolymers show a greater dependence of Raman wavenumber shift upon deformation than the blends (cf. PU-1 and PU-2 with PB-3 and PB-4). These results may be explained in terms of the efficiency of stress transfer between phases [9] and its strong dependence on materials morphology [15] and, in particular, on the connectivity between the phases which is clearly better in the segmented copolyurethanes compared with that in the polyester/polypropylene blends.

Further analysis of the deformation micromechanics data enables the efficiency of stress transfer between phases in the copolymers and blends to be quantified. For example, values of $\Delta\nu$ for the deformed blends were converted into local stresses in the polyester phases using, as a calibration, the corresponding $\Delta\nu$ versus σ curve for the isolated cross-polymerized polyester PE($n = 3$) and the derived value d($\Delta\nu$)/dσ($= -0.11 \pm 0.01$ cm^{-1} MPa^{-1}). In this way the non-linear elastic behaviour of the polyesters was taken into account. The conventional stress–strain curves (from the load cell and strain gauge responses) and the curves of local stress in the polyester phases versus overall applied stress (from

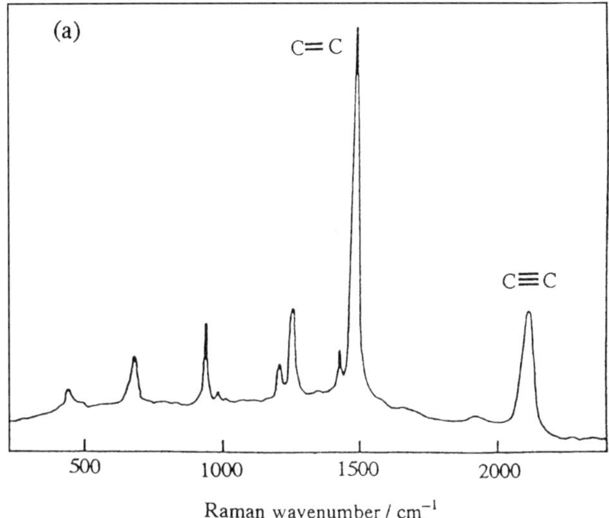

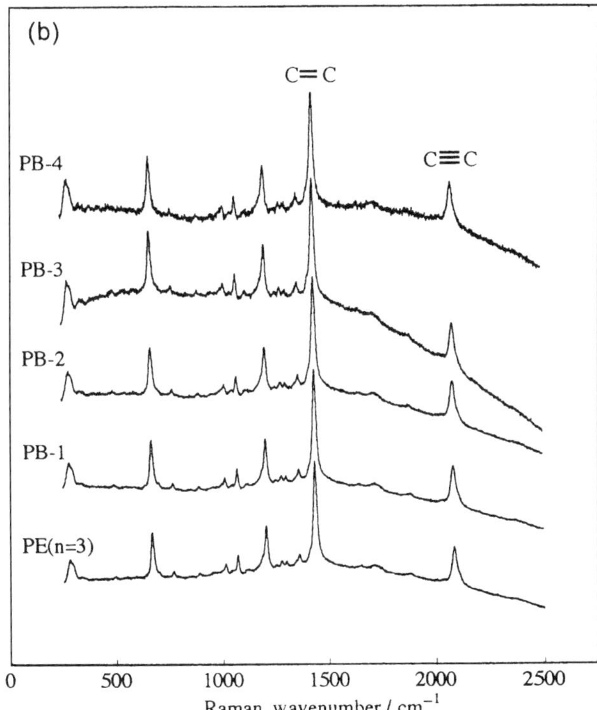

Figure 5.5 Typical resonance Raman spectra for (a) cross-polymerized copolyurethane PU-1 and (b) cross-polymerised polyester, PE($n = 3$) and polyester/polypropylene blends PB-1 to PB-4, containing different percentage weight fractions (80–20%) of PE($n = 3$). All spectra show the strong C=C and C≡C stretching bands characteristic of polydiacetylenes

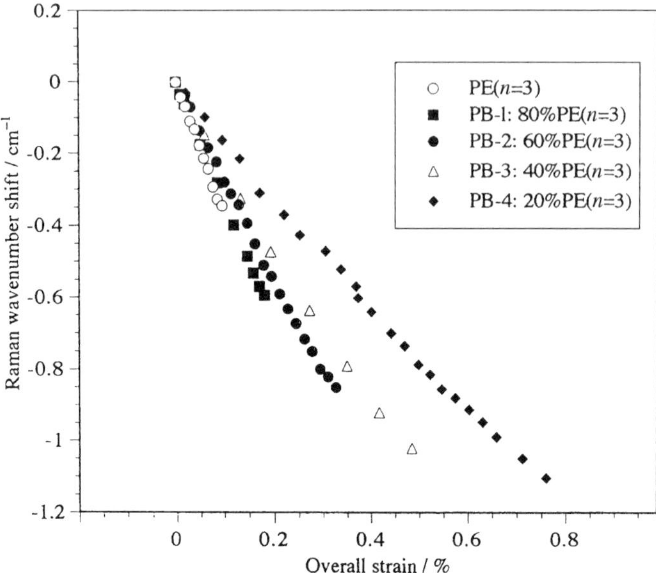

Figure 5.6　Strain dependence of the Raman band position of the C≡C stretching band in cross-polymerized polyester, PE($n = 3$), and polyester/polypropylene blends PB-1 to PB-4, containing different percentage weight fractions (80–20%) of PE($n = 3$)

the shift in Δv and the load cell response) are shown in Figure 5.7 for the PE($n = 3$)/polypropylene blend series. For the blends containing 80 and 60 wt-% PE($n = 3$), the local stress in the polyester is equal to the applied stress, showing that the distribution of stress between the phases is uniform. However, stress transfer to the polyester phase is inefficient in the blends comprising 40 and 20 wt-% PE($n = 3$) and diminishes as the applied stress increases, suggesting partial debonding of the polyester during deformation.

Thus, in conclusion, the results presented here clearly demonstrate that the approach described in this paper is capable of providing unique real-time information about stress transfer between phases during deformation of multiphase polymers such as segmented copolymers and polymer blends.

Figure 5.7　Deformation micromechanics data from simultaneous tensile testing and Raman spectroscopy of cross-polymerized polyester, PE($n = 3$) and polyester/polypropylene blends PB-1 to PB-4, containing different percentage weight fractions (80 –20%) of PE($n = 3$). (a): stress–strain curves with Young's modulus values shown against the curves for each of the materials. (b): local stress in the diacetylene-containing polyester phases plotted against the overall applied stress

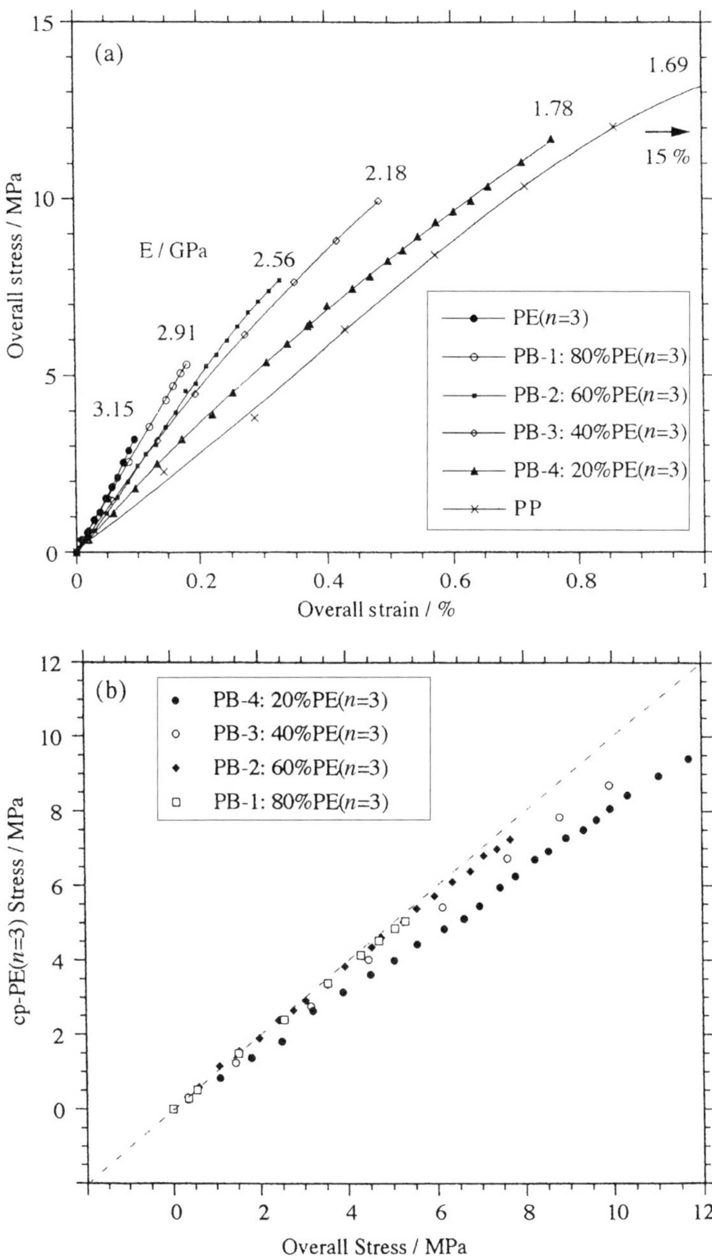
(a)
15
E / GPa
1.69
1.78
2.18
2.56
2.91
3.15
15 %
Overall stress / MPa
Overall strain / %
PE(n=3)
PB-1: 80%PE(n=3)
PB-2: 60%PE(n=3)
PB-3: 40%PE(n=3)
PB-4: 20%PE(n=3)
PP
(b)
PB-4: 20%PE(n=3)
PB-3: 40%PE(n=3)
PB-2: 60%PE(n=3)
PB-1: 80%PE(n=3)
cp-PE(n=3) Stress / MPa
Overall Stress / MPa

ACKNOWLEDGMENTS

The authors express their thanks to the UK Engineering and Physical Sciences Research Council (UK) and the Basque Government (Spain) for funding the research described in this paper. One of us (RC) would like to thank UMIST for a Mohn scholarship.

REFERENCES

1. G. Wegner, *Z. Naturforschung*, **241**, 824 (1969).
2. G. Wegner, *Makromol. Chem.*, **134**, 219 (1970).
3. D. Day and J.B. Lando, *J. Polym. Sci., Polym. Lett. Ed.*, **19**, 227 (1981).
4. M.F. Rubner, *Macromolecules*, **19**, 2114 (1986).
5. J.L. Stanford, R.J. Young and R.J. Day, *Polymer*, **32**, 1713 (1991).
6. X. Hu, J.L. Stanford, R.J. Day and R.J. Young, *Macromolecules*, **25**, 684 (1992).
7. X. Hu, R.J. Day, J.L. Stanford and R.J. Young, *J. Mat. Sci.*, **27**, 5958 (1992).
8. P.A. Lovell, J.L. Stanford, Y-F. Wang and R.J. Young, *Polym. Bull.*, **30**, 347 (1993).
9. P.A. Lovell, J.L. Stanford, Y-F. Wang and R.J. Young, *Macromolecules*, **31**, 842 (1998).
10. A.S. Hay, D.A. Bolon, K.R. Leimer and R.F. Clark, *J. Polym Sci., Poly. Lett. Ed.*, **8**, 97 (1970).
11. P.A. Lovell, J.L. Stanford, Y-F. Wang and R.J. Young, *Polym. Int.*, **34**, 23 (1994).
12. X. Hu, J.L. Stanford, R.J. Day and R.J. Young, *Macromolecules*, **25**, 672 (1992).
13. J.L. Stanford, R.J. Day, X. Hu and R.J. Young, *Prog. Org. Coatings*, **20**, 425 (1992).
14. P.A. Lovell, J.L. Stanford, Y-F. Wang and R.J. Young, *Macromolecules*, **31**, 834 (1998).

6

Novel Intelligent Amphiphilic Conetworks

GÁBOR ERDÖDI[1,2], ÁKOS JANECSKA[1] and BÉLA IVÁN[1,2]

[1]Institute of Chemistry, Chemical Research Center, Hungarian Academy of Sciences, H-1525 Budapest, Pusztaszeri u. 59-67, P.O. Box 17, Hungary
[2]Department of Chemical Technology and Enviromental Chemistry, Eötvös Lóránd Science University, H-1518 Budapest, P.O. Box 32, Hungary

ABSTRACT

Amphiphilic conetworks (APCN) comprising poly(methacrylic acid) and polyisobutylene chain segments have been synthesized by radical copolymerization of trimethylsilyl methacrylate with methacrylate-telechelic polyisobutylene followed by acid catalyzed

Wiley Polymer Networks Group Review Series Vol. 2. Edited by B.T. Stokke and A. Elgsaeter
© 1999 John Wiley & Sons Ltd

deprotection of the trimethylsilyl groups. These new materials exhibit intelligent pH-sensitive response in aqueous media. Drug delivery systems were prepared by loading the networks with theophylline. Release studies indicate sustained release with Fickian diffusion kinetics over a long period of time, i.e. several days.

For the preparation of APCNs with higher structural order, hydroxyl-telechelic three-arm star polyisobutylene with narrow molecular weight distribution was synthesized by quasiliving carbocationic polymerization and subsequent quantitative chain end derivatization. This star polymer was cured with poly(ethylene glycol) to form urethane linkages under suitable conditions. Randomly linked and highly ordered segmented alternating PEG-PIB amphiphilic conetworks with high PEG contents were prepared by a new synthetic strategy. The amphiphilic nature of the resulting networks was proved by swelling in water and n-hexane. DSC measurements confirmed the existence of microphase separation in these new APCNs.

INTRODUCTION

After the discovery of quasiliving carbocationic polymerization [1,2] in the mid-1980s, the synthesis of polyisobutylene (PIB) with controlled microstructure and narrow molecular weight distribution (MWD) has become possible. Telechelic PIBs are among the most important types of polymers synthesized by this method with great current interest for both science and technology. By reacting their reactive termini, these PIBs can be quantitatively converted to various new useful high molecular weight products and networks including amphiphilic conetworks.

Amphiphilic conetworks (APCN) are a new class of crosslinked systems composed of covalently bonded hydrophilic and hydrophobic chains segments [3]. The main property of these networks is their ability to swell uniformly both in hydrophilic and hydrophobic solvents. Because of the large difference in the chemical structure of the macromolecular segments with opposite philicity, these polymer chains are thermodynamically incompatible and exhibit microphase separation. However, due to the covalent bonds between the hydrophilic and hydrophobic chains, the freedom of demixing is significantly prevented in APCNs. Since APCNs usually swell in water as well, they may also be regarded as a special class of hydrogels. Hydrogels are extensively used in biomedical applications, e.g. implants [4–6] and drug delivery devices [6]. Recently a new family of APCNs has been developed by radical copolymerization of methacrylate-telechelic PIB (MA-PIB-MA) and various (meth)acrylic monomers leading to water swellable chains [3,7–11]. These APCNs exhibit excellent biocompatibility and biostability *in vivo* in rats [12] and as such were also studied as potential drug delivery systems [3–7,10].

Polyelectrolyte networks are among the most interesting types of materials. However, only one APCN with polyelectrolyte chain segment, i.e. poly(2-sulfoethyl methacrylate)-*l*-polyisobutylene (*l* stands for linked), has been reported until now [11], the reproducible synthesis of which is questionable. Although the synthesis of such a network is hard to achieve due to synthetic difficulties of linking highly incompatible hydrophilic and hydrophobic polymers, our attention has recently turned toward studying the synthesis and properties of poly(methacrylic acid)-*l*-polyisobutylene (PMAA-*l*-PIB) polyelectrolyte APCNs.

The telechelic macromonomer approach for the synthesis of APCNs leads to a structurally highly undefined network. The distance between branching points (M_c) along the chain crosslinked by the well-defined macromonomer varies in wide ranges. This makes this method uncontrollable regarding the microstructure and morphology of the resulting APCNs. Therefore the synthesis and characterization of a new generation of APCNs with structural order between the components were also attempted by us. These new APCNs are composed of telechelic PIB and poly(ethylene glycol) chain segments.

POLY(METHACRYLIC ACID)-*l*-POLYISOBUTYLENE pH-RESPONSIVE AMPHIPHILIC CONETWORKS

SYNTHESIS AND CHARACTERIZATION

The synthesis of amphiphilic conetworks (APCN) is problematic because it involves the linking of highly incompatible hydrophilic and hydrophobic polymer chain segments. The preparation of poly(methacrylic acid)-*l*-polyisobutylene (PMAA-*l*-PIB) was attempted by radical copolymerization of methacrylate-telechelic PIB (MA-PIB-MA) with a selected monomer. In order to obtain such a network with MA-PIB-MA as crosslinker the following major requirements must be fullfilled:

1. the monomer and the PIB macromonomer must copolymerize in random manner;
2. the kinetic chain must be of sufficient length for the incorporation of at least two MA-PIB-MA units;
3. phase separation during the copolymerization must be prevented.

In our situation the first step involved copolymerization of MA-PIB-MA with trimethylsilyl methacrylate (TMSMA), i.e. a protected methacrylic acid, to obtain the desired APCN. Poly(methacrylic acid) (PMAA) and polyisobutylene (PIB) are incompatible polymers. Therefore direct synthesis of APCNs with these components can only be made by synthesizing a network first with compatible chain segments followed by deprotection (this indirect synthetic route was chosen to prevent phase separation). This generally applicable synthetic strategy for the preparation of APCNs was already applied to prepare poly(2-hydroxyethyl methacrylate)-*l*-polyisobutylene APCNs [8–10]. Similarly, TMSMA was selected as starting monomer for the synthesis of PMAA-*l*-PIB. The copolymerization of MA-PIB-MA with TMSMA and the deprotection process is shown in Scheme 6.1. As exhibited in this Scheme, copolymerization of MA-PIB-MA with the sufficiently nonpolar trimethylsilyl protected methacrylic acid

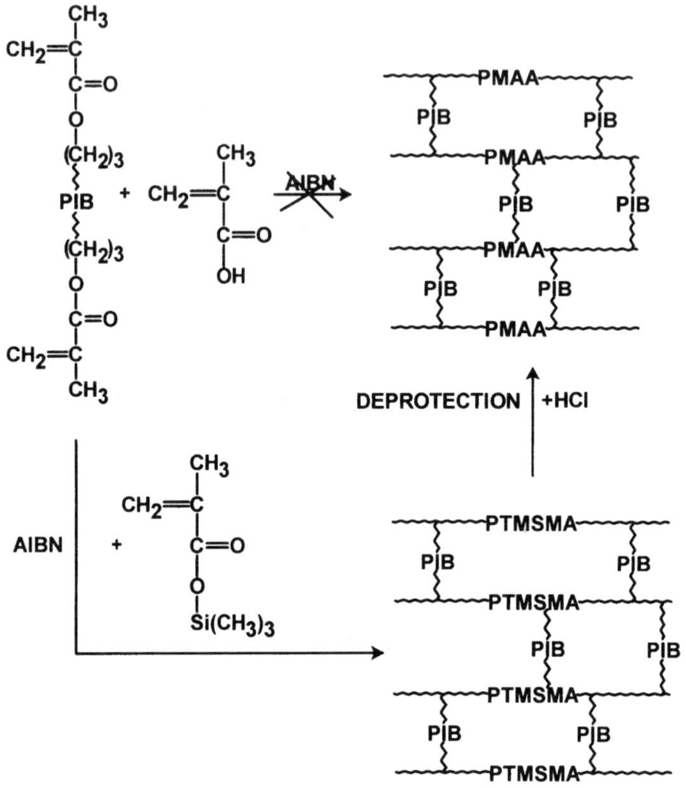

Scheme 6.1 Synthesis of PMAA-*l*-PIB APCN by copolymerization of MA-PIB-MA with TMSMA followed by deprotection

yields a network in a cosolvent (THF) for all the components. Deprotection with HCl in the network leads to the designed PMAA-*l*-PIB APCN.

MA-PIB-MA was prepared by first obtaining allyl-telechelic PIB synthesized by quasiliving carbocationic polymerization of isobutylene with the *tert*-butyldicumyl chloride/TiCl$_4$ initiating system, followed by quantitative functionalization with allyltrimethylsilane [13]. The allyl-telechelic PIB was hydroborated and oxidized [13,14] to prepare hydroxyl termini. The resulting hydroxy-telechelic polymer was converted with methacryloyl chloride to MA-PIB-MA [15] ($M_n = 11\,200$; $M_w/M_n = 1.08$).

The networks were prepared by the radical copolymerization of MA-PIB-MA with TMSMA in THF, which is a common solvent for both PIB and TMSMA, so phase separation during the copolymerization was prevented. The polymerization was carried out in closed Teflon molds. The molds were filled with the reaction mixture containing the PIB macromonomer, the TMSMA comonomer, and AIBN initiator, all solved in THF under nitrogen atmosphere. The molds were sealed by

Table 6.1 Experimental conditions for the synthesis of PMAA-l-PIB APCNs. (THF solvent, total volume 6 ml, 60 °C, 72 h)

Sample	MA-PIB-MA (g)	MAA (g)	MAA $n \times 10^3$ [mol]	TMSMA (g)	AIBN $n \times 10^6$ (mol)
M-11-30	0.3	0.7	8.13	1.29	12.23
M-11-40	0.4	0.6	6.97	1.10	8.63
M-11-50	0.5	0.5	5.81	0.92	6.20
M-11-60	0.6	0.4	4.65	0.74	3.80
M-11-70	0.7	0.3	3.48	0.55	2.17

a teflon-coated rubber lid, and the mold assembly was placed into an oven and heated at 60 °C for three days. After three days of curing the conetworks were removed from the molds and THF was allowed to evaporate. The unreacted MA-PIB-MA was removed by extracting the networks with hexane for 24 h. Deprotection (i.e. the removal of the $-Si(CH_3)_3$ group) was accomplished by swelling the networks in a 5% solution of HCl in methanol, then in HCl + methanol + H_2O followed by 5% HCl in water for 24 h in each step. Then the networks were dried *in vacuo*. The dry networks were transparent tough materials. Table 6.1 summarizes conditions and experimental data. The numbers in the sample column indicate the M_n of the starting MA-PIB-MA (divided by 1000) and the weight percent of PIB in the polymerization charge, whereas the first letter (M) indentifies the hydrophilic chain units, i.e. MAA. The amount of TMSMA was determined by the amount of MAA required for a desired composition. The AIBN concentration ([I]) was selected in relation to the monomer concentration ([M]) in order to provide sufficiently high kinetic chain length (DP_n) for network formation according to the $1/DP_n \sim [I]^{0.5}/[M]$ relation.

The desilylation could be followed visually by observing the gradual advancement of a swelling front across the entire width of the sample until its disappearance indicates complete desilylation. Other facts that proved the perfect desilylation are the DSC traces of the protected (silylated) and deprotected (desilylated) samples. As shown in Figure 6.1, the protected M-11-30 APCN has two T_g value whereas the deprotected sample has only one T_g value indicating satisfactory deprotection. The obtained T_g values for the silylated (M-11-30S),

Table 6.2 T_g values obtained by DSC measurements (conditions as in Figure 6.1)

Sample	T_g (°C)		
	PIB	PTMSMA	PMAA
M-11-30S	−87.7	62.7	
M-11-30DES	−85.4		—–

T_g(PMAA) = 228 °C (literature value [16])

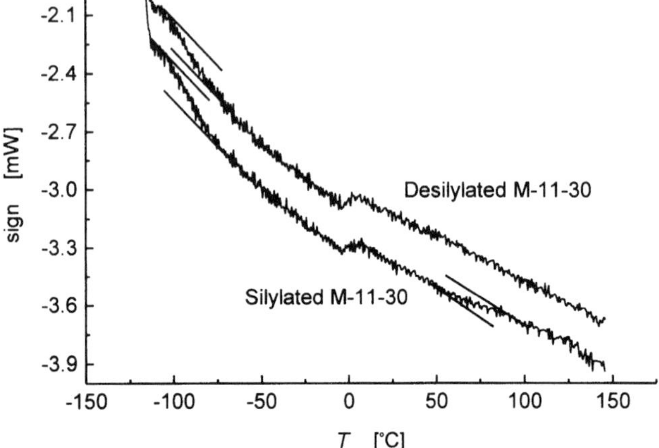

Figure 6.1 DSC thermogram of network M-11-30 before and after desilylation (second heating cycle, heating rate: 10 °C/min)

and desilylated (M-11-30DES) samples are shown in Table 6.2. These data also indicate microphase separated morphology of PMAA-*l*-PIB APCNs. Sequential *n*-hexane extractions after deprotection resulted in less than 10% extractables, indicating high copolymerization yields and close to perfect network formation. This was also confirmed by elemental analysis which provided PIB and PMAA contents nearly to theoretical values.

SWELLING BEHAVIOR

The amphiphilic nature of the resulting conetworks was examined by swelling studies. Figures 6.2 and 6.3 show the swelling ratio (*R*, absorbed solvent per network) as a function of time in water and hexane, respectively. As shown in these Figures the swelling ratio depends on network composition in both solvents. The degree of swelling decreases with increasing PIB content in water, whereas the opposite trend is observed in hexane. The networks swelled isotropically in each solvent, suggesting that the PMAA and PIB domains are cocontinuous in the PMAA-*l*-PIB conetworks. Comparison of Figures 6.2 and 6.3 also indicates that these new APCNs are able to absorb an order of magnitude higher amount of water than hexane. This is due to the presence of the very hydrophilic PMAA chain segments in these networks. PMAA-*l*-PIB with 70% PMAA content absorbs more than 250% water, i.e. these new APCNs can be considered as new superabsorbents. Interestingly, the swelling curves indicate Fickian type water absorbtion at pH = 7, and non-Fickian (anomalous transport) in hexane [16].

The swelling behavior of the new superabsorbent APCNs was also investigated in buffer solutions with low and high pH. The samples were initially swollen to equilibrum swelling ratios at pH = 7. As shown in Figure 6.4, placing these

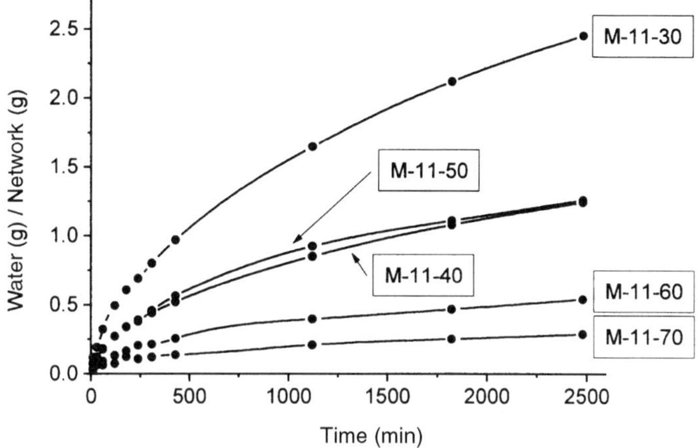

Figure 6.2 Swelling of PMAA-*l*-PIB amphiphilic conetworks in water at room temperature

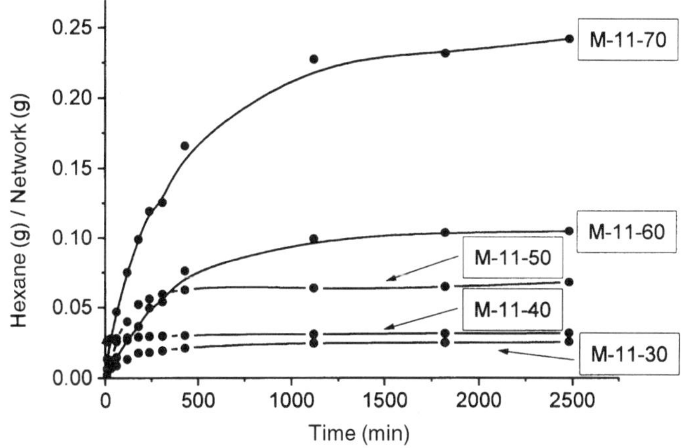

Figure 6.3 Swelling of PMAA-*l*-PIB amphiphilic conetworks in n-hexane at room temperature

samples in a buffer solution with pH $= 2$ leads to rapid deswelling of 15–60%. Changing the pH to 12 results in instantaneous increase in the swelling ratios by 20–80%. The reversibility of the swelling–deswelling phenomenon is presented by the last pH cycle in Figure 6.4, i.e. similar swelling ratios are observed in pH $= 2$ than those in the first cycle. This rapid pH-response resulted also in simultaneous volume change. The 'intelligent' behavior of PMAA-*l*-PIB APCNs can be explained by the polyelectrolyte nature of the PMAA segments. With

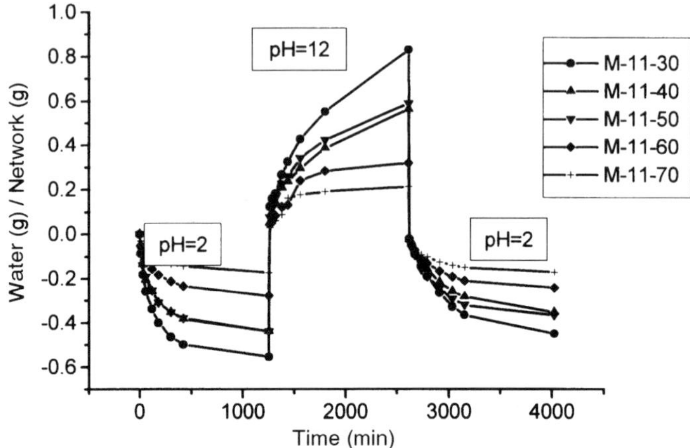

Figure 6.4 Deswelling and swelling of PMAA-*l*-PIB amphiphilic conetworks at pH = 2 and 12

the increasing pH the carboxyl groups of the PMAA rapidly dissociate in the network and the presence of the ionic groups enchances the hydrophilic character of the PMAA phase which can absorb more water. The high rate of the swelling induced by the pH-change indicates that the concentration of ionic groups is especially significant in respect to the swelling properties of amphiphilic conetworks [17,18]. This smart immediate pH-response can be most likely utilized in several biomedical and other advanced applications.

DRUG RELEASE FROM PMAA-l-PIB AMPHIPHILIC CONETWORKS

Amphiphilic conetworks, because of their unique hydrophilic–hydrophobic microphase separated morphology, offer many potential biomedical applications. The hydrophilic–hydrophobic structure may also find unique applications as drug delivery devices for both hydrophilic and lipophilic drugs. In order to test this idea, drug release experiments were carried out with theophylline from PMAA-*l*-PIB APCNs. After loading the PMAA-*l*-PIB with the drug in saturated water solution, it was dried to constant weight. Then the dry sample was immersed in water and the release of theophylline was followed by UV-visible spectroscopy. Figure 6.5 shows the relative amount of theophylline released at time $t(M_t)$ over the total amount loaded (M_o) (i.e. M_t/M_o) as a function of time. Evidently PMAA-*l*-PIB conetworks loaded with theophylline are efficient drug delivery systems with sustained release. Relatively long time (240 hours) is required for nearly complete drug release. The rate and type of release can be analyzed by the use of the expression $M_t/M_o = kt^n$ [19]. In the case of pure Fickian diffusion $n = 0.5$, whereas $n > 0.5$ indicates anomalous transport, i.e. in addition to diffusion another process (or processes) also occurs. If $n = 1$ (zero-order release),

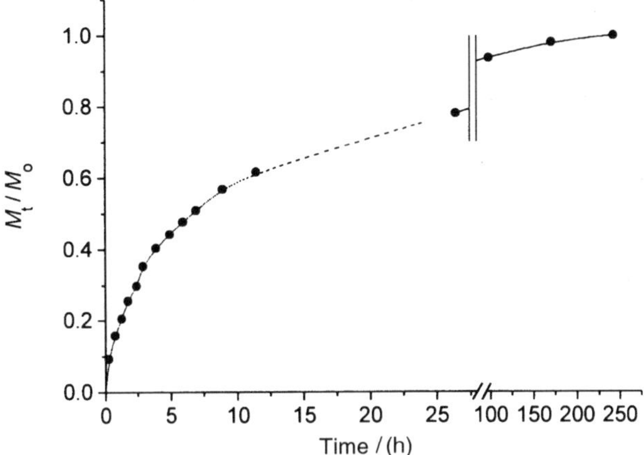

Figure 6.5 Release of theophylline as a function of time from PMAA-*l*-PIB APCNs

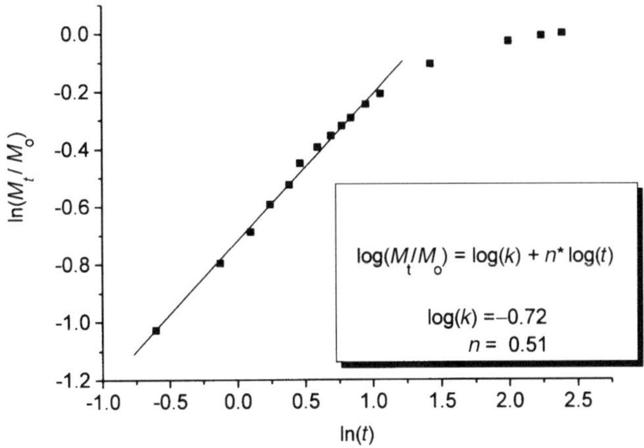

Figure 6.6 The ln(M$_t$/M$_o$) versus ln(t) plot for the release of theophylline from PMAA-*l*-PIB APCNs

the transport is usually polymer relaxation controlled ('Case II transport'). The $\ln(M_t/M_o)$ versus $\ln(t)$ plot is shown in Figure 6.6. This plot yields a straight line for the first phase of the release with a slope of nearly 0.5 ($n = 0.51$). These findings indicate that the slow release of theophylline from the PMAA-*l*-PIB polyelectrolyte APCNs is diffusion controlled. This is most likely connected with the Fickian-type of swelling.

RANDOMLY AND PERFECTLY SEGMENTED AMPHIPHILIC CONETWORKS

SYNTHETIC STRATEGY

In order to obtain high degree of structural order in APCNs, coupling of telechelic hydrophilic and hydrophobic polymer chains with suitable functional groups can be theoretically applied for the synthesis of such networks. As shown in

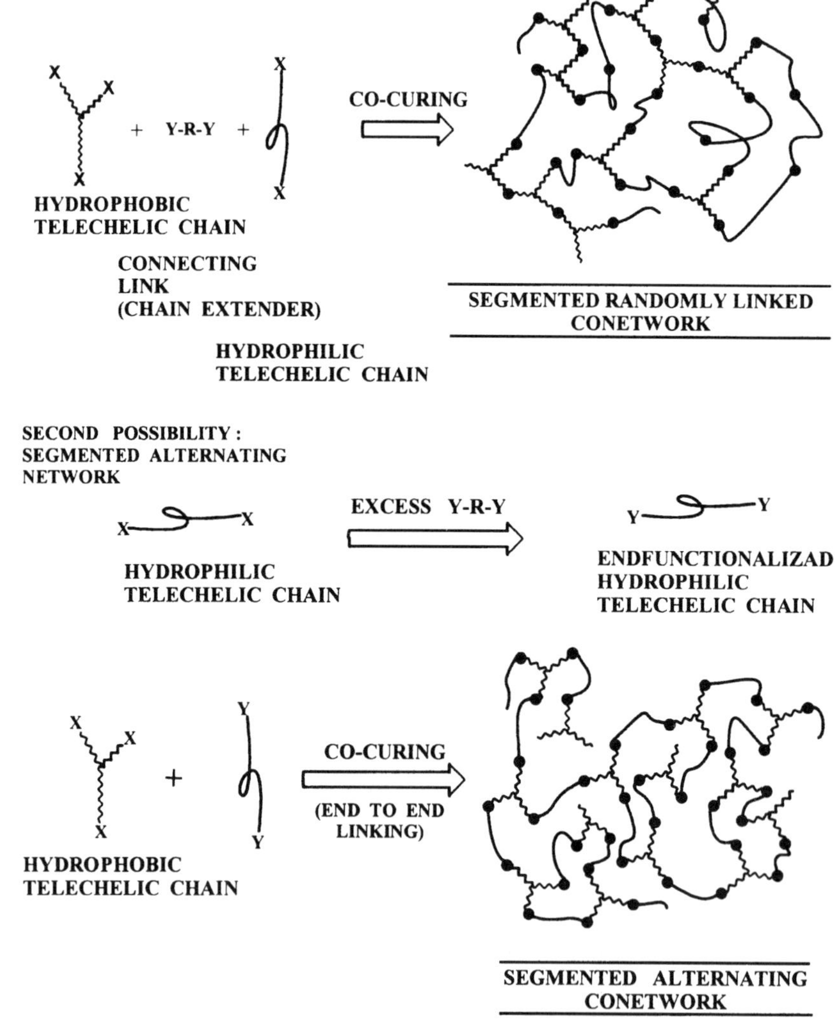

Scheme 6.2   Synthetic strategy for a new generation of APCNs by coupling of telechelic polymer chains

Scheme 6.2, there are two possibilities: (i) co-curing of telechelic chains having the same functional groups with a linking agent when both types of chains bear the same functional endgroups, and (ii) co-curing of telechelic polymers possessing functional groups which are able to react with each other in high yields.

The first approach leads to randomly segmented conetworks. In this case the bifunctional linking agent is able to connect both similar and dissimilar chain segments. Networks are formed when one of the chains is trifunctional. As a consequence the segment length of a polymer with the same philicity will depend on the number of similar chains coupled together.

Structurally a much more regular APCN can be obtained by the second synthetic route. This requires efficient coupling of chains with reactive endgroups, X and Y. The requirement for obtaining segmented perfectly alternating APCN is that X and Y can react only with each other as exhibited in Scheme 6.2. This process involves two steps. First one of the chains are converted to a Y-telechelic polymer by reacting with the connecting link. In the second step the X- and Y-telechelic polymers are reacted with each other.

SEGMENTED AMPHIPHILIC CONETWORKS COMPOSED OF POLYISOBUTYLENE AND POLY(ETHYLENE GLYCOL)

For the synthesis of the new generation APCN three-arm star telechelic PIB was selected as the hydrophobic component. By using quasiliving carbocationic polymerization of isobutylene with a trifunctional initiator (tricumyl chloride) followed by quantitative functionalization with allyltrimethylsilane, three-arm star allyl-telechelic PIB [13] was obtained with narrow molecular weight distribution and predetermined molecular weights. Hydroboration/oxidation led quantitatively to three-arm star hydroxyl-telechelic PIB with exact functionality ($F_n = 3.0$). Using poly(ethylene glycol) as hydrophilic component provides a unique opportunity for the preparation of APCNs with desired structural control. PEG was previously reported as a building block of APCN with polybutadiene but due to phase separation problems the PEG content was limited to $\sim$20% [20].

The process applied for the synthesis of APCNs containing PIB and PEG segments are shown in Scheme 6.3. Segmented randomly linked APCNs were prepared by co-curing three-arm star hydroxyl-terminated PIB ($M_n = 1830$g/mol, $M_w/M_n = 1.07$, $F_n = 3.0$) with PEG in the presence of stoichiometric amount of diisocyanatohexane. Polymers and the connecting link were dissolved in toluene and closed in an ampule under N_2. The mixture in the ampule was allowed to react at $100\,^\circ$C for 3 days. Then it was extracted with THF for 48 h at room temperature and diethyl ether was used to shrink the network. The resulting material was dried in vacuum at $80\,^\circ$C for 3 days. Segmented perfectly alternating conetworks were prepared by reacting isocyanato-telechelic PEG (OCN-PEG-NCO) with three-arm star hydroxyl-terminated PIB. For the synthesis of OCN-PEG-NCO PEG 1450 was dissolved in toluene and diisocyanatohexane was added. It was refluxed for 6 h. Toluene was evaporated and the product was purified from the excess

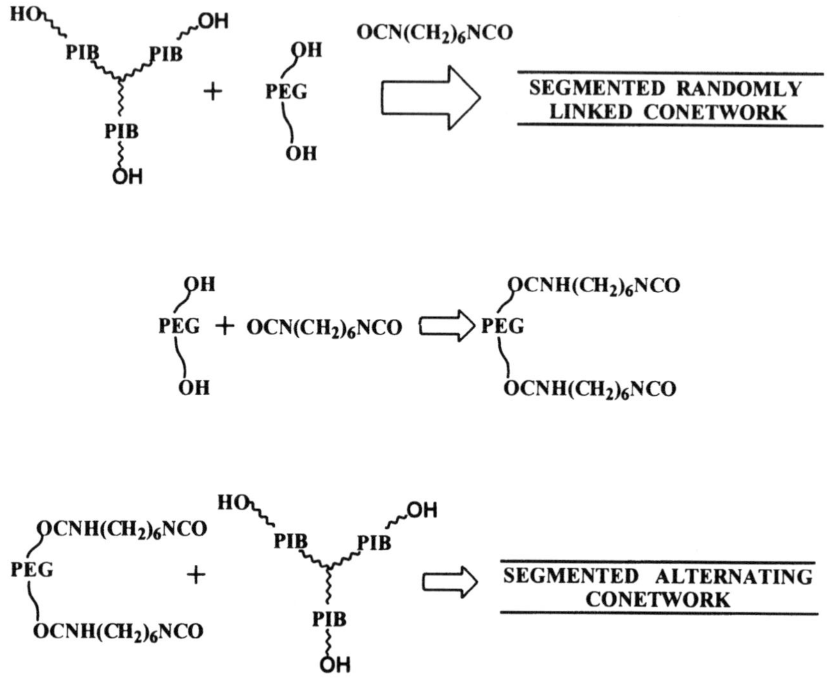

Scheme 6.3 Curing reactions for the segmented randomly linked and the perfectly alternating conetworks

of diisocyanatohexane by repeated precipitation from toluene to n-hexane. The isocyanate-telechelic PEG and the three-arm star hydroxyl-telechelic PIB was closed in an ampule and was cured for 5 days at 100 °C. Then it was extracted with THF for 48 hours followed by a treatment with Et_2O for 6 h and dried in vacuum at 60 °C for 3 days.

The hydroxyl-telechelic chains react with the diisocyanatohexane connecting link even without catalyst at 100 °C. Higher temperature cannot be applied because of the boiling point of toluene (109 °C) and the slow depolymerization

Table 6.3 The characteristics of PEG–PIB APCNs (SA = segmented perfectly alternating conetwork; R = randomly linked conetwork)

Samples	M_w(PIB)	M_w(PEG)	Extr. amount (%)	PEG in network (%)	PIB in network (%)
SA 400	1800	400	11	20	62
R 400	1800	400	42	22	61
SA 1000	1800	1000	21	37	47
R 1000	1800	1000	38	36	48
SA 1450	1800	1450	18	46	38
R 1450	1800	1450	39	45	40

of PEG. In this way six samples were prepared. In these experiments the molecular weight (MW) of the PIB was the same. The MW of PEG ranged from 400 g/mol to 1450 g/mol. For the synthesis of the perfectly alternating conetwork the only way to increase PEG content is to raise the molecular weight of the PEG (or to decrease the M/W of PIB). In the random conetwork this restriction is not necessary but we used the same stochiometry and same type of PEG for the random networks than that for the perfectly alternating conetworks, so reliable comparison is possible. The soluble part ranged from 10 to 40%. PIB and PEG content were determined by elemental analysis. The components, extractable amounts and composition of the different conetworks are summarized in Table 6.3.

Figure 6.7 shows the proof that these materials really have amphiphilic nature, i.e. they are swellable both in hydrophilic and hydrophobic solvents. In the upper

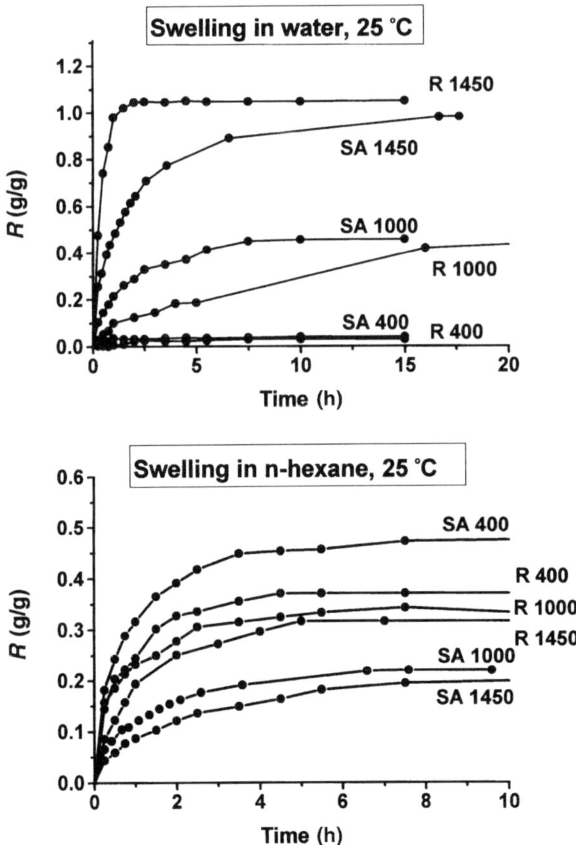

Figure 6.7 Swelling experiments in water and hexane with PEG-PIB conetworks (see Table 6.4 for identification)

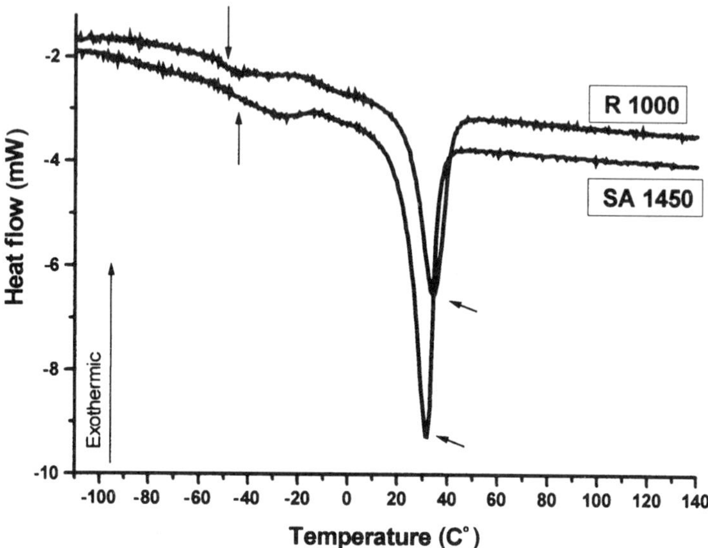

Figure 6.8 DSC traces of PEG-PIB APCNs (see Table 6.4 for identification; second heating cycle, heating rate: 10 °c/ min)

plot, the swelling of samples can be seen in water. It is easy to recognize the straight tendency that the larger the PEG content the higher the degree of swelling and the rate of swelling as well. More surprising is the fact that significant difference exists between the segmented perfectly alternating and the random networks. This may be due to the different curing rate but it can also indicate morphological differences. Swelling experiments in hexane showed similar behavior. The degree of swelling and the rate of swelling are strongly PIB content dependent. Differences between the two type of networks also exist in the course of swelling these conetworks in hexane.

DSC measurements indicate phase-separated morphology for both networks. As shown in Figure 6.8 glass transitions at $-50\,°C$ for PIB and melting peaks at $32\,°C$ and $34\,°C$ are observed. The latter are much lower than the melting point of the starting PEG ($49\,°C$).

In summary, a new synthetic approach has been developed by us for the preparation of novel APCNs containing randomly and alternatingly coupled PEG and PIB chain segments. The method is simple and can be extended to other types of APCNs. The swelling properties can be controlled by composition. Considering that both PIB and PEG are well known for their excellent biocompatibility, the new APCNs are potential candidates in several applications in the area of biomaterials, such as drug release matrices, implants, adhesives etc.

REFERENCES

1. R. Faust and J.P. Kennedy, *J. Polym. Sci., Part A: Polym. Chem.*, **25**, 1847 (1987).
2. J.P. Kennedy and B. Iván, *Designed Polymers by Carbocationic Macromolecular Engineering: Theory and Practice*; Hanser Publishers, Munich, New York, 1992.
3. B. Iván, J. Feldthusen and A.H.E. Müller, *Macromol. Symp.*, **102**, 81 (1996).
4. A.S. Hoffman, in *Polymers in Medicine and Surgery*, (R.N. Kronenthal, Z. Oser and E. Martin, eds), Plenum, New York 1975, p. 33.
5. D.T. Lyman and S.M. Rowland, in *Polymers: Biomaterials and Medical Applications*, (T.I. Kroschwits, ed.), John Wiley and Sons, New York, 1989, p. 52–71.
6. H. Benson, B. Markey and E.E. Schmidt, in *Polymers: Biomaterials and Medical Applications*, (T.I. Kroschwits, ed.), John Wiley and Sons, New York, 1989, p. 131–150.
7. D. Chen, J.P. Kennedy, and A.J. Allen, *J. Macromol. Sci.-Chem.*, **A25(4)**, 389 (1988).
8. B. Iván, J.P. Kennedy and P. Mackey, in *Polymeric Drugs and Drug Delivery Systems*, (R.L. Dunn and R.M. Ottenbritte, eds), ACS Symp. Ser., Vol. 469, American Chemical Society, Washington, D.C., 1991, p. 194–202.
9. B. Iván, J.P. Kennedy and P. Mackey, in *Polymeric Drugs and Drug Delivery Systems*, (R.L. Dunn and R.M. Ottenbritte, eds), ACS Symp. Ser., Vol. 469, American Chemical Society, Washington, D.C., 1991, p. 203–212.
10. B. Iván, J.P. Kennedy and P. Mackey, US Patent, 5,073,381 (1991).
11. B. Keszler and J.P. Kennedy, *J. Polym. Sci., Part A: Polym. Chem.*, **32**, 3153 (1994).
12. D. Chen, J.P. Kennedy, M.M. Kory and D. Ely, *J. Biomed. Mater. Res.*, **23**, 1327 (1989).
13. B. Iván and J.P. Kennedy, *J. Polym. Sci., Part A: Polym, Chem.*, **28**, 89 (1990).
14. B. Iván and J.P. Kennedy and V.S.C. Chang, *J. Polym, Sci.: Polym Chem. Ed.*, **18**, 3177 (1980).
15. J.P. Kennedy and M. Hiza, *J. Polym. Sci.: Polym. Chem. Ed.*, **21**, 1033 (1983).
16. Á. Janecska and B. Iván, to be published.
17. M. Sen and O. Güven, *Polymer*, **39**, 1165 (1998).
18. S.H. Yuk, S.H. Cho and S.H. Lee, *Macromolecules*, **30(22)**, 6856 (1997).
19. N.A. Peppas and R.W. Korsmeyer, in *Hydrogels in Medicine and Pharmacy*, Vol. III, (N.A. Peppas, ed.) CRC Press, 1987, p. 109–135.
20. M. Weber and R. Stadler, *Polymer*, **29**, 1071 (1989).

7

Synthesis and Characterization of Cross-linked Amphiphilic Poly(ethylene oxide) Hydrogels

ZENG-RONG ZHANG and MOSHE GOTTLIEB
Department of Chemical Engineering, Ben-Gurion
University of the Negev, Beer-Sheva 84105, Israel

Wiley Polymer Networks Group Review Series Vol. 2. Edited by B.T. Stokke and A. Elgsaeter
© 1999 John Wiley & Sons Ltd

ABSTRACT

Three types of poly(ethylene oxide) (PEO) networks crosslinked by silicone based compounds have been formed. One type of network prepared from allyl substituted poly(ethylene glycol) is stable, whereas the other two made by direct crosslinking of poly(ethylene glycol) are degradable. The degree of swelling of the resulting hydrogels depends strongly on the PEO/silicone ratio and on temperature. From the weight fraction of soluble material in the networks, the extent of crosslinking reaction has been estimated. The molecular weight between crosslinks has been calculated independently from the weight fraction of soluble material in the network and from swelling data.

INTRODUCTION

Hydrogels may be conveniently described as hydrophilic polymers that are swollen by, but do not dissolve in, water. This is usually achieved by a low degree of crosslinking, as in the case of conventional elastomers, which means that hydrogels are effectively water-swollen polymer networks. Although many naturally occurring polymers may be used to produce this type of material, the structural versatility available in synthetic hydrogels has given them distinctive properties, which in turn have enhanced their practical utility. Due to characteristic properties such as swellability in water, hydrophilicity, biocompatibility, and lack of toxicity, hydrogels have been utilized in a wide range of biological, medical, and pharmaceutical applications [1–4]. Also, hydrogels have become the material of choice as drug carries in controlled release applications [5,6].

Synthetic and semisynthetic hydrogels based on physical [7,8] chemical [9], or photochemical crosslinking [10] have been developed which offer improvements in the barrier and degradation properties of gels. A considerable amount of work has been carried out on hydrogels based on hydroxyethyl methacrylate (HEMA) [11–14]. Measurements of the physical properties and water-binding behavior of various copolymer hydrogels prepared by hydroxyalkyl acrylates and methacrylates with nonhydrophilic monomers have been reported [15]. Copolymers and terpolymers containing *N*-vinyl-2-pyrrolidone find a wide range of applications in the field of hydrogels [16,17]. Less attention has been paid to poly(ethylene glycol) (PEG)-based hydrogels. The routes leading to the formation of endlinked PEO gels discussed in the literature usually involve a reaction between PEO terminal hydroxyl groups and different aromatic or aliphatic multifunctional isocyanates [18–20 and references therein]. Cima [21] has prepared nondegradable crosslinked poly(ethylene glycol) star hydrogels with high ligand capacity as matrices for coupling to cell-binding ligands. Lactidebased poly(ethylene glycol) polymer networks (GL-PEG) prepared by UV photopolymerization using two nontoxic macromers, triacrylated lactic acid oligomer emanating from a glycerol center (GL) and monoacrylated PEG, have also been reported [22]. These materials have been developed for use as polymer scaffolds in tissue engineering, which have cell-adhesion resistant, ligand-immobilizable, and biodegradable characteristics. Another potential route to gel formation is through enzymatic crosslinking of synthetic macromolecular precursors. For example, a formation of hydrogel network by crosslinking functionalized poly(ethylene glycol) and a lysine-containing polypeptide through the action of a natural tissue enzyme, transglutaminase, has been demonstrated recently [23].

There is considerable interest in the preparation of new hydrogels based on polymers with selective water solubility. Poly(ethylene oxide)/poly(dimethylsiloxane) copolymers are interesting possible candidates for such networks. Poly(ethylene glycol) (PEG) or poly(ethylene oxide) (PEO) are hydrophilic polyethers which have received much attention for use in biomaterials due to their low interfacial energy with water, relative structural stability, lack of binding sites for reactive proteins, high chain mobility, and steric stabilization effects [24]. PEO materials are of interest also as result of its low degree of protein adsorption and cell adhesion [25–28]. The reactions leading to the formation of endlinked PEO gels discussed in the literature usually involve the handling of sensitive, highly hygroscopic crosslinkers. Several side reactions are known to impede the desired main reaction and unwanted by-products tend to contaminate the resulting gel. Recently, PEO was modified with either a quaternary tetraalkyl ammonium salt [29] or acrylate [30] to construct polymer networks. In the present work, we introduce a new method for producing crosslinked PEO networks. Poly(dimethylsiloxane) (PDMS) compounds are known for their biocompatibility, high oxygen permeability, low surface tension and surface activity [3,4]. The hydrophobic nature of the silane compounds combined with the hydrophilic nature of the PEO result in an amphiphilic network with tunable properties dependent upon the siloxane/PEO ration. Variable solubility may be obtained by utilizing the strong temperature dependence of PEO solvation and by the introduction of a silane crosslinker. Three types of network will be addressed.

EXPERIMENTAL SECTION

MATERIALS

Poly(ethylene glycol)s PEG-400 (Merck), PEG-600 (Sigma) and PEG-8000 (Sigma) were dried at 105 °C under vacuum for a few hours before use. The crosslinkers tetrakis(dimethylsiloxy)silane $(HSiMe_2O)_4Si$

(United Chemical Technologies, 99% functional material by GC/MS) and tetraethoxysilane $(EtO)_4Si$ (United Chemical Technologies) were used as received. Either stannous-2-ethyl hexanoate (Sigma, 95%) or platinum complex (cisdichlorobis(diethylsulfate)platinum(II)) (Stream, 2% in toluene) were used as the catalyst. Allyl bromide (Aldrich, 99%) was distilled before use. Sodium hydride (Aldrich, 60% dispersion in mineral oil) and anhydrous THF (Aldrich, 99%) were used without further purification. Analytical grade toluene was used as a solvent for the cross-linked reaction.

SYNTHESIS OF ALLYL-SUBSTITUTED POLY(ETHYLENE GLYCOL)

PEO diallyl ethers were prepared from the parent PEG according to a Williamson synthesis [31]: reaction of allyl bromide over the sodium alcoholate in THF solution at room temperature (Scheme 7.1). In a typical synthetic experiment, a few grams of NaH (60% in mineral oil) were added under nitrogen to 100–150 ml solution of PEG in dry THF (cf. Table 7.1). The reaction mixture was stirred for approximately 1 h at room temperature. A solution of allyl bromide in dry THF was then added dropwise (ca. 50 ml) to the reaction mixture. The resulting mixture was further stirred at room temperature for a few hours. After separation in a centrifuge (4000 r.p.m.), the solvent was evaporated under reduced pressure to give the product (in the case of PEG-8000, after centrifugal separation, the solution phase was removed and chloroform was added to dissolve most of the solid. Centrifugal separation was run once again, then the solvent was evaporated under reduced pressure to give the product).

$$HO \sim\!\sim\!\sim OH \xrightarrow{\text{Step 1}} NaO \sim\!\sim\!\sim ONa$$

$$NaO \sim\!\sim\!\sim ONa \xrightarrow{\text{Step 2}} CH_2{=}CHCH_2O \sim\!\sim\!\sim OCH_2CH{=}CH_2$$

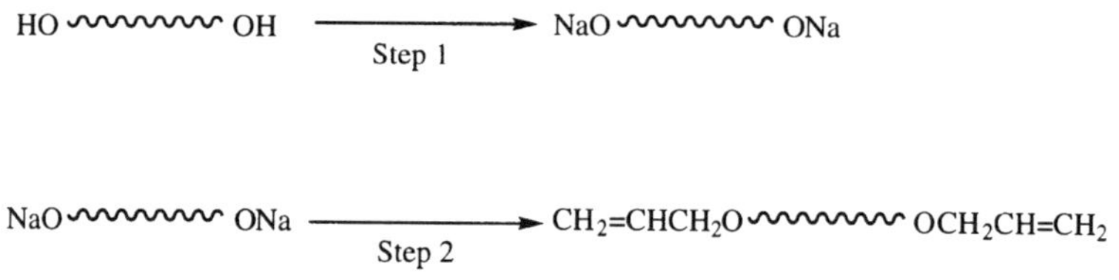

Scheme 7.1 Synthesis of allyl substituted poly(ethylene glycol)-substitution of hydroxyl end groups by allyl groups

Table 7.1 Conditions for the synthesis of allyl-substituted poly(ethylene glycol)s

PEV_i	PEG (g)/ THF (ml)	NaH (g)	T_1^a (h)	Allyl bromide (g)/THF(ml)	T_2^a (h)	Yield(%)	Form of final product
480	20.51/100	5.08	0.5	14.8/50	40	90	light yellow liquid
680	16.78/150	3.00	0.5	9.00/50	48	98	light yellow liquid
8080	14.06/150	0.48	1.5	1.45/50	48	98	white powder

^{a} T_1: reaction time for the first step; T_2: reaction time for the second step (room temperature).

The exact specific conditions for the synthesis of the different allyl-substituted poly(ethylene glycol)s are listed in Table 7.1.

FORMATION OF THE NETWORKS

Type I: In a typical experiment, platinum-based catalyst (9 µl) was added to a mixture of allyl-substituted poly(ethylene glycol) 8080 (2.8336 g), crosslinker tetrakis(dimethylsiloxy)silane $(HSiMe_2O)_4Si$ (0.0633 g) and toluene (1.2508 g). The concentration of polymer in the reaction mixture was 70%. The stoichiometric ratio of reactants defined as the ratio of initial molar concentration of silane groups to that of vinyl groups:

$$r = [SiH]_0/[Vi]_0 \tag{7.1}$$

is equal to one in all these cases ($r = 1$). The reaction mixture was stirred at $80\,^\circ$C under argon for 7 h. The dry network was obtained by vacuum removal of the toluene solvent. Four different networks based on allyl-substituted poly(ethylene glycol) were obtained. These are specified in Table 7.2.

Type II: Platinum-based catalyst (9 µl) was added to a mixture of poly(ethylene glycol) 8000 (2.4612 g), crosslinker tetrakis(dimethylsiloxy)silane $(HSiMe_2O)_4Si$ (0.0579 g) and toluene (1.0409 g). The reaction mixture was stirred at $80\,^\circ$C under argon for 32 h. The dry network was obtained after vacuum removal of the toluene solvent.

Type III: Poly(ethylene glycol) 400 (2.6690 g) was mixed with the catalyst stannous-2-ethyl hexanoate (0.1542 g), and with the stoichiometric amount (0.7068 g) of the cross-linking agent tetraethoxysilane $(EtO)_4Si$. The reaction

Table 7.2 Summary of networks formed from poly(ethylene oxide)

Network	Pre-polymer	Crosslinker	Type of network	Reaction time[a]	Other
A	PEVi-480	$(HMe_2SiO)_4Si$	type I	2 hrs (40 mins)	stable no-degradation
B	PEVi-680	$(HMe_2SiO)_4Si$	type I	3 hrs (80 mins)	stable no-degradation
C	PEVi-8080	$(HMe_2SiO)_4Si$	type I	7 hrs (30 mins)	stable no-degradation
D	PEVi-8080	$(HMe_2SiO)_4Si$	type I	24 hr (30 mins)	stable no-degradation
E	PEG-8080	$(HMe_2SiO)_4Si$	type II	33 hrs (24 hrs)	unstable degradation upon swelling
F	PEG-400	$(EtO)_4Si$	type III	48 hrs	Unstable degradation upon swelling

[a]Total time the system was allowed to react (actual time for completion may be smaller). The number in parentheses is the time after which the reaction mixture can not be stirred.

mixture was allowed to react under vacuum at room temperature for a total of two days.

The specific details of the conditions used to prepare the different types of networks are given in Table 7.2.

SWELLING OF THE NETWORKS

The swelling experiments were based on weight measurement of dry and swollen samples of undefined shapes. The weights of the dry test samples varied between 0.3 and 0.7 g. The equilibrium solvent content was measured by weight difference as follows. After swelling of the crosslinked network in distilled water (or benzene), any water (benzene) on the surface of the hydrogel sample was removed by careful blotting with absorbent paper before the sample was transferred to a pre-weighed sample dish. The sample was weighed, then dried under vacuum at room temperature. The equilibrium solvent content was calculated from the weight differences of the swollen and dry gel.

THE WEIGHT FRACTION OF SOLUBLE MATERIAL IN THE NETWORKS

The weight fraction of soluble material in the networks was obtained from the weight difference of the dry network before and after swelling.

RESULTS AND DISCUSSION

SYNTHESIS OF ALLYL-SUBSTITUTED POLY(ETHYLENE GLYCOL)

Allyl-substituted PEG bis-macromonomers may be prepared from the parent PEO glycols according to a Williamson synthesis [31]. As shown in Scheme 7.1, the first step is to prepare sodium alcoholate from the PEG prepolymer. There are several ways to convert the hydroxyl group to sodium alcoholate. One of them is to react PEG with sodium in the presence of naphthalene in THF solution. Another method is to react poly(ethylene glycol) with sodium hydride in THF solution under nitrogen at room temperature. In our work, the latter method was used. For poly(ethylene glycol) 400 and 600, this reaction was very fast and no remaining hydroxyl groups are detectable by FTIR when the reaction is carried out for 30 min. (Table 7.1). However, in the case of PEG 8000, it took 90 min for the reaction to reach completion. Since the reaction rate depends on the concentration of hydroxyl groups, it takes longer time for the reaction involving the higher molecular weight PEG to be completed.

In the second step, a solution of allyl bromide in dry THF was added dropwise to the reaction mixture which resulted from the first step. The reaction was completed only after stirring the reaction mixture at room temperature for about two days. Three ally-substituted poly(ethylene glycol), PEVi-480, PEVi-680 and PEVi-8080 were obtained with 90–98% of yields (Table 7.1).

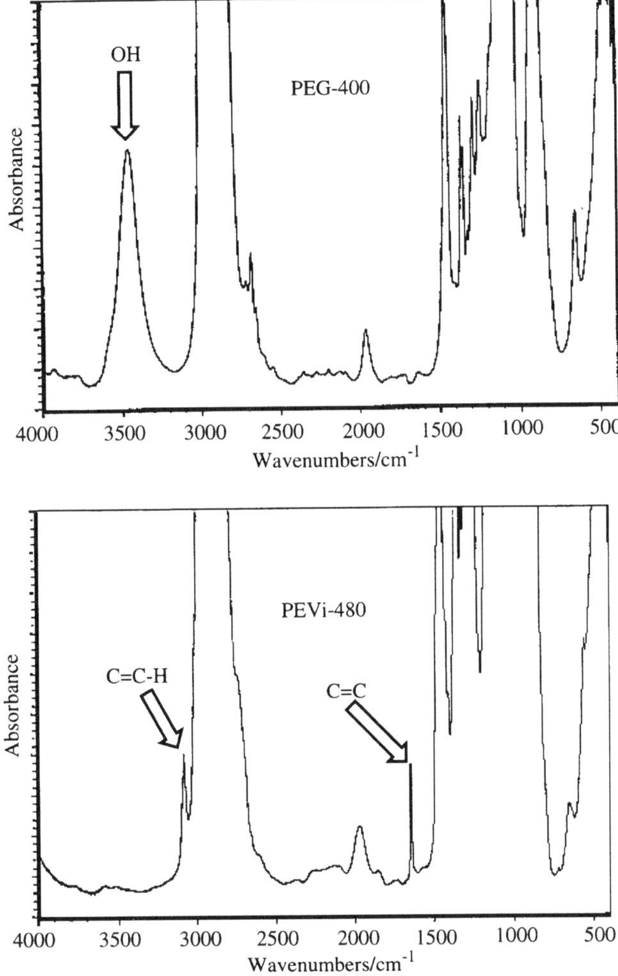

Figure 7.1 IR spectra of poly(ethylene glycol) 400 and allyl-substituted poly(ethylene glycol) 480

A typical IR spectrum is shown in Figure 7.1. As one can see, two peaks at 1648 and 3078 cm^{-1} assigned to the allyl group are observed in the product (PEVi-480), whereas the peak assigned to the hydroxyl group in the poly(ethylene glycol) (PEG-400) has completely disappeared. PEVi-680 gave a similar IR spectrum. These results demonstrate that the conversion of hydroxyl groups of allyl groups is quantitative. Due to the low concentration of terminal groups in the case of PEG-8000, it was impossible to detect either the OH or the allyl groups by FTIR. The substitution in the case of this polymer has been verified by means of NMR as discussed below.

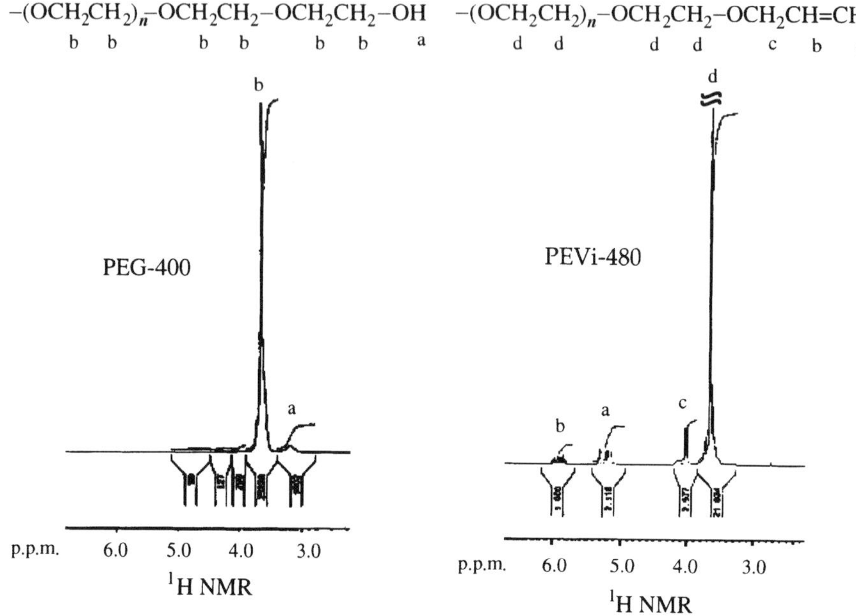

Figure 7.2 ^{1}H NMR of poly(ethylene glycol) 400 and allyl-substituted poly(ethylene glycol) 480

Figure 7.2 shows an example of ^{1}H NMR of original PEG-400 and the allyl-substituted poly(ethylene glycol) (PEVi-480) bis-macromonomer. It is clearly seen from ^{1}H NMR spectra that the hydroxyl group to allyl group substitution was completed quantitatively. This is shown by the complete disappearance of the hydroxyl group peak in PEG-400, peak *a* in the left Figure 7.2, and by the newly formed peak of allyl group in the macromonomer (PEVi-480), which can be seen in the right portion of Figure 7.2 and in Table 7.3: *a* 5.13–5.32 p.P.M.; *b* 5.81–6.00 p.p.m. This indicates that each terminal hydroxyl group in the PEG was substituted by an allyl group. The ^{1}H NMR results are summarized in Table 7.3.

Table 7.3 ^{1}H NMR of allyl-substituted poly(ethylene glycol) (in p.p.m., ref. TMS)

| | $-(OCH_2CH_2)_n-OCH_2CH_2-OCH_2CH=CH_2$ | | | |
| | d   d   d   d   c   b   a | | | |
	a	b	c	d
PEVi-480	5.13–5.32	5.81–6.00	4.00–4.04	3.56–3.76
PEVi-680	5.15–5.32	5.82–6.01	4.01–4.04	3.57–3.66
PEVi-8080	5.14–5.33	5.82–6.03	4.00–4.04	3.52–3.71

The chemical structure of the PEG and PEVi was also characterized by ^{13}C NMR. Figure 7.3 shows representative ^{13}C NMR spectra of PEG and PEVi. The macromonomer PEVi-480 structure was confirmed by the presence of allyl groups at 116 and 134 p.p.m. and by the disappearance of terminal hydroxyl carbons, which is assigned by the peak *a* in the left portion of Figure 7.3. The ^{13}C NMR results are summarized in Table 7.4

As discussed above, due to the low concentration of allyl groups in PEVi-8080, no peak related to the doubt bond (allyl group) was detected in its IR spectrum. However, the NMR results (Table 7.3 and 7.4) demonstrate that the conversion

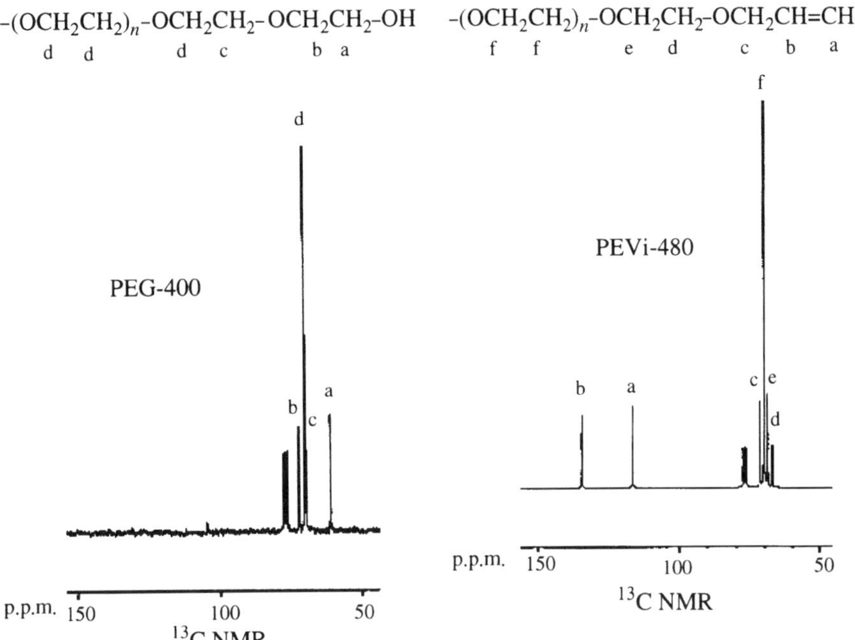

Figure 7.3 ^{13}C NMR of poly(ethylene glycol) 400 and allyl-substituted poly(ethylene glycol) 480

Table 7.4 ^{13}C NMR of allyl-substituted poly(ethylene glycol) (in p.p.m., ref. TMS)

| | $-(OCH_2CH_2)_n-OCH_2CH_2-OCH_2CH=CH_2$ | | | | | |
| | f f e d c b a | | | | | |
	a	b	c	d	e	f
PEVi-480	116.41	134.34	71.66	67.37	68.95	70.10
PEVi-680	116.96	134.69	72.12	68.57	69.33	70.49
PEVi-8080	117.03	134.72	72.16	/	69.37	70.51

of the hydroxyl groups in the PEG to allyl groups in the bis-macromonomer is complete. As shown in Table 7.3, ^{1}H NMR measurements show two groups of peaks at 5.14–5.33 p.p.m. and 5.82–6.03 p.p.m., which are assigned to the double bonds in allyl-substituted poly(ethylene glycol) PEVi-8080. This is also confirmed by the ^{13}C NMR spectrum, in which two peaks at 117 and 135 p.p.m. assigned to the double bonds in the PEVi-8080 are observed.

FORMATION OF THE NETWORKS

A polymer network could be formed by crosslinking via chemical bonds, ionic interactions, hydrogen bonds, hydrophobic interactions, or physical bonds [32]. A network formed in a random way has a highly complex structure in that there is generally a relatively broad distribution of network chain lengths, a significant fraction of imperfections such as dangling chain ends, and presumably some entanglements between neighboring chains which act to some extent as physical crosslinks [33–36]. It wold be desirable to be able to introduce crosslinks in a less random manner so that network structure and the average distance between crosslinking (mesh size) is more uniform. This type of network has several advantages in applications related to membrane performance and diffusion barriers. Of the new synthetic techniques being developed for this purpose, the most promising involve the joining of polymer chains exclusively at their ends. In this work, three possible ways to construct poly(ethylene glycol) based networks have been investigated.

Type I: As shown in Scheme 7.2, the networks are constructed by mixing the allyl-substituted poly(ethylene glycol) and crosslinker tetrakis(dimethylsiloxy)-silane $(HSiMe_2O)_4Si$ in toluene under argon atmosphere. The stoichiometric ratio of reactants was unity ($r = 1$) and the network was formed at $80\,^\circ$C by the hydrosilation reaction.

This reaction is commonly used to form poly(dimethylsiloxane) (PDMS) networks [37,38]. The evolution of structure in this system is well characterized, and the chemistry of its cure has been studied extensively [39–42]. The extend of side reactions of crosslinker hydride silane groups was found to be negligible. Four networks of this type, namely network A, B, C and D were prepared (Table 7.2).

Type II: If the allyl-substituted poly(ethylene glycol) used in type I reaction is replaced by poly(ethylene glycol), another type of network could be formed (Scheme 7.3). It should be stressed that, this time, the PEO chains are linked to the crosslinker by an Si–O–C bond, instead of the Si–C bond in type I. The latter bond is stronger and less likely to undergo hydrolysis to which the former is quite sensitive. One network of this type (Table 7.2, network E) has been prepared.

Type III: Scheme 7.4 shows the formation of another network. In this case, poly(ethylene glycol) is allowed to react with the crosslinker tetraethoxysilane $(EtO)_4Si$ under vacuum at room temperature. This reaction is also commonly used

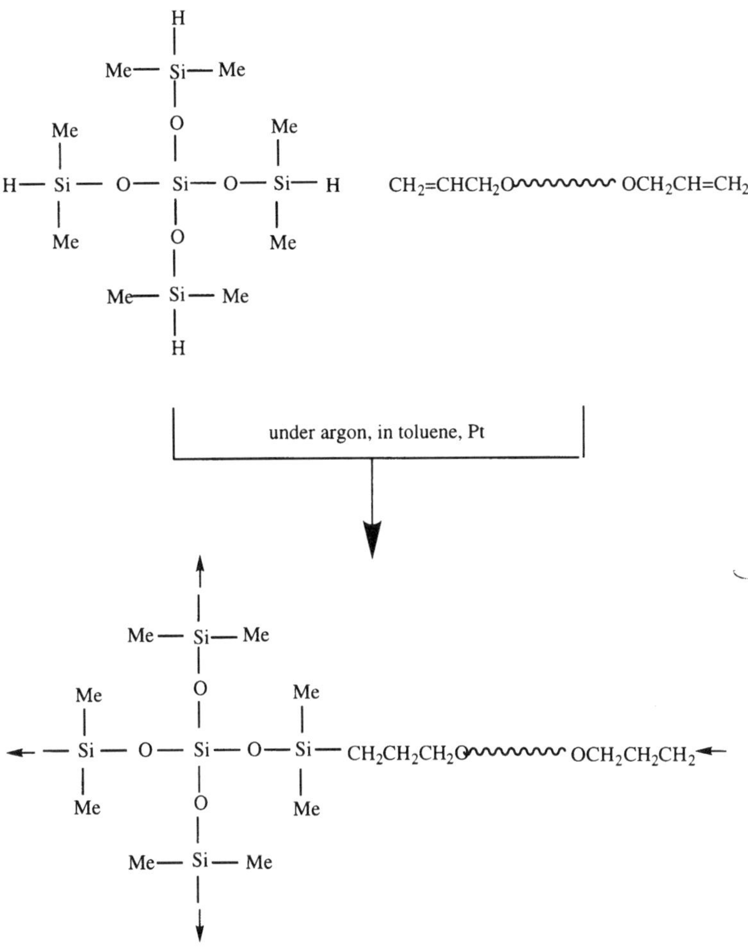

Scheme 7.2 Formation of type I networks

to form PDMS networks from silanol terminated polymers [43]. The networks were tested for unreacted EtO groups using the Zeisel alkoxy group analysis [44]. The fact that no EtO groups were detected, bearing in mind the sensitivity of the method, indicates that the desired end-linking reaction achieved at least 90% completion. One network of this type (in Table 7.2, network F) has been prepared.

It is noteworthy that gels formed by type I reaction are stable and show no sign of degradation even after repeated swelling by different solvents. However, gels formed by type II or type III reactions are not stable and show quick degradation when they are swollen in water. This difference between type I and type II (or III) is a result of the bond type, by which the poly(ethylene glycol) chains are linked to the crosslinker (Si–O–C vs Si–C). Networks type II and III may be useful

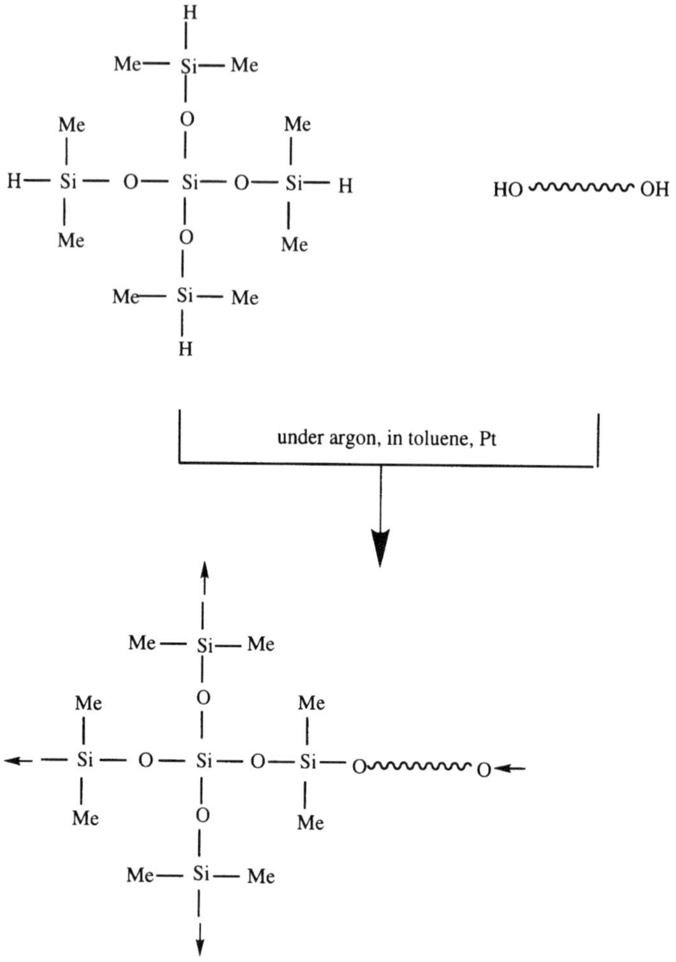

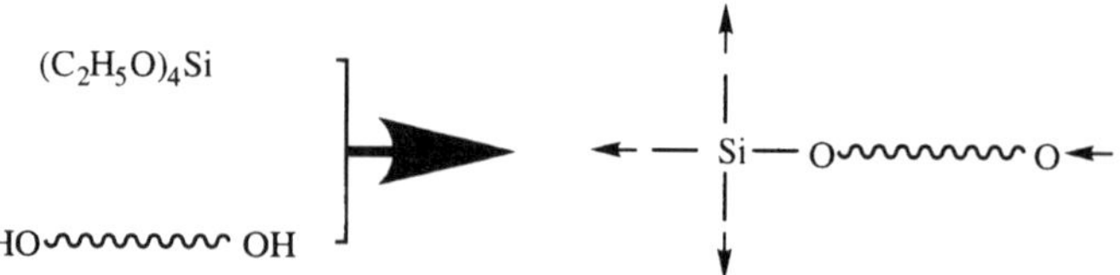

Scheme 7.3 Formation of type II network

Scheme 7.4 Formation of type III network

for water-swellable degradable devices. The following sections in this work will focus on type I networks, i.e. networks A to D.

THE EQUILIBRIUM SWELLING BY WATER OF THE HYDROGELS

The amount of water absorbed by a hydrogel network is quantitatively represented by the equilibrium water content, EWC, or by w_1, the equilibrium water fraction which is the weight fraction of water in the swollen hydrogel:

$$w_1 = \frac{\text{weight of water in the gel}}{\text{total weight of hydrated gel}} \tag{7.2}$$

The EWC is the most important single property of a hydrogel, influencing as it does the premeability, mechanical, surface and other properties of the gel.

The equilibrium water content (EWC) was measured at room temperature for the hydrogels formed by the crosslinked allyl-substituted PEG (type I networks). The EWC value given is an average from three determinations. As shown in Table 7.5, for networks A, B and C, the EWC for the hydrogels at room temperature increases from 0.14 to 0.62 and 0.96 respectively.

The effect of temperature on the equilibrium water content in hydrogels is also investigated. The equilibrium water content (EWC) was measured at three different temperatures and the results are summarized in Table 7.5. The EWCs in these hydrogels decrease with increasing temperature. This is expected due to the LCST type phase diagram of PEG in water. It also demonstrates that the swelling temperature has a remarkable effect on equilibrium water content in hydrogels. By increasing the temperature, a highly swollen hydrogel could be deswollen gradually. This high degree of temperature dependence may be used for sensitive control of swelling.

Three species of network B were repeatedly swollen in water, benzene and water. Table 7.6 shows the EWC in the hydrogels before and after swelling in

Table 7.5 The effect of temperature on the EWCs

Networks		A	B	C
25 °C	1	0.4126	0.6344	0.9572
	2	0.4256	0.6124	0.9594
	3	0.4028	0.6171	0.9581
	ave.	0.4137	0.6213	0.9582
45 °C	1	0.3507	0.5010	0.9490
	2	0.3298	0.5382	0.9542
	3	0.3399	0.5201	0.9507
	ave.	0.3401	0.5198	0.9513
65 °C	1	0.2709	0.4310	0.9400
	2	0.2626	0.4575	0.9421
	3	0.2582	0.4398	0.9380
	ave.	0.2639	0.4428	0.9400

Table 7.6 Swelling of network B

Weight fraction	1	2	3	ave.
Water	0.6319	0.6357	0.6306	0.6327
Benzene	0.8538	0.8550	0.8585	0.8558
Water	0.6252	0.6379	0.6467	0.6366

benzene. It was found that there was hardly any difference between the EWC before swelling in benzene and that after swelling in benzene. This demonstrates that the networks are stable and there is no degradation even after repeated swelling by different solvents.

WEIGHT FRACTION SOLUBLES (w_s) IN THE NETWORKS

The amount of soluble material in a network at any given extent of reaction is of practical interest. It could be used, for instance, to estimate the extent of the reaction, by which the network is formed [45]. Also, the weight fraction of soluble material in the networks can be used to calculate the molecular weight between crosslinks in a network [46]. The weight fraction of soluble material in the networks is defined by the following equation:

$$w_s = 1 - \frac{\text{weight of dry polymer network after swelling}}{\text{weight of dry polymer network before swelling}} \qquad (7.3)$$

Up to the gel point all molecules are finite, thus the weight fraction of solubles, w_s, is unity. Beyond the gel point molecules are rapidly incorporated into the network and w_s decreases quickly. The weight fraction of soluble material in the network was measured for networks A, B and C, and the results are summarized in Table 7.7. As we can see, going from networks A to B and C, w_s increases from 0.062 to 0.15 and 0.38. These results indicate that the formation of network A from PEVi-480 is more complete than network B from PEVi-680 which in turn is more complete than network C formed from PEVi-8080.

Table 7.7 Weight fraction of soluble material in the network (w_s)

Networks	A	B	C
1	0.0645	0.1501	0.3800
2	0.0607	0.1449	0.3899
3	0.0624	0.1442	0.3753
ave.	0.0625	0.1464	0.3817

ESTIMATION OF THE CONVERSION (p) FOR THE CROSSLINKING REACTION

According to the theory of Miller and Macosko [45], the following three equations are used to calculate w_s at a given degree of the reaction, p (where p is defined as fraction of end groups on the polymer chain that have reacted and the stoichiometric ration between polymer reactive groups and crosslinker reactive groups is taken as $r = 1$):

$$P(F_A^{out}) = (1/p^2 - 3/4)^{1/2} - 1/2 \tag{7.4}$$

$$P(F_A^{out}) = pP(F_B^{out}) + (1 - p) \tag{7.5}$$

$$w_s = w_{A4}P(F_A^{out})^4 + w_{B2}P(F_B^{out})^2 \tag{7.6}$$

in which, w_{B2} and w_{A4} are the weight fractions of polymer and crosslinker respectively. The probabilities $P(F_A^{out})$ and $P(F_B^{out})$ have the same meaning as in the original paper [45].

The weight fraction of soluble material in the network (w_s) at different p are calculated by assuming that the molecular weight between crosslinkable sites is 480, 680 and 8080, respectively. The relationship between w_s and p is shown in Figure 7.4. From the measured w_s (cf. Table 7.7), the conversion of reactants (p) is estimated as 79%, 73% and 66% for networks A, B and C, respectively.

Two reasons could in principle lead to imperfection of the networks in our system: the first is the conversion of hydroxyl groups of the PEG to allyl groups in the pre-polymers, the second is the cross-linking reaction. As discussed above, NMR and IR spectra have confirmed that each terminal hydroxyl group in PEG was completely derivatized into allyl group. Thus, one has to conclude that the imperfection of the networks results from the cross-linking reaction itself. During the course of the crosslinking reaction a problem is encountered. After approximately 30 to 60 min. gel has been formed and it is impossible to continue stirring the reaction mixture. It is possible that due to high viscosity and low mobility the reactants have difficulty in diffusing towards each other. To check this point we formed network D which is similar in all respects to network C except the reaction time was three times longer than for network C. Only a minor change in p (68% vs 66% for C) is observed.

THE CALCULATION OF MOLECULAR WEIGHT BETWEEN CROSSLINKS (M_c) FROM THE SWELLING DATA

Another important network parameter is the average molecular weight between crosslinks (M_c) which is useful in understanding the properties of networks [35]. M_c is directly related to the degree of conversion of the network-forming reaction described above.

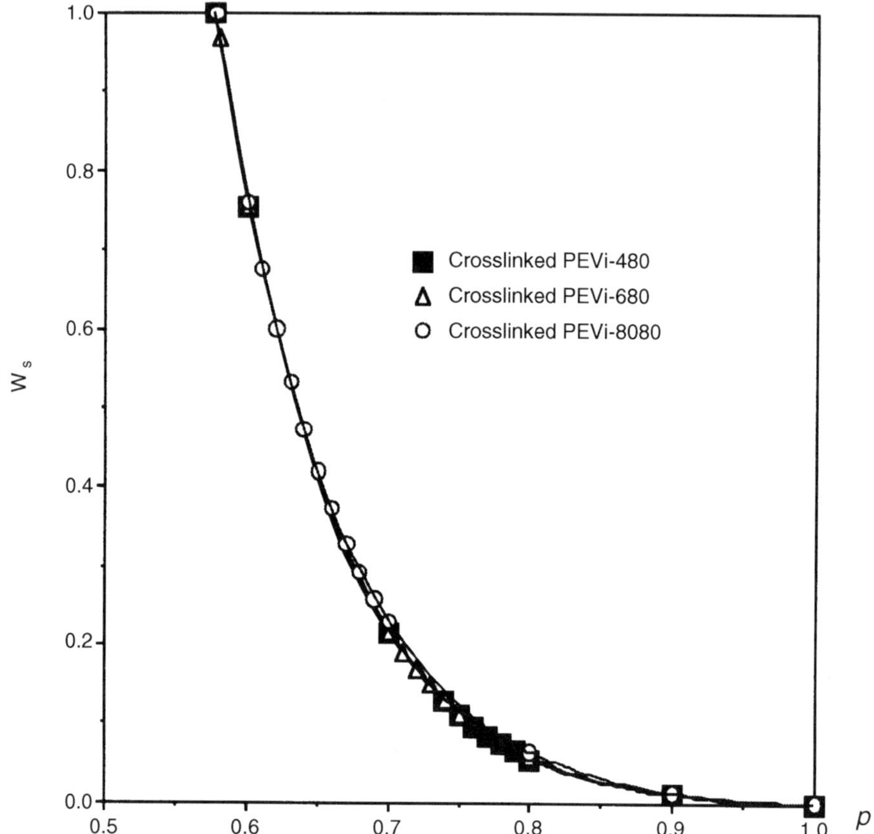

Figure 7.4 The relationship between fraction of solubles (w_s) and conversion of end functional group on the PEO (p) for stoichiometric ratio value $r = 1$

According to Flory [33], the following relationship is used in order to calculate M_c:

$$\ln(1 - \phi_2) + \phi_2 + \chi\phi_2^2 + \nu_e\nu_1(\phi_2^{1/3} - 2\phi_2 f^{-1}) = 0 \tag{7.7}$$

where ν_1 is the molar volume of water (18.05 cm^3 mol^{-1} at 25 °C), f is the functionality of crosslinks (we assume $f = 4$ in our case), ν_e is the effective crosslinking density, ϕ_2 is the volume fraction of polymer in the swollen hydrogel and χ is the polymer–solvent interaction parameter.

From ν_e the effective molecular mass between crosslinks (M_c) can be calculated via the equation:

$$M_c = \rho/\nu_e \tag{7.8}$$

in which ρ is the polymer density.

Table 7.8 The calculated M_c values

Networks	A	B	C
density of polymer (g/cm^3)	1.0473	1.0649	1.2118
ϕ_2	0.5751	0.3641	0.0347
χ^a	0.7751	0.5641	0.2347
M_c	421	739	20239 (20735)^b

[a]The χ values are calculated according to the equation $\chi = 0.2 + \phi_2$. (This equation is obtained by fitting the experimental χ data from the *CRC Handbook of Polymer-Liquid Interaction Parameters and Solubility Parameters* (ED Allan. F.M. Barton, CRC Press, Florida, 1990)).
[b]The number in (parentheses) is calculated from the measured weight fraction of soluble material in the network (w_s).

Obviously the molecular weight between crosslinks (M_c) could also be calculated by the recursion method [46] from the measured weight fraction of soluble material in the networks (w_s). The molecular weights between crosslinks in the networks are calculated by equations (7.7), (7.8) and listed in Table 7.8.

As shown in Table 7.8, the calculated M_c for network C is much higher than the molecular weight of the pre-polymer chain. This is obviously a result of the imperfection of the networks. If each chain is successfully reacted with two crosslinking groups, the result is an 'ideal' or 'model' network and the network chains would have the same average length and distribution of lengths as the sample of noncrosslinked chains from which it was prepared. In other words, M_c will decrease continuously from infinity at the gel point to the weighted sum of the pre-chain and two arms of the crosslinker upon $p = 1$. Table 7.8 shows that networks A and B have M_c close to expected. For network C we also computed M_c from the Macosko–Miller relations based on the value of p. The two methods give almost identical values for M_c.

It should be noted that in addition to the assumptions inherent to the Flory model [47] we have also neglected the contribution of the crosslinker to χ.

CONCLUSIONS

In this work, three types of crosslinked networks based on poly(ethylene oxide) have been synthesized. Type I consists of gels for which the networks are prepared by crosslinking of allyl-substituted PEG with tetrakis(dimethylsiloxy)-silane (HSiMe$_2$O)$_4$Si crosslinker. These gels are stable and show no sign of degradation even after repeated swelling by different solvents. Networks of type II and type III, which are prepared by crosslinking of the PEG with either (HSiMe$_2$O)$_4$Si or with (EtO)$_4$Si, are not stable and gradually degrade when they are swollen in water.

Selective water swellability of the networks could be obtained by controlling the PEO content in the networks and by controlling the temperature. By increasing the temperature, a well-swollen hydrogel could be deswollen gradually.

The measured weight fraction of soluble material in the networks (w_s) and the calculated molecular weight between crosslinks (M_c) show that the synthetic networks are not perfect. The imperfection of the networks results from the crosslinking reaction itself, but the exact source of the problem is not clear. Current work is under way to improve the conversion of the crosslinking reaction.

ACKNOWLEDGMENTS

Financial support for this work was provided by the German Israel Foundation and by the Ministry of Science under the Infrastructure Research program. We gratefully acknowledge Yafa Yagen for her help and advice.

REFERENCES

1. C.P. Pathak, A.S. Sawhney and J.A. Hubbell, *J. Am. Chem. Soc.*, **114**, 8311 (1992).
2. A.S. Sawhney, C.P. Pathak and J.A. Hubbell, *Macromolecules*, **26**, 581 (1993).
3. J.D. Andrade (ed.), *Hydrogels for Medical and Related Applications*, ACS Symposium, Vol. 31, ACS, Washington DC, 1976.
4. N.A. Peppas (ed.), *Hydrogels in Medicine and Pharmacy*, CRC Press, Boca Raton, Florida, 1987.
5. J. Kost and R. Langer, in *Hydrogels in Medicine and Pharmacy*, (N.A. Peppas, ed.), CRC Press, Boca Raton, Florida, 1987, Vol. III, p. 95.
6. N.A. Peppas and R.W. Korsmeyer, in *Hydrogels in Medicine and Pharmacy*, (N.A. Peppas, ed.), CRC Press, Boca Raton, Florida, 1987, Vol. III, p 109.
7. V. Dave, M. Tamagno, B. Focher and E. Marsano, *Macromolecules* **28,** 3531 (1995).
8. H. Yu and D.W.Grainger, *Macromolecules*, **27**, 4554 (1994).
9. T. Roberts, U. De Boni and M.V. Sefton, *Biomater.*, **17**, 267 (1996).
10. J.L.West and J.A. Hubbell, *Proc. Natl. Acad. Sci. USA.* **93**, 13188 (1996).
11. O. Wichterle and D. Lim, *Nature*, **185**, 117 (1960).
12. T. Okano, M. Katayama and I. Shinohara, *J. Appl. Polym. Sci.*, **22**, 369 (1978).
13. I.K. Varma and S. Patnaik, *Eur. Polym. J.*, **12**, 259 (1976).
14. J. Lebduska, J. Snuparek, K. Kaspar and V. Cermak, *J. Polym. Sci.*, **A24**, 777 (1986).
15. P.H. Corkhill, A.M. Jolly, C.O. Ng and B.J. Tighe, *Polymer*, **28**, 1758 (1987).
16. T.P. Davis, M.B. Huglin and D.C.F. Yip, *Polymer*, **29**, 701 (1988).
17. T.P. Davis and M.B. Huglin, *Polymer*, **31**, 513 (1990).
18. Y. Gnanou, G. Hild and P. Rempp, *Macromolecules*, **17**, 945 (1984).
19. N.B. Graham and M.E. McNeill, *Biomaterials*, **5**, 27 (1984).
20. N. B. Graham, in *Hydrogels in Medicine and Pharmacy*, (N.A. Peppas, ed.), CRC Press, Boca Raton, Florida, 1987, Vol. II, p. 95.
21. L.G. Cima, *J. Cell. Biochem.*, **56**, 155 (1994).
22. D.K. Han and J.A. Hubbell, *Macromolecules*, **30**, 6077 (1997).
23. J.J. Sperinde and L.G. Griffith, *Macromolecules*, **30**, 5255 (1997).
24. M. Amiji and K. Park, *J. Biomater. Sci., Polym. Ed.*, **4**, 217 (1993).
25. E.A. Merrill and E.W. Salzman, *ASAIO J.*, **6**, 60 (1984).

26. W.R. Gombotz, W. Guanghui, T.A. Horbett and A.S. Hoffman, *J. Biomed. Mater. Res.*, **25**, 1547 (1991).
27. K. Bergstrom, K. Holmberg, A. Safranj, A.S. Hoffman, M.J. Edgell, A. Kozlowski, B.A. Hovanes and J.M. Harris, *J. Biomed. Mater. Res.*, **26**, 779 (1992).
28. E.L. Chaikof, E.W. Merrill, A.D. Callow, R.J. Connolly, S.L. Verdon and K. Ramberg, *J. Biomed. Mater. Res.*, **26**, 1163 (1992).
29. M. Doytcheva, R. Stamenova, V. Zvetkov and Ch. B. Tsvetanov, *Polymers Networks 98*, Trondheim, Norway, 1998, Paper 2.
30. F.E. Du Prez, D. Christova and E.J. Goethals, *Polymers Networks 98*, Trondheim, Norway, 1998. This book, Chapter 22.
31. M. Newcomb, J.M. Timko, D.M. Walba and D.J. Cram, *J. Am. Chem. Soc.*, **99**, 6392 (1977).
32. N.A. Peppas and A.G. Mikos, in *Hydrogels in Medicine and Pharmacy: Fundamentals*, (N.A. Peppas, ed.), CRC Press, Boca Raton, FL., 1986, Vol. I, pp. 1.
33. P.J. Flory, *Principles of Polymer Chemistry*, Cornell University Press, Ithaca, NY, 1953.
34. L.R.G. Treloar, *The Physics of Rubber Elasticity*, Clarendon Press, Oxford, 1958.
35. K. Dusek and W. Prins, *Adv. Polymer Sci.*, **6**, 1 (1969).
36. W.W. Graessley, *Adv. Polymer Sci.*, **16**, 1 (1974).
37. A. Shefer and M. Gottlieb, *Macromolecules*, **25**, 4036 (1992).
38. M. Adam, D. Lairez, M. Karpasas and M. Gottlieb, *Macromolecules*, **30**, 5920 (1997).
39. M. Gottlieb, C.W. Macosko, G.S. Benjamin, K.O. Meyers and E.W. Merrill, *Macromolecules*, **14**, 1039 (1981).
40. M. Gottlieb, C.W. Macosko and T.C. Lepsch, *J. Polym. Sci., Polym. Phys. Ed.*, **19**, 1603 (1981).
41. S.K. Venkataraman, L. Coyne, F. Chambon, M. Gottlieb and H.H. Winter, *Polym. Prepr.*, **29**, 571 (1988).
42. S.K. Venkataraman, L. Coyne, F. Chambon, M. Gottlieb and H.H. Winter, *Polymer*, **30**, 2222 (1989).
43. J.E. Mark and J.L. Sullivan, *J. Chem. Phy.*, **66**, 1006 (1977).
44. See note 28 in reference 43.
45. D.R. Miller and C.W. Macosko, *Macromolecules*, **9**, 206 (1976).
46. D.R. Miller, E.M. Valles and C.W. Macosko, *Polym. Eng. Sci.*, **19**, 272 (1979).
47. P. Pekarski, Y. Rabin and M. Gottlieb, *J. Phys. II France*, **4**, 1677 (1994).

8

Interpretation of Experimentally Measured Network Moduli on the Basis of Calculated Network Topologies

R.F.T. STEPTO and D.J.R. TAYLOR
Polymer Science and Technology Group, Manchester Materials Science Centre, UMIST and University of Manchester, Grosvenor Street, Manchester M1 7HS, United Kingdom

ABSTRACT

A new Monte Carlo algorithm, developed to simulate non-linear, network-forming polymerizations, is used to calculate the distributions of loop structures in completely reacted networks. Intramolecular reaction is correctly weighted according to the known reactant

Wiley Polymer Networks Group Review Series Vol. 2. Edited by B.T. Stokke and A. Elgsaeter

structures and, more specifically, the flexibilities of the sub-chains connecting pairs of groups reacting intramolecularly. The simulations are based on the polymerization reactions used to form a series of elastomeric, polyurethane networks, whose mechanical properties at complete reaction were measured experimentally. The shear moduli were found to be considerably lower than expected on the basis of a perfect network structure (in which all network chains are assumed to be elastically effective). Smallest loop structures are known to contribute significantly to the decreases in moduli, as they render integer numbers of chains elastically ineffective. However, in this paper the experimentally determined shear moduli are correlated with the calculated proportions of loop structures of *all* sizes in order to estimate the contributions of *larger* loop structures to the decreases in moduli. Dependent upon the reactant structures, the average fractional loss of network-chain elasticity in larger loops is found to lie in the range 0.5–0.8, for trifunctional and tetrafunctional network structures.

LIST OF SYMBOLS

λ_{a0}, λ_{b0}	ring-forming parameters
ν	number of skeletal bonds in a sub-chain forming the smallest loop structure
ρ	density of bulk network
C_{a0}	initial concentration of A groups
C_{b0}	initial concentration of B groups
f	branch-point functionality
p	extent of reaction (i.e. proportion of functional groups reacted)
$p_{\mathrm{re,total}}$	total extent of intramolecular reaction at the end of the polymerisation
$p_{\mathrm{re},1}$	proportion of intramolecular reactions yielding smallest-loop structures
$p_{\mathrm{re},i>1}$	proportion of intramolecular reactions yielding larger-loop structures
$\langle r^2 \rangle^{1/2}$	root-mean-square end-to-end distance of a sub-chain forming a smallest loop
x	fractional loss in elasticity per chain in a larger loop structure
G	experimentally measured shear modulus
G^{o}	shear modulus of a hypothetical, perfect network structure
M_{c}	network-chain molar mass calculated from measured shear modulus, G, via rubber elastic theory
$M_{\mathrm{c}}^{\mathrm{o}}$	network-chain molar mass calculated from known reactant structures
N	number of branch points
N_{k}	number of inelastic network chains
N^{o}	number of elastically effective network chains
$P(\boldsymbol{r} = \boldsymbol{0})$	probability density of a zero end-to-end vector for a Gaussian chain

P_{ab} mutual concentration of pairs of A and B groups on the same molecule which can react intramolecularly

T absolute temperature

INTRODUCTION

It is well established [1] that the occurrence of intramolecular reaction during nonlinear, network-forming polymerizations gives rise to inelastic loop structures, and loop structures of reduced elasticity,in the resulting network topology. It is evident that a *distribution* of loop sizes will exist in a network structure, the details of which are largely inaccessible to experimental analysis. Such loop structures are associated with a reduction in the shear modulus of the network material relative to that expected for the hypothetical, perfect network, in which all network chains are assumed to be elastically effective. However, the relative contributions to the observed modulus-reduction, from loop structures of various sizes, are still not known [2]. It is clear that the formation of the smallest possible loop structure will render a fixed number of chains elastically ineffective [2] (the number being dependent upon the branch-point functionality), whereas the influence of larger loops is, at present, unclear.

The present paper describes the use of a new Monte Carlo (M-C) polymerization algorithm [3] to simulate the formation of a series of polyurethane (PU) network materials. The connectivities in the simulated network structures were recorded as functions of extent-of-reaction, p. By correlating the detailed structural information from the M-C simulations with *experimentally measured* reductions in moduli [4], the loss in network elasticity due to larger loop structures can be estimated.

EXPERIMENTAL PU SYSTEMS

Six series of PU-network materials were formed, via stoichiometric $RA_2 + R'B_f$ polymerizations of hexamethylene diisocyanate (HDI) and star polyoxypropylene (POP) polyols, and using various initial dilutions of reactive groups (in nitrobenzene) [4,5]. The structure of the polyol ($R'B_f$) was varied, in order to test the effects of branch-point functionality (f) and network-chain length (ν) on the reductions in moduli of the resulting networks. The six reaction systems are listed in Table 8.1, where M_c^o is the network-chain molar mass (that for a chain of ν skeletal bonds between branch points) and is defined by the reactant structures. ν is also the number of skeletal bonds in a sub-chain forming the smallest possible loop structure [5].

A precise knowledge of the reactant structures can be used to calculate the shear modulus of the bulk, hypothetically perfect network material, G^o, in which

Table 8.1 Reactant characteristics for the series of PU polymerization systems [4,5]

System ref.	f	ν	M_c^o/g mol^{-1}
1	3	35	635
2	3	62	1168
3	4	28	500
4	4	32	586
5	4	43	789
6	4	65	1220

every chain is assumed to be elastically effective:

$$G^o = N^o kT = \frac{\rho RT}{M_c^o} \tag{8.1}$$

where ρ is the density of the network, T the absolute temperature and N^o the number of elastically effective chains per unit volume of the network. Due to the existence of loop structures in the PU systems listed in Table 8.1, the experimentally measured shear moduli, G, were less than G^o, and the relative *reduction in modulus*, defined as G^o/G (or, equivalently, the relative increase in elastic-chain molar mass, M_c/M_c^o), is thus greater than unity. For each experimental system, an increase in the initial dilution of reactive groups resulted in greater reductions in moduli, consistent with increased incidence of intramolecular reaction and the formation of loop structures. An example of the experimentally observed variation of modulus with dilution, for PU system 1, is shown Figure 8.1.

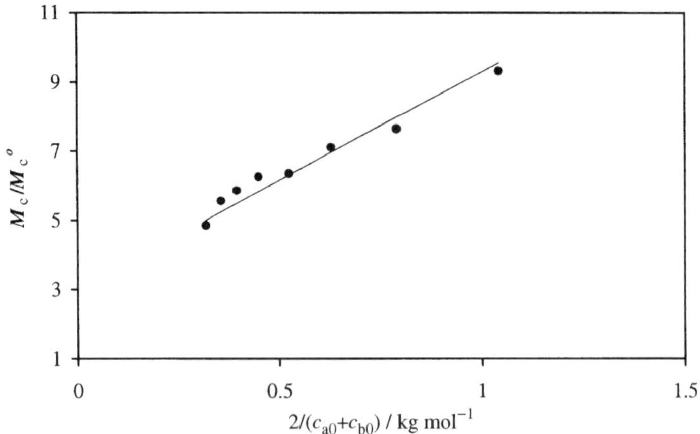

Figure 8.1 Reduction in modulus, expressed as M_c/M_c^o, as a function of the initial dilution of reactive groups, $2/(c_{ao} + c_{bo})$, for PU system 1 (see Table 8.1). The line serves as a guide for the eye

MONTE CARLO POLYMERIZATION ALGORITHM

A new M-C network polymerization algorithm, for single- and two-monomer reactions, has been developed [3] that is based on a perturbation to Flory–Stockmayer (F–S) random-reaction statistics. For a finite population of reactant molecules, the algorithm simulates, as a function of extent-of-reaction, the formation of *all of the connections* in a reaction mixture. Intramolecular reactions are allowed by correctly assigning probabilities for intermolecular and intramolecular reaction, weighted, respectively, according to the initial concentrations of reactive groups in the system, and the detailed conformational statistics of sub-chains separating pairs of reactive groups on the same molecule. Shorter, more flexible sub-chains between reactive groups will enhance the probability of intramolecular reaction. In addition, intramolecular reaction will become increasingly favoured as the reactive-group concentration decreases [1].

The effects of reactant structure are parameterised in terms of P_{ab} for the smallest loop [1]:

$$P_{ab} = \frac{P(r = \mathbf{O})}{N_{Av}} \tag{8.2}$$

where $P(r = \mathbf{O})$ represents the probability density of a zero end-to-end vector between reactive groups. P_{ab} thus represents the mutual concentration of A and B groups at the ends of the smallest sub-chain, which can react intramolecularly. The structure of this sub-chain, whose root-mean-square end-to-end distance is $\langle r^2 \rangle^{1/2}$, is shown in Figure 8.2. If it assumed that the end- to-end distance distributions of network chains can be represented by Gaussian functions with

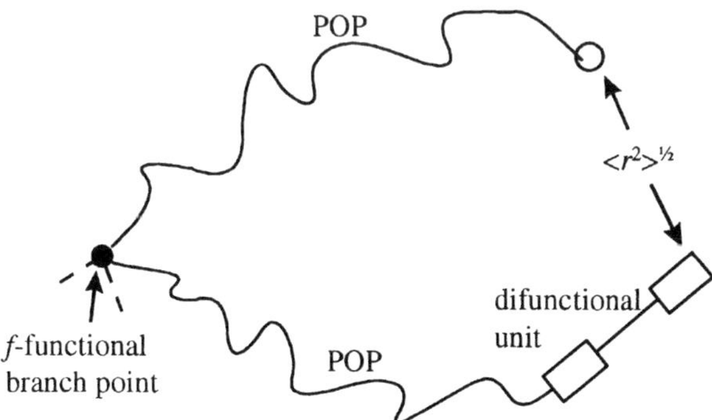

Figure 8.2 Schematic of the sub-chain forming the smallest loop structure; the diagram shows the two arms of the POP polyol, one arm having reacted with the difunctional monomer (a diisocyanate in the case of PU); the root-mean-square distance between the terminal groups is shown as $\langle r^2 \rangle^{1/2}$

mean-square end-to-end distances of $\langle r^2 \rangle$, P_{ab} is given by

$$P_{ab} = \frac{1}{N_{Av}} \left\{ \frac{3}{2\pi \langle r^2 \rangle} \right\}^{3/2} \tag{8.3}$$

The combined effects of reactant structure and reactive-group dilution are accounted for by defining [1] a ring-forming parameter, λ_{a0}

$$\lambda_{a0} = \frac{P_{ab}}{c_{a0}} \tag{8.4}$$

in terms of c_{a0}, the initial concentration of A groups (an analogous expression exists for λ_{b0}, in terms of c_{b0}). For stoichiometric polymerization reactions, $c_{a0} = c_{b0}$, and

$$\lambda_{a0} = \lambda_{b0} = \frac{2P_{ab}}{(c_{a0} + c_{b0})} \tag{8.5}$$

M-C SIMULATION RESULTS

M-C simulations were performed for a series of generic, stoichiometric two-monomer ($RA_2 + R'B_f$) polymerization systems, with trifunctional or tetrafunctional branch units, using a range of $\lambda_{a0}(= \lambda_{b0})$ values up to $\lambda_{a0} = 0.3$. Typical system sizes were not large, and were in the range 3600–4000 reactive groups. At complete reaction ($p = 1$), the proportion of intramolecular reaction, p_{re}, was recorded. As the M-C method is capable of providing the complete distribution of loop sizes, the calculated total p_{re} was broken down into two components: $p_{re,1}$, the proportion of reactions yielding smallest-loop structures, and $p_{re,i>1}$, the fraction of reactions leading to larger loop structures.

The resulting 'master curves', for the stoichiometric trifunctional and tetrafunctional two-monomer polymerizations, are shown in Figure 8.3(a) and Figure 8.3(b), respectively. In both cases it can be seen that the *total* proportion of ring-forming reaction at the end of the polymerization, $p_{re,total}$, remains constant over the whole range of λ_{a0}, at a value dependent only upon the branch-point functionally. Calculated values of $p_{re,total}$ are related to the cycle rank [6] of the completely reacted network structure, and can be predicted analytically [7]:

$$p_{re,total} = \frac{1}{2} \left(1 - \frac{2}{f} \right) \tag{8.6}$$

The factor 1/2 appears due to the presence of difunctional (RA_2) monomer in an $RA_2 + R'B_f$ polymerization. When forming any loop structure, one A group on the closing RA_2 unit must react intermolecularly before the loop can be formed via subsequent intramolecular reaction of the other A group [7]. Hence, $p_{re,total} = 1/6 = 0.1667$ for the trifunctional networks, and $p_{re,total} = 1/4 = 0.25$

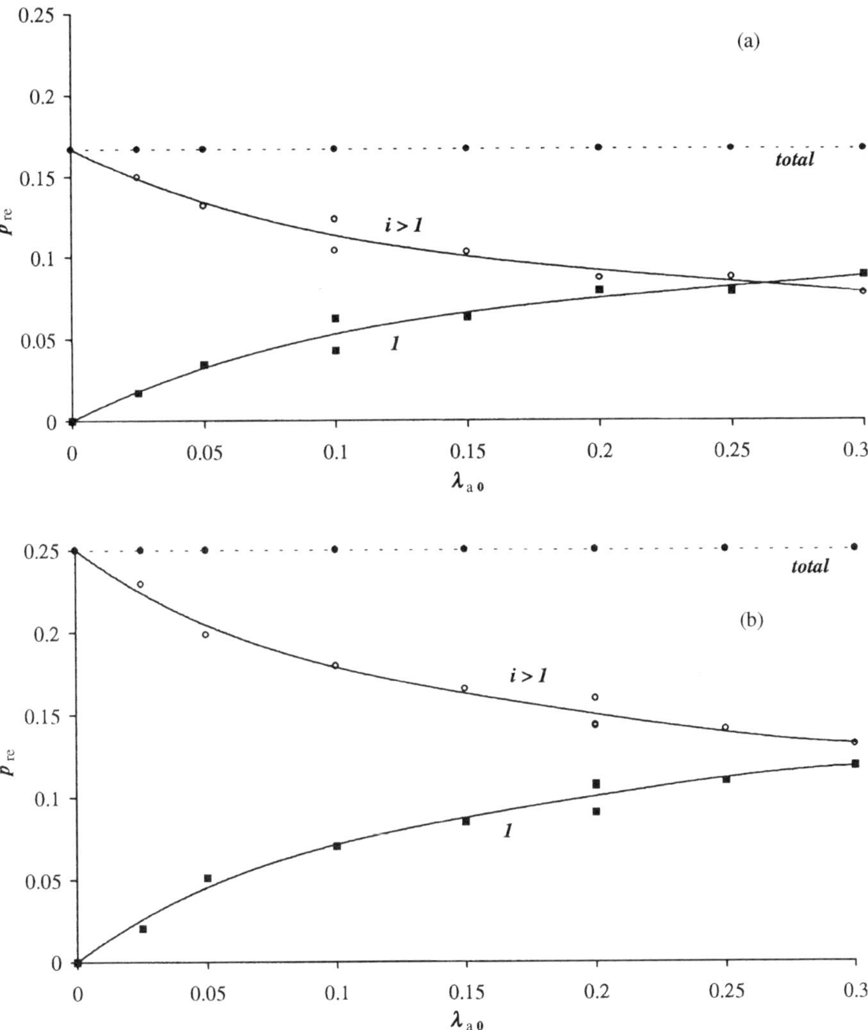

Figure 8.3 $RA_2 + R'B_3$: the proportion of loop-forming, intramolecular reactions, p_{re}, calculated at complete reaction for a series of simulated M-C polymerizations as functions of λ_{a0}; plots are shown for smallest loops (1), and loops of all other sizes ($i > 1$): (a) $RA_2 + R'B_3$; (b) $RA_2 + R'B_4$

for tetrafunctional structures, in agreement with the M-C results shown in Figure 8.3(a) and Figure 8.3(b).

The plots also shown that the incidence of smallest loops ($p_{re,1}$) is negligible when $\lambda_{a0} = 0$. Essentially, in agreement with F–S statistics, all intramolecular reaction then occurs *after* the gel point, resulting in predominantly large loop structures. However, as λ_{a0} increases, $p_{re,1}$ increases, at the expense of the proportion of larger loop structures ($p_{re,i>1}$), as $p_{re,total}$ remains constant. For a given value of λ_{a0}, the tetrafunctional systems give rise to more loop structures than the trifunctional systems, simply due to the greater number of opportunities for loop-formation in the former (equation (8.6)). However, relative to the total number of loop structures, the proportion of smallest loops appears to be fairly insensitive to the branch-point functionality.

CALCULATED REDUCTIONS IN MODULUS

Generally, there are $f/2$ network chains emanating from an f-functional branch point. In a perfect network, all such chains would be elastically effective, i.e. they would be connected to fully elastically effective branch points at both ends. However, in the case of real networks, the formation of loops causes a reduction in the elasticity of the structure, reducing the modulus relative to that of the perfect network. In the case of network structures with trifunctional and tetrafunctional branch points, the formation of a *smallest* loop renders integer numbers of branch points elastically ineffective [2]. This is shown diagrammatically in Figure 8.4(a), and it can be seen that a single smallest-loop structure renders two *trifunctional* branch points (shown as $\square$), or three network chains, elastically ineffective, whereas a single *tetrafunctional* branch point is lost per smallest loop (i.e. two chains are lost).

When larger loop structures are considered, the associated loss in elasticity is not immediately evident. Two instances of two-membered loops for networks with trifunctional and tetrafunctional branch points are shown in Figure 8.4(b). For $f = 3$, the elasticity of the sub-structure shown is equivalent to that of a liner chain between two elastically effective branch points ($\blacksquare$), *plus* the elasticity of the loop structure. Thus, it is assumed that the two remaining branch points in the loop ($\blacksquare$), are of *reduced* elasticity (i.e. giving three chains of reduced elasticity). Similarly, for $f = 4$, the sub-structure shown is equivalent to one in which the two junctions labelled $\blacksquare$ effectively become pseudo-trifunctional, due to the partial loss in the elasticity of the two chains in the loop structure joining them.

It is immediately obvious that the occurrence of loop structures in trifunctional networks will be far more detrimental to their elastic properties, relative to loops in tetrafunctional structures.

Previous work [3] has shown that the effects of the larger loop structures must be taken into account in order to reproduce the experimental reductions in moduli. Hence, the reduction in modulus in each case can be estimated according to the numbers to elastically effective chains corresponding to the values of $p_{re,1}$ and $p_{re,i>1}$ (taken from the M-C simulations), relative to the numbers of chains in the hypothetical 'perfect' network. For N branch points, the latter quantity, denoted N^o, is simply given by

$$N^o = \frac{Nf}{2} \tag{8.7}$$

If the number of chains rendered elastically ineffective, N_k, can also be estimated, the reduction in modulus, or increase in M_c, can be calculated using

$$\frac{G^o}{G} = \frac{M_c}{M_c^o} = \frac{N^o}{N^o - N_k} \tag{8.8}$$

(a) $f = 3$, smallest loop ($i = 1$)

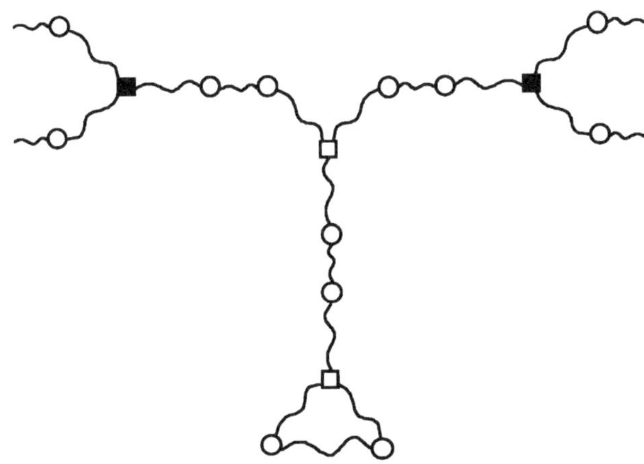

$f = 4$, smallest loop ($i = 1$)

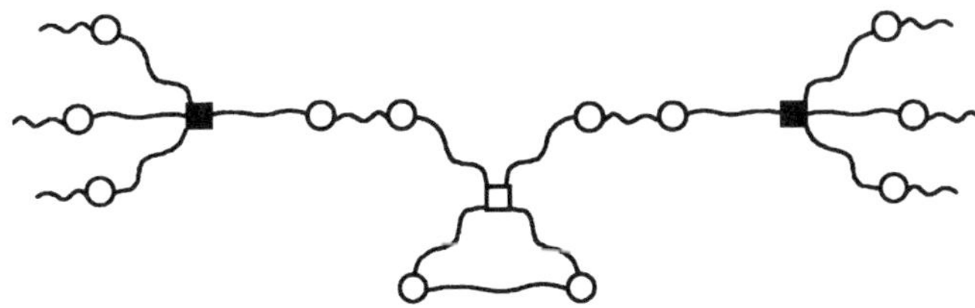

Figure 8.4 (a) Smallest-loop structures, and (b) two-membered loops structures, for networks with trifunctional and tetrafunctional branch points. O denotes a pair of reacted end-groups (i.e. a urethane group, in the case of PU networks); ■ denotes a fully elastically effective branch point; ▨ denotes a *partially* elastically-effective branch point; □ denotes an elastically ineffective branch point

(b) $f = 3$, two-membered loop ($i = 2$)

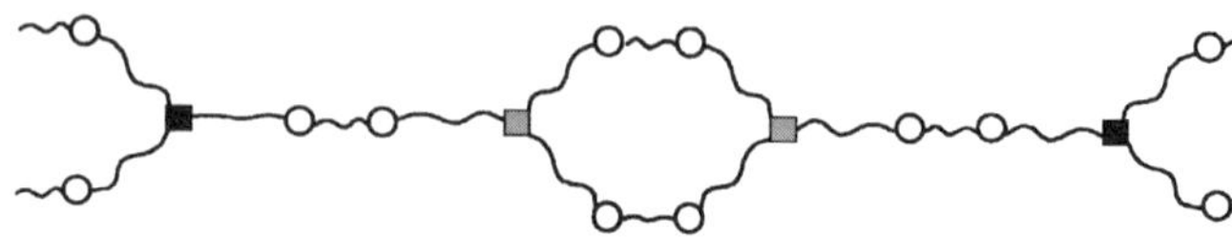

$f = 4$, two-membered loop ($i = 2$)

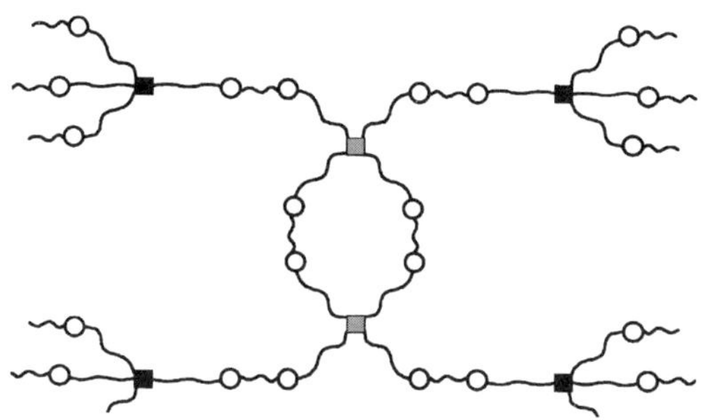

Figure 8.4 (*continued*)

The quantity N_k was evaluated as follows. For the simulated trifunctional networks, the numbers of smallest loops are given by $3N p_{re,1}$, which equate to losses of elastic chains equal to $3 \times 3N p_{re,1}$. However, the $3 \times 3N p_{re,i>1}$ chains in *larger* loops are subject only to a partial loss in elasticity, and are thus equivalent to $x \times 3 \times 3N p_{re,i>1}$ chains, where x is the fractional loss in elasticity per chain in a larger loop structure. ($x = 1$ corresponds to a *total* loss in elasticity, seen in the case of smallest loops only.) Hence, for $f = 3$,

$$N_k = \frac{3N}{2}\{6N p_{re,1} + x.6 p_{re,i>1}\} \tag{8.9}$$

and substituting the results from equation (8.7) and (8.9) into equation (8.8) yields

$$\frac{G^o}{G} = \frac{M_c}{M_c^o} = \frac{1}{1 - 6 p_{re,1} - x.6 p_{re,i>1}} \tag{8.10}$$

A similar expression can be derived for tetrafunctional network structures:

$$\frac{G^o}{G} = \frac{M_c}{M_c^o} = \frac{1}{1 - 4 p_{re,1} - x.4 p_{re,i>1}} \tag{8.11}$$

CORRELATION OF SIMULATED LOOP DISTRIBUTIONS AND EXPERIMENTAL MODULUS MEASUREMENTS

The experimentally determined values M_c/M_c^o for the series of PU networks listed in Table 8.1 were used in conjunction with simulated values of $p_{re,1}$ and $p_{re,i>1}$ to estimate x, the fractional loss of elasticity for chains with larger loop structures, using equations 8.10 and 8.11. However, since $p_{re,1}$ and $p_{re,i>1}$ calculated via M-C are dependent upon P_{ab}, values of M_c/M_c^o depend on both x and P_{ab}. Correlations between M-C calculations and experimental data were therefore performed in two ways:

(i) bivariate least-squares fitting, to evaluate x and P_{ab} simulataneously;
(ii) monovariate least-squares fitting, to evaluate x using P_{ab} values calculated *ab initio*, via chain-conformational analyses [8].

The calculated values of x and P_{ab}, using monovariate and bivariate analyses, for PU systems 1–6 are listed Table 8.2. Using the bivariate fitting method, it can be seen that values of x in the range 0.67–0.60 are required to reproduce the experimental modulus reductions for the trifunctional PU networks (systems 1 and 2), and 0.68 to 0.48 for the tetrafunctional PU networks (systems 3–6). In both cases, an increase in the size of the smallest loop structure results in a decrease in x indicating less elasticity lost. However, the values of P_{ab} estimated from bivariate fitting are much higher than those calculated via conformational analyses of the sub-chain structures (Figure 8.2) and thus the monovariate fitting required larger values of x (0.82–0.69, for $f = 3$, and 0.78–0.56 for $f = 4$). The unrealistically high values of P_{ab} from the bivariate fitting were required in order to match the slopes of the experimental M_c/M_c^o plots, which were larger than those calculated using P_{ab} values from chain-conformational analyses.

Two examples of the fitted experimental M_c/M_c^o data are shown in Figure 8.5(a) and Figure 8.5(b), for PU systems 1 and 3. In both cases, the bivariate fitting procedure results in very good fits to the experimental data,

Table 8.2   values of x and P_{ab} based on correlations of experimentally measured reductions in moduli for PU networks, and extents of intramolecular reaction calculated from M-C simulations

PU system	f	x^a	P_{ab}^a/mol 1^{-1}	x^b	P_{ab}^b/mol 1^{-1}
1	3	0.667	0.538	0.816	0.076
2	3	0.599	0.103	0.690	0.030
3	4	0.684	0.547	0.775	0.120
4	4	0.638	0.457	0.730	0.079
5	4	0.582	0.129	0.637	0.055
6	4	0.481	0.088	0.559	0.032

[a] Bivariate fitting. [b] Monovariate fitting; P_{ab} values were calculated *ab initio*, via chain-conformational analyses

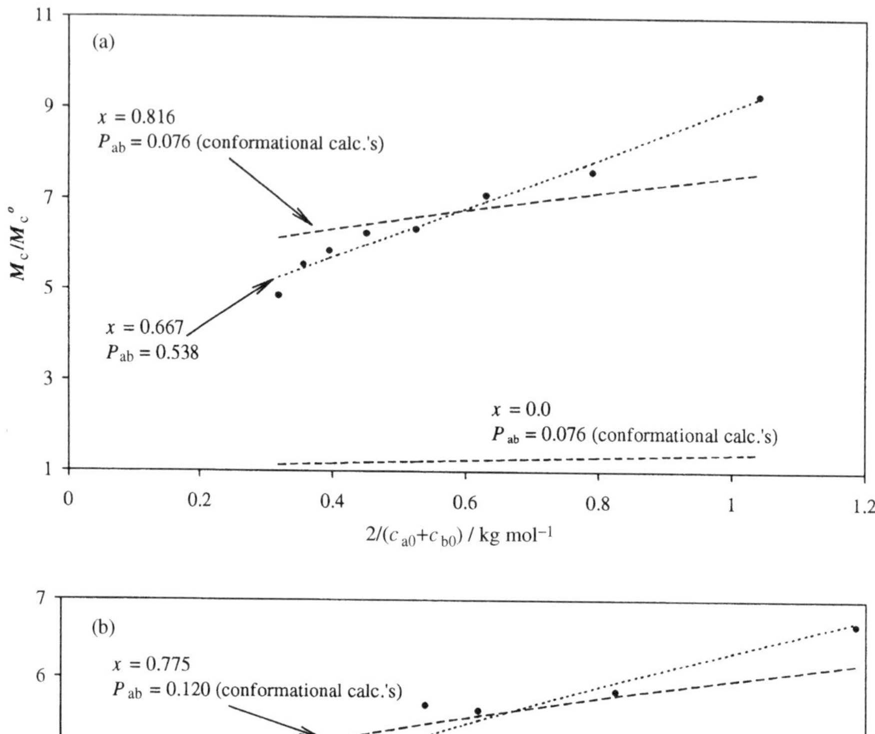

Figure 8.5 Experimental reductions in moduli (●) compared with the predictions of M-C simulations; the values of x and P_{ab} used to calculate the M-C plots are as shown. (a) $RA_2 + R'B_3$, PU system 1: $x = 0.816$, $P_{ab} = 0.076$ (monovariate); $x = 0.667$, $P_{ab} = 0.538$ (bivariate). (b) $RA_2 + R'B_4$, PU system 3: $x = 0.775$, $P_{ab} = 0.120$ (monovariate); $x = 0.684$, $P_{ab} = 0.547$ (bivariate)

as expected. M_c/M_c^o was also calculated using the *ab initio* values of P_{ab}, but assuming that $x = 0$ (i.e, no additional loss in elasticity from larger loop structures), and these values are plotted, for comparison. It is obvious that the neglect of the elasticity loss in larger loop structures results in a gross underestimation of the experimentally observed reductions in moduli. The effects of the values used for P_{ab} are secondary compared with the effects of assuming $x = 0$.

CONCLUSIONS

Detailed structural information, including the distribution of loop sizes, for polymer network materials is generally inaccessible via experiment. However, correlations between experimentally measured network moduli and extents of loop-forming reaction calculated via M-C simulation suggest that larger loop structures ($i > 1$) contribute significantly to the observed loss of elasticity (relative to that of a perfect network). Current work is focusing on the direct calculation of the entropies of larger loop structures, and more exact calculations of P_{ab} for sub-chains in branched structures.

REFERENCES

1. R.F.T. Stepto, 'Non-linear polymerisation, gelation and network formation, structure and properties', in *Polymer Networks Principles of Their Formation, Structure and Properties* (R.F.T. Stepto, ed.), Blackie Academic & Professional, London, 1998; Chapter 2.
2. R.F.T. Stepto and B.E. Eichinger, 'Interpreting the properties of model HDI-based and MDI-based polyurethane networks', in *Elastomeric Polymer Networks*, (J.E. Mark and B. Erman, eds), Prentice-Hall, Englewood Cliffs, New Jersey, 1992, Chapter 18.
3. S. Dutton, R.F.T. Stepto and D.J.R. Taylor, *Angew. Makromol. Chem.* **240**, 39 (1996).
4. J.L. Stanford and R.F.T. Stepto, 'Experimental studies of the formation and properties of polymer networks', in *Elastomers and Rubber Elasticity, Amer. Chem. Soc. Symp. Series 193*, (J.E. Mark and J. Lal, eds), American Chemical Society, Washington DC (1982), Chapter 20.
5. R.F.T. Stepto, 'Intramolecular reaction and network formation and properties, in *Biological and Synthetic Polymer Networks*', (O. Kramer, ed.), Elsevier Applied Science, Baking, 1988, Chapter 10.
6. See P.J. Flory, *Principles of Polymer Chemistry*, Cornell University Press, Ithaca (1953).
7. R.F.T. Stepto, *Polym. Bull. (Berlin)*, **24**, 53 (1990).
8. D.J.R. Taylor and R.F.T. Stepto, to be published.

9

Modeling of Nonequilibrium Nucleation and Growth at Phase Decomposition in Curing Multicomponent Blends

GRIGORI M. SIGALOV and BORIS A. ROZENBERG
Institute of Problems of Chemical Physics, Russian Academy of Sciences, Chernogolovka, 142432 Moscow region, Russia

ABSTRACT

Numerical simulation of the cure-induced phase separation proceeding via nucleation and growth mechanism was performed using novel models of the process. It was found that the particle size distribution in the material formed would be very narrow if the growing particles did not influence each other. The evolution of the concentration profile between neighboring particles was characterized. The ultimate particle size is shown to depend on the bimolecular reaction rate constant according to the power law $r_{\max} \propto K^{-1/2}$.

Wiley Polymer Networks Group Review Series Vol. 2. Edited by B.T. Stokke and A. Elgsaeter
© 1999 John Wiley & Sons Ltd

INTRODUCTION

To improve the characteristics of the network polymers or to obtain new materials based thereon, some modifiers — reactive or nonreactive — are often added to the thermosetting blend. In the course of the cure reaction the solubility of the additive in the thermosetting blend decreases, and it may separate to form a new phase. Finally, as the gelation or vitrification of the matrix phase occurs, this yields a heterophase material. The total process, which includes both chemical reactions and diffusion, is referred to as 'cure reaction induced microphase separation' (CRIMPS) [1]. For example, if a rubber is added to an epoxy/amine system, one can get the material in which the fracture toughness is much higher than that of the unmodified brittle epoxy matrix. This is due to inclusions of viscoelastic rubber dispersed in the matrix. CRIMPS processes have been extensively studied for the past decade [2].

The efficiency of network polymer modification via CRIMPS is mainly determined by the additive volume fraction and the particle size distribution (PSD) of the dispersed phase. It should be also kept in mind that, while the fracture toughness of the modified material generally increases with the volume fraction of the modifier, the elastic moduli and glass transition temperature, T_g, decrease. Therefore, the process must be carried out in a smart way to gain as much as possible in the fracture toughness and, at the same time, to lose as little as possible in the elastic moduli and T_g. This may only be done controlling the PSD by optimizing the blend composition and process conditions.

The most critical parameter which determines PSD is the cure kinetics (i.e., the cure rate and its variation during cure). A spectacular illustration of this idea is provided by the fact that the volume fraction of dispersed phase in the final material, ϕ_{disp}, is not just the volume fraction of the modifier added, ϕ_{20} (even if its solubility in the cured thermoset is negligible). On the contrary, ϕ_{disp} may be much less that ϕ_{20} or even vanish if the cure reaction is fast enough [3]. Thus, merely adding more rubber modifier does no guarantee higher fracture toughness (but may lead to unwanted decrease in elastic moduli and T_g due to plastification effect) if the reaction kinetics is disregarded.

If the cure rate is small enough, then the phase separation proceeds via (quasi) equilibrium nucleation and growth (NG) which is well described by classical theories for both early [4,5] and late stages [6]. If the chemical quench is fast enough, then spinodal decomposition takes place. This case is also covered by many papers (see [2] for references). The intermediate case of moderate reaction rates is most vague and least studied. In this case the phase separation is delayed and, after some induction period, occurs in an essentially nonequilibrium way. Many CRIMPS processes found in practice run this way [2]; this is why this problem is of a considerable importance from both the academic and the practical point of view. Therefore, regulation of the ultimate phase structure in such a kind of CRIMPS process requires adequate theory specific to this case.

In the present paper we consider moderately nonequilibrium phase separation in a model quasibinary blend corresponding to a characteristic system exhibiting CRIMPS. To catch the peculiarities of the process in question, we use a combination of classical and new models. In particular, we adopt the classical nucleation rate [7] and use the conventional linear diffusion equation to describe the concentration changes [8]. However, we take into account that nucleation and growth of a particle lead to nonuniformity of the system's composition both inside and outside the particle. Instead of using average phase compositions, real concentration profiles are calculated with the help of the model of particle growth with equilibrium boundary conditions [1]. Although the consideration is still restricted to the case of noninteracting particles, numerical modeling allows revealing of important features of the PSD and its dependence on the cure rate.

MODEL OF NUCLEATION

Classical models of nucleation [4,5] allow determination of the critical nucleus size, r_c, the free energy of a critical nucleus, ΔG^*, and the probability of nucleation, $P \sim \exp(-\Delta G^*/RT)$. However, the case of solute demixing is quite different from the case of vapor–liquid or liquid–solid transitions, which have been considered conventionally. The difference is that the total quantity of the additive in the system may not change. Since the appearance of a nucleus is a sudden event, the rearrangement of the additive may only take place in the close vicinity of the nucleus. The conclusion is that the newborn nucleus (the phase reach in the additive relative to its average concentration) must be surrounded by a shell relatively poor in the additive. Such consideration, together with the matter conservation principle and some other natural assumptions, lead to the expressions for r_c and ΔG^*:

$$r_c = 2\sigma\omega\varepsilon^{-1}, \quad \Delta G^* = (16\pi/3) \cdot \sigma^3\omega^2\varepsilon^{-2}$$

where σ is the surface tension, ε is the specific energy of phase separation,

$$\varepsilon = -\Delta g_m(\phi_{le}(\alpha)) + \Delta g_m(\phi_{le}(\alpha_0)) - \omega \left[\Delta g_m(\phi_{2e}(\alpha)) - \Delta g_m(\phi_{le}(\alpha))\right]$$

and $\Delta g_m(\phi)$ is the specific free energy of mixing for the given additive concentration, ϕ [9]. The parameter ω is the measure of deflection of the current additive concentration from its equilibrium value for the α-phase,

$$\omega = (\phi_{le}(\alpha_0) - \phi_{le}(\alpha))/(\phi_{2e}(\alpha) - \phi_{le}(\alpha))$$

where α_0 is the reaction conversion at which the binodal — the boundary of metastable state of homogeneous solution — has been reached, α is the conversion at which the given nucleus was born, and ϕ_{le}, ϕ_{2e} are equilibrium concentrations of the additive at given conversion in α-phase and β-phase, respectively. More details of this approach and some of its consequences not mentioned below may be found elsewhere [10].

It is easy to see that the critical nucleus size, birth threshold and probability are determined, among others, by the 'chemical quench depth,' $\Delta\alpha = \alpha - \alpha_0$. Since the cure conversion α may be calculated as a function of time provided that the reaction kinetics is known, $\Delta\alpha$ may be expressed through Δt_n, the nucleation delay. Particles born with different delay will then grow in a similar but somewhat different way. In the next Section we show how PSD may be analyzed provided that the real binodal and, therefore, absolute nucleation delay times are known.

MODEL OF PARTICLE GROWTH

The model of particle growth with equilibrium boundary conditions used in this study [10,11] is a version of the model of diffusion controlled growth [12] completed with proper account for the continuous changing of equilibrium concentrations ϕ_{ie} and interdiffusion coefficients D_i in both phases due to cure reaction, which keeps shifting the phase equilibrium until the gel point is reached. The rate of particle growth, v, is proportional to the difference of the diffusional flows at both sides of the interface,

$$v = \frac{dr}{dt} = 4\pi r^2 \left[D_1 \left(\frac{\partial\phi}{\partial\rho} \right)_{\rho=r+0} - D_2 \left(\frac{\partial\phi}{\partial\rho} \right)_{\rho=r-0} \right] (\phi_{2e} - \phi_{1e})^{-1} \qquad (9.1)$$

where ϕ_{ie} and D_i are functions of conversion and, therefore, time. To describe the growth of a particle, diffusion problems should be solved within both phases together with equation (9.1) and proper initial and boundary conditions. In no case of practical importance does this problem seem to be solvable analytically except for very short times. Accordingly, we have resorted to numerical modeling to study the nucleation and growth of dispersed phase particles in a diepoxy/diamine/nonreactive rubber system. Whenever possible, real values of the parameters involved have been used, so that the results are physically meaningful, though direct fitting of experimental data was not tried. The method presented here may be charaterized as semiquantitative.

RESULTS AND DISCUSSION

We have simulated the CRIMPS process in the quasibinary system described above with different bimolecular reaction rate constants, k. Every program run yielded a growth plot, $r(t)$, t being varied up to gel point, for a single particle. In the first numerical experiment the nucleation delay, Δt_n, was varied within a range such that the nucleation probability was reasonable (not too small or too large). The data obtained for a set of particles born at different moments for the same k are shown in Figure 9.1. It can be easily seen that PSD is very narrow for

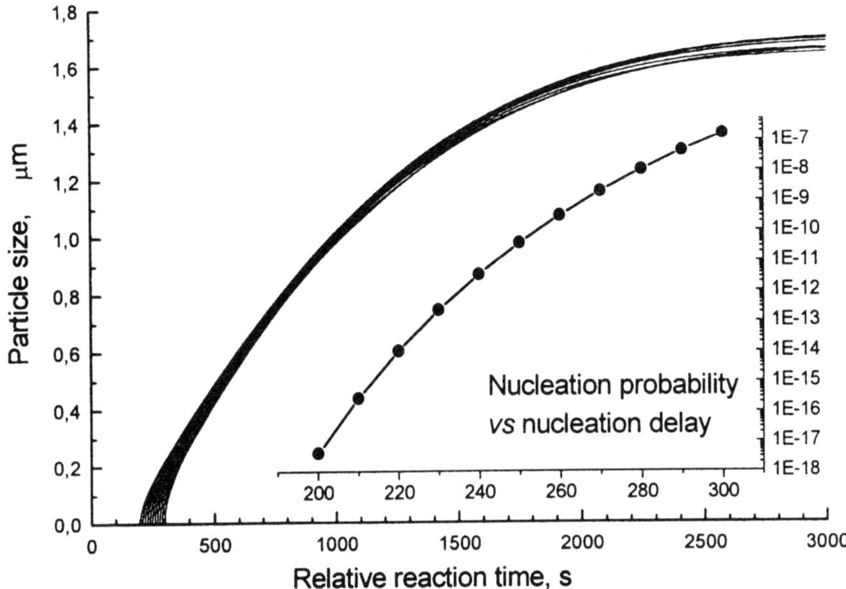

Figure 9.1 Particle growth plots for different nucleation delay times. Times are counted from the moment when the binodal point was reached. The inlay plot shows the corresponding nucleation probabilities. $\phi_{20} = 0.05$, $k = 10^{-6}$ mol^{-1} s^{-1}

this ensemble, while the birth probability changes considerably within the given range of Δt_n. It may be argued that bigger particles ($\Delta t_\mathrm{n} < 200$) will virtually not appear because their birth probability is too small. Moreover, smaller particles ($\Delta t_\mathrm{n} > 300$) will not appear since all the additive will be consumed by particles born earlier. Therefore, the statistical ensemble involved is representative, and the final PSD would be very narrow if the neighboring particles did not interact with each other.

The model allows taking into account such interaction approximately, in particular to estimate the concentration profiles between the centers of neighboring particles. In the profile evolution two stages may be distinguished. At first, the 'concentration fronts' are propagating from the particles' boundaries independently, so that the additive concentration in the center of the segment connecting the particles' centers, ϕ_mid, virtually does not change (Figure 9.2). Then ϕ_mid and, consequently, the supersaturation, start to decrease. This must result in hindering of further growth of particles due to depletion of their vicinity. The latter effect, however, cannot be quantitatively described within the framework of the given model.

At higher reaction conversions the interdiffusion coefficient may decrease so that the diffusion radius may become less than the half-distance between

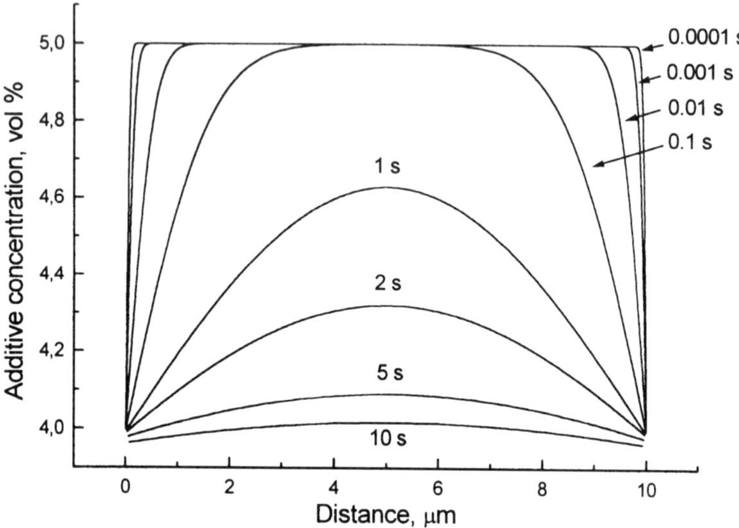

Figure 9.2 Concentration profiles between two particles with centers at 0 and 10 μm at different times indicated on the graph. Particles start to interact at $t \approx 0.1$ s. At larger times the curves are rather approximate and give estimates from below of the real profiles

particles. Then ϕ_{mid} will be virtually fixed, while ϕ_{le} will be still decreasing due to continuing reaction. Thus, the supersaturation ($\phi_{mid} - \phi_{le}$) will be increasing, which may lead to a new wave of nucleation. The final size of such particles will be much less than the mean size of the first generation particles because of smaller supersaturation, diffusion coefficient, and time left to grow. In fact, bimodal and even trimodal PSDs with particle sizes differing by an order of magnitude for successive generations have been observed earlier [3].

When the reaction rate constant was varied and the particles were chosen to be born at the same reaction conversion (though different times), growth plots were found to be very similar but differing from one other by a factor depending on k (Figure 9.3). As expected, the faster is the reaction, the smaller is the particle. The final particle sizes, r_{max}, plotted against k exhibit the obvious power law: $r_{max} \sim k^x$, where $x = (-0.49938 \pm 0.00052)$ as found by the best-fit method (Figure 9.4). Therefore, it may be argued that $x = -0.5$, the error being within ca. 1.2 standard deviations. In the case of bimolecular reaction the total time of the process $t_{proc} \sim 1/k$, therefore, $r_{max} \sim t_{proc}^{1/2}$. The equation looks like the well-known $r \sim t^{1/2}$ law found earlier for diffusion controlled growth [7]. However, note an important difference: in our case *neither of the individual particles* obey the latter law as they grow, whereas the *ultimate radii* of the particles formed with different ks depend on t_{proc} in a misleadingly similar way. The reason for such dependence is still not absolutely clear.

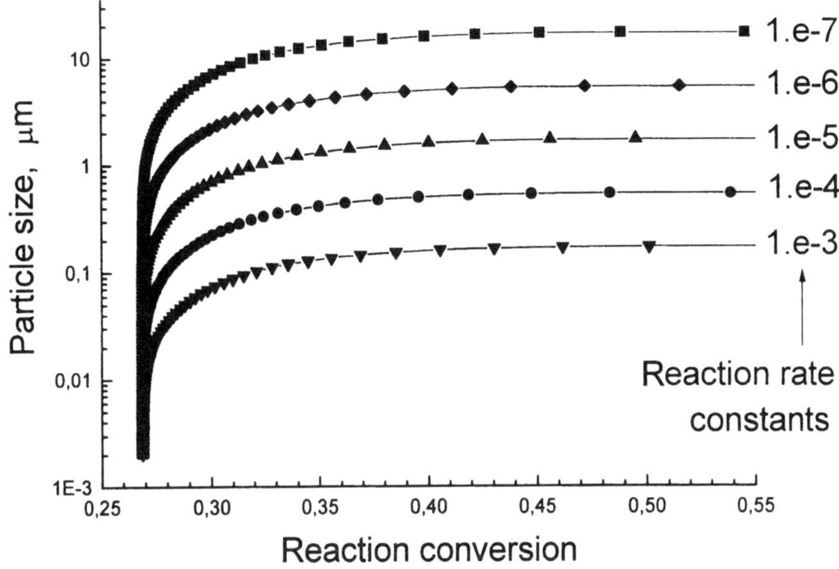

Figure 9.3 Particle growth plots for $\phi_{20} = 0.05$ and different reaction rate constants

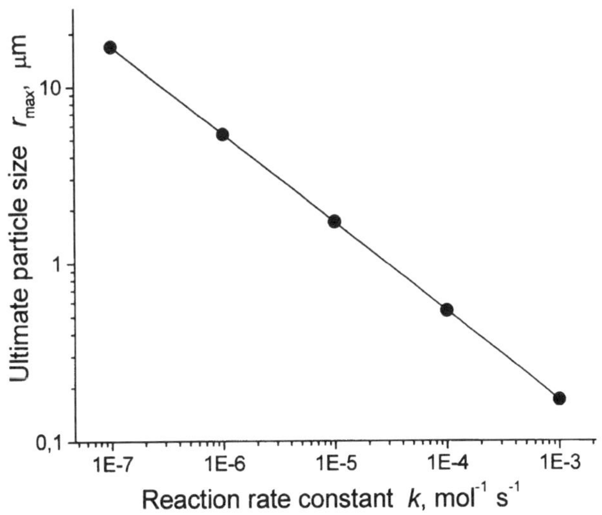

Figure 9.4 Log-log plot of r_{max} against k exhibiting a power law. Same conditions as on Figure 9.3. The line is the best-fit plot (see parameters in text)

The results obtained demonstrate the strong dependence of the morphology of materials formed via CRIMPS on the reaction kinetics and encourage further research which should be directed, first of all, to explicit consideration of interacting particle ensembles. Better understanding of the origin of the above scaling law is also desirable.

ACKNOWLEDGMENTS

This work was supported by the International Science and Technology Center (Project #358) and Russian Foundation for Basic Research (Project 96-03-32027a).

REFERENCES

1. B.A. Rozenberg and G.M. Sigalov, *Polym. Adv. Technol.*, **7**, 356 (1996).
2. R.J.J. Williams, B.A. Rozenberg and J.-P. Pascault, *Adv. Polym. Science*, **128**, 95 (1997).
3. B.A. Rozenberg, *Makromol. Chem., Macromol. Symp.*, **41**, 165 (1991).
4. M. Volmer, *Kinetik der Phasenbildung.* Steinkopff, Dresden, 1939.
5. J.B. Zeldovich, *Acta Physicochim. URSS*, **18**, 17 (1943).
6. I.M. Lifshitz and V.V. Slyozov, *Sov. Phys. JEPT*, **8**, 331 (1959).
7. J.W. Christian, *The Theory of Transformations in Metals and Alloys*, Pergamon Press, Oxford, 1975, Part I.
8. J. Crank, *The Mathematics of Diffusion*, Clarendon Press, Oxford, 1957.
9. P.C. Hiemenz, *Polymer Chemistry: The Basic Concepts*, Marcel Dekker, New York, 1984.
10. G.M. Sigalov and B.A. Rozenberg, *Polym. Sci.*, **40**, 884 (1998).
11. B.A. Rozenberg, G.M. Sigalov and V.O. Nikitin, *Proc. ACS, Div. PMSE*, **74**, 125 (1996).
12. C. Zener, *J. Appl. Phys.*, **20**, 950 (1949).

10

The Role of Strong Intermolecular Interaction in the Formation of Network Polymers

BORIS A. ROZENBERG, EMMA A. DZHAVADYAN and VADIM I. IRZHAK
Institute of Problems of Chemical Physics, Russian Academy of Sciences
Chernogolovka, 142432, Moscow Region, Russia

ABSTRACT

Rheokinetics of epoxy-amine network formation has been studied in the curing of bisphenol A diglycidyl ether by mixtures of primary di- and monoamine within a wide

Wiley Polymer Networks Group Review Series Vol. 2. Edited by B.T. Stokke and A. Elgsaeter
© 1999 John Wiley & Sons Ltd

temperature range using viscosimetry and torsion pendulum braid analysis techniques. The critical conversion was found to depend on cure temperature and to be essentially lower than the gelation point. It was proved that formation of labile physical network due to hydrogen bonding is responsible for these phenomena. An unconventional diagram of relaxation states, containing a formerly unknown area of quasi-rubbery state, was found for the investigated system. The generality of the phenomena discovered for polymeric networks formation has been discussed.

INTRODUCTION

One of the basic concepts of polymer physics is an idea of cross-links of physical networks that are formed due to labile intermolecular bonds of different nature. This concept is the background for description of deformation and glass transition of polymers, formation of thermoreversible polymeric gels, thermoplastic elastomers, etc. [1].

Intermolecular interactions are often considered as an important and sometimes determinant factor in kinetics and catalysis of organic reactions [2,3], including those of polymer formation [4–7]. This approach provides a deeper insight into the mechanism of the reactions, the influence of the chemical structure of reagents and the nature of solvents on the reaction rate. It also allows the processes of auto-acceleration and auto-inhibition and other complicated kinetic features of reactions in a wide range of varying conditions to be described adequately [4–6].

However, only a few studies are available [7–9] that correlate the effects of the association of components in the reacting system with its rheological behavior. Varying of the system viscosity in the process was interpreted as a confirmation of intermolecular interaction responsible for the labile physical network formation. Apparently, the formation of such labile physical networks should appear as a change of the relaxation behavior of the polymeric system during its formation. This aspect of physical network development in curing epoxyamine systems has been considered recently at virtually the same time by two groups of researchers [10,11] using different experimental techniques. The role of strong intermolecular bonding during curing of epoxies by diamine is the focus of this chapter. The following aspects of the problem are considered:

- the effect of cure temperature on the critical conversion (the gelation point observed at rheokinetic measurements);
- the effect of mono-/diamine ratio on critical conversion at the stoichiometric ratio of amino/epoxy groups;
- the diagram of the relaxation state;
- theoretical consideration of critical conversion in epoxy-amine curing, taking into account strong intermolecular interaction.

EXPERIMENTAL

The cure kinetics of a system of diglycidyl ether of bisphenol A (DGEBA) +4, 4'-diaminodicyclohexylmethane (DA) + cyclohexylamine (MA) was studied using an isothermal calorimeter DAK-1-1 in a temperature range 30–80 °C. To calculate the cure kinetics, the value of the reaction thermal effect was taken to be 104.5 kJ/mole.

Rheokinetic investigation was performed using torsion pendulum braid analysis. A technique described in [12] was used for calculating a relative hardness of the cured fiber impregnated with a binder. The time for reaching the critical conversion was estimated from the section cut by the tangent at the point of maximum rate on the plot of relative hardness dependence vs. reaction time. The critical conversion corresponding to this time was estimated from kinetic curves as the given cure temperature.

Viscosity change during the reaction was measured with a rotation viscosimeter Rheotest-2 using a cone–plane type measuring cell in the range of shear rates from 1 to 4860 s^{-1}. Critical conversion corresponding to a drastic growth of viscosity during the reaction was estimated similar to measuring rheokinetics using the torsion pendulum braid analysis technique.

Direct determination of the gelation point of polymer samples was performed using sol–gel analysis in THF. The gelation point is considered to be the conversion at which the presence of an insoluble fraction in THF is first observed.

The glass transition temperature was determined from thermomechanical curves plotted using a UIP-70M device with heating rate of 2.5 K/min and pressure on the sample of 10 MPa.

RESULTS AND DISCUSSION

During formation of a network polymer two relaxation transitions can occur: transition from viscous to rubbery state, and from viscous or rubbery state into a glassy one. These transitions are referred to as gelation and vitrification, respectively. Depending on curing temperature (T_{cure}), either both transitions or one of them can be realized during the reaction. If $T_{cure} > T_{g,\infty}$, only the former transition occurs. If $T_{g,gel} < T_{cure} < T_{g,\infty}$, two transitions occurs; if $T_{cure} < T_{g,gel}$, only the latter transition is possible. Here $T_{g,\infty}$ is the ultimate glass transition temperature of completely cured thermoset, $T_{g,gel}$ is cure temperature that provides maximum conversion at which the curing system reaches gelation point and vitrifies simultaneously. In this case $T_g = T_{g,gel}$. It should be noted that for the two latter cases, when curing occurs at moderate ($T_{g,gel} < T_{cure} < T_{g,\infty}$) or low ($T_{cure} < T_{g,gel}$) temperatures, it is not complete because of diffusion retardation of the reaction and, finally, it stops completely due to vitrification of the curing system [4, 13–15].

Various experimental techniques can be used to characterize the relaxation state in curing systems. In this work we have used two of them, viz., viscosimetry and torsion pendulum braid analysis, widely acceptable for these purposes. Diepoxide cured by a mixture of diamine (DA) and monoamine (MA) was used as the object of investigation. Changing the DA/MA ratio while keeping the stoichiometric ratio of the reacting groups allowed us to vary the cross-linking density of the polymer formed in a wide range, and, hence, to vary relaxation transitions and gelation points which accompany the formation of the network polymer.

CRITICAL CONVERSION AND CURE TEMPERATURE

Figure 10.1 presents rheokinetic curves measured using viscosimetry and torsion pendulum braid analysis techniques at different curing temperatures. Using kinetic curves of curing, changes of rheokinetic parameters can be considered as function of the conversion.

Two unconventional facts deserve particular consideration (Figure 10.2):

1. Critical conversions (α_{cr}) measured by both methods at curing temperatures in the range of 30–80 °C (curve $T_{cure}(\alpha_{cr})$) are lower than the gelation point for the stoichiometric mixture of diepoxide and diamine ($\alpha_g = 0.57$), which was measured by the sol–gel analysis method and was completely consistent with the value calculated provided that no complicating conditions upon the polymeric network formation [16] exist (no substitution effect, i.e., different reactivity of the first and second addition of epoxy-group to amino-group, and no etherification reaction or reactions of ineffective cyclization).

2. The critical conversion value (α_{cr}) depends on the cure temperature and increases upon its increase. The gelation point (α_g) for such systems does not depend on the curing temperature [16]. Moreover, even when the substitution effect exists for epoxy-amine systems, it is usually rather weak [4,17]. It somewhat shifts the gelation point to higher conversions. However, the gelation point remains independent of cure temperature, because there is no

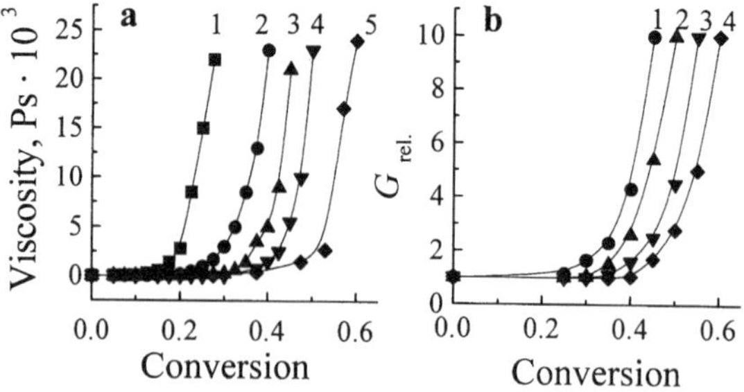

Figure 10.1 (a) Viscosity and (b) relative modulus changes *vs.* conversion at curing DGEBA + DA system at temperatures, °C: 1 − 30; 2 − 40; 3 − 50; 4 − 60; 5 − 80

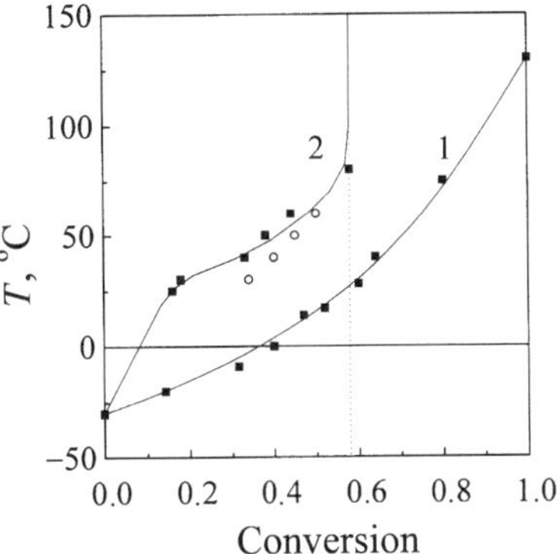

Figure 10.2 *(1)* T_g vs. conversion in DGEBA + DA system curing (experimental results are plotted by dots, solid line is calculated according to equation (10.1)), and (10.2) critical conversion vs. cure temperature for the same system. The viscosimetric and torsion pendulum data are plotted by ■ and O, respectively

temperature dependence for relative constant rate for the first and second addition of epoxy-group to amino-group [17]. It should also be noted that polymers completely preserve their solubility in THF while reaching critical conversions lower than $\alpha_g = 0.57$ at all studied temperatures. Thus, it is evident that a drastic change of rheokinetic curves observed upon curing is not due to the gelation in the curing system with a network formed by covalent chemical bonds.

One may put forward an alternative suggestion that a reason of the observed rheokinetic regularities is the vitrification of the curing system. However, the following data show that it is not the fact. Figure 10.2 presents experimental measurements of the glass transition temperature of the curing system vs. conversion. This dependence is described, as it could be expected, by the Di Benedetto equation [18].

$$\frac{T_g - T_{g0}}{T_{g\infty} - T_{g0}} = \frac{\lambda \cdot \alpha}{1 - (1 - \lambda) \cdot \alpha} \tag{10.1}$$

where T_{g0} and $T_{g\infty}$ are glass transition temperatures at $\alpha = 0$ and $\alpha = 1$, respectively, and $\lambda = \Delta C_{p\infty}/\Delta C_{p0} = 0.34$. Here $\Delta C_{p\infty}$ and ΔC_{p0} are changes of heat capacity at constant pressure on vitrification of a completely cured mixture of reaction product and initial monomer, respectively. As seen in Figure 10.2, the curve for $T_{cure}(\alpha_{cr})$ lies essentially higher than that for $T_g(\alpha)$. Reaching of critical conversions at $\alpha < \alpha_g$ is not accompanied by any specific features (retardation of the reaction) on the kinetic curves. The facts mentioned above definitely suggest that critical conversion is not related to vitrification of the curing system. Later we will discuss the physical nature of the regularities observed.

DEPENDENCE OF CRITICAL CONVERSION ON MONO-/DIAMINE RATIO

Figure 10.3 shows rheokinetic curves of DGEBA curing with MA and DA mixtures at 50 °C, with the ratio of the reacting groups being kept stoichiometric.

The reaction rate is the same in the whole range of MA to DA ratio, and differences begin only when conversation is close to the ultimate one. As could be expected, the ultimate conversion increases as the MA fraction increases. Critical conversion grows with the increase of the MA fraction as well, this being consistent with out expectations (Figure 10.4).

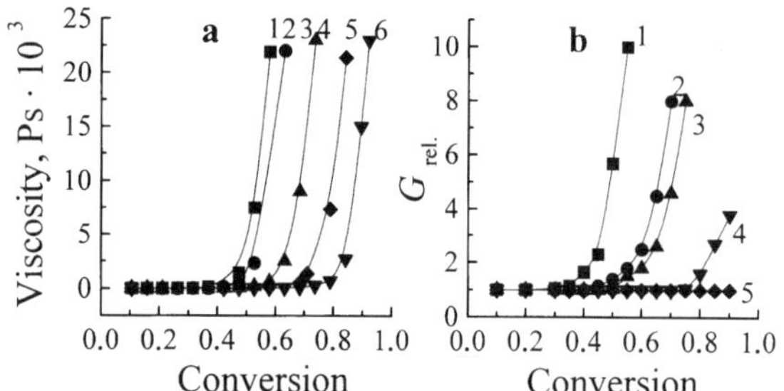

Figure 10.3 (a) Viscosity and (b) relative modulus changes vs. conversion at curing DGEBA + DA + MA system at 50 °C. Weight fractions of MA in amine mixture: (a) 1 − 1.0; 2 − 0.2; 3 − 0.5; 4 − 0.8; 5 − 1.0 and (b) 1 − 1.0; 2 − 0.2; 3 − 0.4; 4 − 0.5; 5 − 1.0

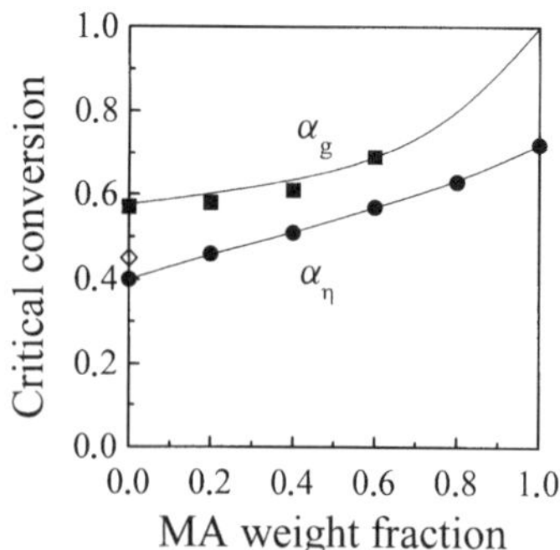

Figure 10.4 Critical conversion vs. MA weight fraction in amine mixture in the DGEBA + DA + MA curing system at 50 °C measured by sol–gel analysis (curve α_g) and viscosimetry (curve α_η). Calculated data for α_g are presented by solid line. The critical conversion determined by the torsion pendulum technique is represented by a rhombus

Experimental values of gelation point as a function of amine ratio, n_2, are in good agreement with those calculated according to the formula:

$$\alpha_g = \frac{1}{\sqrt{3 - 2n_2}} \qquad (10.2)$$

which has been derived using the theory of branching processes for the system studied. Here $n_2 = 2[\text{MA}]/(2[\text{MA}] + 4[\text{DA}])$ is the mole fraction of the monoamine functional groups in the mixture of mono- and diamines.

The experimental curve of critical conversion vs. amine ratio lies below the analogous curve for gelation point.

It should be noted that values of critical conversion measured by the viscosimetric method are somewhat lower than those measured by the torsion pendulum braid analysis technique.

All the results obtained can be unambiguously explained based on the concept of intermolecular interaction in the curing system.

GEL-POINT IN THE CURING SYSTEM WITH STRONG INTERMOLECULAR INTERACTIONS

Condensation of glycidyl ether with amines is accompanied by the formation of hydroxyl groups:

$$
\text{RN}
\begin{matrix} H \\ H \end{matrix}
+ 2\,\text{CH}_2\text{-CHCH}_2\text{OAr (epoxide)} \longrightarrow
\text{RN}
\begin{matrix} \text{CH}_2\text{-CHCH}_2\text{OAr} \ (\text{CH}) \\ \text{CH}_2\text{-CHCH}_2\text{OAr} \ (\text{CH}) \end{matrix}
\qquad (10.3)
$$

The latter are rather strong proton donors, and they are able to form both stable homoassociates with each other, and heteroassociates with all electron-donating groups of the reaction system. The phenomenon of hydrogen bonding in curing epoxy-amine systems is well studied and thermodynamic characteristics of such a type of the reaction are measured (see the table). Formation of the final reaction product even in the model system of monoamine and monoglycidyl ether is accompanied by drastic growth of viscosity due to formation of a labile network of hydrogen bonds between molecules [4,9,10]. A reason for appearance of such labile network is that the reaction product consists of two proton-donating hydroxyl groups and five electron donating atoms (see Scheme (10.3)).

Increase of average functionality of the reacting molecules must result in the formation of a similar labile network of hydrogen bonds if even conversion of the reacting groups is essentially smaller. It is obvious as well that hydrogen bonds concentration decreases as the reaction temperature increases. These conclusions are completely consistent with the results of rheokinetic experiments.

They show that for systems characterized by changing of intermolecular bonds concentration during curing (formation of polyepoxides, polyurethanes, polyurethane-isocyanurates, etc.), the TTT-diagram of relaxation states of the curing system [15] should be complemented by a temperature-conversion region that is characterized by quasi-rubbery behavior of the system (Figure 10.5). This region lies above the $T_g(\alpha)$ curve. Its upper limit is $T = T_{g,\infty}$ (by temperature) and $\alpha = \alpha_g$ (by conversion). This region is the most distinct during low-temperature curing ($T_{g0} < T_{cure} < T_{g,gel}$). When the cure reaction takes place at moderate temperatures ($T_{cure} > T_{g,gel}$), which is common at the first stage of the stepped curing, this region approaches asymptotically $\alpha = \alpha_g$.

Alili *et al.* [11], who used low-frequency rheological and dielectric spectroscopy techniques combined with sol–gel analysis during cure of epoxy–amine system also came to the conclusion that physical gelation does occur in such system.

It should be noted that the more stable intermolecular bonds forming the physical network during cure are, the more the quasi-rubbery region extends to higher temperatures. This facilitates its observation by low-frequency techniques.

Using theory of branching processes and taking into account both chemical reaction and intermolecular bonds formed during cure reaction, the critical conditions of the polymer network formation were obtained [10]:

$$n_2 = \frac{\dfrac{1 - \alpha(3 - 2\alpha)\gamma_0 + \beta_0 + \beta_4}{\alpha(1 + \sum \beta_i)} - Z}{(1 - \alpha)\gamma_2 + \alpha(1 + \gamma_3) - Z} \tag{10.4}$$

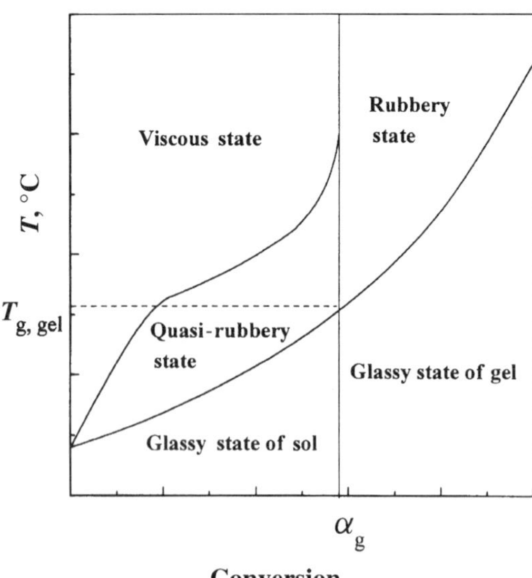

Figure 10.5 Diagram of relaxation states in curing DGEBA + DA system

where

$$Z = (1 - \alpha)^3 (\gamma_1 + \gamma_2) + \alpha(1 - \alpha)^2 (3 + \gamma_1 + 4\gamma_2 + \gamma_3)$$
$$+ 3\alpha^2 (1 - \alpha)(2 + \gamma_2 + \gamma_3) + \alpha^3 (3 + 2\gamma_3)$$

This equation was derived using reasonable assumption that associates formation rates are much higher than those of the chemical reaction and, therefore, the reactions of intermolecular interactions are always in equilibrium in the course of cure.

In the equation (4) $\alpha = \alpha_{cr}$, β_i and γ_i are probabilities, i.e., fractions of functional groups bound into certain associates. For example, probabilities of a hydroxyl group to be bound with an oxygen atom (β_0) into associate, and of an oxygen atom to be found with a hydroxyl group (γ_0) into associate can be expressed as follows:

$$\beta_0 = \frac{K_0[OH][O]}{[OH](1 + K_0[O] + K_1[A_1] + K_2[A_2] + K_3[A_3] + K_4[OH])}$$

$$\gamma_0 = \frac{K_0[OH][O]}{[O](1 + K_0[OH])} \tag{10.5}$$

Probabilities of formation of associates of the OH group with primary (β_1), secondary (β_2), and tertiary (β_3) amines, and of the self-associate (OH–OH) (β_4), as well as probabilities of formation of complexes of primary (γ_1), secondary (γ_2), and tertiary (γ_3) amines with the hydroxyl group are calculated in a similar way. Here K_i is equilibrium constants for corresponding associates formation ($i = 0 - 4$), $[A_1]$, $[A_2]$ and $[A_3]$ are current concentrations of primary, secondary and tertiary amine groups, respectively; $[O]$ is concentration of ether bonds (basicities of the ring and out-of-ring oxygen atoms are assumed to be equal); $[OH]$ is the current concentration of hydroxyl groups.

Expression (10.4) links experimental critical values α and the initial composition of the system for the selected values of equilibrium constants K_i (see Table 10.1).

Qualitative consistency is observed while comparing the calculated values of critical conversion with experimental data (curve for α_η in Figure 10.4).

Table 10.1 Thermodynamic parameters of formation of associates at the interaction of phenylglycidyl ether and aniline

Associate	$K_i(22\,°C)$, 1/mol	$K_i(90\,°C)$, 1/mol	$-\Delta H$, kJ/mol	$-\Delta S$, J/mol· K
OH…O	1.34	0.45	14.2	46.0
OH…A$_1$	1.85	0.44	18.8	66.9
OH…A$_2$	1.35	0.55	11.7	41.8
OH…A$_3$	0.63	0.47	3.8	8.4
OH…OH	0.35	0.04	27.2	98.6

Calculated data can not be directly compared to experimental results because equilibrium constants for aromatic amine (aniline) were used for calculation, while much more basic aliphatic amines were used in this work.

CONCLUDING REMARKS

Thus, for polycondensation systems that are characterized by formation of functional groups able to participate in strong intermolecular interactions (polyepoxides, polyurethanes, polyisocyanurate-urethanes etc.), formation of the covalent network is accompanied by formation of the network with labile physical bonds, as well. This is similar to a continuous growth of apparent functionality of the curing system with conversion. Obviously, the critical conversion that corresponds to a drastic change in rheokinetic curve turns out to be smaller than the conversion at the gel-point, the latter being due to participation of covalent chemical bonds only. The value of critical conversion depends on the cure temperature due to participation of new functional groups, which appeared during curing, in the hydrogen bonding and network formation. This results in essential modification of the diagram of relaxation states that the curing polymeric system passes through. The diagram of relaxation states of the curing system, $T(\alpha)$, is characterized in this case by a special region referred to as the quasi-rubbery state region in Figure 10.5. In this region the curing system has quasi-rubbery properties due to the formation of polymeric network with labile physical bonds.

ACKNOWLEDGMENTS

The authors are indebted for the financial support to the Russian Foundation for Basic Research (Project 96-03-32027a) and to the International Science and Technology Center (Project 358-96).

REFERENCES

1. V.I. Irzhak, G.V. Korolev and M.E. Solovyov, *Usp. Khim.*, **66**, 179 (1997).
2. S.G. Entelis and R.P. Tiger, *Kinetics of Reactions in Liquid Phase. Quantitative Estimate of the Medium Influence.* Khimiya, Moscow, 1973 (in Russian).
3. V.A. Savelova and N.M. Oleinik, *Mechanisms of Organic Catalysts Activity. Bifunctional and Intramolecular Catalysis.* Naukova Dumka, Kiev, 1990 (in Russian).
4. B.A. Rozenberg, *Adv. Polym. Sci.*, **75**, 113 (1986).
5. B.A. Rozenberg, Usp. Khim., **60**, 1473 (1991).
6. R.P. Tiger, D.N. Tarasov and S.G. Entelis, *Khim. Fiz.*, **15**, 11 (1996).
7. B.I. Nakhmanovich and A.A. Arest-Yakubovich, *Vysokomolek. Soedin., Ser. B*, **38**, 359, (1996).
8. B.A. Rozenberg, V.I. Irzhak, and N.S. Enikolopyan. *Interchain Exchange in Polymers*, Khimiya, Moscow, 1975 (in Russian).
9. Kh.A. Arutyunyan, E.A. Dzhavadyan, A.O. Tonoyan, S.P. Davtyan, B.A. Rozenberg and N.S. Enikolopyan, *Zh. Fiz. Khim.*, **50**, 2016 (1976).

10. E.A. Dzhavadyan, V.I. Irzhak and B.A. Rozenberg, *Polym. Sci., Ser.A*, **41**, no. 4 (1999).
11. L. Alili, J. van Turnout and K. te Nijenhuis, in *The Wiley Polymer Networks Group Series. Volume One. Chemical and Physical Networks. Formation and Control of Properties*, (K. te Nijenhuis and W.J. Mijs, eds), Wiley, Chichester, 1998, p. 255.
12. J. Hejboer, *Polym. Eng. Sci.*, **19**, 664 (1979).
31. E.F. Oleinik, *Adv. Polym. Sci.*, **80**, 49 (1986).
14. K. Dušek and I.Havliček, in *Crosslinked Epoxies* (B. Sedlachek and J. Kahove, eds), Walter de Gruyter, New York, 1987, p. 417.
15. J.K. Gillham and J.B. Enns, *Trends in Polymer Science*, **2**, 406 (1994).
16. V.I. Irzhak, B.A. Rozenberg and N.S. Enikolopyan, *Network Polymers: Synthesis, Structure, Properties*. Moscow: Nauka, 1979 (in Russian).
17. E. Girard-Reydet, C.C. Riccardi, H. Sautereau and J.-P. Pascault, *Macromolecules*, **28**, 7599 (1995).
18. J.-P. Pascault and R.J.J. Williams, *J. Polym. Sci.: Part B*, **28**, 85 (1990).

11

Formation Process of Internal Structures in Chemically Cross-linked N-Isopropylacrylamide Gel

YOSHITSUGU HIROKAWA[1]*, TAKUYA OKAMOTO[1]* and TAKEJI HASHIMOTO[1,2]

[1]Hashimoto Polymer Phasing Project, ERATO, JST 15 Morimoto-cho, Shimogamo, Sakyo-ku, Kyoto 606-0805, Japan

[2]Department of Polymer Chemistry, Graduate School of Engineering, Kyoto University, Kyoto 606-8501, Japan

ABSTRACT

The formation process of the hierarchical structures found in the opaque N-isopropyl-acrylamide chemical gels prepared at high temperatures was studied by the quenching method. On the basis of the observation of the optical appearance of the gelation system and the internal structures by laser scanning confocal microscopy, the formation process

---

* Present address: Nippon Zeon R&D Center, 1-2-1 Yako Kawasaki-ku, Kawasaki-shi, 210-8507 Japan

---

Wiley Polymer Networks Group Review Series Vol. 2. Edited by B.T. Stokke and A. Elgsaeter
© 1999 John Wiley & Sons Ltd

was deduced as follows: the microgels were made by cross-linking inside the high polymer concentration parts which are formed by the gelation-reaction-enhanced thermal concentration fluctuations; the microgels were trapped and interconnected by a subsequent formation of loose networks. The increase of gelation temperature seemed not to affect the formation process of the internal structures except for an increase in the number of the microgels.

INTRODUCTION

Polymer gels prepared by chemical cross-linking reaction are three-dimensional polymer networks with fluid. It has been reported that the changes of external conditions, such as solvent compositions, temperatures, pH, and so on, induce volume phase transition of the gels [1]. N-Isopropylacrylamide (NIPAAm) chemical gel is an attractive material which shows volume phase transition in pure water at $32\,^{\circ}$C [2]. There are many studies of the NIPAAm gel concerning not only the fundamental aspects but also the applications of the gel [3,4]. Inhomogeneities in the gel, i.e. internal structures, are also very important to understand and utilize the gel, and recently the number of reports of inhomogeneities has been increasing [5–19].

It has been reported that the optical appearance of NIPAAm gel depended on the preparation temperatures [20]. The gels prepared at low temperatures such as those prepared below $20\,^{\circ}$C are transparent, while the gels prepared at high temperatures above $25\,^{\circ}$C showed permanent opacity. Since the difference in the appearance of the gel is probably related to the difference in the internal structures, the NIPAAm gel gives a particularly suitable system for studying the internal structures of gels.

We have been studying the internal structures of opaque NIPAAm gel by means of laser scanning confocal microscopy (LSCM) [21], laser light scattering (LS), and small-angle neutron scattering (SANS) [22] and found hierarchical structures in the length scale ranging from nm to μm. Opaque NIPAAm gel was found to have bicontinuous domain structures consisting of domains dense and sparse in polymer networks, and the spacing between the domains is about 10 μm. Furthermore, small-angle neutron scattering revealed that the dense polymer network domain was made of a large number of highly cross-linked microgels connected by loose polymer networks.

In this report, we present the study of the formation process of such interesting hierarchical internal structures of the opaque gel by the temperature drop (quenching) technique of the gelation system. In this technique, the transient internal structures appearing during the gelation process at a given temperature will be fixed by the gelation induced after the temperature drop. By the observation of the fixed internal structures by LSCM, we can deduce the formation process of the internal structures of the gel at the given temperature.

EXPERIMENTS

N-Isopropylacrylamide (NIPAAm) was purified by recrystallization from toluene to *n*-hexane. *N,N'*-methylene-bis(acrylamide) (BIS) (electrophoresis grade), *N,N,N',N'*-tetramethylethylenediamine (TEMEDA) (electrophoresis grade) and ammonium peroxodisulfate (AP) (electrophoresis grade) were used without any further purification.

NIPAAm (0.7524 g), BIS (7.71 mg) and TEMEDA (12.0 µl) were dissolved into deionized water (4.5 ml) purged with nitrogen. After addition of 0.4% AP aqueous solution (0.5 ml), the pregel solution was loaded into the cell which was assembled with two round glass plates (22 mmϕ, 0.17 mm thickness), a ring spacer (0.5 mm thickness) and a holder. Then the cell was put into the temperature controlled water bath for a certain period.

The time change in the transmittance of the gel during gelation was recorded by use of a laser power meter. For the determination of NIPAAm conversion, the gel (or solution) was taken out from the cell after a given gelation time and put into the amount of chilled water needed for the unreacted monomer to diffuse out from the gel. After 1 day, the residual monomer concentration in the outside of NIPAAm gel was measured by UV spectroscopy to determine the conversion. For the observation of time dependence of internal structures by LSCM at a given gelation temperature (T_{gel}), the cell was quickly transferred from the temperature controlled water bath at T_{gel} into an ice water bath at a temperature denoted by T_{quench} and kept for overnight, during which further gelation is expected to take place to result in the formation of a uniform network which would trap the gel structures formed at T_{gel}. In this manner, hereafter designated as a 'quenching method' for convenience, the internal structures formed at T_{gel} will be fixed in the gel.

RESULTS AND DISCUSSION

INTERNAL STRUCTURES OF NIPAAm GELS

N-Isopropylacrylamide (NIPAAm) gels were prepared at 20.0, 27.0 and 38.0 °C. The gel obtained at 20.0 °C was transparent while the gels obtained at 27.0 and 38.0 °C showed opacity. Laser scanning confocal microscope (LSCM) reflective images were observed only for the gels having opacity. The representative LSCM image of gel prepared at 27.0 °C and its binarized image after the image process were shown in Figure 11.1, where (a) is a sliced image subjected to image analyses by use of three-dimensional filters such as a median filter, a mean filter and a smoothing filter and (b) is the binarized image. The bright and the dark areas were identified to be dense and sparse network polymer domains, respectively, by means of fluorescent technique [23]. Three-dimensional structures consisting of dense and sparse polymer domains were found to be bicontinuous by the aid of computational stacking of the two-dimensional LSCM images. A close-up of the

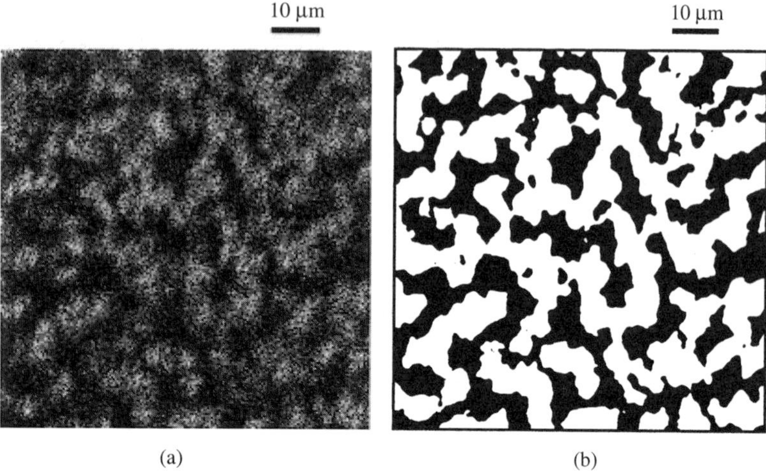

Figure 11.1   LSCM image (a) of the gel prepared at 27.0 °C and its binarized image (b). In (b), light and dark areas represent dense and sparse polymer domains, respectively

bright area in the LSCM image showed that the dense polymer domain consists of a lots of light spots as shown in Figure 11.2. These light spots were considered to be highly cross-linked microgels connected by loose networks on the basis of small-angle neutron scattering analysis [22]. Thus the opaque NIPAAm gels have hierarchical structures in the length scale ranging from 1 nm to 10 μm.

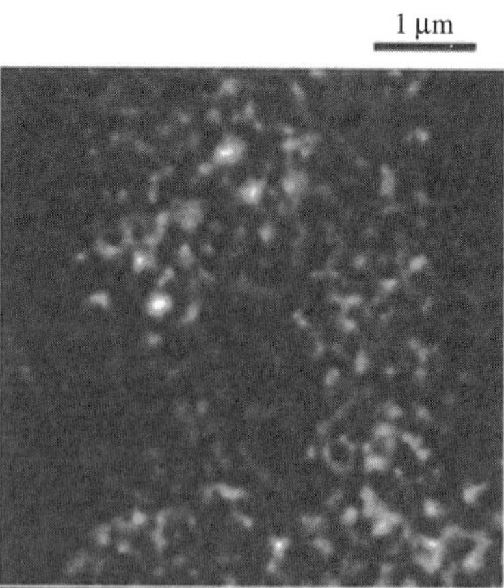

Figure 11.2   Close-up image of dense polymer domain shown in Figure 11.1(a). Many light spots present inside the domain dense in polymer are considered to be highly cross-linked microgels

FORMATION PROCESS OF INTERNAL STRUCTURES

In order to investigate the process forming the internal structures of the gel, the gelation of NIPAAm was carried out at 27.0 °C in the presence of the cross-linker. The time conversion curve of NIPAAm is shown in Figure 11.3. NIPAAm was consumed almost 100% within 60 min. When the conversion was reaching about 50%, the system started to become turbid, which did not disappear even after the completion of gelation. The gelations at 27.0 °C were quenched at 180, 240, 352, 444, 523 and 3600 s by immersing the cells into ice water bath. The former two systems (i.e. those quenched at 180 and 240 s did not show any turbidity before quenching, and the quenched gels were also clearly transparent, showing no meaningful LSCM images. The systems quenched at 180 and 240 were still liquid, which is similar to the epoxy gelation systems [7,8], although the polymer structures formed in the two systems at this stage are very different.

With the system quenched at 352 s, slight opacity was observed before the quenching but after quenching the resultant gel was transparent. This quenched gel did not give any meaningful LSCM image. These facts indicate that the observed opacity during the gelation is probably due to the dynamical concentration fluctuations induced by the increasing concentration of linear or branched polymers produced by the reaction. The thermal concentration fluctuations are suppressed and hence the opacity disappears upon lowering temperature, namely, quenching, simply because the linear or branched polymer solution has the lower critical solution temperature (LCST) type phase diagram. Because if the opacity is due to the static frozen fluctuations, the quenched gel should show the LSCM image characterizing the internal structures. Phase separation of the polymer also produces the opacity, but the heat generated by the radical polymerization is

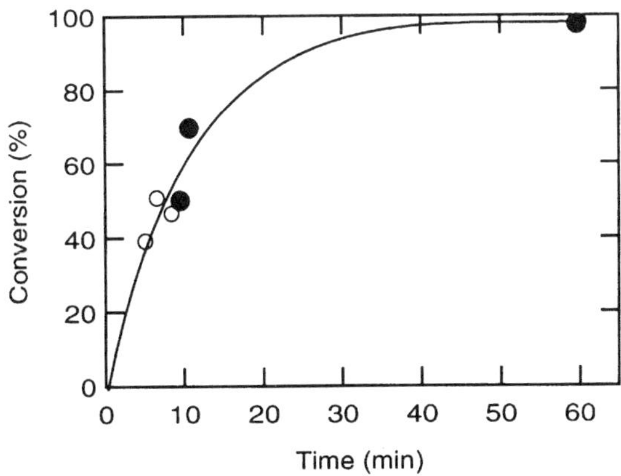

Figure 11.3 Time conversion curve of the gelation at 27.0 °C. Open or filled circles mean that the appearance of the quenched gels is transparent or opaque, respectively

not enough to raise the temperature of the gelation system to induce the phase separation of poly(*N*-isopropylacrylamide) aqueous solution which occurs above at 32 °C [2].

With the latter three systems quenched at 444, 523 and 3600 s., the gels showed opacity after quenching as well as during the gelation. Figure 11.4 showed the LSCM images of the quenched gels. The image of the gel quenched at 444 s showed that the small number of bright spots present inside the domains dense in polymer networks were rather sparsely distributed in space. These bright spots correspond to the highly cross-linked microgels. As the gelation proceeded, the number of bright spots increased. Three images shown in Figure 11.4 seem to have the rather similar spacing between the domains dense in polymer networks although numerical analysis is necessary.

Even when gelation was conducted at 38.0 °C which is above the phase separation temperature of the linear poly(*N*-isopropylacrylamide), the scheme of the

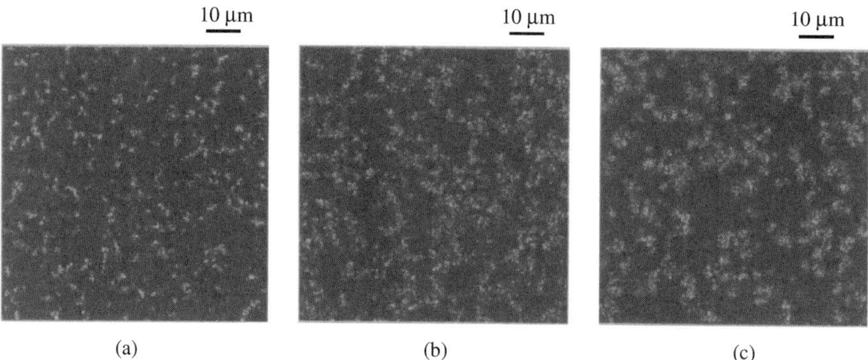

(a)   (b)   (c)

Figure 11.4 LSCM images of the quenched gels. Gelation temperature was 27.0 °C. Quenching time: (a): 444 sec; (b): 523; (c): 3600

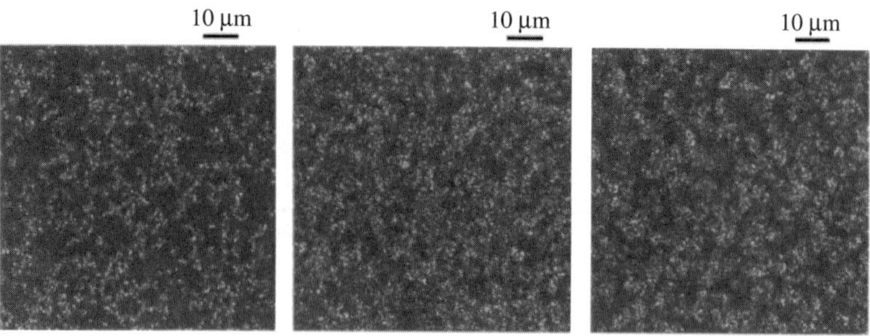

Figure 11.5 LSCM images of the quenched gels, Gelation temperature was 38.0 °C. Quenching time: (a): 400 sec; (b): 550; (c) 3600

internal structure formation seems to be similar to the system at 27.0 °C, although the reaction rate at 38 °C was higher than that conducted at 27 °C. The gel quenched at the time when the slight opacity was observed during gelation (quenched at 315 s after immersion in water bath setting at 38.0 °C) was transparent and did not show any meaningful LSCM images. On the other hand, the gels quenched at 400, 550 and 3600 s showed the LSCM images as shown in Figure 11.5. As the gelation time goes on, the bright area in the LSCM images increased. Almost all of the areas in the images of the gels quenched at 550 and 3600 s was covered with bright area consisting of a large number of bright spots. These facts may indicate that the formation of microgels takes place efficiently in the polymer-rich regions induced by the phase separation of the linear or branched polymer.

CONCLUSION

The gelation forms the more or less linear or branched polymer at the first stage. If the temperature T_{gel} for the gelation reaction is higher than or equal to 27.0 °C, this process is followed by large concentration fluctuations of the resultant linear or branched polymers. In the high concentration regions, cross-linking takes place effectively to make the highly cross-linked microgels. Soon after the microgels are formed, they are trapped by the formation of the loosely cross-linked network. The network formation hinders further enhancement of the concentration fluctuations or progress of the phase separation.

In order to reveal the more precise mechanism of the formation of the hierarchical structures, further studies are necessary. We are now conducting investigation by means of the time-resolved scattering methods.

REFERENCES

1. T. Tanaka, *Science American*, **249**(1), 124(1981).
2. Y. Hirokawa and T. Tanaka, *AIP Conference Proceedings No. 107, Physics and Chemistry of Porous Media* (D.L. Johnson and P.N. Sen, eds) 1984, p. 203; Y. Hirokawa and T. Tanaka, *J. Chem. Phys.*, **81**, 6379(1984).
3. M. Shibayama and T. Tanaka, *Advances in Polymer Science*, **109**, 1(1993).
4. S. Hirotsu, *Phase Transitions*, **47**, 183(1994).
5. N. Weiss and A. Silberberg, *Brit. Polymer, J.*, **9**, 144(1977).
6. T-P. Hsu, D.S. Ma and C. Cohen, *Polymer*, **24**, 1273(1983).
7. B. Chu and C. Wu, *Macromolecules*, **21**, 1729(1988).
8. C. Wu, J. Zuo and B. Chu, *Macromolecules*, **22**, 838(1989).
9. S. Mallam, F. Horkay, A-M. Hecht and E. Geissler, *Macromolecules*, **22**, 3356(1989).
10. T. Tokuhiro, T. Amiya, A. Mamada and T. Tanaka, *Macromolecules*, **24**, 2936(1991).
11. E. Mendes Jr., P. Lindner, M. Buzier, F. Boue and J. Bastide, *Phys. Rev. Lett.*, **66**, 1595(1991).
12. M. Tokita and T. Tanaka, *J. Chem Phys.*, **95**, 4613(1991).

13. Y. Suzuki, K. Nozaki, T. Yamamoto, K. Itoh and I. Nishio, *J. Chem. Phys.*, **97**, 3808(1992).
14. M. Rubinstein, L. Leibler and J. Bastide, *Phys. Rev. Lett.*, **68**, 405(1992).
15. M. Shibayama, T. Tanaka and C.C. Han, *J. Chem. Phys.*, **97**, 6842(1992).
16. M. Shibayama, T. Tanaka and C.C. Han, *J. Chem. Phys.*, **97**, 6842(1992).
17. M. Shibayama, M. Morimoto and S. Nomura, *Macromolecules*, **27**, 5060(1994).
18. M. Tokita, T. Miyoshi, K. Takegoshi and K. Hikichi, *Phys. Rev. E*, **53**, 1823(1996).
19. A. Suzuki, M. Yamazaki and Y. Kobiki, *J. Chem. Phys.*, **104**, 1751(1996).
20. C.M. Rathjen, C-H. Park, P.R. Goodrich and D.D. Walgenbach, *Polymer Gels and Networks*, **3**, 101(1995).
21. Y. Hirokawa, H. Jinnai, T. Okamoto, and T. Hashimoto, *Polymer Preprints, Japan*, **46**, 1918(1997).
22. T. Okamoto, Y. Hirokawa, K. Kimishima, H. Jinnai, S. Koizumi, and T. Hashimoto, *Polymer Preprints, Japan*, **46**, 1920(1997).
23. A.C. Saucier, S. Mariotti, S.A. Anderson, D.L. Purich, *Biochemistry*, **24**, 7581(1985).

PART 2

Network Characterization

12

'Butterfly' Light Scattering Pattern from Stretched N-Isopropylacrylamide Gel

TAKUYA OKAMOTO[1]*, YOSHITSUGU HIROKAWA[1]*, KATSUO MATSUZAKA[1], and TAKEJI HASHIMOTO[1,2]

[1]Hashimoto Polymer Phasing Project, ERATO, JST, 15 Morimoto-cho, Shimogamo Sakyo-ku, Kyota 606-0805, Japan

[2]Department of Polymer Chemistry, Graduate School of Engineering, Kyota University, Kyoto 606-8501, Japan

ABSTRACT

The effect of stretching on internal structures of translucent chemically crosslinked N-isopropylacrylamide (NIPAAm) gel has been studied by laser light scattering (LS). The anisotropic small-angle light scattering pattern, known as the 'butterfly pattern', was observed when the stretching ratio (λ) of the gel exceeded about 1.50. The Fourier spectra obtained from the fast Fourier transform (FFT) of the transmission optical microscope image also showed the butterfly pattern. This fact indicates the presence of frozen inhomogeneities at mesoscopic scale (on the order of micrometres) in the macroscopically uniform gel.

* Present address: Nippon Zeon R&D Center, 1-2-1 Yako Kawasaki-ku, Kawasaki-shi, 210-8507 Japan

Wiley Polymer Networks Group Review Series Vol. 2. Edited by B.T. Stokke and A. Elgsaeter
© 1999 John Wiley & Sons Ltd

INTRODUCTION

We have been studying internal structures of chemically crosslinked *N*-isopropylacrylamide (NIPAAm) gel with slight opacity by a laser scanning confocal microscope (LSCM), laser light scattering (LS), and small angle neutron scattering (SANS). The results obtained from LSCM and LS indicated that the NIPAAm gels had internal structures consisting of domains having dense and sparse polymer networks at mesoscopic scale (μm order). These domain structures were found to be frozen inhomogeneous structures on the basis of the time resolved LSCM images [1,2].

Many studies on inhomogeneities of the gels have been reported so far [3–13]. SANS experiments have demonstrated that the uniaxially deformed gels show an anisotropic scattering pattern oriented along the stretching direction (so called 'butterfly pattern'). The origin of the butterfly pattern is considered to be the frozen inhomogeneity existing at a length scale of nm order. Therefore, when the NIPAAm gel having frozen inhomogeneities at the length scale of μm order is stretched, we would expect to observe the butterfly pattern by LS. In this report, we wish to present the observation of the butterfly pattern by LS from stretched NIPAAm gels with slight opacity in order to confirm the conjecture described above.

MATERIALS AND METHODS

NIPAAm monomers purified by recrystallization and ammonium peroxodisulfate (AP) (initiator) were used as received. Sheet-like NIPAAm gels were prepared by free radical polymerization in the presence of crosslinker, i.e. *N,N'*-methylene-bis(acrylamide) (BIS). For this purpose, a pregel solution was first prepared by dissolving NIPAAm monomers, BIS, and *N,N,N',N'*-tetramethylethylenediamine (TEMED) (accelerator) in pure water. The concentrations of NIPAAm monomers, BIS, and TEMED were 1.33 M, 10 mM, and 21 mM, respectively, to a final volume. After addition of 0.4 wt.% AP solution, the pregel solution was quickly transferred into a cell (1.1 mm thickness) made of glass plates, and the cell was immersed in the temperature controlled water bath for 1 h for gelation.

The stretching of gel samples prepared in this way was carried out by using a specially designed stretching apparatus in which both edges of the samples were held with clamps. The stretching ratio (λ) of the gels was determined by the distance between the clamps. The LS measurements were performed with an He–Ne laser as an incident beam and the LS patterns were recorded with a CCD camera.

RESULTS AND DISCUSSION

Figure 12.1 shows the appearance of NIPAAm gels prepared at different preparation temperatures. The outer white ring is the cell used for gelation and the

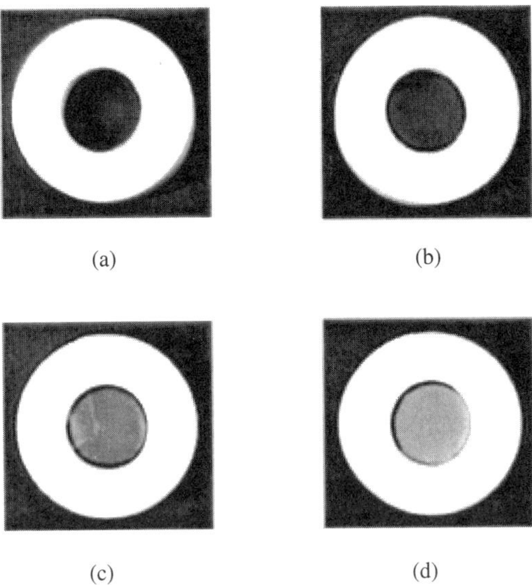

(a)   (b)

(c)   (d)

Figure 12.1 Optical appearance of NIPAAm gels prepared at (a) 20 °C, (b) 24.5 °C, (c) 32 °C, and (d) 38 °C

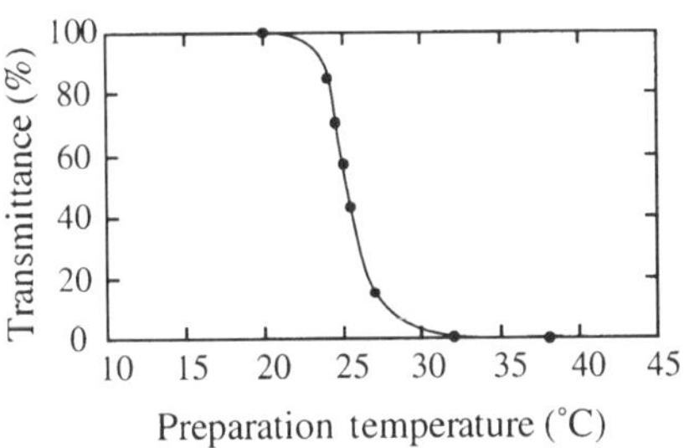

Figure 12.2 Transmittance of NIPAAm gels prepared at various temperatures

center part is the gel. The gel prepared at 20 °C was quite transparent, while slight opacity was observed for the gel prepared at 24.5 °C. The gels prepared at 32 °C and 38 °C were obviously opaque. The observed opacity of the gels did not disappear even when the temperature was lowered. This may indicate that the inhomogeneities were permanently fixed in the gel by the crosslinking.

To quantitatively study the differences of the opacity, we measured the transmittance of these samples at wavelength 488 nm as a function of the preparation temperature (Figure 12.2). The transmittance of the gel prepared at 20 °C was 100%. With increase of preparation temperature, the transmittance suddenly decreases at around 25 °C and eventually goes to zero at temperature higher than 32 °C.

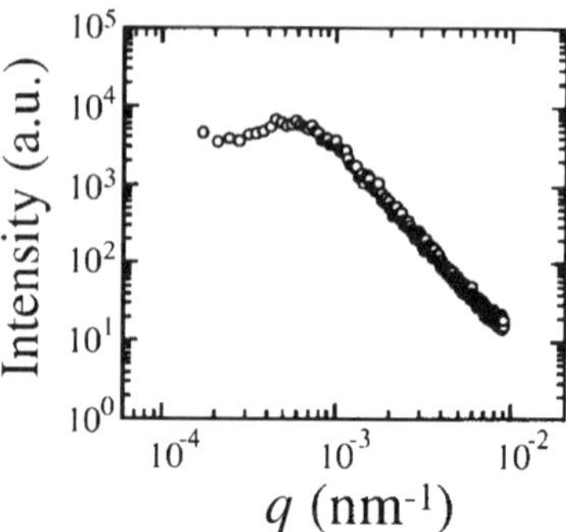

Figure 12.3 LS profile of NIPAAm gel prepared at 24.52 °C. q is magnitude of the scattering vector defined by $q = (4\pi/\lambda_1)\sin(\vartheta/2)$ with ϑ and λ_1 being the scattering angle and wavelength of the incident beam in the medium

In order to investigate the internal structures of the gel, LS measurements were carried out. Figure 12.3 shows the double logarithmic plot of the scattered intensity as a function of the wave number q for the gel prepared at 24.5 °C. The profile showed a maximum at $q = q_m \sim 0.0005$ nm^{-1}, corresponding to 12 μm in spacing D between the dense (or sparse) polymer network domains ($D = 2\pi/q_m$). The spacing agrees well with that observed by the LSCM [1,2].

In order to confirm that this internal structures consisting of dense and sparse polymer network domains at mesoscopic scale (μm order) is the frozen inhomogeneous structure, we further investigated the LS patterns from the stretched gels. The transparent gel prepared at 20 °C appeared to show only directionally independent (circular) scattering pattern, even when the gel was stretched up to the stretching ratio (λ) 1.82. This indicates that the gel does not have inhomogeneities at mesoscopic scale (μm order). The gel is uniform before stretching, as evidenced by no appreciable LS maximum at $q = q_m$ before stretching, and the more or less uniform gel is uniformly stretched at the length scale of μm, though it might be inhomogeneous at the length scale of nm as reported by Mendes *et al.* [11]. The gel prepared at 24.5 °C showed the circular scattering pattern with a scattering maximum at $q = q_m$ before stretching (see Figure 12.3). However, the butterfly pattern of the light scattering began to appear, when the stretching ratio (λ) of the gel is greater than $\sim$1.50 as shown in Figure 12.4.

Figure 12.5 shows the transmission optical micrographs of NIPAAm gel prepared at 24.5 °C at different λ. The contrast in the pictures was supposed to reflect the concentration fluctuations between the dense and sparse polymer network domains. The fluctuations at λ = 1.00 was small and directionally independent, while those at λ = 2.14 appeared to be enhanced along the stretching direction in terms of its amplitude but remain unchanged normal to the stretching direction, similarly to that observed for the sheared semidilute solutions of high molecular weight polymers [14–16].

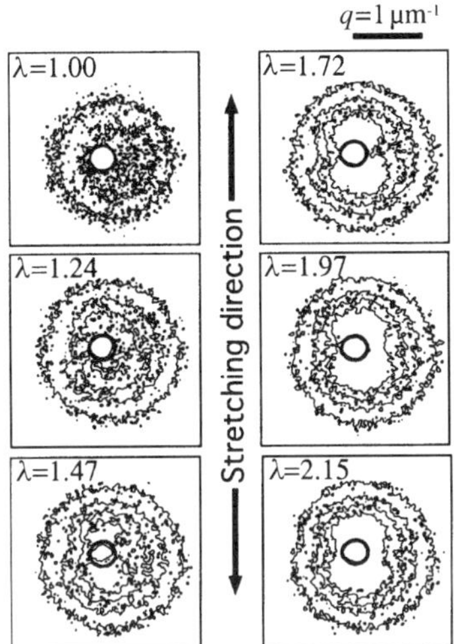

Figure 12.4 Isointensity curves of LS patterns of NIPAAm gel at various stretching ratios (λ)

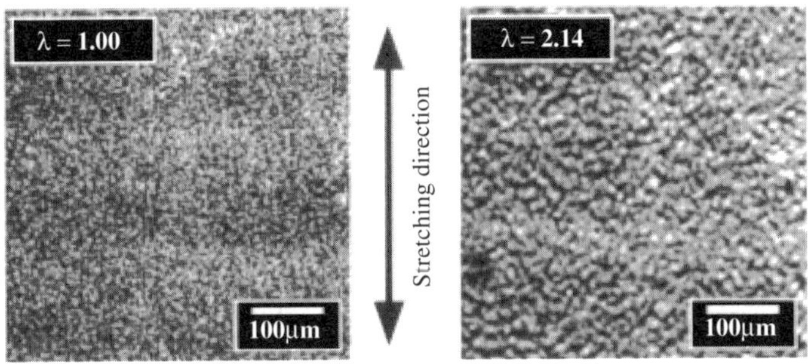

Figure 12.5 Transmission optical micrographs of NIPAAm gel prepared at 24.5 °C at different stretching ratio (λ)

In order to ascertain that the optical images as shown in Figure 12.5 are a true representation of structural entities occurring in the gel, we carried out a fast Fourier transform (FFT) of the optical images. The isointensity curves of FFT spectra corresponding to the images shown in Figure 12.5 are presented in Figure 12.6. FFT spectra showed the same patterns as those obtained by LS, i.e., the circular pattern at $\lambda = 1.00$ and the butterfly pattern at $\lambda = 2.14$. Hence the images shown in Figure 12.5 represent the true structural entities.

This fact elucidated the presence of the frozen inhomogeneities in the gel at mesoscopic scale (µm order). A dense polymer network domain is more diffi-cult to deform than a sparse one. Therefore, the butterfly scattering pattern can

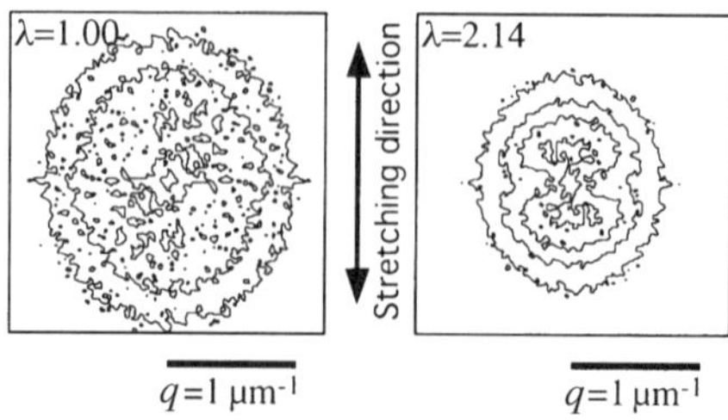

Figure 12.6 Isointensity curves of FFT spectra calculated from the images shown in Figure 12.5

be observed due to the enhanced frozen concentration fluctuation consisting of dense and sparse polymer network domains upon stretching. These concentration fluctuations become anisotropic upon stretching: the fluctuation amplitude is enhanced along the stretching direction but remains essentially unchanged normal to the stretching direction, giving rise to a butterfly pattern with strong scattered intensity along the stretching direction.

REFERENCES

1. Y. Hirokawa, H. Jinnai, T. Okamoto and T. Hashimoto, *Polymer Preprints, Japan*, **46**, 1918 (1997).
2. T. Okamoto, Y. Hirokawa, K. Kimishima, H. Jinnai, S. Koizumi and T. Hashimoto, *Polymer Preprints, Japan*, **46**, 1920 (1997).
3. J. Bastide, L. Leibler, *Macromolecules*, **21**, 2647 (1988).
4. A. Baumgartner and C.E. Picot, *In Molecular Basis of Polymer Networks* (A. Baumgartner and C.E. Picot, eds), Springer, Berlin, 1989.
5. P.N. Pusey and W. van Megen, *Physica A*, **157**, 705 (1989).
6. E. Mendes, P. Lindner, M. Buzier, F. Boue and J. Bastide, *Phys. Rev. Lett.*, **66**, 1595 (1991).
7. A. Onuki, *J. Phys. Paris*, **2**, 45 (1992).
8. A. Onuki, *Adv. Polym. Sci.*, **63**, 109 (1993).
9. C. Rouf, J. Bastide, J.M. Pujol, F. Schosseler and J.P. Munch, *Phys. Rev. Lett.*, **73**, 830 (1994).
10. S. Panyukov and Y. Rabin, *Phys. Rep.*, **1**, 269 (1996).
11. E. Mendes, R. Oeser, C. Hayes, F. Boue and J. Bastide, *Macromolecules*, **29**, 5574 (1996).
12. J. Bastide and S.J. Candau, in *Physical Properties of Polymeric Gels* (J.P. Cohen-Addad, ed.), John Wiley, New York, 1996.
13. M. Shibayama, *Maclomol. Chem. Phys.*, **1**, 198 (1998).
14. E. Moses, T. Kume and T. Hashimoto, *Phys. Rev. Lett.*, **72**, 2037 (1994).
15. H. Murase, T. Kume, T. Hashimoto, Y. Ohta and T. Mizukami, *Macromol. Symp.*, **104**, 159 (1996).
16. T. Kume, T. Hashimoto, T. Takahashi and G.G. Fuller, *Macromolecules*, **30**, 7232 (1997).

13

Osmotically Active and Osmotically Passive Counter Ions in Polyelectrolyte Gels

K.B. ZELDOVICH, O.E. PHILIPPOVA and A.R. KHOKHLOV
Physics Department, Moscow State University, Moscow 117234, Russia

ABSTRACT

The counter ion distribution in a realistic inhomogeneous polyelectrolyte gel is considered. It is shown that some of the counter ions are trapped into the local electrostatic potential wells originated by higher local polymer content, and do not contribute to the intranetwork osmotic pressure. Accordingly, the gel swelling ratio should be less, and the collapse less abrupt in comparison with a homogeneous gel. The experimental evidence presented qualitatively supports the proposed theory: the more inhomogeneous the gel, the less is its swelling ratio.

INTRODUCTION

Polyelectrolyte gels contain charged monomer units and counter ions. Suppose that a piece of such gel is immersed in a large amount of water. The counter ions cannot escape outside the gel because of the condition of macroscopic

Wiley Polymer Networks Group Review Series Vol. 2. Edited by B.T. Stokke and A. Elgsaeter
© 1999 John Wiley & Sons Ltd

electroneutrality. Being forced to reside inside the gel, they try to occupy as much volume as possible because in this way they gain some translational entropy. As a result they create an exerting osmotic pressure leading to the gel swelling. The exerting osmotic pressure of counter ions plays a very essential (often dominant) role in the behaviour of polyelectrolyte gels. In particular, it determines the two basic properties of the gels: their superabsorbing ability with respect to water and the pronounced collapse transition.

The first of these properties is connected with the fact that the gel swelling caused by the exerting osmotic pressure of counter ions results in the absorption of a significant amount of solvent, in which the gel is immersed. Just 1 g of the dried gel can acquire up to 1 kg of solvent. This property is widely used in many commercial applications, i.e. in hygiene (disposable diapers, sanitary napkins), agriculture (water retention in soil) etc. [1].

Another striking property of gels which is determined by the osmotic pressure exerted by counter ions is the pronounced collapse transition which occurs when the quality of solvent becomes poorer. When we increase the volume fraction of a poor solvent, at first the volume of the gel does not change too much and then suddenly the gel shrinks in a jump-like fashion (Figure 13.1). The large amplitude and the abrupt character of the collapse transition are consequences of the osmotic pressure exerted by counter ions. This osmotic pressure induces an enormous swelling of the gel in a good solvent. On the other hand, the collapsed gel, which is stabilized by the attraction of monomer units in a poor solvent, is relatively unaffected by counter ions. As a result, there is a large difference between the gel superswollen due to the exerting osmotic pressure (good solvent region) and the collapsed gel (poor solvent region) with the potential barrier between these two states, and this difference can be overcome only in a jump-like

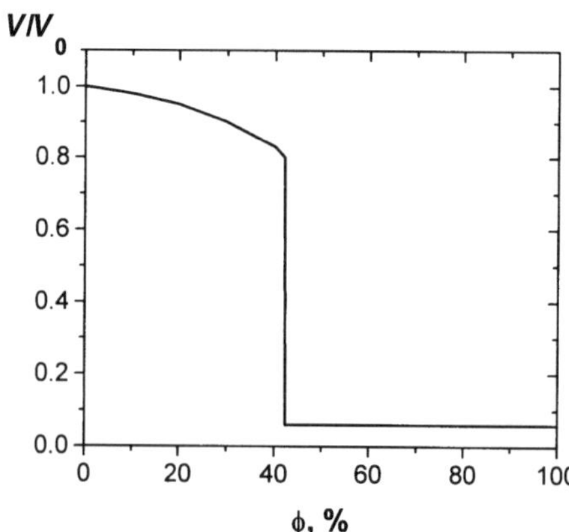

Figure 13.1   Dependence of a volume of polyelectrolyte get V on the volume fraction ϕ of the poor solvent added to water. V_0 is the volume of the gel swollen in pure water

fashion via a first-order phase transition (for a detailed theory, see ref. [2]). Owing to the large amplitude and the abrupt character of the collapse transition, the gels near the transition threshold become extremely sensitive to small variations of the external parameters, which can induce the collapse (temperature, pH etc.). Such gels are called 'responsive' gels and they are widely used as sensors, switching devices, actuators etc. [1].

Thus, the osmotic pressure of counter ions is a very important factor, which governs many of the gel properties, and therefore it is necessary to be able to calculate it correctly. Existing theories of polyelectrolyte gels usually treat the system as spatially homogeneous, with counter ion osmotic pressure given by [3]

$$\pi = nkT \tag{13.1}$$

where n is the total average concentration of counter ions within the microscopic gel sample. The aim of the present paper is to present some experimental results contradicting this formula, and to propose possible theoretical explanations. Namely, it turns out that the gel swelling ratio observed experimentally is significantly lower than that predicted by formula (13.1). We attribute this effect to the existence of osmotically passive counter ions, which are condensed either on gel chains or on microheterogeneities in the gel sample, and do not contribute to the exerted osmotic pressure.

EXPERIMENT

Below we will give two experimental examples in order to illustrate this fact.

The first of these concerns the shrinking of polyelectrolyte gels in low molecular weight salt solutions. When a polyelectrolyte gel is immersed in salt solution, it shrinks because of the decreasing difference in the osmotic pressure inside and outside the gel. Figure 13.2 shows the dependence of the swelling ratio of poly(sodium methacrylate) gel (with respect to pure water) on the concentration of NaCl [4]. It is seen that gel shrinking is observed only starting from some characteristic salt concentration c'. It is clear that this salt concentration should be of the order of the concentration of free counter ions inside the gel which determines the intranetwork osmotic pressure. If we compare the value of c' with the total concentration of counter ions available in the gel volume, c_{pol}, we will see that only a small fraction f of gel counter ions contributes to the osmotic pressure, i.e. is osmotically active (Table 13.1). At the same time, there is a large fraction of osmotically passive counter ions. A first reason for this is Manning condensation of counter ions on the highly charged network chains [5]. However, taking into account this type of condensation, we still obtain a rather large deviation of the calculated concentration of free counter ions (c_M) from the experimental value of c' (Table 13.1).

Moreover, when the monomer concentration at the synthesis conditions decreases, this deviation becomes more pronounced. As is known [6], the lower

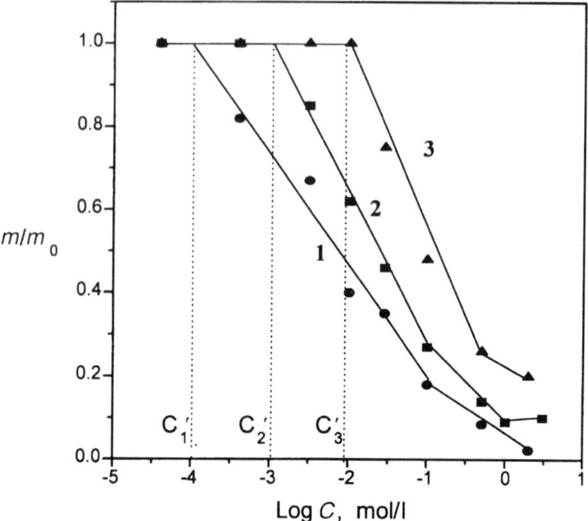

Figure 13.2   Dependence of relative mass m/m_0 of poly(sodium methacrylate) gels on NaCl concentration for the gels 1–3 from Table 13.1. m_0 is the mass of gel swollen in pure water

Table 13.1   Characteristic parameters of poly(sodium methacrylate) gels

Gel number	1	2	3
Monomer concentration at synthesis conditions	20	30	40
m_0/m_{dry}	1429	250	91
c'	0.00012	0.0016	0.01
c_{pol}	0.0065	0.037	0.1
$f = c'/c_{pol}$	0.018	0.043	0.10
c_M	0.0023	0.013	0.036
$f' = c'/c_M$	0.052	0.12	0.28

c' is some characteristic salt concentration, at which the gel starts to shrink (this concentration should be of order of the concentration of free counter ions inside the gel)

c_{pol} is the total concentration of counter ions in the gel volume

c_M is the concentration of counter ions which should be free, when correction is made for Manning condensation.

the concentration of monomer at the synthesis conditions, the more heterogeneous gel is obtained. Therefore, the experimental data suggest that the fraction of osmotically passive counter ions increases with the gel inhomogeneity. Most probably, in the inhomogeneous polyelectrolyte gel in the relatively highly charged polymer-rich regions the local potential wells are formed, which

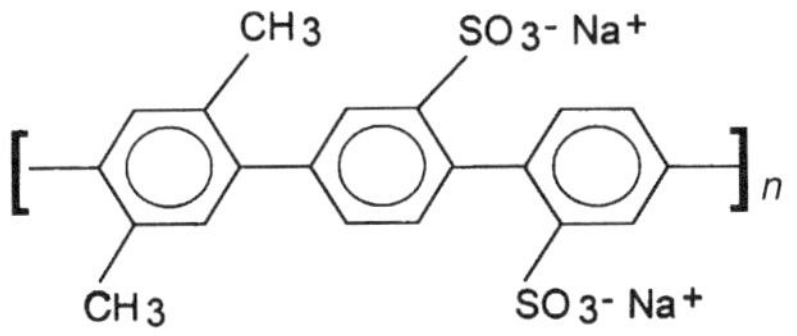

Figure 13.3 Chemical structure of rigid-rod polyelectrolyte PPP3, poly (4,4''-(disodium-2', 5'-dimethyl-1,1' : 4', 1''-terphenyl-3,2''-disulfonate))

additionally entrap some of the counter ions. The data presented above (Table 13.1) evidence that the fraction of counter ions trapped by charge inhomogeneities may be rather high.

Another way to introduce the inhomogeneous distribution of charges in the gel consists in the incorporation of charged polymer rods in the neutral gel. This was done for polyacrylamide hydrogel with polyelectrolyte rods PPP3 of the chemical structure presented in Figure 13.3 [7]. The rods are embedded in the gel, but not covalently attached to it. In this case the charged units belong only to the rods, while the rod-free space in the gel volume does not contain charged units at all. The rods bring with them their counter ions and the latter create the exerting osmotic pressure due to which the gel swells.

Let us compare the swelling of the gel with entrapped rods (inhomogeneous distribution of charges) with that of the model gel with more or less randomly distributed charges. The model gel was prepared by free radical copolymerization of neutral (acrylamide) and charged (sodium salt of 2-acrylamido-2-methyl-1-propanesulfonic acid) monomers. Figure 13.4 depicts the degree of swelling of the gels as a function of the fraction of charged units. A striking effect is evident: the swelling of rod-containing gel is much less than that of the model gel (cf. curves 1 and 3). Therefore, in the gel with entrapped rods a large fraction of counter ions does not contribute to the osmotic pressure, i.e these counter ions are osmotically passive.

Obviously, one part of the osmotically passive counter ions are condensed on the rods via the Manning mechanism. We assume that locally the polymer chains are rigid enough for the Manning theory to be applicable. Then, the effective line charge σ of the chain is given by the expression $\sigma = \sigma_0 a/l_B$, where σ_0 is the initial charge density of the chain, a is the distance between charges along the chain, and $l_B = e^2/\varepsilon kT$ is the Bjerrum length. In our case, $a = 2.5$ Å, $l_B = 7$ Å. Consequently, the osmotic pressure of the gas of free counter ions should be $l_B/a = 2.8$ times smaller than that calculated without the consideration of counter ion condensation. If we take into account this correction, we obtain the curve 2 (Figure 13.4), which is still far from the line for the random copolymer gel. Therefore, the Manning condensation does not explain the whole magnitude of the observed effect. Thus, along with passive counter ions condensed in a molecular vicinity of the rods via Manning mechanism, there is some other fraction of

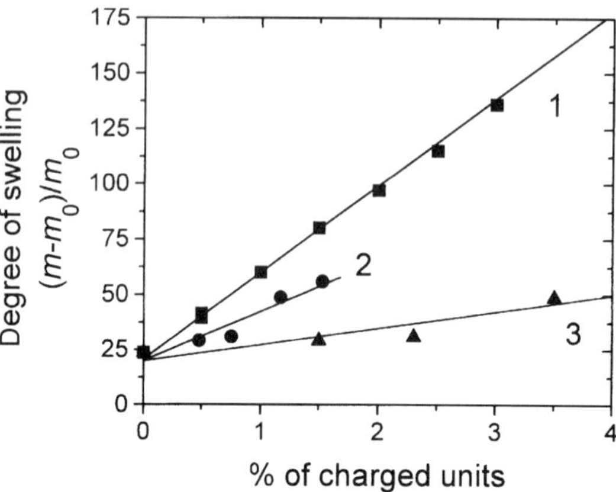

Figure 13.4 Dependence of the degree of swelling of polymer gels on the fraction of charged units for model copolymer gel of poly(acrylamide-co-sodium 2-acrylamido-2-methyl-1-propanesulfonate) with random distribution of charged units in the gel volume (1) and for polyacrylamide gel with embedded rigid-rod polyelectrolyte PPP3 with (2) and without (3) correction for Manning condensation

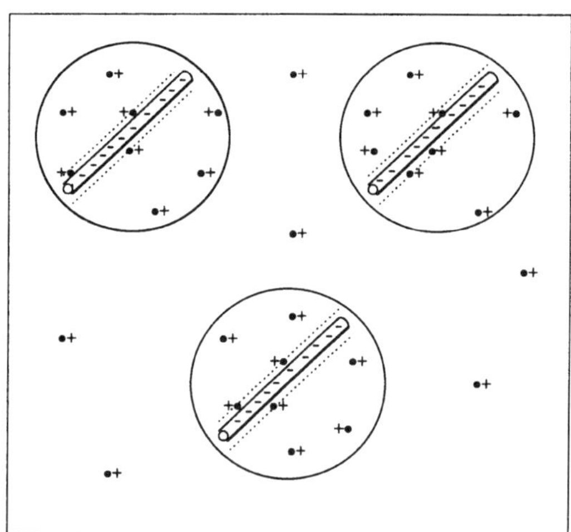

Figure 13.5 Schematic representation of uneven distribution of counter ions in a gel with inhomogeneously arranged polyelectrolyte molecules

osmotically passive counter ions. It seems reasonable to suggest that the existence of these counter ions is related with the inhomogeneous distribution of charged units within the gel. The schematic illustration of this fact is given in Figure 13.5. In addition to counter ions, which are in the molecular vicinity of the rods and which are condensed by the Manning mechanism, there are counter ions trapped

by local potential wells on a larger spatial scale, which also do not contribute to
the osmotic pressure.

Therefore, the experimental data suggest that there can be two types of osmotically passive counter ions: (1) counter ions condensed via Manning mechanism
and (2) counter ions trapped by charge inhomogeneities. Moreover, the contribution of the last type of counter ions can be rather significant.

THEORY

In this section we propose a simple theory which describes the phenomenon of
a decrease of osmotic pressure in inhomogeneous gels. This inhomogeneity may
be due, for example, to the inclusion of charged particles in the gel, or by an
uneven distribution of polymer chains, naturally occurring during gel synthesis.
In our simplified theory we will not distinguish these cases, and we will assume
that small charged aggregates are imbedded in a gel, creating the inhomogeneity.
Let us consider the behaviour of counter ions in such a medium.

Indeed, if the charge of the chains is distributed inhomogeneously, this gives
rise to local potential wells, in which many of the counter ions become trapped
and do not participate in creation of the osmotic pressure. Only truly free counter
ions, which are out of these local potential wells, are osmotically active and
contribute to the osmotic pressure.

The simple model which accounts for this effect can be outlined as follows:
consider a uniform homogeneous neutral gel with small spherical regions of
excess charge, randomly distributed within the network. Some counter ions will
be trapped by these regions, while some other counter ions will float in the
network truly free. We will use the two-phase approximation to describe the
inhomogeneous counter ion distribution. Let Q be the charge of charged regions,
R be their radius, and β the fraction of truly free counter ions that are outside
of the charged region. Correspondingly, the fraction of counter ions outside of
the potential well is $(1 - \beta)$. We assume that charged regions are separated by
a distance much larger than R, so we can neglect the electrostatic interaction
between different regions, as well as the Debye–Hückel corrections to the free
energy of the gas of counter ions.

This system should be compared with a polyelectrolyte gel with the same
overall small ion concentration: if the gel has subchains of length N with σ
uncharged monomer units between charged ones, the charge per subchain, N/σ,
should be equal to KQ, where K is the number of charged particles per one
subchain ($K \ll 1$). We will consider the case where $N/\sigma = KQ = 10$ charges per
subchain, $K = 1/100$, $Q = 1000$, $N = 100$, $\sigma = 10$.

The free energy F of an uncharged polymer network (per one subchain) can
be expressed as follows [2]:

$$\frac{F_{\text{unch}}}{kT} = \frac{3}{2}\left(\alpha^2 + \frac{1}{\alpha^2}\right) + \frac{Bn_0}{\alpha^3} + \frac{Cn_0^2}{\alpha^6} \qquad (13.2)$$

where α is the swelling ratio, $\alpha = (V/V_0)^{1/3}$, B and C are the second and third virial coefficients, respectively, $B = a^3\tau$, and n_0 is the monomer concentration within an unperturbed polymer coil of length N, $n_0 = 3/(4\pi a^3 \sqrt{N})$. The parameter τ characterizes the solvent quality, $\tau = 0$ corresponds to θ-point, and $\tau > 0$ to a good solvent. The equilibrium swelling ratio is found from the condition of the minimum of the free energy,

$$\frac{\partial F}{\partial \alpha} = 0 \tag{13.3}$$

Considering the polyelectrolyte network, we must add to this free energy the free energy of ideal gas of counter ions, assuming that there are N/σ counter ions per one subchain:

$$\frac{F_{\text{gas}}}{kT} = \frac{N}{\sigma} \ln \left(\frac{n_0}{\alpha^3 \sigma} \right) \tag{13.4}$$

(The Debye–Hückel correlational contribution to this free energy can be shown to be very small.) The condition for the equilibrium state of the gel remains the same. The analysis of the gel behaviour in this case [2] shows that at very high σ (low charge content) the gel undergoes continuous shrinking when the solvent quality becomes poorer, while at higher charge content the collapse is jump-like at a certain value of $\tau < 0$.

In our case, another contribution should be added to F_{unch}: the free energy corresponding to the redistribution of counter ions in an electric field created by charge inhomogeneities:

$$\frac{F}{kT} = KQ(1 - \beta) \ln \frac{(1 - \beta)KQ}{R^3} + KQ\beta \ln \frac{\beta n_0 KQ}{\alpha^3} + (KQ\beta)^2 \frac{e^2}{\varepsilon kT} \frac{1}{R} \tag{13.5}$$

Here the first term is the translational entropy of counter ions within the charged sphere, the second term is the translational entropy of truly free counter ions, not trapped by a potential well, and the last term is electrostatic energy of a charged sphere of radius R. Here we have assumed that charged regions are far apart from each other, and consequently, their energy is proportional to $1/R$. The values of the equilibrium swelling ratio α and counter ion redistribution coefficient β are found from the conditions

$$\frac{\partial F}{\partial \alpha} = 0, \quad \frac{\partial F}{\partial \beta} = 0 \tag{13.6}$$

We assumed $N = 100$, $e^2/\varepsilon akT = 1$, $R = 0.5aN^{1/2}$. It turns out that in this case β is close to unity, i.e most of the counter ions are trapped within the potential wells and do not contribute to the osmotic pressure. The corresponding collapse curves for a gel with homogeneous distribution of small ions and a gel with embedded charged regions are shown in Figure 13.6. From this Figure it is evident that for the gel with embedded charged regions the picture of the collapse changes

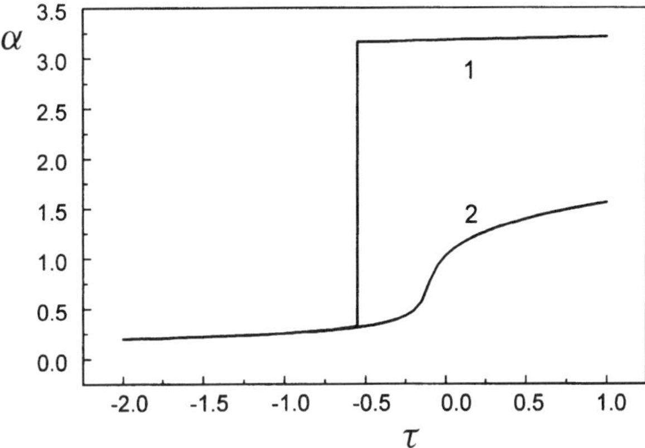

Figure 13.6 Collapse curves: (1) for a homogeneous polyelectrolyte gel (10 charges per chain) and (2) for a gel with inhomogeneously distributed charged regions (1 aggregate per 100 chains, 1000 charges per aggregate)

drastically. Instead of a pronounced collapse transition with a big jump we have a continuous transition with much lower swelling in a good solvent. Therefore, if the fraction of osmotically passive ions trapped by charge inhomogeneities is high enough, the osmotic pressure is lower, and the gel collapse is less abrupt and with lower amplitude compared with the same gel, but with homogeneous charge distribution.

CONCLUSION

In realistic inhomogeneous polyelectrolyte gels the electric charge is unevenly distributed in space. As a result, some counter ions become trapped by the local potential wells around highly charged domains and lose the ability to contribute to the intranetwork osmotic pressure. Such counter ions become osmotically passive. This paper gives some experimental evidence for the existence of osmotically passive counter ions trapped by charge inhomogeneities. A theory is proposed to calculate the drop of osmotic pressure for a certain distribution of charge inhomogeneity, namely, a rare inclusion of charged particles. The predictions of theory for the decrease of the swelling ratio of the gels are in qualitative agreement with the experimental data.

REFERENCES

1. K.S. Kazanskii and S.A. Dubrovskii, *Adv. Polym. Sci.* **104**, 97 (1992).
2. A.Yu. Grosberg and A.R. Khokhlov, *Statistical Physics of Macromolecules*, AIP Press, New York, 1994.

3. L.D. Landau and E.M. Lifshitz, *Statistical Physics*, Pergamon Press, Oxford, 1988.
4. C.H. Jeon, E.E. Makhaeva and A.R. Khokhlov, *Macromol. Chem. Phys.*, **199**, 2665 (1998).
5. G.S. Manning, *Annu. Rev. Phys. Chem.*, **23**, 117 (1972).
6. K. Dusek and W. Prins, *Adv. Polym. Sci.*, **6**, 1 (1969).
7. O.E. Philippova, R. Rulkens, B.I. Kovtunenko, S.S. Abramchuk, A.R. Khokhlov and G. Wegner, *Macromolecules*, **31**, 1168 (1998).

14

Visualization of Adsorption and Diffusion Processes of Probe Molecules Interacting to Polymer Networks by Anti-Stokes Fluorescence Imaging

EIJI KATO[1] and TOMOHIRO MURAKAMI[2]

[1]Marine Engineering, Kobe University of Mercantile Marine, 1-1,5, Fukae-Minami, Higashinada, Kobe 658, Japan

[2]Yuge National College of Maritime Technology, Ehime 794-25, Japan

ABSTRACT

The penetration process of probe molecules (rhodamine B) into N-isopropylacrylamide gels was visualized by an anti-Stokes fluorescence imaging method. By analyzing the intensity profiles by a theory based on Fickian diffusion, the diffusion coefficient was determined for the different swollen state of the gel under the volume phase transition. The diffusion coefficient drastically decreased at the phase transition from the swollen state to the shrunken state. The result was discussed by using the free volume theory.

Wiley Polymer Networks Group Review Series Vol. 2. Edited by B.T. Stokke and A. Elgsaeter
© 1999 John Wiley & Sons Ltd

INTRODUCTION

The diffusion of solute and solvent molecules in polymer gels is not only of scientific interest but also important in biomedical applications. Particularly, an understanding of the diffusion process of water-soluble molecules in hydrogels is relevant to the application to drug delivery systems.

Direct investigations of the diffusion process have been carried out for various polymeric systems by using an NMR imaging technique [1–5]. However, there are fewer investigations for gels in spite of the importance in applications [5–7].

In this work, we visualized penetration and release processes of probe molecules (rhodamine B) into nonionic poly(N-isopropylacrylamide)(NIPA) gels by using an anti-Stokes fluorescence imaging technique. This technique is very simple and very effective for hydrogels. Rhodamine B is a water-soluble dye. It emits intense anti-Stokes fluorescence with a peak of 580 nm when illuminated by incident light with a longer wavelength, an He–Ne laser (632.8 nm) [8]. The great advantage of the anti-Stokes fluorescence imaging method is that the incident light can travel through the sample gels without significant absorption. Therefore, we can get the whole image of intensity (concentration) distribution of the probe molecules in the gel which represents the penetration process of the molecules.

Analyzing the obtained intensity profiles by a formula based on Fickian diffusion, we calculated the diffusion coefficients of the probe molecules for different swollen states of the gel under a volume phase transition. The results were discussed by using the free volume theory [9,10].

EXPERIMENTAL

A detailed description of the experimental apparatus was given in a previous paper [8]. A monochromatic beam from an He–Ne laser was transformed into light in the form of a sheet by three cylindrical lenses. The light was introduced vertically to the side of the cylindrical sample gel to get a sectional planner image of the gel. Rhodamine B, of which the molecular weight is 478, emits a broad anti-Stokes fluorescence spectrum with a peak around 580 nm when excited by an He–Ne laser of 632.8 nm. The sectional planar profiles of the fluorescence emitted from the probe molecules in the gel were imaged by the CCD camera (1316×1034 pixels). To eliminate the undesirable light brought by elastic scattering from polymer networks, a holographic notch filter was placed in front of the CCD camera. This filter absorbed effectively the light of 632.8 nm and transmitted only the fluorescence. Before the measurements, the sample gels were kept in an equilibrium state in pure water at a constant temperature. Then, the surrounding water was replaced by the solutions of rhodamine B. The sectional planar profiles of the fluorescence were measured as a function of time.

The anti-Stokes fluorescence intensity was assumed to be linear with the concentration of the probe molecules in the measured concentration range ($\sim 10^{-4}$ mol/l). In a higher concentration region, the influence of absorption and quenching must be taken into account.

Sample nonionic poly(N-isopropylacrylamide) gels were prepared as follows. NIPA monomer (3.96 g) and crosslinker N,N'-methylenebisacrylamide (0.067 g) were dissolved in distilled water (total volume 50 ml). After purging with nitrogen, we added ammonium persulfate (20 mg) and tetramethylethylenediamine (120 µl) as an initiator and an accelerator for polymerization. Gelation was carried out in a day in glass tubes of 4.29 mm in diameter at 20 °C. The gels were removed from the glass tubes and washed in distilled water, cut into segments about 7 mm in length, and placed in a glass cell of $10 \times 10 \times 110$ mm for the fluorescence measurements. The glass cell was immersed in a transparent water bath. The temperature of the bath was controlled within ± 0.1 °C.

RESULTS

Figure 14.1 shows representative sets of the fluorescence image and intensity profile in a radial direction of the gel for the different cases in which the NIPA gel was surrounded (a) by aqueous solution of rhodamine B(10^{-4} mol/l) (Figure 14.1(a)), and (b) by methanol solution of rhodamine B(8×10^{-4} mol/l) (Figure 14.1(b)), after a typical penetration time. In both cases, before the measurements, the gel was kept swollen in pure water. Then the water was replaced by each solution. Figure 14.1(a) shows that a sharp intensity boundary was formed across the solution–gel interface. This indicates that NIPA gel selectively adsorbed the probe molecules in the aqueous solution. It is seen the adsorbed molecules diffuse into the interior of the gel. Meanwhile, in the methanol solution (Figure 14.1(b)), such an intensity boundary was not formed. The intensity profile indicates that the probe molecules diffuse from the surrounding methanol solution into the gel. From these results, it is supposed that hydrophobic interaction acting between rhodamine B and NIPA is responsible for the adsorption of the molecules.

Figure 14.2 shows a representative set of the fluorescence images and the intensity profile for NIPA gel in aqueous solution after different penetration times. The figure represents clearly the adsorption and the diffusion processes of the probe molecules. The intensity at the surface increases with increasing time and the molecules diffuse from the surface into the interior of the gel. The intensity at the surface is plotted as a function of time in Figure 14.3. It shows that the intensity rapidly increases with increasing time and reaches a saturated value after ~ 70 min. The intensity profiles after the saturation time are shown superimposed in Figure 14.4.

In order to determine the diffusion coefficient, we compared the intensity (concentration) distribution with those calculated from a theory based on Fickian

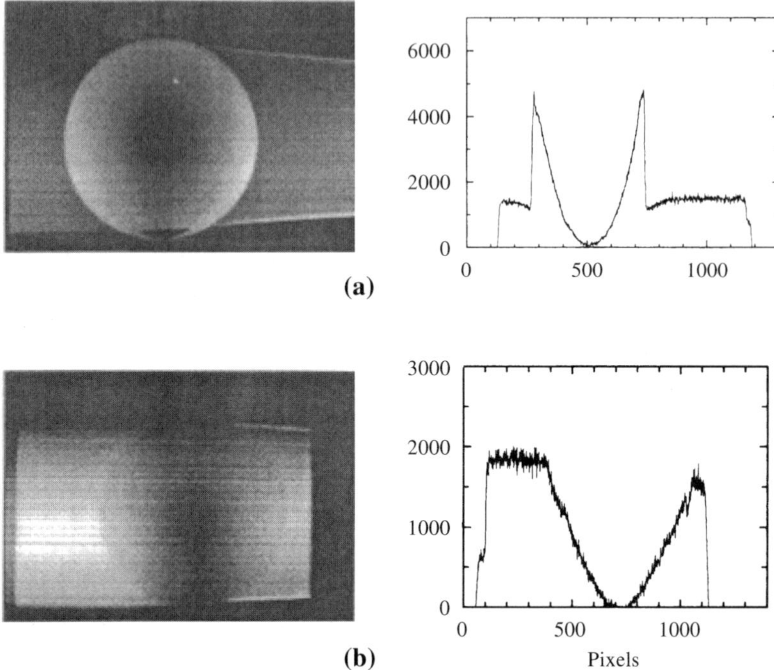

(a)

(b)

Figure 14.1 Typical fluorescence images and intensity profiles representing the penetration process of probe molecules into nonionic NIPA gels in aqueous solution and methanol solution.

The gels were immersed in each solution. The circles represent sectional planar images of the anti-Stokes fluorescence from the gel.

(a): Penetration of rhodamine B from the surrounding aqueous solution (penetration time, 51 min; temperature, 33.4 °C; diameter of the gel, 4.54 mm)

(b): Penetration of rhodamine B form the surrounding methanol solution (penetration time, 30 min; temperature, 36.0 °C; diameter of the gel, 6.15 mm)

diffusion [11]. When we consider the diffusion in a long cylinder of radius a, the concentration $C(r, t)$ at a position r in a radial direction and in time t is given as for the conditions ($C = C_0$, $r = a$, $t \geq 0$; $C = 0$, $0 < r < a$, $t = 0$),

$$C(r, t) = C_0 \left\{ 1 - \frac{2}{a} \sum_{n=1}^{\infty} \frac{\exp(-D\alpha_n{}^2 t)J_0(r\alpha_n)}{\alpha_n J_1(a\alpha_n)} \right\} \tag{14.1}$$

Figure 14.2 A representative set of the fluorescence images and the intensity profiles for the NIPA gel in aqueous solution (10^{-4} mol/L) after different penetration times. The times are inserted in the figure (temperature, 33.4 °C; diameter of the gel, 4.54 mm)

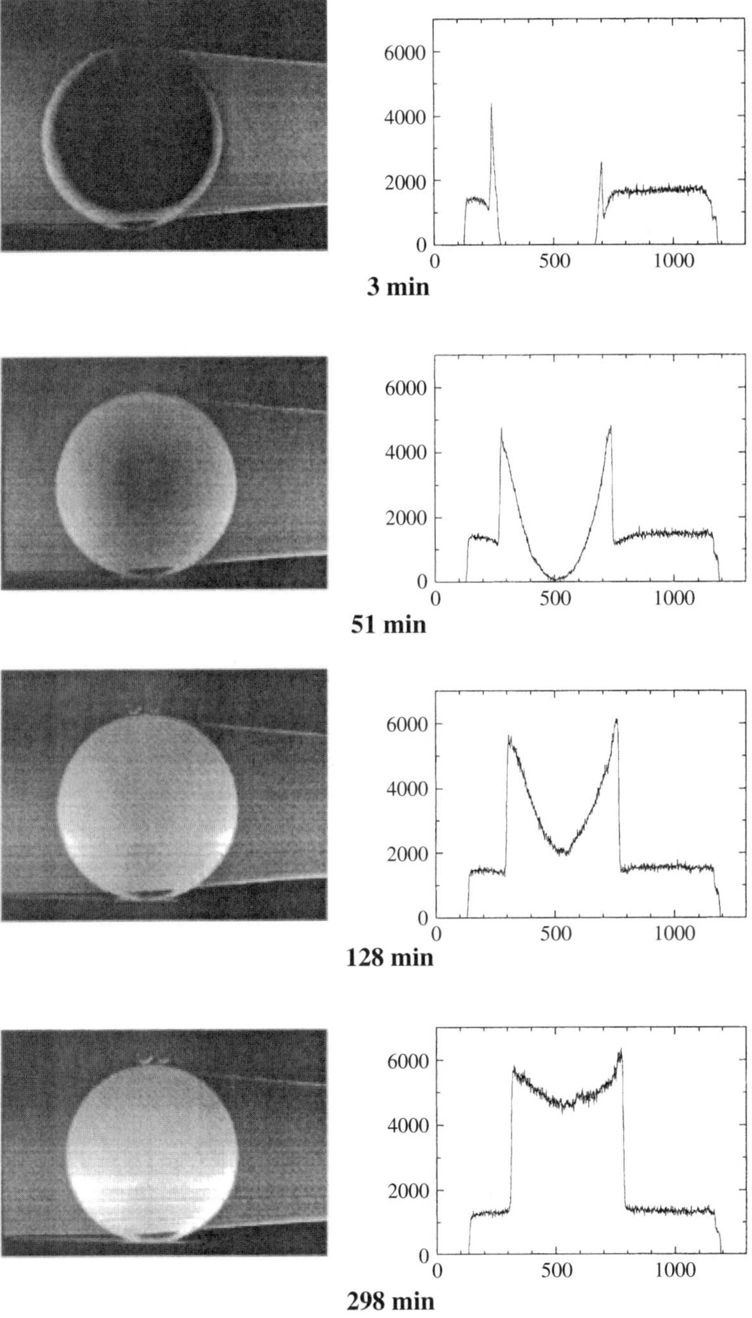
6000
4000
2000
0
0
500
1000
3 min
6000
4000
2000
0
0
500
1000
51 min
6000
4000
2000
0
0
500
1000
128 min
6000
4000
2000
0
0
500
1000
298 min

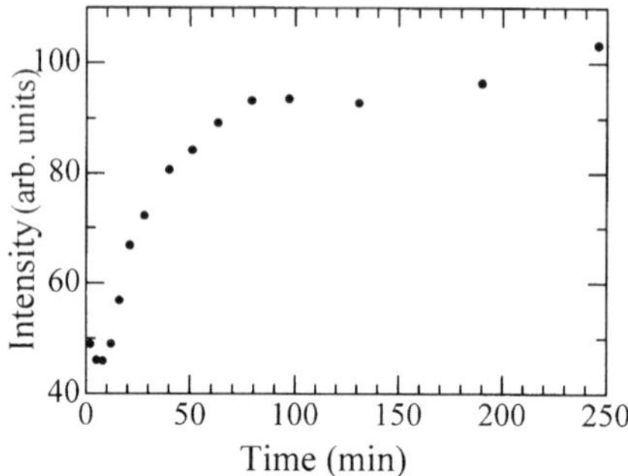

Figure 14.3 Dependence of the intensity at the surface on the penetration time

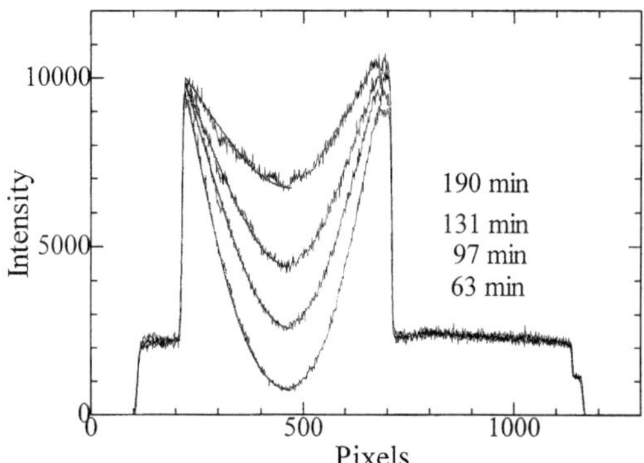

Figure 14.4 Intensity distribution in the gel for the different penetration times. The times are inserted in the figure. Solid curves are there calculated by Fickian diffusion (temperature, 32.5 °C; diameter of the gel, 4.85 mm)

where D is the diffusion coefficient, J_0 and J_1 are the Bessel functions of the first kind of order zero and the first order, respectively, and α_n are the positive roots of $J_0(a\alpha_n) = 0$. And, $\alpha_n = C_n/a$, C_n is a constant depending on the parameter n.

In order to compare the concentration (intensity) distributions in Figure 14.4 with theoretical ones, we calculated equation (14.1) as a function of r for various values of Dt/a^2. The solid curves in the figure were those obtained when we select 0.078, 0.13, 0.184, and 0.275 as the values of Dt/a^2. It is seen that these curves fit the experimental distributions well. These selected values are plotted as a function of the corresponding penetration time t in Figure 14.5. Assuming that the relation between these values and the time is linear, we can determine the diffusion coefficient from a gradient of the line. The obtained value is $D = 1.53 \times 10^{-6}$ cm^2/s for the swollen NIPA gel in the aqueous solution at 32.5 °C.

Similar measurements visualizing the penetration process of the probe molecules were carried out for the different swollen state of the gel under

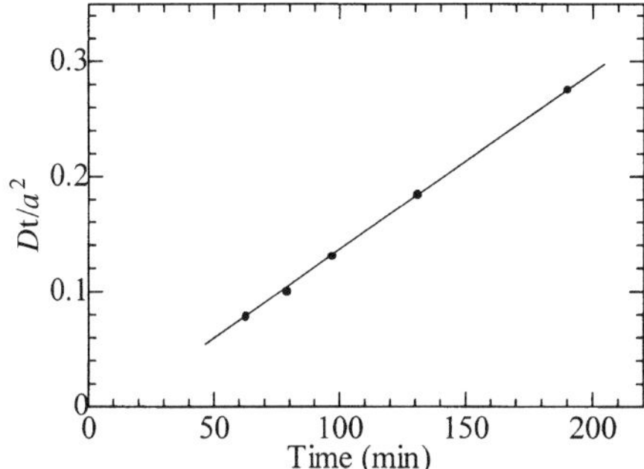

Figure 14.5 Relation between Dt/a^2 and the corresponding penetration time

the volume phase transition. It is known that a nonionic NIPA gel undergoes a continuous volume phase transition by the change in temperature [12,13]. From the measurement of swelling curve, it was found that the present NIPA gel continuously changed its volume from the swollen state to the shrunken state at 34.5 °C. When the gel shrank the gel became opaque. However, when more than a week passed the gel became transparent. The measurements for the shrunken gel were done after then.

Comparing the obtained intensity (concentration) distributions with ones calculated from Fickian diffusion (equation (14.1)), we determined the diffusion coefficients for the different swollen states. They are plotted as a function of temperatures, together with the swelling curve, in Figure 14.6. It is seen that the diffusion coefficient drastically decreases in the phase transition from the swollen state to the shrunken state. The transition temperature coincides with that determined from the swelling curve. It was found that in the phase transition, the diffusion coefficient decreased in more than two orders of magnitude.

The free volume theory [9,10] has been adopted to elucidate the dependence of the diffusion coefficient on the concentration of polymer in solutions or gels [14–16]. At lower polymer concentrations (<50%), the theory expresses the dependence by a simple relation as

$$\log(D/D_0) = Aw_2/(1 - w_2) \tag{14.2}$$

where D_0 is the diffusion coefficient in the absence of polymer, A is a system-dependent parameter, and w_2 is the weight fraction of the polymer.

The obtained diffusion coefficient is plotted as a function of $\phi/(1 - \phi)$, in Figure 14.7, where ϕ is the volume fraction of polymer. The values of ϕ can be calculated from the swelling data in Figure 14.6, by taking account of the fact that $\phi/\phi_0 = V_0/V$ and $\phi_0 = 0.072$. Since the density (1.10 g/cm^3) of NIPA polymer, which was roughly estimated in the present experiment, does not differ much from that of water, we can regard as $\phi/(1 - \phi) = w_2/(1 - w_2)$. Equation (14.2) predicts

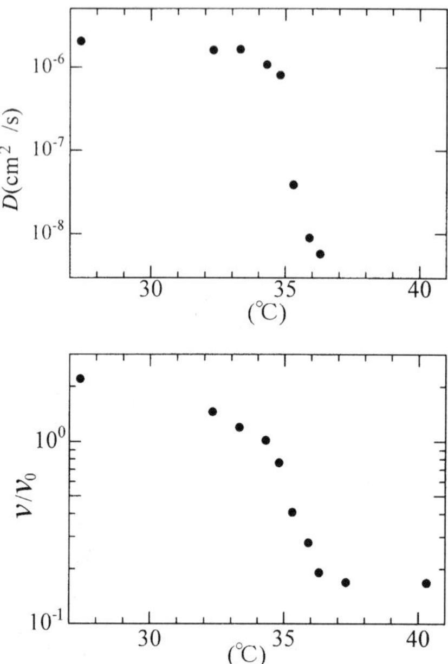

Figure 14.6 The diffusion coefficient as a function of temperature, and the swelling curve of the present NIPA gel

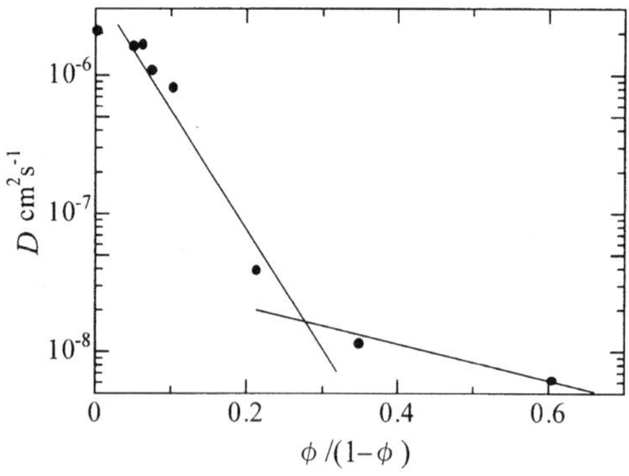

Figure 14.7 Dependence of the diffusion coefficient D on $\phi/(1 - \phi)$, where ϕ is the volume fraction of network

that $\log D$ linearly depends on $w_2/(1 - w_2)$, that is, $\phi/(1 - \phi)$. Figure 14.7 shows that $\log D$ vs $\phi/(1 - \phi)$ can be expressed by two lines with very different gradients. The gradient of the line for the swollen state is much larger than that for the shrunken state. This means that the system-dependent parameters in equation (14.2) vary greatly with states of the gel.

From these results, it is found that the diffusion of solvent in the shrunken gel qualitatively differs from that in the swollen gel. We suggest that this is due to the formation of inhomogeneous structures of polymer network in the shrunken state.

REFERENCES

1. L.A. Weisenberger and J.L. Koenig, *Macromolecules*, **23**, 2445 (1990).
2. M.Valtier, P. Tekely, L. Kiéné and D. Canet, *Macromolecules*, **28**, 4075 (1995).
3. M. Ercken, P. Adriaensens, D. Vanderzande and J. Gelan, *Macromolecules*, **28**, 8541 (1995).
4. J.-M. Petit, X.X. Zhu and P.M. Macdonald, *Macromolecules*, **29**, 70 (1996).
5. T. Shibuya, H. Yasunaga, H. Kurosu and I. Ando, *Macromolecules*, **28**, 4377 (1995).
6. S. Pajevic, R. Bansil and C. Konak, *Macromolecules*, **26**, 305 (1993).
7. S. Schlick, J. Pillar, S.-C. Kweon, J. Vacik, Z. Gao and J. Labsky, *Macromolecules*, **28**, 5780 (1995).
8. E. Kato and T. Murakami, *Polymer Gels and Networks*, **6**, 179 (1998).
9. J. S.Vrentas, J. L. Duda and H.-C. Ling, *J. Polym. Sci., Polym. Phys. Ed.*, **23**, 275 (1985).
10. J.S. Vrentas, J.L. Duda and H.-C. Ling and A.-C. Hou, and *J. Polym. Sci., Polym. Phys. Ed.*, **23**, 289 (1985).
11. J. Crank, *The Mathematics of Diffusion*, Clarendon Press, Oxford, 1975.
12. Y. Hirokawa and T. Tanaka, *J. Chem. Phys.*, **81**, 6379 (1984).
13. S. Hirotsu, Y. Hirokawa and T. Tanaka, *J. Chem. Phys.*, **87**, 1392 (1987).
14. R.A. Waggoner, F.D. Blum and J.M.D. MacElroy, *Macromolecules*, **26**, 6841 (1993).
15. S. Schlick, J. Pillar, S.-C. Kweon, J. Vacik Z. Gao and J. Labsky, *Macromolecules*, **28**, 5780 (1995).
16. S. Matsukawa and I. Ando, *Macromolecules*, **29**, 7136 (1996).

15

Nanostructures in Swollen Polymer Networks

FERENC HORKAY[1], GREGORY B. MCKENNA[2] and ERIK GEISSLER[3]

[1]General Electric Corporate Research and Development, Schenectady, NY 12301, USA

[2]Polymers Division, National Institute of Standards and Technology, Gaithersburg, MD 20899, USA

[3]Laboratoire de Spectrométrie Physique CNRS UMR5588, Université J. Fourier de Grenoble, B.P.87, 38402 St Martin d'Hères, France

Wiley Polymer Networks Group Review Series Vol. 2. Edited by B.T. Stokke and A. Elgsaeter
© 1999 John Wiley & Sons Ltd

ABSTRACT

The thermodynamic properties of cross-linked polyisoprene networks swollen in toluene are investigated by scattering and osmotic methods. Small-angle neutron scattering measurements indicate that in these gels static scattering is caused by (i) local elastic constraints frozen in by cross-links, and (ii) dense clusters present in the uncross-linked system. A scattering function of the form $I(Q) = \Delta\rho^2\{A(1 + Q^2\xi^2)^{-1} + B_1(1 + Q^2\Xi_1^2)^{-2} + B_2(1 + Q^2\Xi_2^2)^{-2}\}$ is fitted to the neutron data, where Q is the transfer wave vector, ξ is the thermodynamic correlation length, Ξ_1 and Ξ_2 are static correlation lengths, $\Delta\rho^2$ is a contrast factor, A, B_1 and B_2 are constants. The intensity of the dynamic component of the gel scattering spectrum is in satisfactory agreement with the result obtained from independent osmotic and mechanical measurements.

INTRODUCTION

Recently, significant progress has been made in the understanding of the relationship between the structure and the thermodynamic properties in cross-linked polymer gels [1–3]. By combining scattering measurements of different length scales and macroscopic elastic and osmotic observations, it was found that a certain part of the network polymer is restrained from participating in the dynamic concentration fluctuations that control the thermodynamic properties of gels [4–5]. Since this static structure governs the macroscopic mechanical performance of these materials, its characterization is of great importance.

The subject of the present work is the investigation of a complex gel system that exhibits structural nonuniformities at both molecular and supermolecular (colloidal) levels. We measure the concentration dependence of the osmotic swelling pressure and the elastic modulus of polyisoprene (PIP) networks swollen in toluene. We also determine the scattering response of the same gels and compare them with that of the corresponding uncross-linked solution.

THEORETICAL BACKGROUND

THERMODYNAMICS

According to the Flory–Huggins theory [6], the osmotic pressure of a polymer solution can be expressed as

$$\Pi = -(RT/v)[\ln(1 - \varphi) + (1 - P^{-1})\varphi + \chi\varphi^2] \qquad (15.1)$$

where v is the molar volume of the solvent, φ is the volume fraction of the polymer, P is the degree of polymerization, and χ is the interaction parameter. In general χ depends on the polymer volume fraction, i.e., $\chi = \chi_0 + \chi_1\varphi + \chi_2\varphi^2$.

The osmotic (mixing) contribution of the cross-linked polymer is assumed to have the same form as equation (15.1) with $P = \infty$. In addition, however, an elastic contribution arises from the deformation of the chains and acts in opposition to the mixing pressure. The elastic contribution, G, has the form

$$G = G_0 \varphi^m \tag{15.2}$$

where G_0 is the elastic modulus of the unswollen network, and $m = 1/3$.

The difference between these two yields the swelling pressure of the gel,

$$\omega = \Pi_{\mathrm{mix}} - G \tag{15.3}$$

where Π_{mix} is the mixing pressure of the cross-linked polymer.

SCATTERING

The intensity of radiation scattered elastically from a polymer solution is inversely proportional to the osmotic compressibility [7]

$$I(Q) = a(\Delta\rho^2 kT\varphi^2/K_{\mathrm{os}})(1 + Q^2\xi^2)^{-1} \tag{15.4}$$

where a is a constant, $K_{\mathrm{os}}[= \varphi(\partial\Pi/\partial\varphi)]$ is the osmotic compression modulus, ξ is the polymer–polymer correlation length, $\Delta\rho^2$ is a contrast factor. Q is the transfer wave vector, $Q = (4\pi/\lambda)\sin(\theta/2)$, where λ is the incident wavelength and θ is the scattering angle.

In gels, the scattered intensity is governed by the longitudinal osmotic modulus [8]

$$M_{\mathrm{os}} = K_{\mathrm{os}} + (4/3)G \tag{15.5}$$

The scattering function of a gel is the combination of at least two components [8,9]: a solution-like (dynamic) component described by equation (15.4) and, a static component that depends on the detailed mechanism of the cross-linking process. In the simplest case the static component can be approximated by a Debye–Bueche form factor [9] and the gel scattering spectrum takes the form [10]

$$I(Q) = \Delta\rho^2\{(kT\varphi^2/M_{\mathrm{os}})(1 + Q^2\xi^2)^{-1} + B(1 + Q^2\Xi^2)^{-2}\} \tag{15.6}$$

where Ξ is the radius of the static concentration fluctuations governed by the frozen-in elastic constraints. The amplitude of the second term in equation (15.6) can be shown to be $B = 8\pi\Xi^3\langle\delta\varphi^2\rangle$ where $\langle\delta\varphi^2\rangle$ is the mean square amplitude of the static concentration fluctuations.

EXPERIMENTAL [11]

GEL PREPARATION

The poly(isoprene) gels were made according to a method described by McKenna and Crissman [1]. A poly(*cis*-isoprene) sample (Polysciences, Inc.) was dissolved in toluene and mixed with differing amounts of dicumyl peroxide: 0.5, 2.0, 5.0 and 10.0% (w/w). The specimens were dried and then cured at 150 °C for 2 h.

OSMOTIC AND MECHANICAL MEASUREMENTS

The swelling pressure of the gels was determined as a function of the polymer concentration using a modified deswelling method [12]. Gels were equilibrated with poly(vinyl acetate) solutions ($M_w = 140\,000$) of known osmotic pressure. A semi-permeable membrane was inserted between the gel and the solution to prevent the diffusion of the polymer molecules into the swollen network.

Swollen network specimens were uniaxially compressed (at constant volume) between two parallel flat plates. The stress–strain data were determined in the range of deformation ratio $0.7 < \Lambda < 1$. The absence of volume change and barrel distortion was checked by determining the dimensions of the deformed and undeformed gel cylinders.

SMALL-ANGLE NEUTRON SCATTERING MEASUREMENTS

The SANS measurements were performed on the NG3 instrument at NIST, Gaithersburg, MD, using an incident wavelength of 6 Å and selector bandwidth of 10%. The detector was placed at three distances, 1.8 m, 6 m and 13 m from the sample. The Q range explored was $0.005\ \text{Å}^{-1} < Q < 0.23\ \text{Å}^{-1}$, with counting times of between 10 min and 1 h. The ambient temperature during the experiments was $25° \pm 1°$C. The solvent was deuterated toluene. The sample was placed between two 1 mm thick quartz windows separated by a 1 mm thick spacer, sealed by a Viton O-ring. Corrections for incoherent background, detector response and cell window scattering were made. Calibration of the scattered neutron intensity was performed using the signal from standard samples of silica aerogel [13].

RESULTS AND DISCUSSION

SWELLING AND MECHANICAL RESULTS

In Figure 15.1 the shear moduli of the polyisoprene/toluene gels are displayed as a function of polymer volume fraction in a double logarithmic plot. In this representation, according to equation (15.2), the slopes of the straight line through data sets from each sample (dotted line in Figure 15.1) should display a slope of 1/3. The experimental results are in agreement with this value.

The mixing pressure, Π_{mix}, in the swollen network can be expressed as $\Pi_{\text{mix}} = \omega + G$, where the measured values of G are shown in Figure 15.1. In Figure 15.2, the values of Π_{mix} are plotted as a function of φ. Within the experimental error, all these points fall on a single master curve, i.e., the mixing contribution is

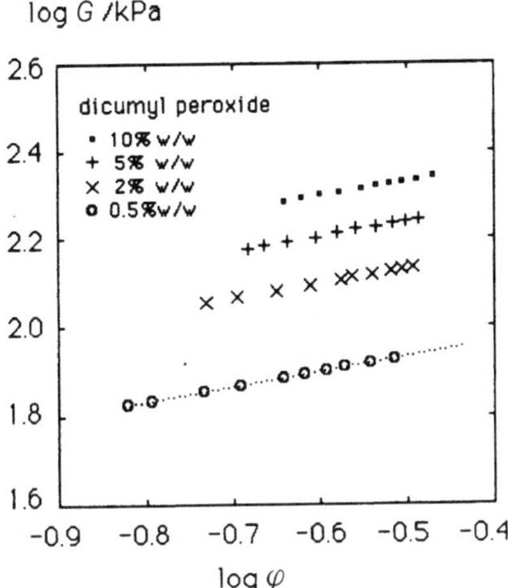

Figure 15.1 Shear modulus G of polyisoprene/toluene gels as a function of polymer volume fraction. The straight line is the best fit through the lowest data set. The slope is 0.332 ± 0.002

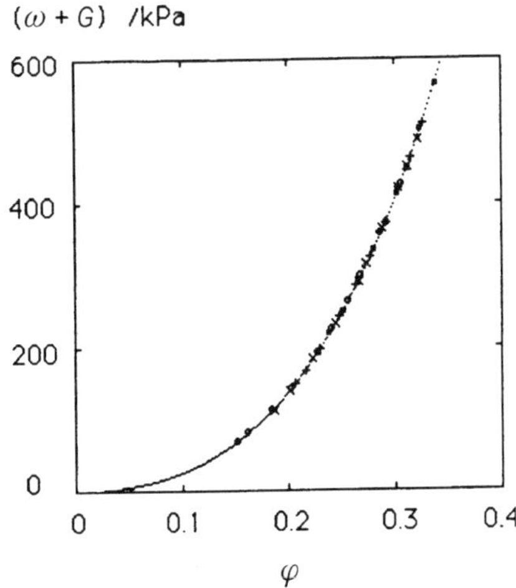

Figure 15.2 Mixing pressure, $\Pi_{mix} = \omega + G$ as a function of polymer volume fraction for polyisoprene/toluene gels. (Different symbols refer to gel samples prepared with different amounts of cross-linker)

practically independent of the cross-linking density. The continuous curve shows the least squares fit of $(\omega + G)$ to the Flory–Huggins equation which yields the following values for the interaction parameters: $\chi_0 = 0.427$, $\chi_1 = 0.000$, and $\chi_2 = 0.112$.

SANS MEASUREMENTS ON POLYISOPRENE/TOLUENE SYSTEMS

The scattering response of a semi-dilute solution of a non-associating polymer is described by equation (15.4). Polyisoprene/toluene solutions, however, display slight haze, indicating the presence of clusters (nanostructures). This is probably due to oxidation of polyisoprene. The size of these clusters is comparable with the wavelength of visible light. In Figure 15.3 is shown the SANS spectrum of a polyisoprene/toluene solution. In addition to the thermodynamic term (line a) described by equation (15.4), strong extra scattering is observed at low Q (line b). The continuous line shows the fit of equation (15.6). The average size of the large objects, Ξ, is 38.3 nm.

In Figure 15.4 the SANS spectrum is shown for a polyisoprene/toluene gel. The nanostructures that are formed prior to cross-linking in polyisoprene/toluene solutions are present in the gels (line b_2), and influence the SANS spectrum. The gel spectrum, however, contains another structural element due to frozen-in elastic constraints [4,5,14]: the latter can be described by concentration fluctuations of mean square amplitude $\langle \delta\varphi^2 \rangle$ with a spatial range Ξ_1 (line b_1). Since static contributions can be described by the Debye–Bueche formalism, it is natural to extend equation (15.6) by writing

$$I(Q) = \Delta\rho^2\{A(1 + Q^2\xi^2)^{-1} + B_1(1 + Q^2\Xi_1^2)^{-2} + B_2(1 + Q^2\Xi_2^2)^{-2}\} \quad (15.7)$$

where $A = kT\varphi^2/M_{os}$, $B_1 = 8\pi\langle\delta\varphi^2\rangle\Xi_1^3$, and B_2 is taken to be the intensity scattered by the clusters formed in the solution prior to cross-linking.

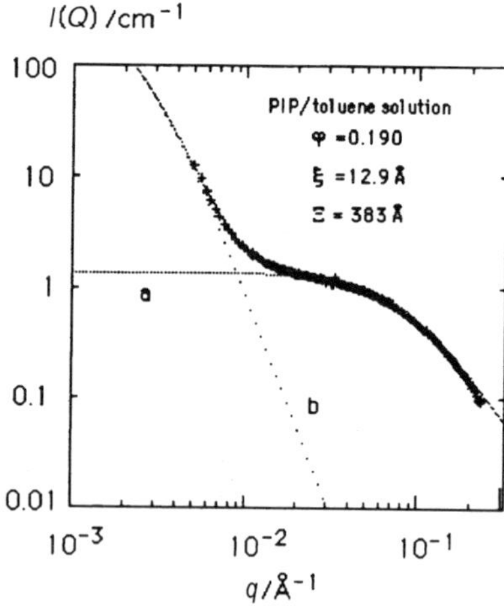

Figure 15.3 Small-angle neutron scattering spectrum of a polyisoprene/toluene solution. The heavy dotted line (a) is the first term in equation (15.6); the light dotted line (b) is the second term

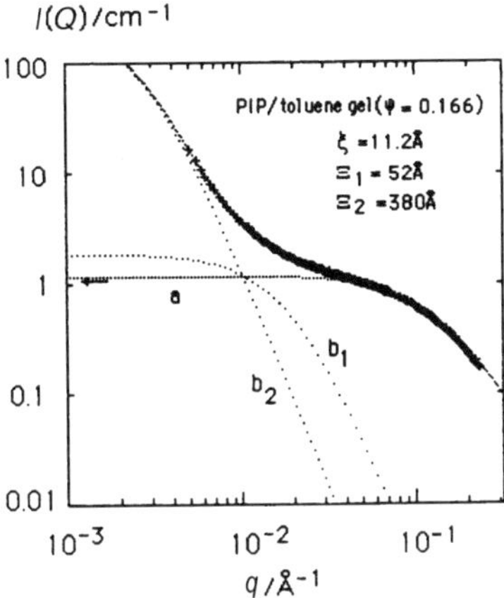

Figure 15.4 Small-angle neutron scattering spectrum of a fully swollen poly-isoprene/toluene gel. Dotted lines *a*, *b₁* and *b₂* show respectively the first, second and third terms of the least squares fit to equation (15.7). The arrow designates the intensity calculated from the swelling pressure measurements, corresponding to dotted line *a*

In Figure 15.4 the components of equation (15.7) are shown as dotted lines. The parameters obtained from the fits to the spectra of two gels and a solution are listed in Table 15.1.

From measurements of the swelling pressure and the elastic modulus the longitudinal osmotic modulus, M_{os}, can be determined independently

$$M_{os} = (\varphi^2 RT/v_1)\{1/(1 - \varphi) - 2\chi_0 - 3\chi_1\varphi - 4\chi_2\varphi^2\} + (4/3)G_0\varphi^{1/3} \qquad (15.8)$$

This allows us to compare the thermodynamic component in the SANS spectrum with the result derived from macroscopic observations. Substitution of the parameters χ_0, χ_1, χ_2 and G_0 into equation (15.8), along with the contrast factor, yields the theoretical intensity scattered by the thermodynamic concentration fluctuations in the swollen network.

The arrow placed against the ordinate axis shows the value of the intensity of the thermodynamic contribution calculated from the osmotic and mechanical measurements with equation (15.8). The agreement between the values of the

Table 15.1 Parameters from fits of equation (15.7) to SANS spectra

Sample	vol.fract. φ	$\Delta\rho^2 A$ /cm⁻¹	$\Delta\rho^2 B_1$ /cm⁻¹	$\Delta\rho^2 B_2$ /cm⁻¹	ξ /Å	Ξ_1 /Å	Ξ_2 /Å	$\langle\delta\varphi^2\rangle/\varphi$
gel 1	0.166	1.2	2.0	290	11.2	52	380	0.083
gel 2	0.224	1.0	1.8	260	8.8	43	370	0.079
solution 1	0.192	1.1	—	233	12.9	—	383	—

thermodynamic component obtained from the scattering spectra and that evaluated from the macroscopic measurements is satisfactory. We note that the gel signal can also be described by equation (15.6) (i.e. using only two terms), but both the fit to the scattering data and the agreement between the thermodynamic components are unacceptably poor. Furthermore, as described earlier, a third component is in any case expected on physical grounds because of the presence of cross-links.

CONCLUSIONS

Neutron scattering in polyisoprene/toluene gels reveals a hierarchy of characteristic lengths: the thermodynamic correlation length ξ, the elastic correlation length Ξ_1, and the correlation length that characterizes the clusters formed in the solution prior to cross-linking Ξ_2. The large scale structures in these gels thus consist of two contributions: the first is caused by the competition of the osmotic pressure and the residual elastic constraints frozen into the network, while the second is associated with the presence of clusters in this particular system.

ACKNOWLEDGMENTS

We acknowledge the support of the National Institute of Standards and Technology, US Department of Commerce, in providing the neutron research facilities used in the experiment. This work is based upon activities supported by the National Science Foundation under Agreement No. DMR-9423101. We are grateful to B. Hammouda for his invaluable help F.H. acknowledges grant number OTKA T016872 from the Hungarian Academy of Sciences.

REFERENCES

1. G.B. McKenna and J.M. Crissman, *J. Polym. Sci. Polym. Phys. Ed.*, **35**, 817 (1997).
2. J.F. Douglas and G.B. McKenna, in *Elastomeric Polymer Networks*, (Mark, J.E. and Erman, B., eds), Prentice Hall, Englewood Cliffs, NJ, (1992), p. 327.
3. F. Horkay and G.B. McKenna, Gels, in *Handbook of Polymer Properties* (J.E. Mark, ed.) American Institute of Physics (1996), p. 363.
4. F. Horkay, A.M. Hecht, S. Mallam, E. Geissler and A.R. Rennie, *Macromolecules*, **24**, 2896 (1991).
5. E. Geissler, F. Horkay and A.M. Hecht, *Phys. Rev. Lett.*, **71**, 645 (1993).
6. P.J. Flory, *Principles of Polymer Chemistry*, Cornell University Press, Ithaca, 1953.
7. P.G. de Gennes, *Scaling Concepts in Polymer Physics*, Cornell University Press, Ithaca, 1979.
8. L.D. Landau and E.M. Lifshitz, *Theory of Elasticity*, (2nd edn), Pergamon Press, Oxford, 1970.
9. P. Debye and R.M. Bueche, *J. Appl. Phys.*, **20**, 518 (1949).
10. E. Geissler, F. Horkay, A.M. Hecht, C. Rochas, P. Lindner, C. Bourgaux and G. Couarraze, *Polymer*, **38**, 15 (1997).

11. Certain commercial materials and equipments are identified in this article to specify adequately the experimental procedure. In no case does such identification imply recommendation or endorsement by the National Institute of Standards and Technology, nor does it imply necessarily that the product is the best available for the purpose.

12. F. Horkay and M. Zrínyi, *Macromolecules*, **15**, 1306 (1982).

13. NIST Cold Neutron Research Facility. NG3 and NG7 30-meter SANS Instruments Data Acquisition Manual, January 1996.

14. C. Rouf, J. Bastide, J.M. Pujol, F. Schosseler and J.P. Munch, *Phys. Rev. Lett.*, **73**, 830 (1994).

16

Structural Changes in Hydrophilic Networks in Water–Salt Solutions Studied by SANS

GENNADII EVMENENKO[1], VLADIMIR ALEXEEV[1], TATIANA BUDTOVA[2]*, ALEXANDER BUYANOV[2] and SERGEI FRENKEL[2]†

[1]Petersburg Nuclear Physics Institute, 188350 Gatchina, Russia

[2]Institute of Maromolecular Compounds, Russian Academy of Science, Bolshoi prosp. 31, 199004 St-Petersburg, Russia

---

* To whom correspondence should be addressed

† Deceased in November 1998

Wiley Polymer Networks Group Review Series Vol. 2. Edited by B.T. Stokke and A. Elgsaeter
© 1999 John Wiley & Sons Ltd

ABSTRACT

The structural changes in hydrogels immersed in water and water–salt solutions are studied by small angle neutron scattering technique. The sizes of heterogeneities calculated from scattering curves were analysed as a function of the chemical nature of the cross-linking, agent and of the gel capability to absorb water: degree of cross-linking, degree of ionization and salt concentration. The changes of the contrast (relative density between the inhomogeneity and the matrix) during gel swelling was also investigated. The results show that gels with potential high degrees of swelling have an inhomogeneous structure (dense areas near the cross-linking junction and the swollen matrix around it). This inhomogeneity becomes more pronounced when the gel is immersed in a salt solution. Hydrogels with not a very high absorption capability have a homogeneous structure and in a salt solution they contract as a unit.

INTRODUCTION

Polyelectrolyte gels are soft and gentle stimulus-responsive materials, so-called 'intelligent' polymers. In order to understand and predict their properties and swelling behaviour it is necessary to study their structure changes during swelling. In particular, the structural changes of strongly charged gels which are able to swell more than 1000 times are of a special interest. The mechanisms of gel swelling and contraction in different aqueous solutions are still under discussion (gel swelling/contracting as a unit or via microphase separation? [1]). Small angle neutron scattering is a convenient tool to measure structural characteristics in the range of 10–1000 Å [2,3].

In this paper we study how the structure of highly swelling hydrophilic networks changes when they are immersed in water and water–salt solutions. The goals of the work are:

- to study changes of hydrogel structure during swelling in water as a function of the chemical nature of the cross-linking agent and of the gel capability to absorb water: degree of cross-linking and degree of ionization;
- to investigate structural changes in gels immersed in aqueous salt solution.

MATERIALS AND METHODS

The samples were hydrogels based on polyacrylamide of different degrees of ionization cross-linked by macromolecular compounds: allylhydroxy-ethylcellulose (AHEC), allylcarboxymethylcellulose (ACMC) and allyldextran (AD). The sample synthesis is described in detail by Buyanov *et al.* [4]. We also used gels cross-linked by N, N'-methylenebisacrylamide (MBA), in order to check if the results obtained depend on the molecular weight and functionality of the cross-linking agent. The sample characteristics are given in Tables 16.1 and 16.2.

Table 16.1 Samples for studies of gel swelling in D_2O

Cross-linking agent	Degree of cross-linking, wt%	Maximal degree of swelling, g/g	Degree of ionization, mole%
Allylhydroxyethylcellulose (AHEC):			
AHEC, $M = 10^5$	0.7	500	50
AHEC, $M = 10^5$	0.7	150	10
Allyldextrane (AD):			
AD1, $M = 1.5\ 10^4$	1.0	1400	100
AD2, $M = 5\ 10^5$	1.0	1200	100
AD3, $M = 5\ 10^5$	5.0	400	100
AD4, $M = 5\ 10^5$	25.0	50	100
N, N'-Methylenebisacrylamide (MBA):			
MBA[a]	0.1	550	75
MBA[a]	0.1	33	75
gel swollen in 1 M NaCl			

[a]This gel was kindly provided by Atochem, a French company

Table 16.2 Samples for studies of gel swelling in NaCl solutions

Cross-linking agent	Degree of cross-linking (wt%)	Degree of swelling at equilibrium g/g		
		Neutral gels in D_2O	Ionized gels (100%)	
			inD_2O	in 0.2 M NaCl
Allylhydroxyethylcellulosc (AHEC):				
AHEC1	0.7	12	300	50
AHEC2	0.3	18	700	70
Allylcarboxymethylcellulose (ACMC):				
ACMC	0.3	14	1300	60
N, N'-Methylenebisacrylamide (MBA):				
MBA	0.01	11	430	50

Hydrogels of different degrees of swelling were prepared by direct mixing of dry gel particles and heavy water (or solutions of NaCl in heavy water). The degree of swelling was calculated simply as $Q = W/W_0$, where W and W_0 are the weights of a swollen and dry gel, respectively.

The SANS measurements were performed on the small angle neutron diffractometer 'Membrana-2', St-Petersburg Nuclear Physics Institute, Russia. The average wavelength of the incident beam was 3 Å and the spectral half-width was varied from 0.1 to 0.3. The experimental data were processed taking into account spectral and collimation distortions of the diffractometer, detector efficiency, sample transmission and incoherent background scattering.

The scattering intensities were converted to the absolute differential scattering cross-sections per unit sample volume using H_2O calibration data.

RESULTS AND DISCUSSION

THEORETICAL APPROACH

The scattering patterns from chemically cross-linked gels were analysed from the point of view of additive contribution from a homogeneous, liquid-like (dynamic) fluctuating gel matrix, and embedded heterogeneities having a high cross-linking density. The scattering equation for an isotropic substance may be written as the spatial Fourier transform of the correlation function $\gamma(r)$ [5]:

$$I(q) = \int_0^\infty \gamma(r)\frac{\sin(qr)}{qr}r^2 \, dr \qquad (16.1)$$

where $q = (4\pi \sin\theta)/\lambda$ is the scattering vector, 2θ is the scattering angle.

To analyse the scattering intensity in hydrogels we used the correlation function proposed by Mallam *et al.* [6]:

$$\gamma(r) = \langle c \rangle^2 \frac{\xi}{r} \exp\left(-\frac{r}{\xi}\right) + \langle \Delta c^2 \rangle \exp\left(-\frac{r^2}{2\Xi^2}\right) \qquad (16.2)$$

where Ξ is the characteristic mean size of the static (frozen) heterogeneity in a gel; ξ is the correlation length of polymer–polymer interactions, which take place between the fluctuating chains of the gel network, $\langle c \rangle$ is the ensemble-averaged value of a local polymer concentration, $\langle \Delta c^2 \rangle$ is the mean square amplitude of the concentration fluctuations in the gel.

The Fourier transform of equation (16.2) gives, for the scattering function,

$$I(q) = I_G(0)\exp\left(-\frac{q^2\Xi^2}{2}\right) + \frac{I_L(0)}{1+q^2\xi^2} \qquad (16.3)$$

where $I_L(0) \cong \langle c \rangle^2 \xi^3 (2/\pi)^{0.5}$ and $I_G(0) \cong \Xi^3 \langle \Delta c^2 \rangle$ are linear coefficients of the Lorentzian and Gaussian terms, respectively. Thus, the model of the scattering system contains four independent variables, $I_L(0)$, $I_G(0)$, Ξ and ξ, which were calculated from the best agreement between the experimental SANS curves and simulated ones (equation (16.3)).

The data treatment showed that for highly swelling gels each term in equation (16.3) contributed to different parts of a scattering curve [7]. The information about the excess scattering due to the presence of cross-links can be extracted from the region of low q values $(q < 0.02 \text{ Å}^{-1})$ and variation of ξ from 5 to 25 Å in the second term does not influence (within the experimental errors) the values of Ξ. Thus we assumed that for obtained scattering patterns the second term may be considered as a 'background' and we followed only the changes of the size of solid-like heterogeneities. A mean value of Ξ was calculated for each degree of swelling.

Additional information on how the dense areas around cross-linking junctions are changing during gel swelling can be obtained from the dependence of a contrast $\Delta\rho$ (a relative density between the inhomogeneity and the matrix) on the degree of swelling:

$$\Delta\rho^2 \sim I_G(0)/(V^2 N) \tag{16.4}$$

where N is the number of scattering centres in the volume unit (roughly proportional to $1/Q$), V is the volume of a heterogeneity ($\sim \Xi^3$).

HYDROGEL SWELLING IN WATER (D_2O)

The scattering curves for gels cross-linked by AHEC (gel with 50% of ionised groups), AD1, AD2 and MBA (see Table 16.1) and swollen in D_2O at 20 g/g are given in Figure 16.1. The solid lines correspond to the calculated dependences according to equation (16.3). In Figure 16.2 the SANS data for hydrogel cross-linked by AD2 swollen at different degrees in D_2O are presented. Similar results were obtained for other samples. In all the cases the curves were smooth, not showing any order peak.

The changes of the size of heterogeneities Ξ as a function of gel degree of swelling Q are presented in Figure 16.3 for gels of different degrees of ionization,

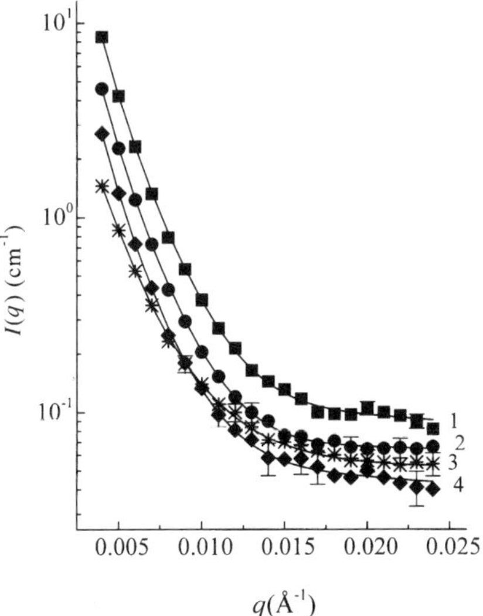

Figure 16.1 SANS data for hydrogels cross-linked by AD1 (1), AD2 (2), MBA (3) and AHEC (gel degree of ionization 50%) (4) and swollen at 20 g/g in D_2O. The experimental errors are shown by error bars or coincide with the size of the points

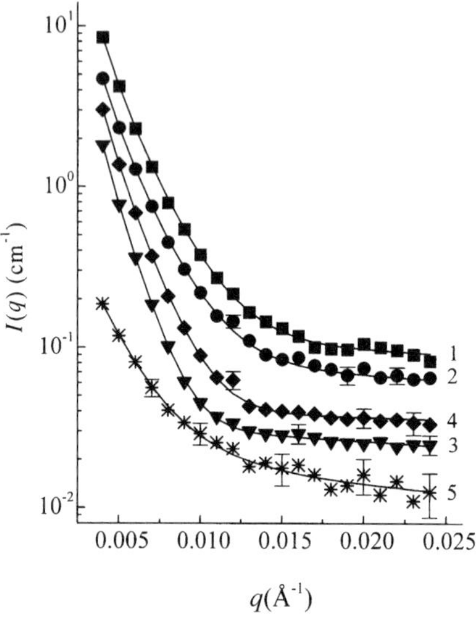

Figure 16.2 SANS curves for hydrogel cross-linked by AD2 and swollen in D_2O at 20 g/g (1), 70 g/g (2), 200 g/g (3), 500 g/g (4) and 1000 g/g (5)

cross-linked by AHEC and in Figure 16.4 for gels cross-linked by different low and high molecular weight compounds. The characteristics of gels used in this part are given in Table 16.1.

The dependence of the size of heterogeneities Ξ on Q shows a maximum for

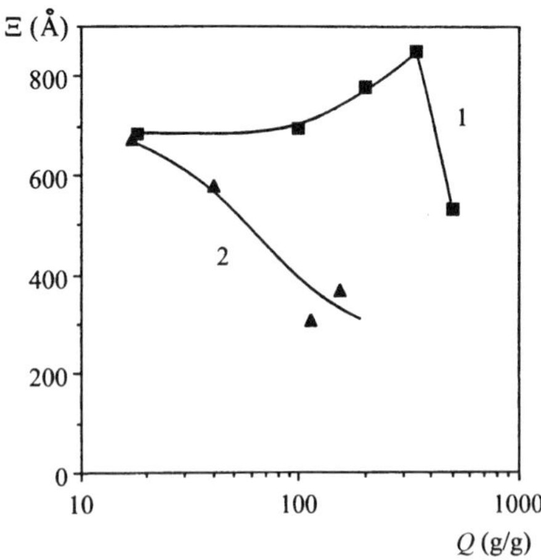

Figure 16.3 Size of heterogeneities vs degree of swelling for hydrogels cross-linked by AHEC with 50% (1) and 10% (2) ionized groups

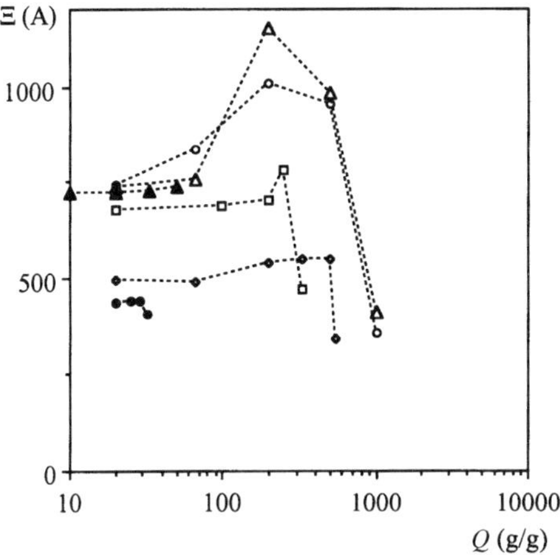

Figure 16.4 The same for gels cross-linked by AD1 (O), AD2 (△), AD3 (□), AD4 (▲), MBA (◇) and by MBA, in 1 M NaCl solution (●)

gels with high absorption ability (>200 g/g). There is no maximum for gels with low degrees of swelling (which is a result of either high degree of cross-linking, or low degree of ionization, or gel swollen in a salt solution).

The vicinities of cross-linking junctions are enriched with entangled sodium polyacrylate macromolecules and the increase of heterogeneities size is due to the polyelectrolyte swelling mechanism (similar to the viscosity-concentration dependence for polyelectrolytes). It becomes apparent when a gel is strongly charged and has a low degree of cross-linking. The result does not depend on the chemical nature or the molecular weight of the cross-linking agent.

The influence of the chemical nature of the cross-linking agent on hydrogel structural properties can be seen from the dependence of the contrast on gel degree of swelling (see Table 16.3). We took the value of the contrast for each gel swollen at 20 g/g as a point of reference ($(\Delta\rho)_{Q=20}$) and calculated how $\Delta\rho$ is changing during swelling. The contrast is increasing with the increase of gel swelling for gels cross-linked by macromolecular compounds. It is practically not changing for gel cross-linked by MBA. This means that hydrogels cross-linked by compounds of low functionality and low molecular weight are homogeneously swelling in water. On the contrary, the structure of hydrogels cross-linked by macromolecular agents becomes less and less homogeneous with gel swelling. At equilibrium it is a network with dense areas around junction points and a swollen matrix. The more is the gel absorption capability, the less homegeneous is its structure.

HYDROGEL SWELLING IN NaCl SOLUTIONS

In this part we also used gels with different absorption capability, but cross-linked

Table 16.3 Changes of contrast for hydrogels swelling in D_2O

Cross-linking agent	Degree of swelling (g/g)	Relative contrast $\Delta\rho/(\Delta\rho)_{Q=20}$
AD1 ($Q_{max} = 1400$ g/g)	20	1
	70	1.6
	200	1.6
	500	3.3
	1000	3.6
AD2 ($Q_{max} = 1200$ g/g)	20	1
	70	1.3
	200	1.1
	500	2.3
	1000	2.3
AD3 ($Q_{max} = 400$ g/g)	20	1
	100	1.1
	200	1.9
	250	2.0
	330	1.5
AD4 ($Q_{max} = 50$ g/g)	20	1
	33	1.4
	50	1.4
MBA ($Q_{max} = 550$ g/g)	20	1
	70	0.9
	200	1.0
	330	0.8
	500	0.8
MBA (gel swollen in 1M) NaCl, $Q_{max} = 33$ g/g)	20	1
	25	0.9
	29	0.9
	33	0.8

by allylhydroxyethylcellulose (AHEC), allylcarboxymethylcellulose (ACMC) and N,N'-methylenebisacrylamide (MBA) (see Table 16.2). First, neutral gels were prepared and the sizes of heterogeneities in D_2O were calculated from the scattering data. Then gels were neutralized, swollen at their maximum and SANS measurements were performed. Finally, neutralized hydrogels were immersed in $NaCl–D_2O$ solutions of different concentrations.

In all the cases gels were swollen up to the equilibrium state. The SANS measurements were performed and the scattering data were analysed in the same way as for gels swollen in D_2O solutions. The scattering curves did not show any order peak for gels swollen at the studied salt concentrations.

The dependencies of the size of heterogeneities on NaCl concentration C_{NaCl} are presented in Figure 16.5. In the same graph the values of Ξ for neutral gels are also given. For all the samples the values of Ξ for ionized gels in D_2O are several times higher than for neutral ones. It is known that the presence of

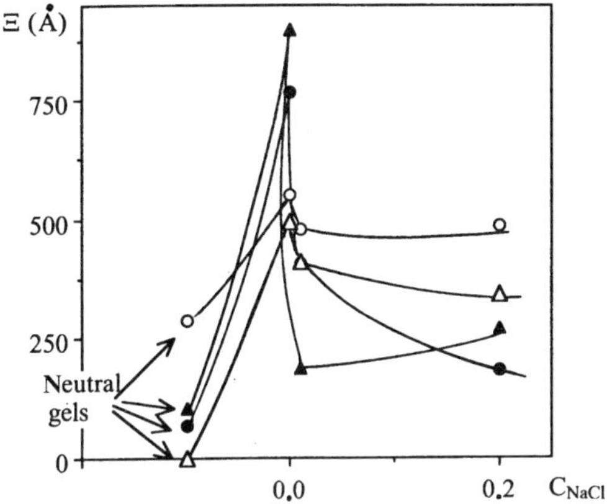

Figure 16.5 Dependence of the size of heterogeneities on the salt concentration

charged units on a polymer chain induces several new physico-chemical properties. SANS results show that areas enriched with polymer chains around junction points are acting as 'microgels', swelling more when they are charged. When a polyelectrolyte gel is immersed in a salt solution the polyelectrolyte effect is suppressed, the gel contracts and the size of inhomogeneities decreases.

The summary of how gel degree of swelling, volume of heterogeneities and the contrast change in NaCl solution as compared with pure D_2O is given in Table 16.4. Two remarkable trends can be noted:

1. In gels having *not a high absorption capability* (cross-linked by AHEC1 and MBA) the volume of heterogeneities is changing less than the degree of swelling and the contrast in water is higher than contrast in salt solution.
2. In gels having a *high absorption capability* (cross-linked by AHEC2 and ACMC) the volume of heterogeneities is changing in 40–70 times, much more than the degree of swelling, and the contrast in salt solution is higher than in water.

The interpretation of this phenomena can be as follows. In paper [1] two possibilities of gel contraction in a bad solvent are discussed: (i) when gel contracts

Table 16.4 Changes of gel degree of selling Q, volume of heterogeneity V and contrast $\Delta\rho$ as a function of salt concentration

Sample	Q_{D_2O} (g/g)	$\dfrac{Q_{D_2O}}{Q_{0.2N\ NaCl}}$	$\dfrac{V_{D_2O}}{V_{0.2N\ NaCl}}$	$\dfrac{(\Delta\rho)_{D_2O}}{(\Delta\rho)_{0.2N\ NaCl}}$
AHEC1	300	6	1.5	2.0
AHEC2	700	10	40	0.6
ACMC	1300	21	70	0.5
MBA	430	10	3	1.3

as a unit and (ii) via microphase separation, when microdomains of contracted areas are surrounded by a swollen matrix. It was shown that the changes of translational entropy of mobile ions favor the second mechanism of contraction.

SANS results show that hydrogels not having a high absorption capability seem to contract in a salt solution as a unit. On the other hand, gels with high absorption capability seem to contract through a microphase separation.

CONCLUSIONS

The structural changes in hydrogels swelling in water and water–salt solutions depend mainly on their absorption capability. Those gels which potentially can swell in water more than 400–500 times have an inhomogeneous structure (dense areas in the vicinity of a cross-linking junction and a swollen matrix around) and they contract in salt solutions through a microphase separation. Hydrogels with a not very high absorption capability have a homogeneous structure and they contract in a salt solution as a unit.

ACKNOWLEDGMENTS

The authors are grateful to the Russian Foundation of Fundamental Research, grant N 96-03-33852a, for the financial support of this work.

REFERENCES

1. K.B. Zeldovich, E.E. Dormidontova, A.R. Khokhlov and T.A. Vilgis, *J. Phys. II*, **7(4)**, 627–635 (1997).
2. E. Geissler, in '*Neutron and Synchrotron Radiation for Condensed Matter Studies*, 3, *Application to Soft Condensed Matter and Biology*, (J. Baaruchel, J.L. Hodeau, M.S. Lehmann, J.R. Regnard and C. Schlenker, eds) Les Editions de Physique, Springer Verlag, 1994, p. 7–22.
3. J. Bastide and S.J. Candau, in *Physical Properties of Polymeric Gels* (J.P. Cohen-Addad, ed.), John Wiley, New York, 1996, pp. 143–308.
4. A.L. Buyanov, L.G. Revelskaya, G.A. Petropavlovsky, M.F. Lebedeva, S.K. Zakharov, V.A. Petrova and Nud'ga, *Polymer Science* (translation of Vysokomolec. Soed., Russia), **B31**, 883–887 (1989).
5. P. Debye and A.M. Bueche, *J. Appl. Phys.*, **20**, 518–525 (1949).
6. S. Mallam, A.M. Hecht and E.J. Geissler, *J. Chem. Phys.*, **91**, 6447–6451 (1989).
7. G. Evmenenko, V. Alexeev, T. Budtova, A. Buyanov and S. Frenkel, *Polymer.*, **40**, 2981–2985 (1999).

17

Styrene–Butadiene Copolymers: Mechanical Characterization and Molecular Orientation

L. BOKOBZA and C. MACRON
Laboratoire de Physico-Chimie Structurale et Macromoléculaire ESPCI,
10 rue Vauquelin, 75231 Paris, Cedex 05, France

ABSTRACT

Results of equilibrium stress–strain and swelling experiments are reported for styrene–butadiene copolymers of varying crosslink density. The orientation of polymer chains is investigated under uniaxial elongation by birefringence and infrared dichroism spectroscopy which monitor orientation on a segmental scale.

Wiley Polymer Networks Group Review Series Vol. 2. Edited by B.T. Stokke and A. Elgsaeter
© 1999 John Wiley & Sons Ltd

INTRODUCTION

The study of cross-linked elastomeric networks and essentially the analysis of segmental orientation offers unique insights into the physics of rubber elasticity.

While the strain dependence of the stress is the most common quantity to assess elasticity theories, methods used to measure orientation such as deuterium nuclear magnetic resonance (^{2}H-NMR), Fourier transform (FTIR) dichroism or birefringence are particularly informative for understanding the physics of dense polymeric media. These techniques directly probe the orientation behavior of network chains on a segmental scale.

Both the FTIR [1,2] and the ^{2}H-NMR techniques [3–6] directly measure the orientation of specific labels on a chain relative to a laboratory-fixed axis, the orientation being suitably induced by stretching the sample uniaxially.

In this paper, in addition to stress–strain experiments, infrared dichroism and birefringence are used to characterize the orientation behavior of styrene–butadiene copolymers. Swelling measurements are also carried out in order to get an indirect estimation of the cross-link density of the networks.

EXPERIMENTAL PART

SAMPLES

Four networks with styrene–butadiene copolymers exhibiting varying cross-link density were investigated.

The styrene–butadiene copolymer (Cariflex S1502 from Shell) contains 23.4 wt% of styrene units randomly distributed in the chain. The microstructure of the butadiene phase is the following: 9.1% cis, 54.5% trans, 13% 1,2. These networks were cross-linked with sulfur for 50 min at 150 °C under a pressure of 150 bars. The formulation and the vulcanization characteristics of the samples are compiled in Table 17.1. The amount of sulfur and of CBS were varied in order to obtain samples of different cross-link density.

METHODS OF INVESTIGATION

All experiments reported here were performed at room temperature.

Table 17.1 Formulations of the rubber compounds

Ingredients (phr)	SBR 1	SBR 2	SBR 3	SBR 4
Rubber	100	100	100	100
Sulfur	0.75	1.51	1.51	3.01
Diphenyl guanidine (DPG)	1.99	1.99	1.99	1.99
Zinc oxide	2.49	2.49	2.49	2.49
Stearic acid	1.51	1.51	1.51	1.51
Cyclohexyl benzothiazole sulfenamide (CBS)	0.89	1.78	0.89	3.56

Stress–strain measurements reported here were carried out by simply streching strips of $40 \times 10 \times 0.2$ mm^3 between two clamps by means of a sequence of increasing weights attached to the lower clamp. The distance between two marks is measured with a cathetometer after allowing sufficient time for equilibration.

To determine the equilibrium swelling of the vulcanizate, a sample of 20 mm $\times$ 10 mm $\times$ 0.2 mm was put into cyclohexane. After 72 h at room temperature, the sample was taken out of the liquid, the cyclohexane removed from the surface and the weight determined. The weight swelling ratio, Q, was also determined from the lengths of the sample in the unswollen and swollen states.

Infrared spectra were recorded with an FTIR spectrometer (Nicolet Model 210) with a resolution of 4 cm^{-1} and an accumulation of 32 scans.

Birefringence was measured by using an Olympus BHA polarizing microscope fitted with a Berek compensator. The thickness of the sample was obtained with a micrometer comparator and averaged all along the specimen.

THEORETICAL BACKGROUND

The stress–strain behavior can be analyzed according to the affine and phantom network models which are the simplest approaches of rubber elasticity used to relate the state of deformation at a molecular level to the externally applied macroscopic deformation [7,8].

In an affine network, the junction points are assumed to be embedded in the network and undergo affine displacements. The true stress (force divided by the deformed area) A is defined, for dry networks formed in the bulk state, as:

$$\sigma = \frac{\nu K T}{V}(\alpha^2 - \alpha^{-1}) = \frac{\rho k T \mathcal{N}_A}{M_c}(\alpha^2 - \alpha^{-1}) \qquad (17.1)$$

where

ν/V = number of chains per unit volume;
α = extension ratio defined as the ratio of the final length of the sample in the direction of stretch to the initial length before deformation;
T = absolute temperature;
ρ = density of the network;
M_c = average molecular weight between cross-links;
$\mathcal{N}_A$ = Avogadro number.

The other extreme case is the phantom behavior where the junction points fluctuate over time without being hindered by the presence of the neighboring chains. The fluctuations about the mean positions are not affected by the macroscopic state of deformation. The true stress for a phantom network is given by

$$\sigma = \left(1 - \frac{2}{\phi}\right)\frac{\nu k T}{V}(\alpha^2 - \alpha^{-1}) \qquad (17.2)$$

where ϕ is the junction functionality.

Both models assume that all chains are elastically active and the stress of an affine network, which does not depend on the junction functionality, is twice that predicted from the phantom theory for a tetrafunctional network.

Other theories have been developed taking into account the effect of entanglements which result from the uncrossability of network chains. In the constrained-junction model described by Flory and Erman [9], the entanglements are assumed to diminish the size of the fluctuation domains of the junctions, Edwards and Vilgis [10] extend the concept of the tube model to the study of cross-linked systems.

The average chain length M_c can be estimated from swelling measurements by using the following equation:

$$M_c = -\frac{\rho(1 - 2/\phi)V_1 v_{2m}^{1/3}}{\ln(1 - v_{2m}) + \chi v_{2m}^2 + v_{2m}} \tag{17.3}$$

In this expression, similar to the well-known Flory–Rehner equation [11] based on the affine network model, ρ denotes the network density during formation, V_1 is the molar volume of solvent, v_{2m} is the volume fraction of polymer at conditions of equilibrium and χ is the interaction parameter for the solvent–polymer system. The factor $(1 - 2/\phi)$ comes from the fact that at high degree of swelling Q (equal to v_{2m}^{-1}) and defined as the ratio (volume of the network plus solvent/volume of the dry network), the system may be treated essentially as a phantom network.

Measurements of strain birefringence of deformed networks is also an interesting technique which can be used to assess elasticity theories [12,13]. According to the theory, the birefringence for uniaxial extensions is related to the strain function by the expression:

$$\Delta n = \mathscr{P}\frac{\nu k T C}{V}(\alpha^2 - \alpha^{-1}) \tag{17.4}$$

where $\mathscr{P} = 1$ for an affine network and $(1 - 2/\phi)$ for a phantom network and C is the stress-optical coefficient which is related to the optical anisotropy Γ_2 of the network through the following equation:

$$C = \frac{2\pi(n^2 + 2)^2\Gamma_2}{27nkT} \tag{17.5}$$

n being the mean refractive index. C is usually referred to in the literature as the stress-optical coefficient since:

$$C = \Delta n/\sigma \tag{17.6}$$

A quantity characterizing the orientational birefringence of polymers is the intrinsic birefringence, $[\Delta n]_0$, which is defined as follows:

$$[\Delta n] = [\Delta n]_0\langle P_2(\cos\theta)\rangle \tag{17.7}$$

$\langle P_2(\cos\theta)\rangle$ is the second Legendre polynomial given by

$$\langle P_2(\cos\theta)\rangle = (3\langle\cos^2\theta\rangle - 1)/2 \tag{17.8}$$

where θ is the angle between the macroscopic reference axis (usually taken as the direction of strain) and the local chain axis of the polymer. The angular brackets indicate an average over all molecular chains and over all possible conformations of these chains.

$[\Delta n]_0$ may be called the maximum birefringence because the perfect orientation corresponds to $\langle P_2(\cos\theta)\rangle$.

Infrared dichroism spectroscopy is an independent technique to determine the second Legendre polynomial. This technique directly measures the orientation of electric dipole-transition moments associated with particular molecular vibrations. It is based on the determination of the dichroic ration of a selected absorption band for a deformed network. For a network under simple tension, this ration is defined as $R = A\,\|\,/A\perp$ ($A\,\|$ and $A\perp$ being the absorbances of the investigated band, measured with radiation polarized parallel and perpendicular to the stretching direction, respectively) [14, 15].

The orientation of the transition moment vector with respect to the direction of stretch is expressed in terms of the second Legendre polynomial $\langle P_2(\cos\gamma)\rangle$ related to the dichroic ratio by the following expression:

$$\langle P_2(\cos\gamma)\rangle = \frac{R-1}{R+2} \tag{17.9}$$

On the other hand, the orientation of the local chain axis with respect to the stretching direction is given by

$$\langle P_2(\cos\theta)\rangle = \frac{2}{3\cos^2\beta - 1}\frac{R-1}{R+2} \tag{17.10}$$

where β is the angle between the transition moment vector of the vibrational mode considered and the local chain axis of the polymer (Figure 17.1). Equation (17.10) can be rearranged using equation (17.9) as:

$$\langle P_2(\cos\gamma)\rangle = \langle P_2(\cos\theta)\rangle\langle P_2(\cos\beta)\rangle \tag{17.11}$$

where $\langle P_2(\cos\beta)\rangle = (\tfrac{1}{2})(3\cos^2\beta - 1)$.

The orientation function $\langle P_2(\cos\gamma)\rangle$, which characterizes the orientation of the transition moment vector relative to the direction of stretch in uniaxially deformed networks, is obtained form the expression

$$\langle P_2(\cos\gamma)\rangle = \mathscr{P}D_0(\alpha^2 - \alpha^{-1}) \tag{17.12}$$

where the front factor $\mathscr{P}$ equals 1 for the affine network model and to $(1 - 2/\phi)$ for a phantom network model and D_0 is the configurational factor defined by

$$D_0 = (3\langle r^2\cos^2\Phi\rangle_0/\langle r^2\rangle_0 - 1)/10 \tag{17.13}$$

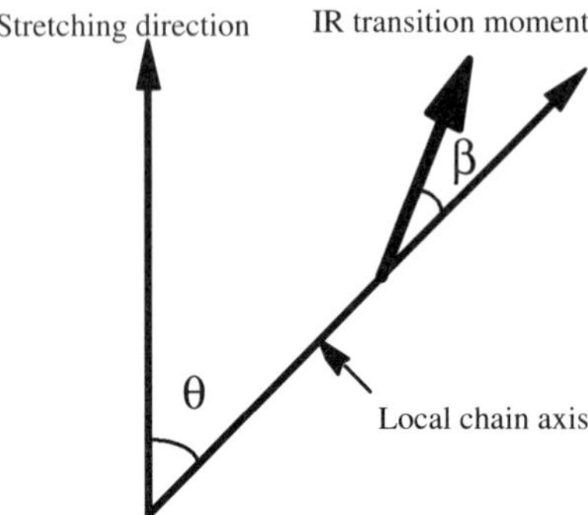

Figure 17.1 Positions of local chain axis and transition moment with respect to the stretching direction

Here ϕ is the angle between the transition moment vector whose orientation is being considered and the chain end-to-end vector r. The averaging is performed for unconstrained chains. The moments $\langle r^2 \rangle_0$ and $\langle r^2 \cos^2 \Phi \rangle_0$ in equation (17.13) can be calculated by using the matrix generation technique of the rotational isomeric state formalism. The configurational factor D_0 which incorporates the structural features of the network chains is inversely proportional to the number n of bonds in the chain between two junctions [16].

The specificity of infrared absorption bands to particular chemical functional groups makes infrared dichroism especially attractive for a detailed study of segmental orientation simultaneously occurring in either component of copolymeric chains composed of monomeric units, provided one or several absorption bands are specific of each component. In a theoretical study dealing with the effects of intrinsinc structural and conformational properties on segmental orientation in uniaxially deformed copolymers, Bahar *et al.* [17] pointed out the importance of the local structure and of the conformational characteristics of the different segments along the chain backbone.

RESULTS AND DISCUSSION

STRESS-STRAIN AND SWELLING MEASUREMENTS

The relation between stress (force divided by the deformed area) and the strain function is shown in Figure 17.2 for the four networks investigated in this study. The points represent equilibrium experimental data.

The quantity most often used to analyze results of stress–strain measurements in uniaxial deformation is the reduced stress $[\sigma^*] = \sigma/(\alpha^2 - \alpha^{-1})$ [18–20]. Real networks are frequently described by the Mooney-Rivlin equation:

$$[\sigma^*] = 2C_1 + 2C_2\alpha^{-1} \tag{17.14}$$

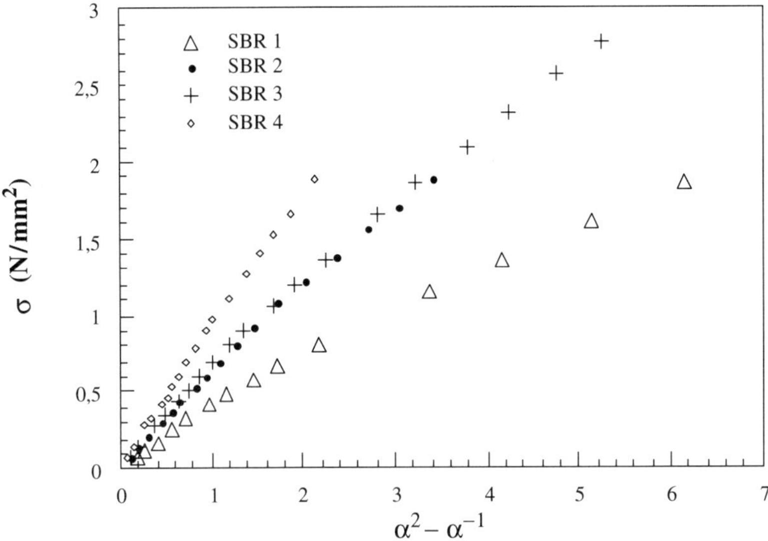

Figure 17.2 Stress–strain curves at room temperature

in which $2C_1$ and $2C_2$ are constants independent of α. The term $2C_1$ has been identified as the high-deformation modulus (the phantom network model limit) while $(2C_1 + 2C_2)$ is an estimate of the small strain modulus (the affine network model limit). So the phantom behavior is approached at large extensions $(\alpha^{-1} \to 0)$

As a typical example, the reduced stress for the SBR 2 sample, is given in Figure 17.3 as a function of inverse extension ratio. The molecular weights between cross-links, M_c, are determined, for the four samples, from the value obtained at $\alpha^{-1} = 0$ of $2C_1 (2C_1 = \frac{1}{2}\rho RT/M_c$ by assuming that cross-links are tetrafunctional).

It was deduced empirically that cross-link densities (proportional to $1/M_c$) are linear with Q^{-2} (Q being the equilibrium weight swelling ratio) [21]. A linear relation has been effectively observed between $1/M_c (M_c$ determined from the phantom modulus given by $2C_1$) and Q^{-2} (Figure 17.4). In Table 17.2,

Table 17.2 Analysis of stress–strain behavior of the SBR samples and equilibrium swelling ratios Q in cyclohexane

Sample	$2C_1 (\mathrm{N/mm}^2)$	$2C_2 (\mathrm{N/mm}^2)$	$M_c^{\mathrm{Ph}} (\mathrm{g/mol})$	Q
SBR 1	0.17	0.33	6564	4.43
SBR 2	0.38	0.36	2981	3.53
SBR 3	0.34	0.46	3374	3.69
SBR 4	0.62	0.43	1843	2.84

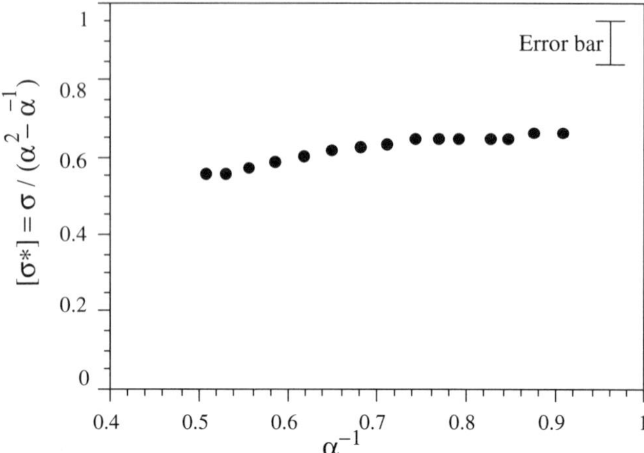

Figure 17.3 Reduced stress versus reciprocal elongation

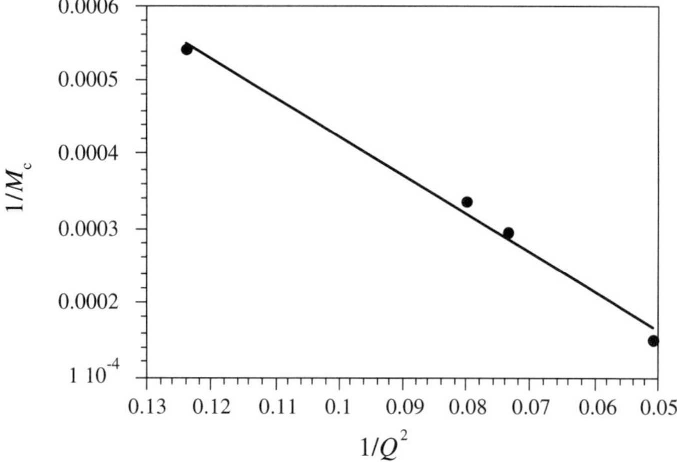

Figure 17.4 Relation between M_c^{-1} and Q^{-2} (Q being the equilibrium swelling ratio for the cyclohexane swollen gels)

the Mooney–Rivlin constants $2C_1$ and $2C_2$ are listed together with the molecular weight between junctions M_c^{Ph} obtained from the phantom limit $2C_1$ (equation (17.14)).

INFRARED DICHROISM AND BIREFRINGENCE MEASUREMENTS

We have examined the dichroic behavior of the bands located at 1640 and 1493 cm^{-1} respectively ascribed to the C=C stretching vibration of the vinyl unit and to the benzene ring vibration ν_{19a} (in Wilson's notation [22]). On account of the large force constant of the double bond, the transition moment direction corresponding to the band at 1640 cm^{-1} is localized along the C=C chemical bond of the vinyl lateral group. On the other hand, in the case of a C_{2v} local symmetry of the ring, the transition moment of the band at 1493 cm^{-1}, related to a vibration of species A_1, lies along the binary molecular axis of the aromatic group. The transition moment vectors as well as the choice of the baselines for the determination of the dichroic ratios, are given in Figure 17.5. The transition moment associated with the band at 1493 cm^{-1} of the benzene ring, is perpendicular to a local chain axis, this local chain axis being defined as the line connecting two successive backbone atoms.

Figure 17.6 displays, for all the samples, the dichroic functions of the two bands against the strain function $(\alpha^2 - \alpha^{-1})$. According to equation (17.9), these dichroic functions given by the experimental quantity $(R - 1)/(R + 2)$ represent the orientation $\langle P_2(\cos \gamma) \rangle$ of the two transition moment vectors relative to the direction of stretch. This way of plotting the data precludes any assumption concerning the local chain axis which is a rather fictitious entity, defined as an axis of cylindrical symmetry with respect to the transition moment vector. It is worthwhile to notice that, in practice, in data interpretation, the orientation of

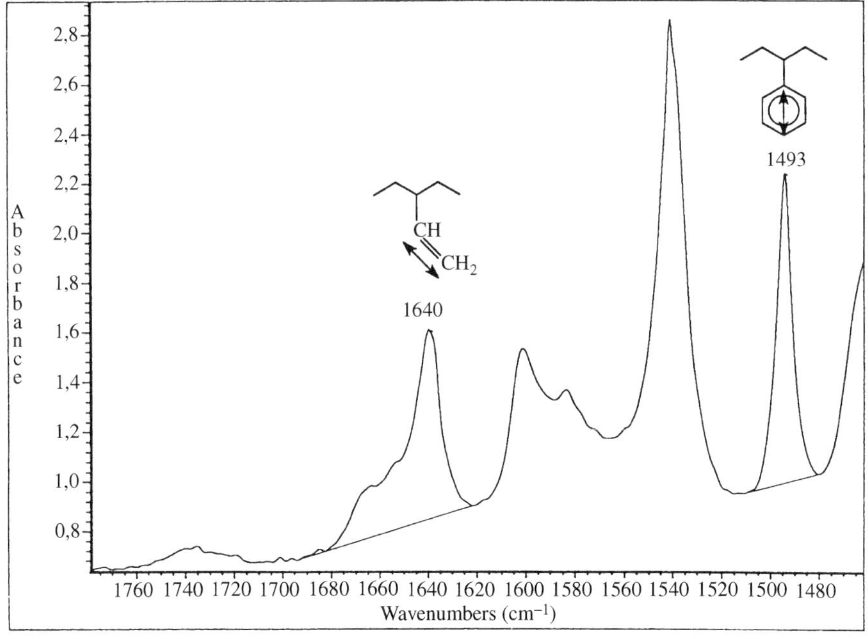

Figure 17.5 Infrared spectrum of an SBR rubber : transition moments associated with the investigated bands and baseline determination

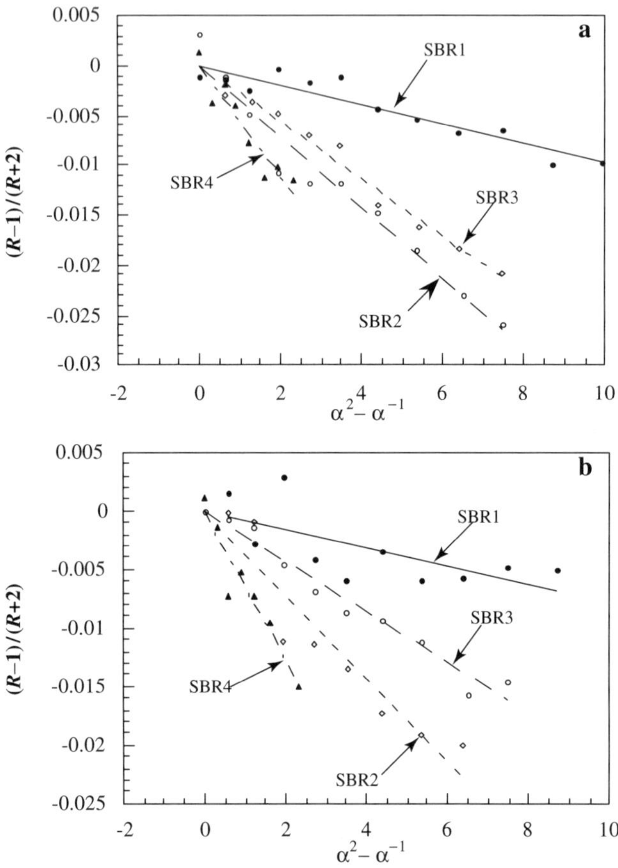

Figure 17.6 Dichroic functions for the bands at 1640 cm^{-1} (a) and 1493 cm^{-1} (b)

the local chain axis is determined rather than the orientation of specific transition moments.

In Figure 17.7 are compared, for the SBR 2 sample, the dichroic behavior of the two bands investigated. It can be seen that both bands exhibit negative orientation (negative D_0 or $\langle P_2(\cos \gamma) \rangle$). On the other hand, within experimental error, the two bands display the same dichroic behavior, thus proving that the corresponding transition moments are at nearly the same position with respect to the stretching direction. This result is consistent with that predicted by the work of Bahar *et al.* [17] where it is shown that the configurational factors D_0 associated with perpendicular transition moment vectors are indistinguishable while those associated with vectors along the backbone may exhibit quite distinct orientations.

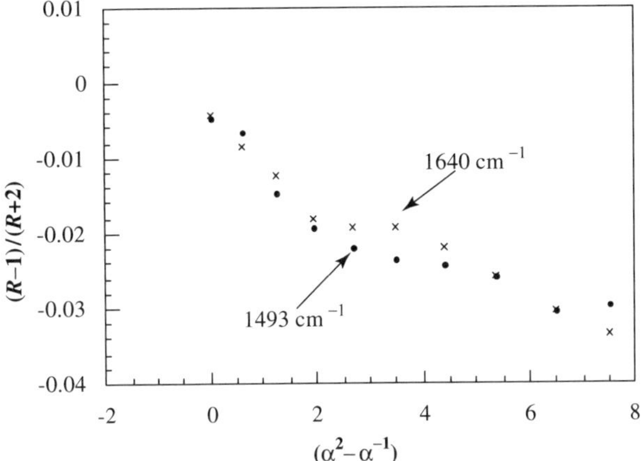

Figure 17.7 Dichroic functions versus $(\alpha^2 - \alpha^{-1})$ for the SBR 2 sample

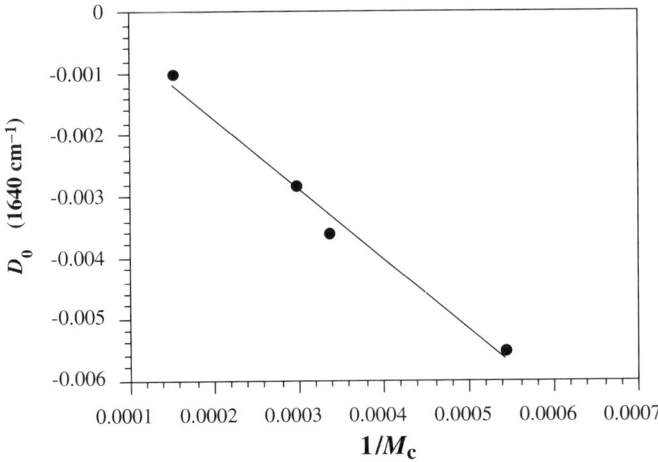

Figure 17.8 Dependence of the configurational factor (obtained for the band at 1640 cm^{-1}) on the molecular weight between cross-links, M_c

As the configurational factor has been shown to be inversely proportional to chain length, and is conveniently expressed as $D_0 \sim 1/n$ (n being the number of bonds in the network chain), it was interesting to plot D_0 (determined from the slopes of the curves representing $\langle P_2(\cos\gamma)\rangle$ versus the strain function) against $1/M_\mathrm{c}$. As expected, a linear relation is obtained (Figure 17.8).

Figure 17.9 shows for the four samples, the birefringence as a function of $(\alpha^2 - \alpha^{-1})$. The four sets of data exhibit, throughout the applied deformation range, a linear relationship between the experimentally observed birefringence

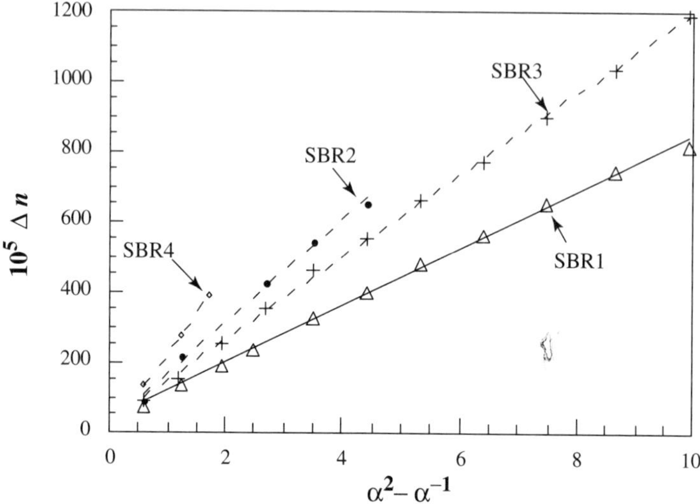

Figure 17.9 Strain dependence of the birefringence

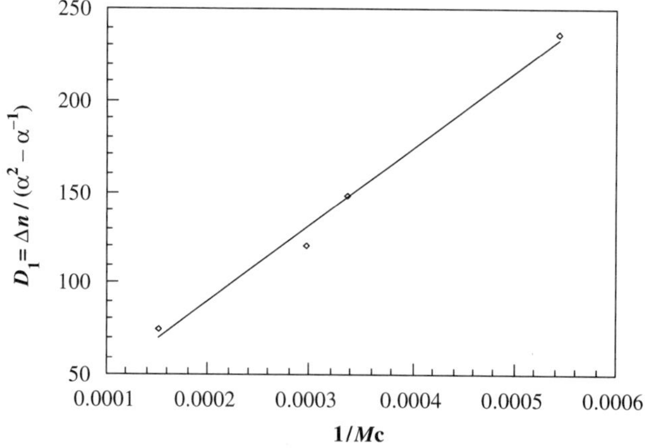

Figure 17.10 Dependence of birefringence on molecular weight between cross-links

and the strain function. Birefringence seems to yield a more reliable measure of orientation than infrared linear dichroism. Effectively, one of the problem in the infrared measurements is the lack of sensitivity in the determination of small dichroic effects, essentially at small strains.

The experimental results are in good agreement with the theory equation (17.4) which states that the slope, D_1, of the curve representing Δn versus the strain function is proportional to the cross-linking density (Figure 17.10).

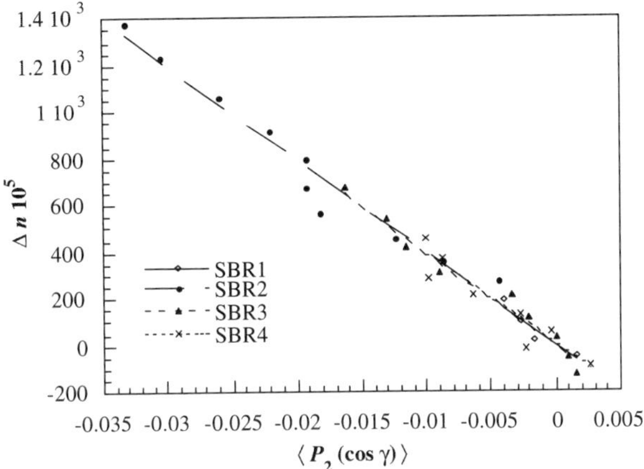

Figure 17.11   Relation between birefringence and orientation

The relation between birefringence and the second order moment of the orientation function, $\langle P_2(\cos \gamma)\rangle$ determined by infrared dichroism, is illustrated in Figure 17.11. The slope of the curve yields the intrinsic birefringence, $[\Delta n]_0$, which is a characteristic of the polymer.

CONCLUSION

In addition to stress–strain and equilibrium swelling analysis, we have applied infrared dichroism and birefringence for the characterization of network behavior. Infrared spectroscopy is able to characterize the orientational behavior of the different sequences of a copolymer and also of the different units constituting the microstructure of the elastomer phase. The local conformational characteristics are expected to play a major role in molecular orientation. These techniques are exploited further in order to obtain the dependence of molecular orientation on chain microstructure.

REFERENCES

1. B. Jasse and J.L. Koenig, *J. Macromol. Sci., Rev. Macromol. Chem.*, **C17**, 16 (1979).
2. L. Bokobza, B. Amram and L. Monnerie, in *Elastomeric Polymer Networks*, (J.E. Mark and B. Erman, eds), Prentice Hall, Engkwood cliffs, NJ, 1992, p. 289.
3. A. Dubault, B. Deloche and J. Herz, *Polymer*, **25**, 1405 (1984).
4. A. Dubault, B. Deloche and J. Herz, *Prog. Coll. Polymer Sci.*, **75**, 45 (1987).
5. P. Sotta and B. Deloche, *Macromolecules*, **23**, 1999 (1990).
6. P. Sotta, B. Deloche and J. Herz, *Makromol. Chem., Macromol. Symp.*, **45**, 177 (1991).

7. J.E. Mark and B. Erman, in *Rubberlike Elasticity. A molecular Primer*, Wiley-Interscience, New York, 1988.

8. J.E. Mark and B. Erman, in *Structure and Properties of Rubberlike Networks*, Oxford, 1997.

9. P.J. Flory and B. Erman, *Macromolecules*, **15**, 800 (1982).

10. S.F. Edwards and Th. Vilgis, *Polymer*, **27**, 483 (1986).

11. P.J. Flory and J. Rehner, *J. Chem. Phys.* **12**, 412 (1944).

12. B. Erman and P.J. Flory, *Macromolecules*, **16**, 1601 (1983).

13. B. Erman and P.J. Flory, *Macromolecules*, **16**, 1607 (1983).

14. B. Amram, L. Bokobza, J.P. Queslel and L. Monnerie, *Polymer*, **27**, 877 (1986).

15. B. Amram, L. Bokobza, L. Monnerie, and J.P. Queslel, *Polymer*, **29**, 1155 (1988).

16. S. Besbes, I. Cermelli, L. Bokobza, L. Monnerie, I. Bahar, B. Erman and J. Herz, *Macromolecules*, **25**, 1949 (1992).

17. I. Bahar, B. Erman and T. Haliloglu, *Macromolecules*, **27**, 1703 (1994).

18. J.P. Queslel and J.E. Mark, in *Encyclopedia of Physical Science and Technology*, **14**, 717 (1992).

19. J.E. Mark, *Polymer J.*, **17**, 265 (1985).

20. J.E. Mark, *Makromol. Chem.*, **202/203**, 1 (1992).

21. B. Saville and A.A. Watson, *Rubber Chemistry and Technology*, **40**(1), 100 (1967).

22. E.B. Wilson, *J. Phys. Rev.*, **45**, 706 (1934).

18

Effects of Polymer Networks on the Bending Electrostriction of Polyurethanes

MASASHI WATANABE[1], NOBUHISA TAKAHASHI[1], TSUTOMU UEDA[2], MAKOTO SUZUKI[2], YOICHI AMAIKE[2] and TOSHIHIRO HIRAI[1]*

[1]Faculty of Textile Science and Technology, Shinshu University, 3-15-1 Tokida, Ueda 386–8567, Japan
[2]Narayama Laboratory, Nitta Industries Corporation, 6-5-6 Sakyo, Nara 631–0801, Japan

---

* To whom correspondence should be addressed.

Wiley Polymer Networks Group Review Series Vol. 2. Edited by B.T. Stokke and A. Elgsaeter
© 1999 John Wiley & Sons Ltd

ABSTRACT

Polyurethane films coated with thin gold electrodes are bent using an applied electric field. When one of the surfaces of the polyurethane film was insulated from gold electrode by a thin layer of polyethylene, the bending deformation was not observed. This suggests that the bending deformation requires charge injection from the electrode.

The influence of chemical structures—especially network structures—of polyurethanes on the electrically induced bending deformation was also investigated. It turned out that the bending deformation was quick and large in the polyurethane which had long chain length between chemical crosslinking points. When the chain length between the crosslinking points was the same, the increase of the number of urethane groups in the chain depressed the bending deformation. It is concluded that the mobility of the soft segment plays a critical role in the bending deformation, and the urethane groups provide friction among the chains.

INTRODUCTION

Recently, various polymer actuators have been studied by many groups. Examples of the organic materials that have a marked ability to respond to an electric field include polymer gels [1], Nafion membranes [2], conducting polymers such as polypyrrole [3] and polyaniline [4]. These materials need to be actuated in the presence of aqueous solutions.

To exclude the effect of electrochemical reaction on the electrode, we investigated the electrical actuation of polyvinylalcohol (PVA) gel swollen with dimethyl sulfoxide (DMSO) [5]. Although the PVA-DMSO gel showed a remarkable improvement, the poor durability still remains as a serious defect. For the practical use of these polymer actuators, it is important to develop materials that can be actuated in air or without the presence of a solvent. Therefore, we have studied the electrically induced deformation of polyurethane films [6–10].

In our previous papers [9,10], we showed that polyurethane films can bend by applying an electric field. The electric-field dependence of the bending deformation was also examined. The degree of the bending deformation was proportional to the square of the applied electronic field. It suggests that the bending deformation is caused by bending electrostriction.

However, as far as we know, only one report has been published regarding bending electrostriction of polymers [11]. The mechanism of the deformation has been studied insufficiently. In addition, the effect of chemical structures on the bending electrostriction in also unclear.

In the work covered by this paper, we prepared various polyurethanes systematically and studied the effect of the network structure on the bending electrostriction.

Table 18.1   Compositions of polyurethanes

Sample code	Composition and molar ratio
PU-1	$MDI/PMPA_{3000}/1, 4 - BD/TMP = 2/1/0.4/0.333$
PU-2	$PPDI/PMPA_{1000}/1, 4 - BD/TMP = 2/1/0.5/0.333$
PU-3	$PPDI/PMPA_{2000}/1, 4 - BD/TMP = 2/1/0.5/0.333$
PU-4	$PPDI/PMPA_{5000}/1, 4 - BD/TMP = 2/1/0.5/0.333$
PU-5	$PPDI/PMPA_{2000}/1, 3 - BD/TMP = 2/1/0.5/0.333$
PU-6	$PPDI/PMPA_{1000}/1, 3 - BD/TMP = 2/1/0.75/0.167$

MDI: 4,4'-diphenylmethane diisocyanate, PPDI: p-phenylene diisocyanate, $PMPA_{1000}$: poly(3-methyl-1,5-pentane adipate) $M_n = 1000$, $PMPA_{2000}$: poly(3-methyl-1,5-pentane adipate) $M_n = 2000$, $PMPA_{3000}$: poly(3-methyl-1,5-pentane adipate) $M_n = 3000$, $PMPA_{5000}$: poly(3-methyl-1,5-pentane adipate) $M_n = 5000$, 1,3-BD: 1,3-Butanediol, 1,4-BD: 1,4-Butanediol, TMP: Trimethylolpropane.

EXPERIMENTAL

PREPARATION OF POLYURETHANES

Polyurethanes were prepared through a prepolymer route. The polymer compositions are summarized in Table 18.1. A typical procedure is described. 40 g of poly (3-methyl-1,5-pentane adipate) (PMPA) (Kuraray Co., Ltd, Kurapol P-2010, Mn = 2000, acid value <0.5 KOH mg/g) was placed in a glass reactor that had nitrogen inlet and outlet, and was dried at 100 °C in vacuo for 1 h before use. A 6.41 g of p-phenylene diisocyanate (PPDI) (E.I. DuPont, HylenePPDI, purity = 99.86%) was charged into the reactor at 30 °C. The reaction mixture was then warmed to 85 °C. under vigorous stirring. The reaction time was approximately 1.5 h, and the reaction was followed by determining the NCO values using a conventional method of dibutylamine back titration [12]. After the theoretical NCO value, which means the completion of NCO terminated prepolymer preparation, was attained, 0.90 g of 1,3-butanediol (1,3-BD) and 0.89 g of trimethylolpropane (TMP) were added to the prepolymer under vigorous agitation. The reaction mixture was then degassed under vacuum and was cast on a Teflon mold. After curing at 120 °C for 24 h, a film of 0.3 mm in thickness was obtained. Using an ion-sputtering method, the prepared polyurethane film was coated with a thin gold layer on each surface as an electrode. Then the film was cut into 5 mm in width and 30 mm in length, and was prepared for measurement. All measurements were carried out in thermostatic atmosphere.

MEASUREMENT OF DISPLACEMENT AND CURRENT

The experimental setup for measuring the bending deformation is shown in Figure 18.1. The film was vertically suspended in air, and the top of the film was fixed. By applying a d.c. voltage, the displacement of the free bottom end

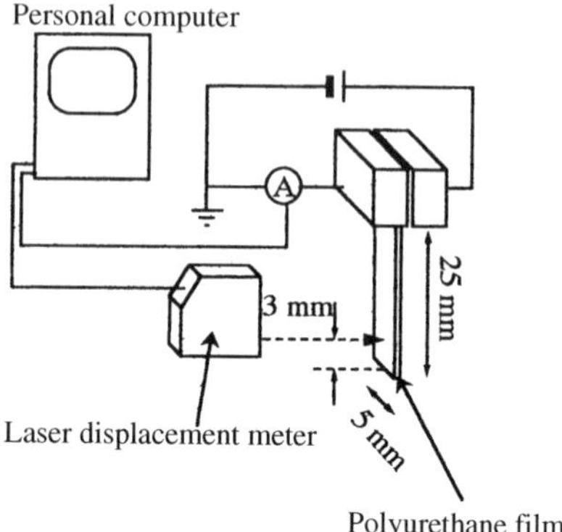

Figure 18.1   Schematic diagram to measure bending deformation and current

of the film was measured using a laser displacement meter (Keyence Corporation, LB-62). The current was measured using an ampere meter (Advantest Corporation, Ultra High Resistance Meter).

DSC MEASUREMENT

Differential scanning calorimetric (DSC) curves were recorded using a Mac-Science DSC3200 apparatus with a heating rate of 10 °C/min, under nitrogen purging.

RESULTS AND DISCUSSION

CHARGE INJECTION FROM THE ELECTRODES

Three samples, Sample 1a, 1b, and 1c, were prepared from a polyurethane (PU-1) (Figure 18.2), In Sample 1a, each surface was coated with a thin gold layer (contacting electrode). Sample 1b was also coated with a thin gold layer on each surface, but a polyethylene thin film was stuck onto one of the gold layers. As shown in Figure 18.2, one surface of Sample 1c was covered with a polyethylene thin film and was insulated from a gold electrode layer (blocking electrode).

On applying the electric field of 2 MV/m to Sample 1a, it bent toward the cathode side. In the case of Sample 1b, although elongation or contraction of the cathode surface was prevented by the stuck polyethylene film, it also bent toward the cathode side (Figure 18.3). These results suggest that the bending deformation of Sample 1a was mainly caused by elongation of the anode surface.

On the other hand, Sample 1c, which had an insulating layer on one of the surfaces, showed much smaller bending deformation than that of Sample 1b. The irreversible current observed in Sample 1b was considered to be caused by charge injection, and the current observed in Sample 1c was much smaller than

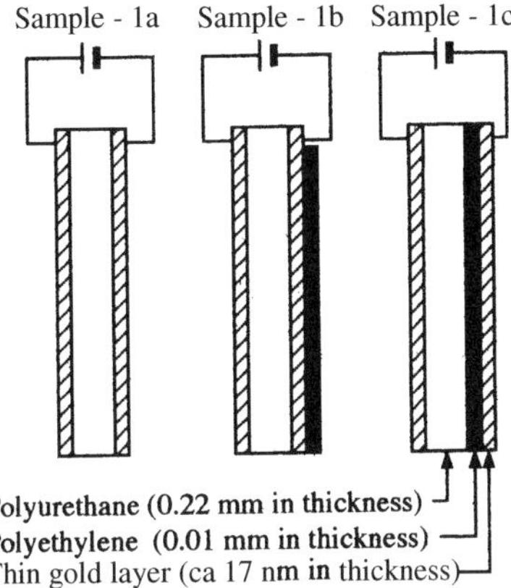

Figure 18.2   Fabrication of polyurethane/polyethylene laminate films

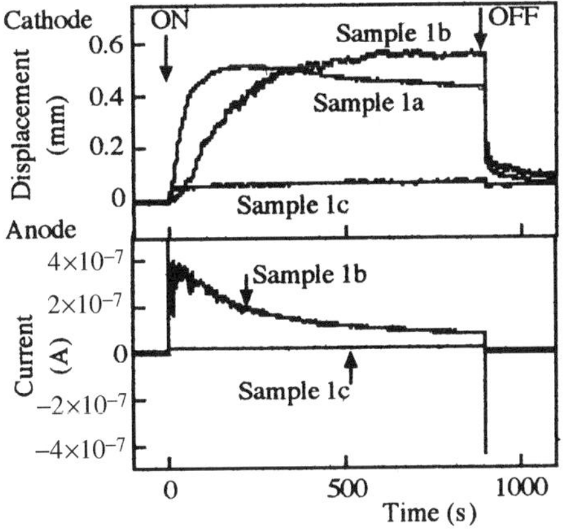

Figure 18.3   Courses of displacement and current of polyurethane/polyethylene laminate films. The measurements were carried out at 30 °C. The applied electric filed was 2 MV/m

that in Sample 1b. The results suggests that the bending deformation requires the charge injection. Therefore, the origin of the expansion of the anode surface is likely to be the repulsion among the charge-injected species, which are under investigation and not clarified yet.

EFFECT OF THE CHAIN LENGTH OF A SOFT SEGMENT

Electrically induced bending deformations of PU-2, PU-3 and PU-4 were examined. As shown in Figure 18.4, these polyurethanes were different from each

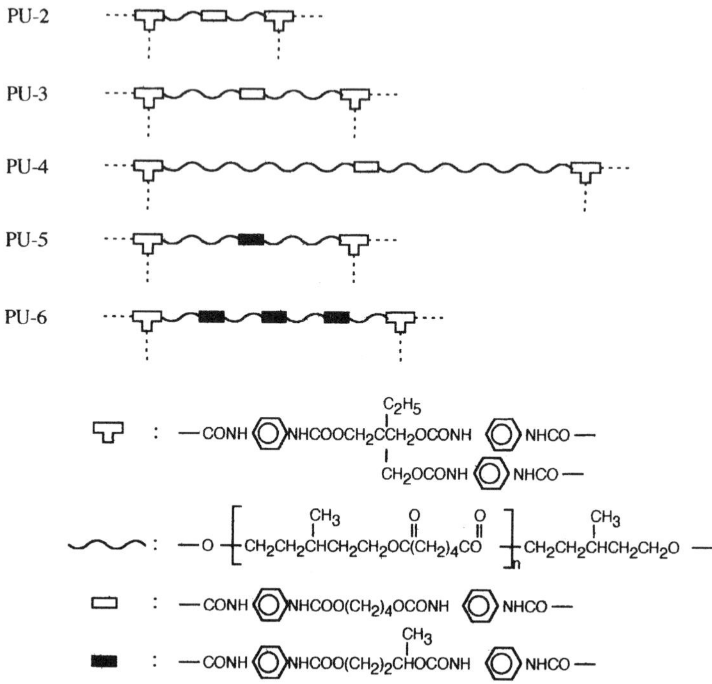

Figure 18.4 Structures of polyurethanes assumed from the molar ratio of the composition

other only in the chain length of the soft segment (PMPA). In this experiment, each surface was directly coated with a thin gold layer.

Figure 18.5 shows the courses of displacement and current induced by applying 1 MV/m. In the Figure, the polyurethane that had longer soft segment shows quicker and larger bending deformation. It can be explained as follows. The mobility of charge in the polyurethane that has a longer soft segment is considered higher. Higher mobility leads to the promotion of charge injection from the electrode, and caused the asymmetric space charge distribution which is large enough for the bending deformation of the film.

INFLUENCE OF FRICTION AMONG THE POLYMER CHAINS

Polyurethanes, PU-5 and PU-6, have the same chain length between chemical crosslinking points but have different number of hard segments between the crosslinking points (Figure 18.4). Figure 18.6 shows the courses of bending deformation and current induced at 1 MV/m.

In PU-6, both the displacement and the current caused by the charge injection were smaller than those in PU-5. This suggests that the hard segments increase

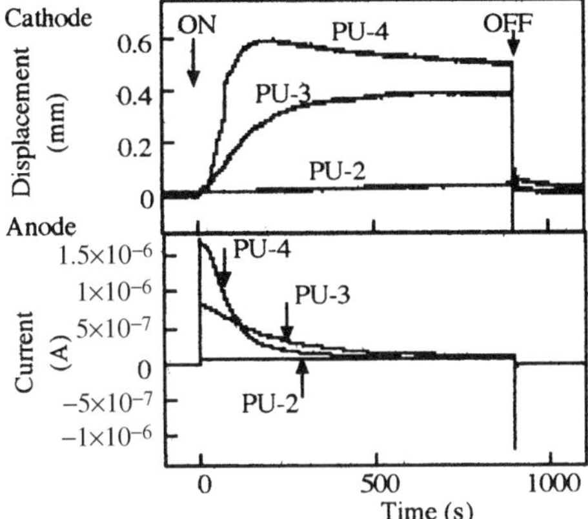

Figure 18.5 Effect of the chain length of soft segment on displacement and current. The measurements were carried out at 30 °C. The applied electric field was 1 MV/m

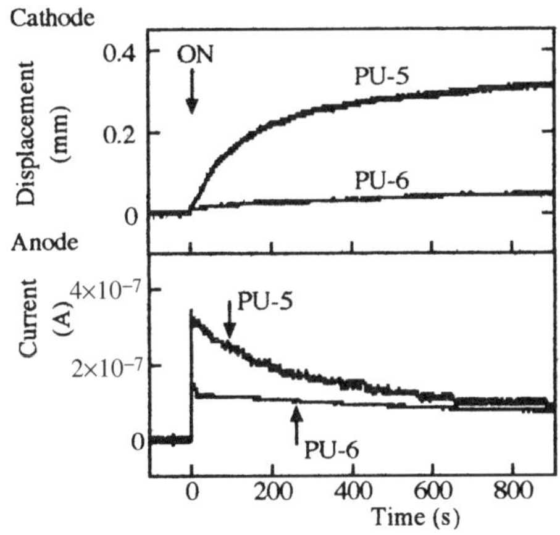

Figure 18.6 Courses of displacement and current of PU-5 and PU-6. The measurements were carried out at 30 °C. The applied electric field was 1 MV/m

the friction between the polymer chains and depress the micro-Brownian motion of polymer chains (through electrostatic interaction between the urethane groups).

TEMPERATURE DEPENDENCE OF THE INCREASING RATE OF THE DISPLACEMENT

The electrically induced bending deformation of PU-6 was examined under various temperatures (Figure 18.7). The polyurethane film bent much more quickly at 50 °C or above than below 45 °C. This can be considered because

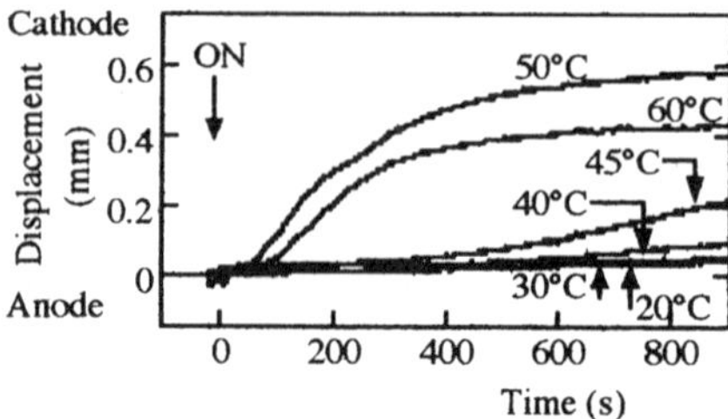

Figure 18.7 Temperature dependence of electrically induced bending deformation of PU-6. The applied electric field was 1 MV/m

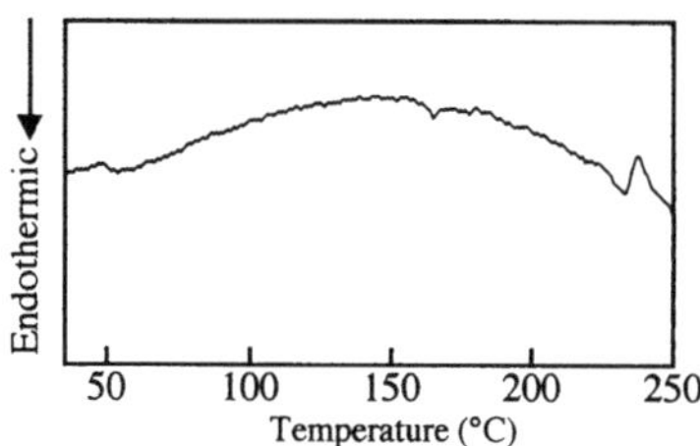

Figure 18.8 DSC curve of PU-6

at higher temperature the mobility of injected charge became larger with the increase in micro-Brownian motion of polymer chain, and the friction among the soft segements loosened in the temperature range.

DSC curve of PU-6 exhibits three endothermic peaks (Figure 18.8). Peak 1 at 234 °C can be attributed to relatively well-formed hard segment domains [13]. Peak 2 at 165 °C can be attributed to somewhat poorly formed hard segment domains [13]. Peak 3, starting at about 48 °C and ending at about 85 °C, is unknown, but the temperature is much higher than the glass transition temperature of the soft segment which is below 0 °C.

Assuming that Peak 3 is assigned to dissociation of the relatively weak electrostatic interaction between urethane groups, the result from DSC agrees with the temperature dependence of the displacements.

CONCLUSION

It was proved that the electrically induced bending deformation of the polyurethane films requires charge injection from the electrode. The polyurethane that has long chain length between chemical crosslinking points shows quick and large bending deformation as far as the electric field application (less than 15 min).

An advantage of the electrically induced bending deformation of polyurethane films is that they can actuate in air. In addition, the polyurethane films are one layer and homogeneous but can bend like bimorph actuators. From these advantages, this material offers great promise in applications as artificial muscles and actuators.

ACKNOWLEDGMENT

This work was partly supported by Grant-in-Aid for COE Research (10CE2003) by the Ministry of Education, Science, Sports and Culture of Japan.

REFERENCES

1. Y. Osada, H. Okuzaki and H. Hori, *Nature*, **355**, 242 (1992).
2. K. Asaka, K. Oguro, Y. Nishimura, M. Mizuhata and H. Takenaka, *Polym. J.* **27**, 436 (1995).
3. T.F. Otero, J. Rodriguez, E. Angulo and C. Santamaria, *Synth. Mett.*, **55–57**, 3713 (1993).
4. K. Kaneto, M. Kaneko and W. Takashima, *Jpn. J. Appl. Phys.*, **34**, L837 (1995).
5. T. Hirai, H. Nemoto T. Suzuki, S. Hayashi, and M. Hirai, *J. Intell. Material Syst. Struct*, **4**, 277 (1993).
6. T. Hirai, T. Kasazaki, T. Sugino, A. Sukumoda, H. Sadatoh, M. Hirai and S. Hayashi, *Rep. Prog. Polym. Phys. Japan*, **36**, 341 (1993).
7. T. Hirai, H. Sadato, S. Hayashi, Y. Amemiya, T. Ueki and M. Hirai, *Rep. Prog. Polym. Phys. Japan*, **240**, 221 (1994).
8. T. Hirai, H. Sadatoh, T. Ueda, T. Kasazaki, M. Hirai and S. Hayashi, *Angew. Makromol. Chem.*, **240**, 221 (1996).
9. M. Watanabe, M. Yokoyama, T. Ueda, T. Kasazaki, Y. Kurita, M. Hirai and T. Hirai, *Chem. Lett.*, 773 (1997).
10. M. Watanabe, M. Yokoyama, T. Ueda, T. Kasazaki, M. Hirai and T. Hirai, *Reports on Progress in Polymer physics in Japan*, **40**, 401 (1997).
11. H. Kawai, *Oyo Butsuri*, **39** (9), 869 (1970).
12. C. Hepburn, in *Polyurethane Elastomers*, (2nd edn), Elsevier Applied Science, London and New York, 1992, p. 293.
13. A. Cunningham, N.C. Hilyard, in *Advances in Urethane Science and Technology*, (K.C. Frisch and D. Klempner, eds), Technomic Publishing Co., Lancaster and Basel, 1996, p. 112.

PART 3

Polymer Networks and Precursor Architectures

19

Hyperbranched Polymers in Thermoset Applications

MATS JOHANSSON

Department of Polymer Technology, Royal Institute of Technology,
SE-100 44 Stockholm, Sweden

ABSTRACT

This paper describes the use of hyperbranched polymers as scaffolds for thermoset resin structures. Structural variations of hyperbranched aliphatic polyesters and how these variations affect the properties of the resins are described. It is shown that very different properties can be obtained for thermosets based on hyperbranched polyester by variation of either reactive or non-reactive end-groups. The T_g and the mechanical properties mainly depend on the crosslink density whereas other properties such as wettability (polarity) depend more on the structure of the non-crosslinkable groups. The reactive end-groups are easily polymerized, indicating a good accessibility of these groups, resulting in low residual unsaturation.

Wiley Polymer Networks Group Review Series Vol. 2. Edited by B.T. Stokke and A. Elgsaeter
© 1999 John Wiley & Sons Ltd

INTRODUCTION

The research in the field of coating technology, polymer networks in thin films, has during the last decades been focused on improvements of the material properties as well as reduction of the environmental hazards. One goal which has obtained increasing attention is the reduction of volatile organic content (VOC) in coating formulations. Introduction and expansion of new techniques such as powder coatings, radiation curable systems and waterborne coatings have to some extent been the solution but a significant fraction of current coating systems still consists of liquid systems employing solvents. One obstacle for the reduction of VOC lies in the demands on the properties of the final dry film. The molar mass of the resin needs to be high enough to obtain good final film properties and to reduce toxic hazards, which leads to minimum viscosity of the resin. One way to reduce the viscosity of a resin, and thus reducing the VOC, is by changing the architecture of the resin instead of changing the molar mass or chemical structure. It has been shown that the introduction of branching in resin structures reduces the viscosity while still retaining good resin properties [1].

Another field of research which has obtained increasing attention during the last decade concerns highly branched dendritic polymers, comprising dendrimers and hyperbranched polymers [2]. These polymers are based on AB_x-functional monomers producing polymers with a potential branching point in every repeat unit. Dendrimers represent the extreme where all monomers are fully reacted, giving a monodisperse polymer with a layerwise, spherical structure. Dendrimers are in most cases synthesized via multi-step procedures including purification between each step, hence they are tedious to produce. Hyperbranched polymers are built of the same type of monomers but with the difference that the monomers are not fully reacted, i.e. the polymers contain linear segments, and are polydisperse. Hyperbranched polymers can be produced more easily in larger amounts, which is why they are more suitable as an alternative to dendrimers for material consuming applications. When looking at a property such a solubility it has been shown that hyperbranched polymers have a much higher solubility compared to linear analogues and dendrimers an even higher solubility [3]. How the structural differences between dendrimers and hyperbranched polymers affect this difference is still not fully clarified, e.g. what is the effect of the degree of branching of the hyperbranched polymer?

The highly branched structure of dendritic polymers gives properties differing from their linear counterparts with respect to, for instance, melt viscosity and solubility [4,5]. Various materials applications, ranging from rheological additives to thermoset resin structures, have been suggested for hyperbranched polymers [6]. Numerous different hyperbranched polymers such as, for example, aromatic polyphenylenes [4], aromatic polyesters [7] to aliphatic polyesters [8] have been synthesized. Hyperbranched polymers have, unlike linear polymers, the special feature that they contain numerous end-groups which greatly affect the properties. The properties of hyperbranched polymers are determined both

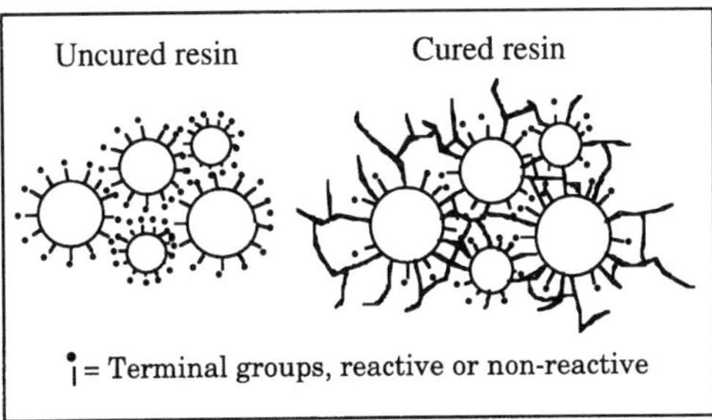

Figure 19.1 Generalized description of hyperbranched polymers as thermoplastic and thermosets

by the backbone structure and by the end-group structure, increasing the possibilities to tailor a polymer for a certain application. Hyperbranched polymers incorporated into networks represent an extreme where all or a fraction of the end-groups are 'locked' in the network. When looking at polymers for thermoset applications both the properties before and after crosslinking are affected (Figures 19.1 and 19.3). This paper aims to describe the properties of some different hyperbranched aliphatic polyesters used in thermoset applications. The properties before crosslinking as well as the crosslinking performance and the properties of the final network will be discussed.

HYPERBRANCHED ALIPHATIC POLYESTERS BASED ON bis-MPA

This paper is focused on hydroxy-functional hyperbranched aliphatic polyesters based on 2,2-bis-methylolpropionic acid, bis-MPA, as AB_2-monomer. Several different polyesters, based on bis-MPA, have been synthesized and functionalized for thermoset applications.

SYNTHESIS

The base polyesters were synthesized via an acid catalyzed condensation reaction performed in bulk at 140 °C, employing a hydroxyfunctional core to determine the molar mass. Figure 19.2 presents a general description of the synthesis of a hydroxyfunctional hyperbranched polyester. A more extensive description of the synthesis has been reported previously [9,10].

The molar mass of the formed polyester depends on the ratio between bis-MPA and the core molecule, a higher ratio producing a higher molar mass. The formed polyesters have a T_g close to 35 °C and a degree of branching of 0.45 [11].

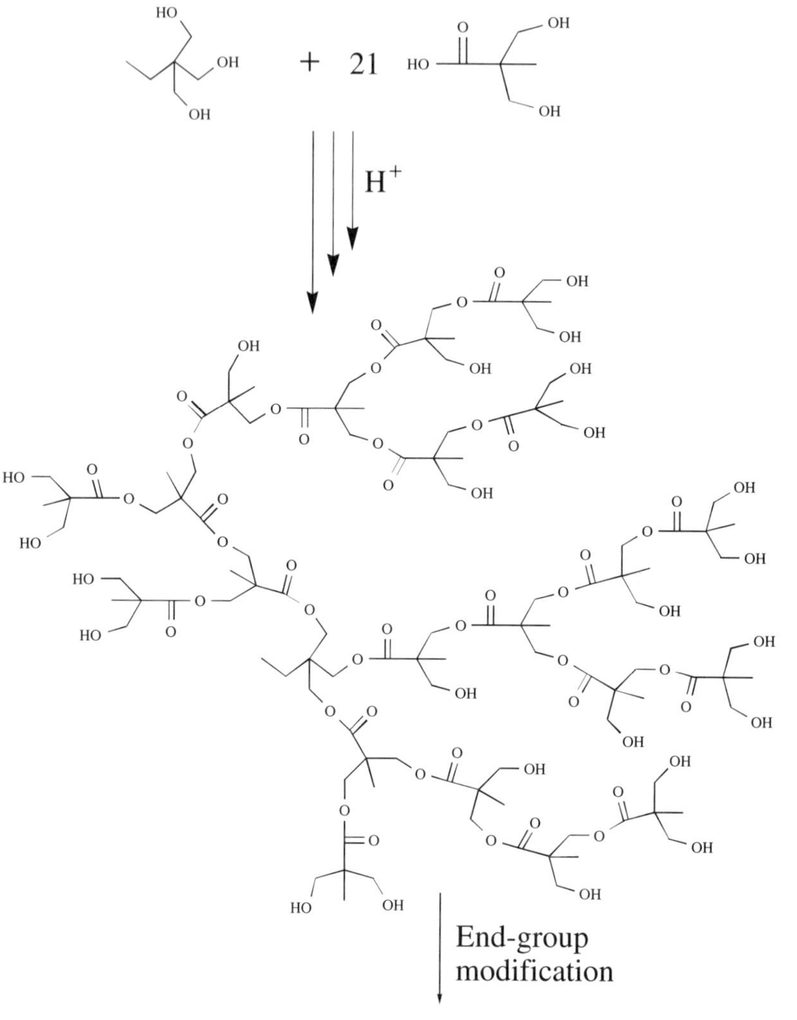

Figure 19.2 Schematic description of the synthesis of a hydroxyfunctional hyperbranched aliphatic polyester based on 2,2-bismethylolpropionic acid

END-GROUP MODIFICATIONS

The numerous end-groups on the hyperbranched polyesters can be chemically modified in different ways. All or a fraction of the end-groups can be reacted to form either reactive (crosslinkable) or non-reactive moieties with difference in chemical structure affecting for instance the polarity of the polymer. More extensive descriptions of modifications of hyperbranched aliphatic polyesters have been published elsewhere [12].

Non-reactive End-groups

Variations of the chemical nature of the end-groups will greatly affect the properties of hyperbranched polymers when considering them as thermoplastic materials (i.e. a resin before cure). The modification of the hydroxyl end-groups on the hyperbranched polyester to groups such as propionates, benzoates, and carboxylic acids have been demonstrated. A series of hyperbranched polyesters based on bis-MPA with alkyl chain end of different length has been synthesized to study possible side chain crystallization. Both the size of the hyperbranched polyester and the length of the alkyl chains have been varied. The effect of these end-group modifications on the material properties is further discussed here.

Reactive End-groups

The end-groups can also be modified to reactive, crosslinkable species. A series of hyperbranched acrylate resins has been synthesized in order to study the effect of the structure of the fraction of non-reactive end-groups on the resin and on the network properties. A hydroxy-functional hyperbranched aliphatic polyester was reacted with acryloyl chloride to form a resin with approximately 30% of the hydroxyl end-groups transformed to acrylate groups. Modified resins, where the residual hydroxyl end-groups were transformed to either propionate or benzoate groups, were also synthesized, resulting in three different hyperbranched resins with the same amount of crosslinkable end-groups but different non-reactive moieties *were obtained* (Figure 19.3).

In a second series, a hydrofunctional hyperbranched polyester was reacted with methacrylic anhydride to various extents, producing three different resins with different methacrylate functionality ranging from 30 to 60% of the end-groups.

PROPERTIES OF NON-CROSSLINKED HYPERBRANCHED POLYESTERS

The rheological behaviour of hyperbranched polymers is of importance when considering the potential use of them for various applications. It has been shown that for example the glass transition temperature can be shifted within a wide range by changing the polarity of the end-groups while still having the same backbone structure [13].

Figure 19.4 shows the melt viscosity for hyperbranched aliphatic polyesters having either propionate or hydroxyl end-groups. The T_g shifts over $50\,^\circ$C for the less polar structure and the viscosity above T_g drops several decades. In the case of reactive resins, for example, the acrylate series, the shifts will be smaller since only a fraction of the end-groups can be varied in the structure.

All the hyperbranched polyesters also exhibit a Newtonian behaviour, i.e. no shear thinning or thickening, within a medium shear rate range ($0.1-100$ rad s^{-1}) indicating a lack of entanglements. The viscosity increases with molar mass for the hyperbranched polyesters up to a certain value where it levels out.

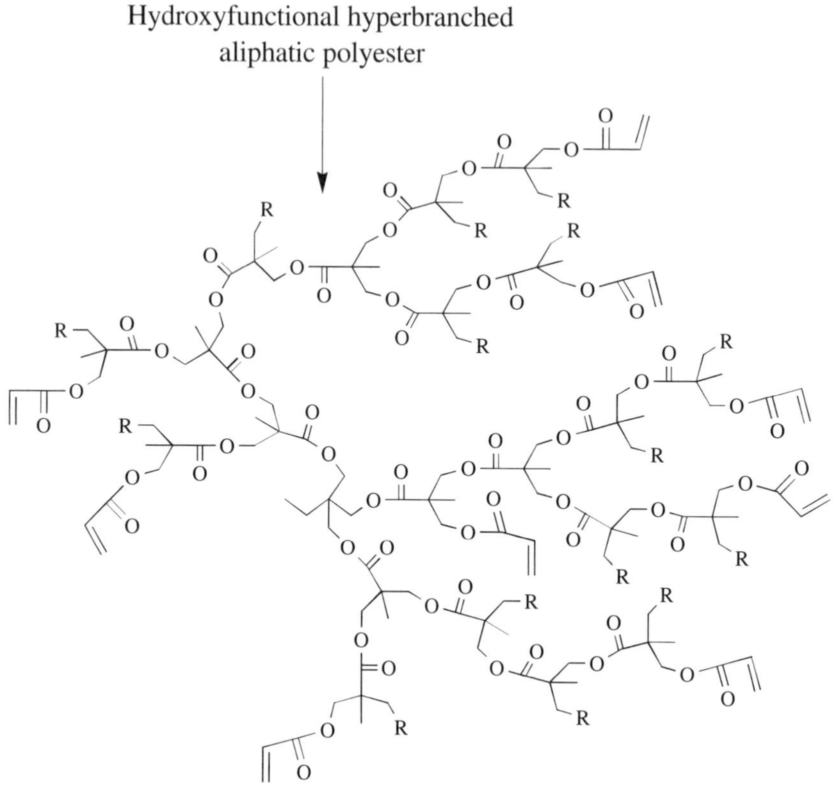

Figure 19.3 Modification of a hyperbranched aliphatic polyester to different acrylate resins. R is either hydroxyl, propionate, or benzoate

Hyperbranched polymers are in most cases considered to be amorphous structures due to the high degree of branching which hinders conventional crystallization. Semi-crystalline hyperbranched polymers can however be obtained if suitable crystallizable end-groups are attached to the polymer, as shown on hyperbranched polyesters with long alkyl chain ends [14].

CURING PERFORMANCE AND NETWORK PROPERTIES

The possibility of adjusting the polarity of the hyperbranched polyester affects not only the viscosity but also the wetting and compatibility properties. This has elegantly been shown by Boogh et. al. who adjusted the polarity of a hyperbranched polyester to cause phase separation from a blend with a conventional composite matrix system during the crosslinking [15]. The hyperbranched polyester was initially miscible with an epoxy/amine resin but due to changes in

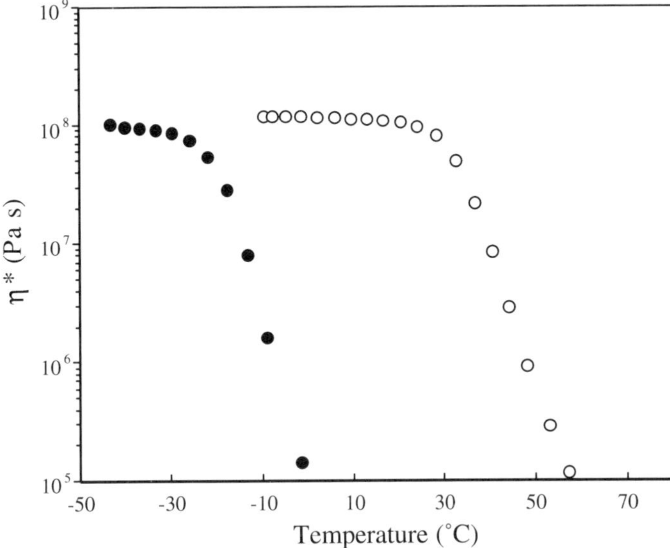

Figure 19.4 Complex dynamic viscosity as a function of temperature for hyperbranched aliphatic polyesters with either hydroxyl (O) or propionate (●) end-groups

polarity of the conventional resin when the crosslinking proceeded, the hyperbranched polyester became immiscible and small phase separated domains were formed. The phase separation greatly improves the toughening properties while retaining the modulus of the system.

Free radically initiated UV-curing of the hyperbranched acrylate resins proceeded rapidly and resulted in low amounts of residual unsaturation as measured with FT-Raman spectroscopy (Figure 19.5). This shows that the reactive species are accessible for polymerization and not trapped inside the hyperbranched polyester skeleton. The amount of residual unsaturation, however, increases for resins with higher functionality probably due to vitrification effects.

Figure 19.6 presents tan δ as a function of temperature for crosslinked acrylate resins which have the same crosslink density but different structures of the non-reacted end-groups. The T_g shifts 10 °C with increasing polarity of the resin, indicating that reduced mobility of the non-reactive end-groups to some extent affects the network, although the effect is small.

A much larger effect was seen when the crosslink density of the system was varied as seen with cured films of the methacrylate resins. The T_g shifted from 100 °C to 170 °C going from 30 to 60% methacrylate end-groups on the hyperbranched polyester. The transition also becomes wider with increasing crosslink density, indicating a more disperse network structure. It can be concluded that it is possible to vary the T_g of a network based on a hyperbranched polyester

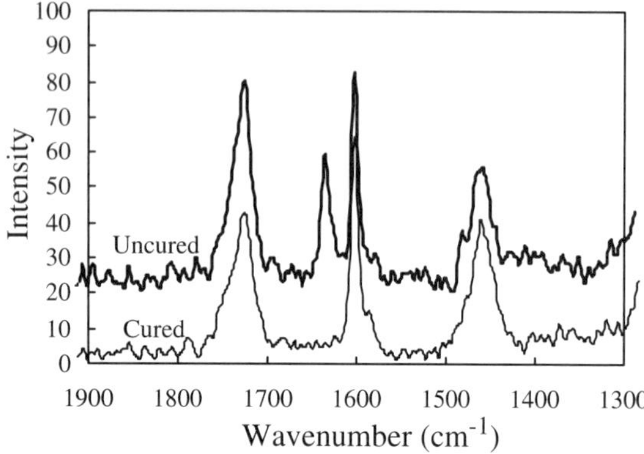

Figure 19.5   FT-Raman spectra of a hyperbranched acrylate resin before and after UV-curing

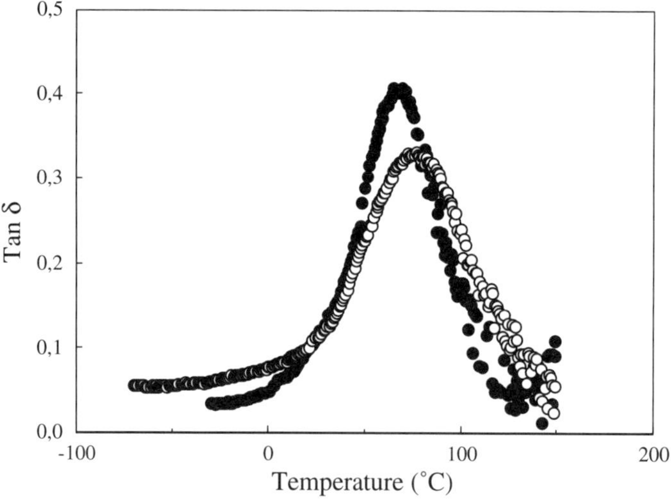

Figure 19.6   Tanδ versus temperature for crosslinked acrylate resins having either hydroxyl (O) or propionate (●) end-groups

by changing the number of crosslinkable end-groups. The structure of the non-reactive end-groups affects the mechanical network properties to some extent, although this effect is small.

Other properties can, however, be affected greatly by the structure of the non-reactive end-groups. Measurements of the contact angle of water on cured films (approximatly 30% acrylate/70% non-reactive end-groups) show changes from 10° to 80° when going from polar carboxylic acid end-groups to less polar silyl end-groups.

CONCLUSIONS

This paper presents the concept of using hyperbranched polyesters as scaffolds for network structures. One specific hyperbranched polyester can be modified to form resin structures with very different properties, both as a resin before crosslinking and as a final crosslinked network. Some properties such as T_g and modulus of the cured film mainly depend on the crosslink density, i.e. functionality of the resin. Other properties such as melt viscosity, solubility and wetting behaviour are more dependent on the chemical structure of the non-reactive end-groups. Crosslinking of acrylate functional hyperbranched polyesters proceeds rapidly, resulting in low amounts of residual unsaturation, indicating good accessibility of the acrylate groups for polymerization.

REFERENCES

1. M. Johansson, M. Trollsås and A. Hult, *J. Pol. Sci., Part A; Polym. Chem.*, **30**, 2203, (1992).
2. (a) D.A. Tomalia, A.M. Naylor and W.A. III Goddard, *Angew. Chem. Int. Ed. Engl.* **102**, 119 (1990). (b) B.I. Voit, *Acta Polymer.*, **46**, 87 (1995). (c) J.M.J. Fréchet, *Science*, **263**, 1710 (1994). (d) M. Johansson, E. Malmström and A. Hult, *Trends in Polymer Science*, **4**(12), 398 (1996).
3. K.L. Wooley, J.M.J. Fréchet and C.J. Hawker, *Polymer.* **35**, 4489 (1994).
4. Y. Kim and O.W. Webster, *Macromolecules*, **25**, 5561 (1992).
5. S.R. Turner, B.I. Voit, T.H. Mourey, *Macromolecules*, **26**, 4617 (1993).
6. M. Johansson and A Hult, *J. Coat. Techn.*, **67**, 35 (1995).
7. K.L. Wooley, C.J. Hawker, R. Lee, J.M.J. Fréchet, *Polym. J.*, **26**, 187 (1994).
8. M. Johansson, E. Malmström and A. Hult, *J. Polym. Sci. Part A: Polym. Chem.*, **31**, 619 (1993).
9. E. Malmström, Ph.D. thesis, Royal Institute of Technology, Stockholm, Sweden (1996).
10. E. Malmström, M. Johansson and A. Hult, *Macromolecules*, **28** 1698 (1995).
11. E. Malmström, M. Trollsås, C.J. Hawker, M. Johansson, and A. Hult, *Proc. Am. Chem. Soc., Div. Polym. Mater.: Sci. Eng.*, **77**, 151 (1997).
12. B. Pettersson and K. Sörensen, *Proceedings of the 21st Waterborne, Higher Solids, & Powder Coatings Symposium*, New Orleans, Louisiana, USA (1994), p. 753. (b) M. Johansson and A. Hult, *PRA's 16th Waterborne, High Solids & Radcure Technolgies, Conference*, Frankfurt, Germany (1996), p. 1.
13. S.R. Turner, F. Walter, B.I. Voit and T.H. Mourey, *Macromolecules*, **27**, 1611 (1994).
14. E. Malmström, M. Johansson and A. Hult, *Macromol. Chem. Phys.*, **197**, 3199 (1996).
15. L. Boogh, B. Pettersson, P. Kaiser and J.-A. Månson, *Proceedings of 28th International SAMPE Technical Conference*, Seattle, USA (1996), p. 236.

20

A Graph-like Method of Calculating Radii of Gyration of Starburst Dendrimers

HENRYK GALINA and GRAŻYNA GROSZEK
Faculty of Chemistry, Rzeszów University of Technology,
35-959 Rzeszów, Poland

ABSTRACT

A method of calculating radii of gyration of Gaussian models of dendrimers is presented. The models consisted of the set of f-functional units ($f = 3$, 4, and 6) connected into the form of dendrimers with links of the properties of Gaussian chains. The number of generations g was up to 10. The explicit formulas for the radii of gyration have been derived by using a straightforward graph-theoretical analysis of the model molecules. The results were very close to those published for similar models by other authors.

INTRODUCTION

The size and shape of regular highly branched molecules became of interest since the development of synthetic routes leading to preparation of dendrimers

Wiley Polymer Networks Group Review Series Vol. 2. Edited by B.T. Stokke and A. Elgsaeter
© 1999 John Wiley & Sons Ltd

or starburst molecules [1]. Some of the studies [2] visualized these molecules as having unusual radial segment density profiles. Such profiles, namely the low segment density in the interior of molecules, further stimulated rapid growth of effort in seeking various synthetic methods of preparing the 'hollow' molecules. Despite, however, several reports [3] where this peculiar property was claimed to be successfully exploited for 'encapsulating' host substances within the dendrimer core, other studies did not confirm the picture outlined by DeGennes and Hervet [2]. Monte-Carlo studies [4,5] and self-consistent mean field calculations [6] have shown that dendrimers are not particularly different than other polymer molecules and the segment density decreases monotonically from the center of the coil outwards [6].

With respect to the size of dendrimer molecules, however, both theoreticians and experimentalists agree that their fractal dimension, D, considerably exceed that of linear polymer chains or even that of randomly branched molecules. The exponent $v = 1/D$ in the relation linking the radius of gyration, R_G, and the molecular weight or polymerization degree, N, varies in the range of 0.2–0.22 in computer simulations [7,8] to 0.36 as deduced from diffusion properties of aliphatic dendrimers reported in a recent paper by Ihre at al. [9].

In this work we study dimensions of the Gaussian model of dendrimers. This model pretty well describes dimensions of unperturbed linear chains. It becomes, however, less and less realistic as the topology of a macromolecule gets complicated due to the presence of branch points or cycles. This limitation is illustrated by, e.g., the simple fact that however complex a Gaussian molecule may be, its mean square radius of gyration is always proportional to the length of the longest path through the molecule. Therefore, for Gaussian molecules with the topology of a dendrimer, the general prediction of the mean square radius of gyration is simply [6,10]

$$R_G^2 \sim g \tag{20.1}$$

where g is the number of generations in the dendrimer molecule ($g \gg 1$).

The advantage of the Gaussian model is that it provides an explicit formula linking the radius of gyration with the number of generations and functionality of units in a dendrimer molecule. Such a formula could be useful at least for the sake of comparing the numerical values of the proportionality constants in equation (20.1) for molecules with different functionalities of branching points.

In fact, the explicit equations linking these structural parameters with R_G has been, to our knowledge, published only very recently, independently by two authors [11,12]. We have derived essentially the same equation by using the graph theoretical approach developed long ago by Eichinger [13] and used to calculate mean square radii of gyration for Gaussian models of molecules having different topologies [14,15] as well as for polymer networks [16]. The approach makes use of Kirchhoff matrices of molecular graphs obtained from the molecule by labeling its units as vertices and bonds (which are Gaussian subchains) as edges. The mean square radius of gyration is expressed through the generalized inverse of the Kirchhoff matrix of the graph of the molecule. The radii of gyration are simply expressed as recurrence equations which start with the summation of elements of matrices consisting of ones.

THE MODEL

We consider a regular f-dendrimer model molecule consisting of f-functional units in generations 0 through $g-1$ and terminal units of functionality 1 in generation g. The entire mass of the molecule is uniformly distributed among the units. The units are connected with identical links which are in fact Gaussian subchains of the (effective) mean square length b_{eff}^2 (cf. Figure 20.1.) The mean square radii of gyration will be expressed as multiplicity of b_{eff}^2. The molecule configuration is unperturbed by any interactions other than the constraints resulting from the connectivity.

Define two integer values functions, $N_g(f)$ and $M_g(f)$ characterizing dendrimers of g generations ($g = 0, 1, 2, \ldots, f = 3, 4, \ldots$).

The first function

$$N_g = N_g(f) = 1 + f + f(f-1) + \cdots + f(f-1)^{g-1}$$

$$= 1 + f \sum_{i=0}^{g-1} (f-1)^i = \frac{f(f-1)^g - 2}{f-2} \tag{20.2}$$

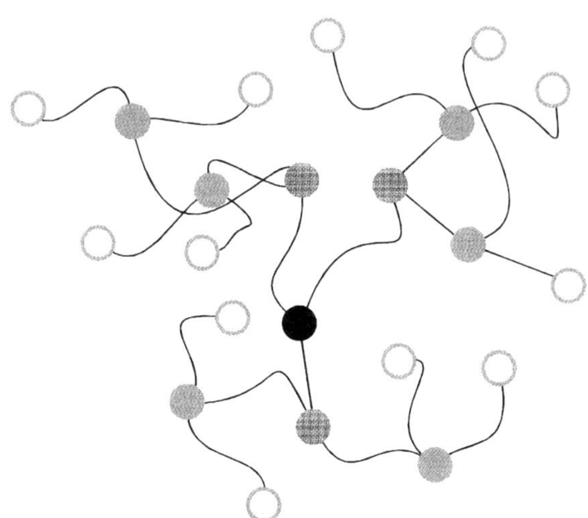

Figure 20.1 A model dendrimer molecule. The phantom links between units represented by circles are Gaussian subchains. The higher generation of the units the lighter is their color

is simply the total number of units in the dendrimer, while the second one

$$M_g(f) = \frac{(f-1)^g - 1}{f - 2} \tag{20.3}$$

is the number of units in each branch of a dendrimer extending from the core, i.e., from the unit in the zero-th generation.

According to the approach outlined in Appendix the unperturbed mean square radius of gyration of a Gaussian molecule is given by

$$R_G^2 = \frac{b_{\text{eff}}^2}{N}(\text{tr } \mathbf{K}^{-1} - N^{-1}\mathbf{j}^T\mathbf{K}^{-1}\mathbf{j}) \tag{20.4}$$

where tr $\mathbf{K}^{-1}$ is the trace (sum of diagonal entries) of the inverse of $\mathbf{K}$, the matrix obtained from the Kirchhoff matrix $\mathscr{K}$ of the corresponding molecular graph. $\mathbf{K}$ is obtained from $\mathscr{K}$ by deleting the first row and the corresponding column of the latter, $\mathbf{j}$ is the column vector of the size matching that of $\mathbf{K}$ whose all elements are ones, and superscript 'T' stands for the vector transpose. N is the size of dendrimer as given by equation. (20.2). The last term, $\mathbf{j}^T\mathbf{K}^{-1}\mathbf{j}$, is simply the sum of all entries of $\mathbf{K}^{-1}$.

The regularity of dendrimer graphs make their Kirchhoff matrices very neat. The inverse $\mathbf{K}^{-1}$ has the same regularity and can be expressed as the block diagonal matrix of the form

$$\mathbf{K}_g^{-1} = \begin{bmatrix} \mathbf{A}_g^{-1} & \mathbf{0} & .. & \mathbf{0} \\ \mathbf{0} & \mathbf{A}_g^{-1} & .. & \mathbf{0} \\ .. & .. & .. & .. \\ \mathbf{0} & \mathbf{0} & .. & \mathbf{A}_g^{-1} \end{bmatrix} \tag{20.5}$$

There are f blocks in the inverse and each has the size $M_g \times M_g$ (M_g is given by equation (20.3)). The matrices $\mathbf{A}$ are in fact the Kirchhoff submatrices corresponding to branches extending from the core unit. The most important property of an $\mathbf{A}$ is that its inverse is given by the following recurrent relationship.

$$\mathbf{A}_g^{-1} = \mathbf{U}_g + \begin{bmatrix} 0 & 0 & .. & .. & 0 \\ 0 & \mathbf{A}_{g-1}^{-1} & 0 & .. & \mathbf{0} \\ .. & 0 & \mathbf{A}_{g-1}^{-1} & .. & \mathbf{0} \\ .. & .. & .. & .. & .. \\ 0 & \mathbf{0} & \mathbf{0} & .. & \mathbf{A}_{g-1}^{-1} \end{bmatrix} \tag{20.6}$$

where $\mathbf{U}$ is the matrix of ones and there are $f - 1$ blocks in the matrix.

For a dendrimer with g generations equation (20.4) can be written in the form

$$R_G^2 = \frac{f b_{\text{eff}}^2}{N_g}\left[M_{g-1} + (f-1)\text{tr}\left(\mathbf{A}_{g-2}^{-1}\right) - N_g^{-1}\left(M_{g-1}^2 + (f-1)\mathbf{j}^T\mathbf{A}_{g-2}^{-1}\mathbf{j}\right)\right] \tag{20.7}$$

Since $\mathbf{A}_1^{-1} = A_1^{-1} = 1$, one finds for any generation g

$$\mathrm{tr}(\mathbf{A}_g^{-1}) = \frac{(f-1)^g[g(f-2)-1]+1}{(f-2)^2} \tag{20.8}$$

and

$$\mathbf{j}^T\mathbf{A}_g^{-1}\mathbf{j} = \sum_{i=0}^{g-1} \left[\frac{(f-1)^{g-i}-1}{f-2}\right]^2 (f-1)^i \tag{20.9}$$

RESULTS AND DISCUSSION

The explicit mean square radii of gyration for model Gaussian dendrimers are presented in Table 20.1 for $f = 3$, 4, and 6 and for generations g up to 10. The values are very close to those one can calculate using the equations published by Carl [11] and La Ferla [12]. The slight differences seem due to different simplifications used in various versions of the same model. As one can see in Table 20.1, although the model applied is purely graph theoretical and describes phantom molecules without any volume ascribed to units or chains, there are differences between mean square radii of gyration for dendrimers of different functionality of units. They are indeed linear functions of the number of generations, see Figure 20.2, as required by equation (20.1). When log-log plotted against the mass of dendrimers represented by N they give curved relationships (Figure 20.3).

Table 20.1 The mean square radii of gyration for Gaussian models of dendrimers with g generations of f-functional units

| Number of | R_G^2/b_{eff}^2 | | |
generations	$f = 3$	$f = 4$	$f = 6$
1	0.562	0.640	0.735
2	1.170	1.384	1.604
3	1.878	2.244	2.559
4	2.675	3.176	3.546
5	3.542	4.146	4.543
6	4.457	5.133	5.542
7	5.405	6.128	6.542
8	6.374	7.126	7.542
9	7.356	8.125	8.542
10	8.346	9.125	9.542

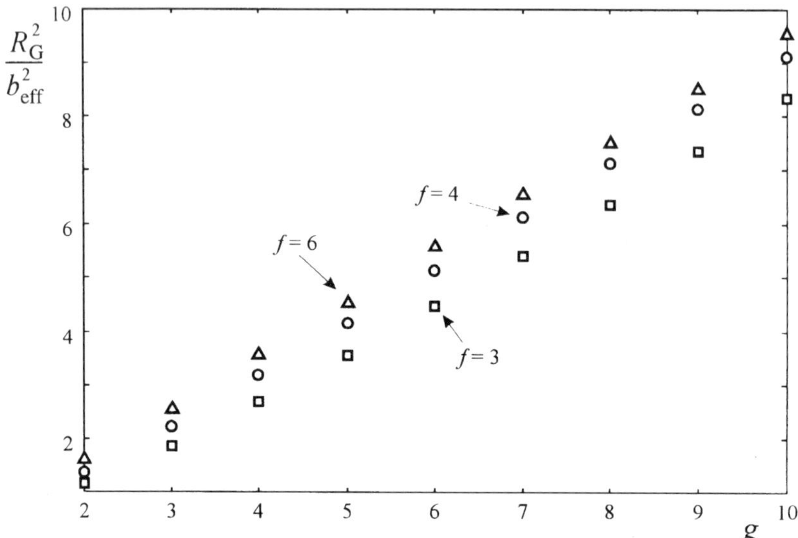

Figure 20.2 The dependence of the mean square radii of gyration of the Gaussian dendrimer molecules on the number of generation, *g*. The mean square radius is expressed in terms of the mean square bond length units

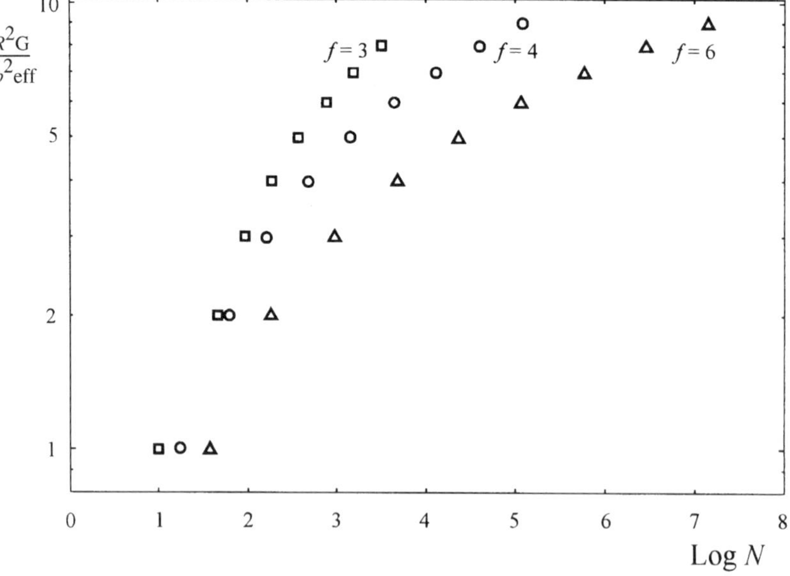

Figure 20.3 The log-log plot of the mean square radii of gyration of the Gaussian dendrimer molecules vs. number of units in the molecule. No single exponent (fractal dimension) seems to exists in the range of structures studied

ACKNOWLEDGMENTS

The financial support of this work from the Polish Committee of Scientific Research, grant no. 3T09A 038 011, is gratefully acknowledged.

APPENDIX

An instantaneous configuration of a Gaussian molecule having N units in a d-dimensional space is fully described by the $d \times N$ matrix $\mathbf{R} = (\mathbf{r}_1, \mathbf{r}_2, \ldots, \mathbf{r}_N)$ where $\mathbf{r}_j$ is the vector linking the center of mass of the molecule with the jth unit. The bonds linking units are considered in the model to be identical springs the length of which has the Gaussian distribution with the mean square length b_{eff}^2. By assuming the contributions of the bonds-springs to the configurational potential to be additive, the latter can be expressed as [13]

$$V(\mathbf{R}) = \frac{d}{2b_{\mathrm{eff}}^2} kT \sum_{\substack{\text{allpairs} \\ i-j}} (\mathbf{r}_i - \mathbf{r}_j)^2 = \frac{d}{2b_{\mathrm{eff}}^2} kT \operatorname{tr}(\mathbf{R}\mathscr{K}\mathbf{R}^T) \tag{A.1}$$

where k is the Boltzmann constant, T is the absolute temperature and $\mathscr{K}$ is the Kirchhoff matrix which contains information on the connectivity of the molecular graph. The Kirchhoff matrix is easily constructed after labeling units of the molecular graph in an essentially arbitrary way. The diagonal entries of the Kirchhoff matrix are functionalities of units and the off-diagonal ones are zeros or negative integers. The (i, j) entry $(i \neq j)$ is the negative of the number of links between the unit labeled i and the unit labeled j. Since for a molecule containing N units there are $N!$ ways one can define its Kirchhoff matrix the method of labeling may be crucial. We labeled the graph of the dendrimer starting from the central (core) unit which had label 1. Then the whole dendron extending from the core was labeled outwards up to the unit of functionality 1 in the highest generation. It was given the label $g + 1$. Labels $g + 2$, $g + 3$, etc., had 'brothers' of this monofunctional unit, i.e., units extending from the same 'father' unit in generation $g - 1$. Then the units of the father's brother were labeled, etc. This method of labeling yielded the block structure of the Kirchhoff matrix which greatly facilitated inverting its submatrices.

The potential (A.1) was used to derive the configurational integral and then the mean square radius of gyration according to the procedure described by Eichinger [13]. The latter is expressed simply as the sum of reciprocals of all eigenvalues λ_i of the Kirchhoff matrix:

$$R_{\mathrm{G}}^2 = \frac{b_{\mathrm{eff}}^2}{N} \sum_{i=1}^{N} \lambda_i^{-1} \tag{A.2}$$

For an ordinary matrix the sum of the reciprocals of its eigenvalues is simply equal to the trace of its inverse (i.e., the sum of the inverse diagonal entries). Unfortunately, the Kirchhoff matrix of a connected graph is singular and has exactly one zero eigenvalue. One can use Moore–Penrose generalized inverse [17] or calculate eigenvalues numerically. The special structure and properties of the Kirchhoff matrix makes it possible to replace its generalized inverse by the ordinary inverse of a submatrix of the Kirchhoff matrix obtained by deleting one row and its corresponding column (we have deleted the row and column of the dendrimer core). The trace of the inverse of $\mathscr{K}$ is then expressed by the term in parenthesis of equation (20.4).

REFERENCES

1. see, e.g., G.R. Newkome, C.N. Moorefield and F. Vögtle, *Dendritic molecules. Concepts. Syntheses. Perspectives*, VCH Verlag GmbH, Weinheim, 1996.
2. P.-G. DeGennes and H. Hervet, *J. Phys. Lett.*, **44**, L351–60 (1983).
3. J.F.G.A. Jansen, E.W. Meijer and E.M.M. de Brabander-van den Berg, *J. Am. Chem. Soc.* **117**, 4417–8 (1995); S. Stevelmans, J.C.M. van Hest, J.F.G.A. Jansen, D.A.F.J. van Boxtel and E.M.M. de Brabander-van den Berg, *J. Am. Chem. Soc.*, **118**, 7398–9 (1996).
4. R.L. Lescanec and M. Muthukumar, *Macromolecules*, **24**, 4892–7 (1991).
5. M.L. Mansfield and L.I. Klushin, *Macromolecules*, **26**, 4262–8 (1993).
6. D. Boris and M. Rubinstein, *Macromolecules*, **29**, 7251–60 (1996).
7. M. Murat and G.S. Grest, *Macromolecules*, **29**, 1278–85 (1996).
8. R.L. Lescanec and M. Muthukumar, *Macromolecules*, **23**, 2280–8 (1990).
9. H. Ihre, A. Hult and E. Soderlind, *J. Am. Chem. Soc.*, **118**, 6388–95 (1996).
10. Z.Y. Chen and S.-M. Cui, *Macromolecules*, **29**, 7943–52 (1996).
11. W. Carl, *J. Chem. Soc., Faraday Trans.*, **92**, 4151–4 (1996).
12. R. La Ferla, *J. Chem. Phys.*, **106**, 688–700 (1997).
13. B.E. Eichinger, *Macromolecules*, **13**, 1–11 (1980).
14. H. Galina, *Macromolecules*, **16**, 1479–83 (1983).
15. B.E. Eichinger, *Macromolecules*, **10**, 671–5 (1977); J.E. Martin and B.E. Eichinger, *J. Chem. Phys.*, **69**, 4588–94 (1978); B.E. Eichinger and J.E. Martin, *J. Chem. Phys.* **69**, 4595–9 (1978).
16. B.E. Eichinger, *Macromolecules*, **5**, 496–505 (1972).
17. C.R. Rao and S.K. Mitra, *Generalized Inverse of Matrices and its Application*, Wiley, New York 1971.

21

Hyperbranched Biopolymers

WALTHER BURCHARD,[1] THOMAS ABERLE[1] and
CATALINA ELENA IOAN[2]

[1]Institute of Macromolecular Chemistry, Hermann-Staudinger-Haus,
University of Freiburg, D-79104; Germany
[2]'P. Poni' Institute of Macromolecular Chemistry, 6600 Jassi, Romania

ABSTRACT

Hyperbranched polymers have a high segment density and offer attractive potentials
for functionalized well defined heterogeneities when covalently bound into a network
of low crosslinking density. The structure formation, even of very large *hyperbranched
biopolymers*, is controlled by the high specificity of enzymes. In contrast to corresponding
synthetic polymers ring formation and other side reactions are excluded. Furthermore the
molar mass and the branching density of these biological macromolecules can be varied in
a wide range by applying special enzymes. The paper describes a number of theoretically
predicted and experimentally observed properties. The knowledge of these parameters is
a fundamental prerequisite for a quantitative study of the influence of heterogeneities on

Wiley Polymer Networks Group Review Series Vol. 2. Edited by B.T. Stokke and A. Elgsaeter
© 1999 John Wiley & Sons Ltd

network properties. The present contribution is considered as complementary to the paper by Frey on chemically synthesized hyperbranched structures presented in this volume.

INTRODUCTION

Polymer networks are commonly considered as being homogeneous in their segment densities. This assumption appears reasonably fulfilled for bulk networks (no solvent). However, the crosslinking density is often enhanced in the neighborhood once a first crosslink has been formed. When these networks are immersed in a suitable solvent, the swelling causes a pronounced spatial heterogeneity in the segment density. The arising problems have been studied extensively by SANS [1] and a consistent interpretation of the results by a suitable theory is now in progress [2]. Recent studies with gels (formed from solutions) revealed that such heterogeneity is much more common than anticipated [3,4]. In all these cases the size and shape of these domains are not sufficiently well known. They are the basis of many not well–understood phenomena. Often just a simple spherical shape of impenetrable objects is assumed which may be not justified. Examples are the anisotropic domains in gel forming polysaccharides and proteins (gelatin, fibrin [5]), based on double helix formation and ensuing lateral aggregation. The domains are often not fully rigid and posses a certain capability for segment interpenetration [6,7].

In order to obtain a better insight into the effect of such heterogeneities on the properties of swollen networks it is desirable to prepare well–defined supramolecular structures which subsequently can be incorporated into a network. The advantage of such a procedure is that the properties of the built-in heterogeneities can be studied separately. Suitable objects are end functionalized regular star-branched macromolecules with many arms, dendrimers, polymacromonomers, hyperbranched materials and monodisperse microgels. The synthesis of regular structures involves laborious preparatory work that has to be done in many well–controlled steps. The preparation of hyperbranched materials on the other hand appears to be much easier. These materials are prepared, for instance, with AB_2 monomers, in which only reactions between functional groups A with B are possible. In a random process also reactions among alike functional groups are possible which soon leads to gel formation. The constraint of exclusive reactions among A and B functional groups shifts the gel point beyond $\alpha = 1$, i.e. gelation does not take place. Hence, exceedingly high branching densities can be obtained, for instance in the above mentioned example, every second repeating unit in the polymer can become a branching point. These samples may be appropriate since they have many functional end groups which could easily be incorporated via covalent bonding in a swollen network. Not all of them are needed for gel formation, the remaining ones could be used for other purposes, for instance as catalysts.

The possibilities of preparative chemistry are outlined in the contribution by Frey [8]. However, the desired structures are not as easily obtained as expected. (1) So far, only fairly low molar masses could be obtained, probably because of the quickly increasing branching density. (2) Ring formation seems to occur to an unexpectedly high extent which makes the desired simple hyperbranched structure unpleasantly complex [9,10]. (3) At present no possibility has been worked out that allows a variation of the branching density in a homopolymerization process. All these restrictions do not exist with hyperbranched biopolymers: ring formation and other side reactions are excluded due to the high specificity of enzymes, and the branching density is controlled by the inequality of the two functional groups B_1 and B_2 in the A $\prec^{B_1}_{B_2}$ monomer. Thus the biological examples provide a much wider range of defined structures than available so far with synthetic monomers.

HYPERBRANCHED POLSACCHARIDES

The two monomers which have the characteristics for hyperbranching are glucose and fructose, often combined as a hetero-dimer in sucrose. Their chemical structures are shown in Figure 21.1. The glucose is the repeating unit in amylopectin, glycogen and in dextran, it has a reducing endgroup in C1 position (A-group) and three non−reducing groups of which only the two in C4 and C6 positions (B_1 and B_2 groups) of the sugar ring are used. In amylopectin the $\alpha(1,4)$ bond is the predominant linkage, and the $\alpha(1,6)$ bonds form the branches, while in dextran the flexible $\alpha(1,6)$ bonds establishes the main linkages and the $\alpha(1,4)$ bonds form the branches. Depending on the microbiological system also $\alpha(1,3)$ linkages are found in dextran, but these result in short chains which are not further branched and do not take part in the hyperbranching process [11]. Fructose has two primary OH groups in the C1 and C6 positions and three secondary OH groups of which only the one in C2 position is used in fructans. It has no reducing end group. Here the secondary OH in C2 represents the functional group A, and reaction can occur with the primary OH groups in C1 and C6 (the B_1 and B_2 groups). Fructans with predominant C2–C1 linkages belong to the *inulin* type while those with predominant C2–C6 linkages are of the *levan* type [12]. The C2–C6 bonds form the branching points in inulins, while those in levans are the C2–C1 linkages. The main difference in the two types of fructans consists in the fact that the inulins resemble derivatized polyethyleneoxides, with the sugar ring as side group, while in levans the sugar ring is built in the chain. Therefore, the inulins should be fairly flexible, whereas the sugar ring in the main chain of levan will cause a more extended conformation. The structures of the main chain in the two fructans are shown in Figure 21.2. (The structures of amylopectin and dextran are well known [13] and are not reproduced here.) In common hyper-branched materials, as defined by Flory [14], the two functional B-groups have the same reactivity. Its branching density increases with increasing molar mass

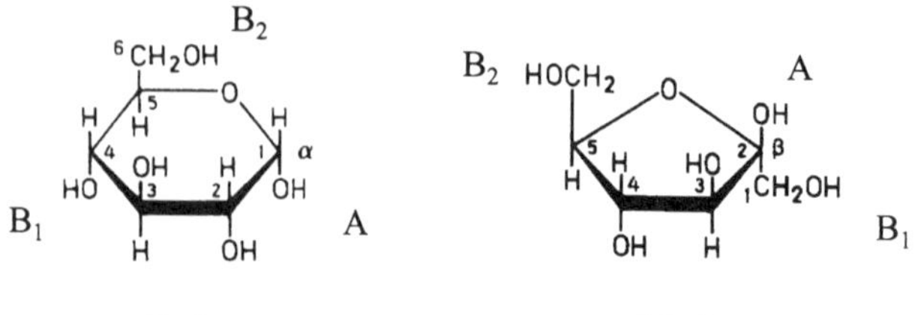

Figure 21.1 α-*D*-Glucose (left) and β-*D*-Fructose (right) as typical monomers for hyperbranching. The reducing end groups in C1 position of the glucose form bonds only with the OH groups in the C4 and C6 positions. Fructose has no reducing end group. The secondary OH group in C2 position establishes here the functional group A while the two primary OH groups in the C1 and C6 positions represent the two functional groups B_1 and B_2, respectively

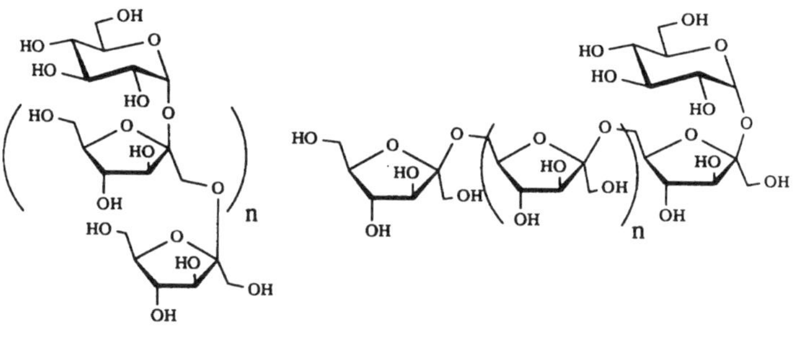

Figure 21.2 Fructans of the inulin and levan type. In the inulin type the sugar forms a side chain of a derivatized polyoxyethylene. In the levan type the sugar is in the main chain

and reaches a limiting value of 50%. In the general case of non-equal reactivity of the B-groups a branching probability p can be defined by $\beta_1 = \alpha(1 - p)$ and $\beta_2 = \alpha p$, where β_1 and β_2 are the probability of reaction of the B_1 and B_2 groups, respectively. The branching probability can range from $p = 0$ (linear chain, no branching) up to $p = 0.5$ (maximum branching, equal reactivity of the B-groups). Note: the branching density is defined as the ratio of the number of branching units to the total number of reacted monomers. The branching density increases when the particle grows in size, i.e. when the probability of reaction α is increased. In hyperbranched polysaccharides the molar mass is mostly very high and $\alpha \approx 1$. Therefore the branching density (or degree of branching) differs only insignificantly from the branching probability p.

PROPERTIES OF HYPERBRANCHED STRUCTURES

MOLAR MASS DISTRIBUTIONS

The weight fraction molar mass distribution for common AB_2 polymers was calculated by Flory more than 50 years ago [14]. Only a few years later the distributions for the general cases were given by Erlander and French [15]. Details of the lengthy equations are given in a recent review and in the original papers [16]. Figure 21.3 demonstrates the difference in the molar mass distributions of the randomly branched A_3 and the non-randomly branched AB_2 polymers, respectively, where the latter form the hyperbranched structure [17]. In addition Figure 21.4 shows the change in the distributions when the branching probability is changed. The two types of distribution differ strongly in their polydispersities. For the randomly branched materials one has in the limit of large molar masses M_w a polydispersity ratio of $M_w/M_n \propto DP_w$, but for the hyperbranched material the polydispersity increases only with the square root of the weight average degree of polymerization $M_w/M_n \propto (DP_w)^{1/2}$. The polydispersity is decreased when $p < 0.5$; the limiting behavior is obtained in all cases (with the exception of $p = 0$), but for small p the limit is reached at much higher degrees of polymerization. Figure 21.5 gives the results from measurements with levans [18] and randomly branched polyester fractions [19]. The data demonstrate the approximate properties of hyperbranching for the polysaccharide and of random branching for the polyester.

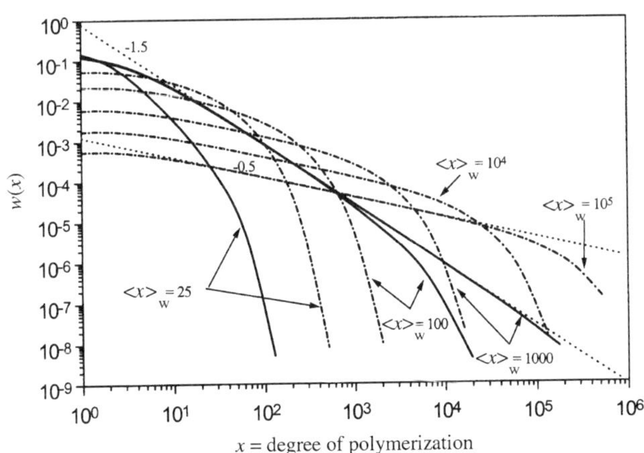

Figure 21.3 Double logarithmic plot of the weight fraction molar mass distributions $w(x)$ for randomly branched three functional polycondensates for three different weight average degrees of polymerization x_w (A_3 monomers, full lines) compared with that of the AB_2 hyperbranched macromolecules (dashed lines, 5 different x_w). Note: the randomly branched polymer approaches a power law with exponent $-(\tau - 1) = -1.5$. Unexpectedly also the hyperbranched distribution approaches a power law behavior with exponent $-(\tau - 1) = -0.5$ [14,17]. (The power law behavior is indicated by dotted lines)

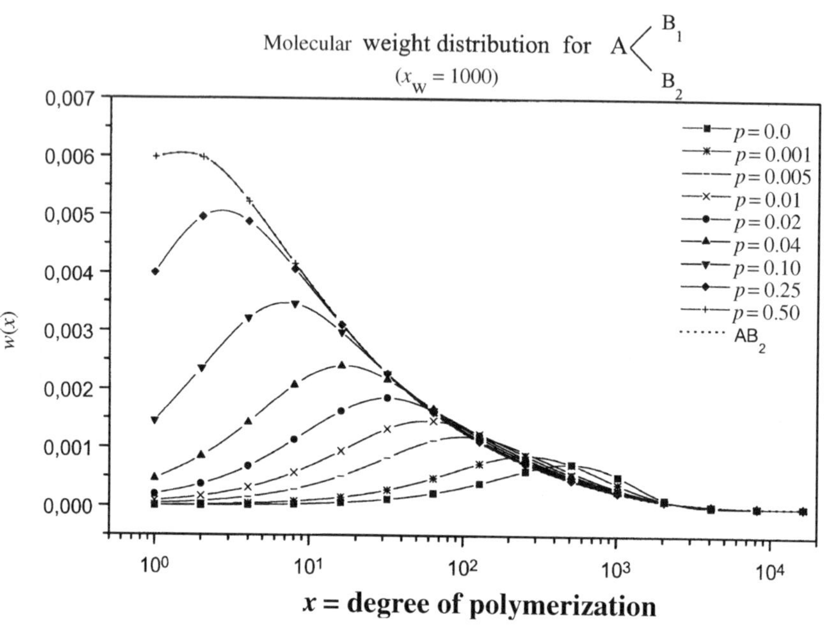

Figure 21.4 Change of the molar mass distribution $w(x)$ with the propability of branching p for hyperbranched macromolecules [15,17]. (Note: $w(x)$ is plotted linearly against $\log x$)

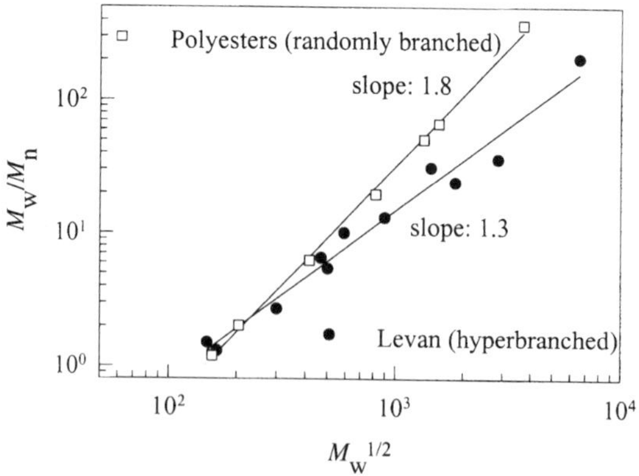

Figure 21.5 Double logarithmic plot of the polydispersity ratios M_w/M_n as a function of $M_w^{1/2}$ levans [18] (filled circles) in comparison with the data obtained from randomly crosslinked polyesters [19]. The hyperbranched samples approximately follow the predicted $M_w^{1/2}$ behavior whereas for the randomly branched samples the predicted M_w dependence is approached

ANGULAR DISTRIBUTION OF SCATTERED LIGHT

The conformational properties of the generalized hyperbranched polymers were calculated in 1972 [20] on the basis of the cascade theory, developed by Gordon in 1962 [21,22]. Most informative is the angular dependence in static light scattering (LS) which was derived on the basis of unperturbed Gaussian statistics. Figure 21.6(b) shows some of the theoretical curves for a branching probability of 0.04, and the curves are contrasted in Figure 21.6(a) with the actual measurements from amylopectin fractions. A fit of the scattering curves with the theoretical

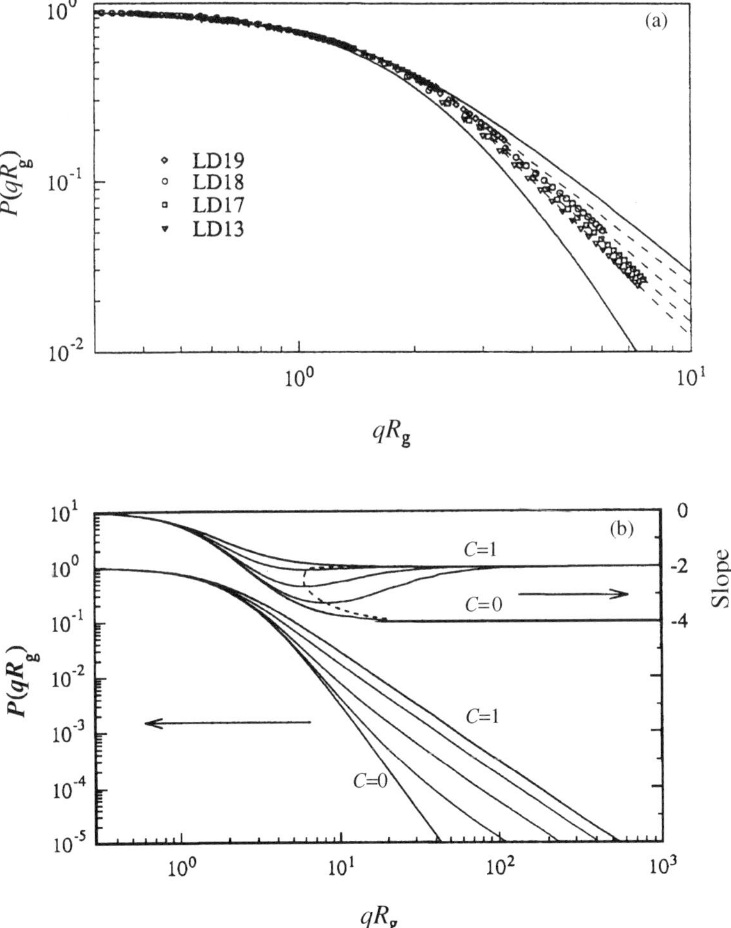

Figure 21.6 Double logarithmic plot of the particle scattering factors $P(qR_\mathrm{g})$ (a) from measurements with amylopectin fractions [6] of different molar masses compared with (b) theoretical predictions [20] when the branching probability is varied. R_g is the radius of gyration and $q = (4\pi/\lambda)\sin(\theta/2)$

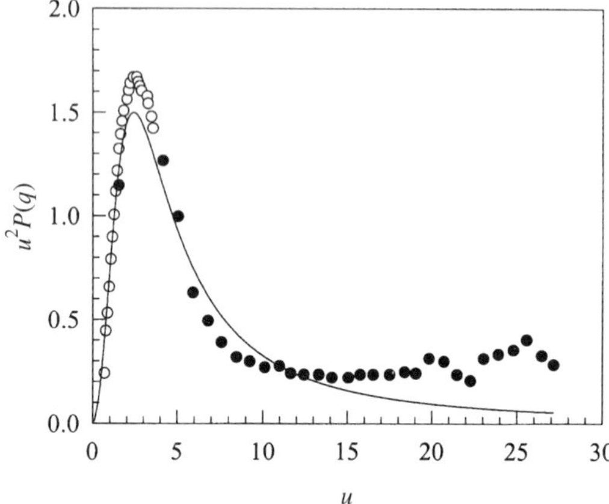

Figure 21.7 Kratky plot of SANS measurements from mussels glycogen [24] in D_2O. The line corresponds to a theoretical curve given in Ref. [24] with a characteristic parameter of $C = 0.00057$ [16], under the assumption of Gaussian chain statistics for hyperbranched polymers. The increase at large qR_g values results from excluded volume interactions and partly from a certain chain stiffness, the steeper decrease is an effect of a dense packing of repeating units at the center of the macromolecules [25]

relationship for the particle scattering factor $P(qR_g)$ (where $q = (4\pi/\lambda)\sin(\theta/2)$) gave a branching density of only 0.016. This deviation was discussed in detail previously [23] and was interpreted to result mainly from the neglect of excluded volume interactions. Similar deviations were also found by SANS from mussels glycogen in D_2O [22,24]. Figure 21.7 shows a Kratky plot of measurements in comparison with theory for hyperbranching assuming Gaussian statistics. An overall agreement in behavior is observed. Deviations occur in two regions of $q = (4\pi/\lambda)\sin(\theta/2)$: (i) the experimental curve decays steeper in an intermediate q-region; this was interpreted as a result of the finite volume of the glucose units which are densely packed around the center of the molecule. (ii) at large q-values an increase occurs again; this increase could be a result of chain stiffness, but more likely it is the effect of excluded volume which should cause a predicted increase of $q^{0.4}$ in the Kratky plot. (The exponent results from the fractal dimension of freely swollen branched clusters $d_f = 1.6$.)

INTERPARTICLE INTERACTION AND RESTRICTED SEGMENT INTERPENETRATION

Since the hyperbranched materials are intended to use as models for the study of heterogeneity in a network it is of special interest to know to which extent a segment interpenetration would be possible [6,7]. To this end we studied with four examples of hyperbranched glucans the inter-particle interactions. These were native waxy maize starch ($M_\mathrm{w} = 76 \times 10^3$ kg/mol) in water [6], partially hydrolyzed amylopectins ($M_\mathrm{w} = 35\text{--}800$ kg/mol) in 0.5 M NaOH [7], hydroxyethyl (waxy maize) starch in water ($M_\mathrm{w} = 200$ kg/mol) [6] and glycogen from mussels ($M_w = 7.1 \times 10^3$ kg/mol) in 0.5 M NaOH. The branching densities were $p = 0.04$ for the amylopectins and $p = 0.08$ for the glycogen [26]. The interparticle interactions could be measured directly by static light scattering from the reciprocal scattering intensity at zero scattering angle. When normalizing these data with respect of the molar mass (at zero concentration) one obtains a relationship that can be approximated by the following equation [6,7,27]:

$$\frac{M_\mathrm{w}}{M_\mathrm{app}(c)} = 1 + 2X + 3g_\mathrm{A}X^2 + 4h_\mathrm{A}X^3 \tag{21.1}$$

$$X = A_2 M_\mathrm{w} c \propto c/c^* \tag{21.2}$$

where $M_\mathrm{app}(c) = R_{\theta=0}/Kc$ is the apparent molar mass at the concentration c, A_2 the second virial coefficient and c^* the overlap concentration. All amylopectin data form a common curve up to $X \cong 4$ that lies in between the theoretical curves for flexible linear chains and hard (impenetrable) spheres. For the glycogen a slightly stronger increase was observed. The difference is better recognized when the forward scattering intensity is used and normalized by multiplication with the second virial coefficient. Also this curve can be represented in a scaled form as follows [27]:

$$A_2\frac{R_{\theta=0}}{K} = \frac{X}{1 + 2X + 3g_a X^2 + 4h_a X^3} \tag{21.3}$$

Figure 21.8 shows the result of these measurements. The behavior corresponds to the expectation since the glycogen with its higher branching density will have, relative to its overall dimension, a larger impenetrable core than the amylopectins and should therefore be more like hard spheres. At $X > 4$ deviations from the expected curves are obtained. In both cases the strong upturn in Figure 21.8 results from the onset of association. For glycogen eventually phase separation occurred, and was accompanied by a strong turbidity. The association was confirmed by the observation of a marked increase in the radius of gyration $R_\mathrm{g}(c)$ [6].

INTRINSIC VISCOSITY

Assuming validity of the Fox–Flory relationship also for branched structures one has

$$[\eta] = \Phi_\mathrm{b}\frac{R_{\mathrm{g,b}}^3}{M_\mathrm{w}} \tag{21.4}$$

Thus, one should find a decrease of the intrinsic viscosity when the branching density is increased. This behavior was indeed observed as is shown in

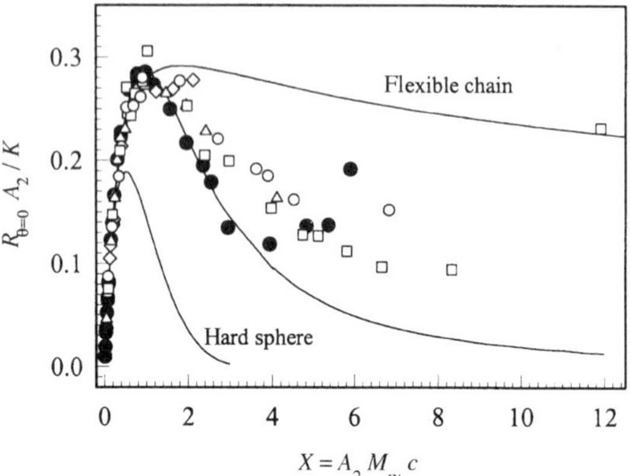

Figure 21.8 Plot of the normalized forward scattering as a function of the scaled concentration $X = c/c^* = (A_2 M_w c)$, where c^* is the overlap concentration and A_2 the second virial coefficient, for hydrolyzed amylopectins [7] and glycogen from mussels. The segment interpenetration is more strongly inhibited with the more extensively branched glycogen than with the amylopectins: therefore the glycogen curve becomes more like a hard sphere. The full lines represent theoretical curves for special model structures

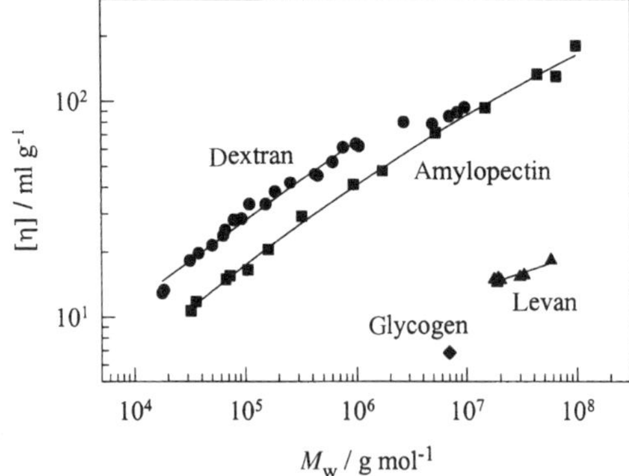

Figure 21.9 Molar mass dependence of the intrinsic viscosity $[\eta]$ for dextran in water ($p \approx 0.04$) [28], hydrolyzed amylopectins in 0.5 M NaOH ($p \approx 0.04$) [23], levan in DMSO ($p \approx 0.09$) [18] and glycogen in 0.5 M NaOH($p \approx 0.08$) [26]. As expected the viscosities are strongly lowered when the branching density is increased. Because of different solvents used and differences in the monomer topology only the trend can be observed

Figure 21.9. A quantitative estimation of the branching density is at present not possible, since the influence of a screened particle draining on the parameter Φ_b is yet not known [17]. Furthermore, polydispersity has some effect. This will be the subject of further studies in future work.

CROSSLINKING

Crosslinking of amylopectins by epichlorhydrine has already been applied in food industry for a long time. Here the interest was focussed on preparing thickeners of especially high efficiency. These examples make clear that covalent binding of water soluble hyperbranched materials into a hydrogel will cause no particular difficulties. Corresponding work is in progress in our laboratory.

REFERENCES

1. J. Bastide and S.J. Candau, 'Structure of gels as investigated by means of static scattering', in *Physical Properties of Polymer Gels*, Ed. (J.P. Cohen-Addad, ed.) Wiley, Chichester, (UK) 1996.
2. Y. Rabin and R. Bruisma, *Europhys. Lett.*, **20**, 79 (1992).
3. Y. Rabin, P. Pekarski and R. Bruisma, *Europhys. Lett.*, **24**, 145 (1993).
4. W. Burchard, T. Aberle, T. Fuchs, W. Richtering, T. Coviello, E. Geissler and L. Schulz, Ch. 2 in *Chemical and Physical Networks*, Vol. 1, eds. K. te Nijenhuis and W.J. Mijs, Wiley, Chichester (UK) 1998.
5. A.H. Clark and S.B. Ross-Murphy, *Adv. Polym. Sci.*, **83**, 58 (1987).
6. T. Aberle and W. Burchard, *Comput. Theor. Polym. Sci.*, **7**, 215 (1997).
7. G. Galinsky and W. Burchard, *Macromolecules*, **29**, 1498 (1996).
8. H. Frey, see contribution in this volume.
9. C. Cameron, A.H. Fawcett, C.R. Hetherington, R.A.W. Mee and F.V. McBride, *J. Chem. Phys.* **19**, 8235 (1998).
10. K. Dusek, private communication
11. L. Kenne and B. Lindberg, *Bacterial Polysaccharides*, in G. Aspinall (ed.) *The Polysaccharides*, Academic Press, New York and London 1983, Vol. 2, pp. 346–347.
12. M.A. Clarke, A.V. Bailey, E.J. Roberts and W.S. Tsang, 'Polyfructose: a new microbial polysaccharide', in *Carbohydrates as Organic Raw Materials.* (F.W. Lichtenthaler, ed.) VCH, Weinheim, Germany, 1991.
13. Ch. Mercier and A. Guilbot, Starch in *The Polysaccharides*, G. Aspinall, (ed.) Academic Press, New York and London, 1985, Vol. 3. pp. 210–283.
14. P.J. Flory, *Principles of Polymer Chemistry*, Cornell University Press, Ithaca., N.Y., 1953.
15. S.R. Erlander and D. French, *J. Polym. Sci.*, **20**, 7 (1956).
16. G. Galinsky and W. Burchard, *Macromolecules*, **30**, 4445 (1997).
17. W. Burchard, *Adv. Polym. Sci.*, **143**, 1 (1999).
18. S.S. Stivala, J.J. Zweig and J. Ehrlich, *'Dilute solution properties of streptococcus salvisarius levan and its hydrolysates'*, in *Solution Properties of Polysaccharides* (A.A. Brant, ed.), *ACS Symp. Ser.*, **150**, 101 (1981).
19. V. Trappe, Ph.D. Thesis, University of Freiburg 1994.
20. W. Burchard, *Macromolecules*, **5**, 604 (1972).
21. M. Gordon, *Proc. Roy. Soc. (London)*, **51**, 240 (1962).
22. W. Burchard, *Adv. Polym. Sci.*, **48**, 1 (1983).
23. G. Galinsky and W. Burchard, *Macromolecules*, **28**, 2363 (1995).
24. W. Burchard, *Macromolecules* **10**, 919 (1977); see also Ref. [22], p. 72.
25. M. Daoud, J.-P. Cotton, *J. Physique (Paris)*, **43**, 531 (1982).
26. P. Mischnick, personal communication; to be published.
27. W. Burchard, *Macromol. Symp.*, **39**, 179 (1990).
28. F.R. Senti, N.N. Hellmann, N.H. Ludwig, G.E. Babcock, R. Tobin, C.A. Glass and B.L. Lamberts, *J. Polym. Sci.*, **17**, 533 (1955).

22

Amphiphilic Multicomponent Polymer NETWORKS: Design, Evaluation and Applications

F.E. DU PREZ,[1] D. CHRISTOVA[2] and E.J. GOETHALS[1]

[1]Department of Organic Chemistry, Polymer Chemistry Division, University of Gent, Krijgslaan 281 S4, B-9000 Gent, Belgium
[2]Bulgarian Academy of Sciences, Institute of Polymers, 1113 Sofia, Bulgaria

ABSTRACT

Polymer blends and copolymer networks were formed between a hydrophilic polymer, i.e. poly(ethylene oxide) [PEO, $(CH_2CH_2O)_n$] or poly(1,3-dioxolane) [polyDXL,

---

Wiley Polymer Networks Group Review Series Vol. 2. Edited by B.T. Stokke and A. Elgsaeter
© 1999 John Wiley & Sons Ltd

$(CH_2OCH_2CH_2O)_n$] and a hydrophobic polymer with a high T_g, i.e. poly(methyl methacrylate), or a low T_g, i.e. poly(butyl acrylate). While a coarse phase separation was observed in the polymer blends, compatibility was strongly increased by the introduction of crosslinks between the polymers. These AB-block copolymer networks were synthesized by the free radical copolymerization of bis-macromonomers of the hydrophilic polymers (e.g. α, ω-acrylate-terminated PEO) with a vinyl monomer (e.g. MMA, BA). This leads to networks with well-defined segmented structures. The nature and the ratio of the two components in the amphiphilic segmented networks determine both the morphology and the hydrophilic/hydrophobic balance while the molecular weight and the fraction of the bis-macromonomer determine the crosslink density. DMA, DSC and swelling experiments revealed the influence of these parameters on the ultimate phase behaviour. The forced compatibility, the mechanical stability and the controllable crosslink density of the segmented networks could lead to applications as polymer membranes for pervaporation and as intermediates in the field of conductive PEO-materials.

INTRODUCTION

It is well-known that combining two polymers enables one to adjust the properties of polymer materials, not only by variation of the nature or the ratio of the polymers but also by changing the way of blending [1–3]. The way of blending determines the ultimate phase morphology of the end product, which in turn is governed by the thermodynamic incompatibility of the polymer segments. In general, the end products are referred to as multicomponent polymer materials. Two thermoset combinations of polymers, schematically presented in Figure 22.1, belong to this category: the interpenetrating polymer networks (IPN) and the AB-crosslinked polymers (ABCP).

The IPNs, which have been extensively studied during the last two decades, are defined as a combination of two or more polymers in network form, at least one of which is synthesized in the immediate presence of the other [4–6]. In an

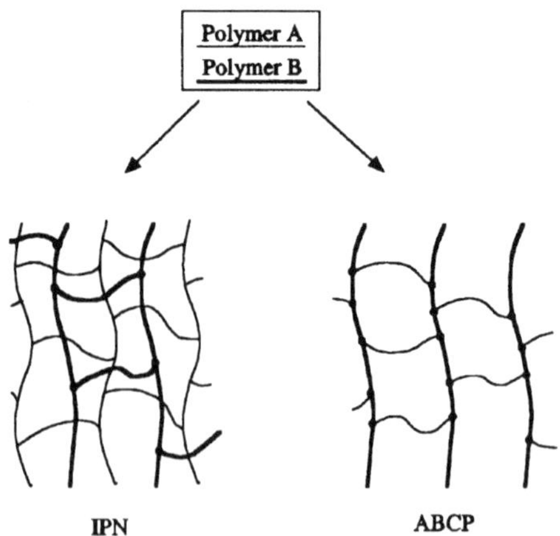

Figure 22.1 Schematic structures of an IPN and an AB-cross-linked polymer

ABCP, on the contrary, a single network is generated by grafting of polymer A to polymer B at both ends or at various points along the chains. In this way, intrinsically immiscible polymers can be combined by permanent fixations. The physicochemical properties of the ABCPs could be expected to be the result of a combination of the properties of segmented structures such as block copolymers on one hand and of polymer networks on the other hand. Although different preparation methods have been described in literature, few of them result in the formation of well-defined networks with good control of the molecular weight between the crosslinks and of the number of branches at the junction points [7–10]. Tezuka *et al.* [11,12], for example, proposed an elegant system for the synthesis of 'two-component model networks' based on a macromolecular ion-coupling reaction between pyrrolidinium and carboxylate-containing prepolymers. Other authors have described amphiphilic network structures that have been obtained by two approaches. In the first approach, a hydrophobic polymer containing polymerizable groups at both chain ends, also called a bis-macromonomer, is copolymerized with a hydrophilic monomer [13]. In the second, a hydrophilic bis-macromonomer is reacted with a hydrophobic polymer [14].

The copolymer networks we focus on in the present investigation, further referred to as segmented networks, were prepared from the combination hydrophilic bis-macromonomer-hydrophobic comonomer [3,15,16]. In this chapter, we report on the description of networks based on the bis-macromonomers (acrylate endgroups) of two water-soluble, semi-crystalline polymers, polyDXL and PEO, for which the procedures of synthesis lead to controllable molecular weights and functionality. PEO was chosen for its wide application in polymer alloys [17,18], while polyDXL was selected as starting material because it was demonstrated earlier that this polymer is a hydrophilic material which can be degraded in acid environment and which could be applied for different reasons [3,19]. In this study, these well-defined pre-polymers were combined with hydrophobic vinyl polymers by free radical copolymerization of their end-groups with vinyl monomers such as methyl methacrylate (MMA) or butyl acrylate (BA). This results in the formation of amphiphilic segmented networks in which the hydrophilic polymer segments act as a polymeric crosslinker for the hydrophobic polymer chains. The first results on the synthesis and properties of these networks have already been reported [16].

The dependence of the morphology and swelling properties of the networks on the nature and ratio of both segments as well as on the molecular weight of the bis-macromonomer will be described. The influence of these parameters on the crystallinity behaviour of the PEO-containing networks will also be discussed. This interest originates from the possible use of PEO as a polymer electrolyte (e.g. as solid batteries) at ambient temperature after complexation with metal salts [20]. As the conductivity is related to the amorphous part of the polymer [21], it will be investigated whether the crystallinity of PEO can be decreased by incorporation of the PEO segments in a copolymer network. The mechanical and dimensional stability of the final polyelectrolyte will be provided by the chemical crosslinks, thus no longer by the crystalline parts of the polymer.

Another application of segmented networks, that has been described recently [19], is the use of the networks as membranes for the dehydration process of ethanol–water solutions by the pervaporation technique.

EXPERIMENTAL

MATERIALS

Poly(ethylene glycol) (PEG, $M_n = 1500$, 4000 and 6500 (Aldrich)) was purified by azeotropic distillation in toluene and by further drying under vacuum for 6 h at 80 °C (water content negligible). 1,3-Dioxolane (DXL) was purified by distillation over CaH_2 and dried on sodium wire under reflux. Dichloromethane was distilled twice over CaH_2, followed by drying under reflux on sodium–lead alloy for several hours. The initiator of the cationic polymerization, methyl trifluoromethanesulfonate (methyl triflate) was purified by distillation over CaH_2 just before use. Methyl methacrylate (MMA), methyl acrylate (MA), butyl acrylate (BA) and acryloyl chloride were refluxed over CaH_2 in the presence of a radical inhibitor, phenothiazine, before distillation. Triethylamine was distilled and refluxed prior to use. The initiators for the free radical copolymerization, bis(4-*tert*-butylcyclohexyl)peroxydicarbonate (Perkadox 16, purity 95%), or 2,4,6-trimethylbenzoyl (diphenyl)phosphine oxide (TMBPO) were used as received (Akzo and Aldrich).

SYNTHESIS OF BIS-MACROMONOMERS

The synthesis of α, ω-acrylate terminated PEO (illustrated for $M_n = 1500$) starts from the commerically available PEG. Into a 500 ml three-necked flask, equipped with a magnetic stirrer, 0.05 mol dried PEG (75 g), 200 ml of dichloromethane and 0.11 mol of triethylamine (proton trapper) were charged under nitrogen atmosphere. To this mixture, 0.11 mol of acryloyl chloride was added dropwise at a temperature below 10 °C. The esterification reaction proceeded overnight at room temperature. The solution was filtrated several times to remove the precipitated triethylamine hydrochloride salt. The clear liquid was washed with 5% NaOH and several times with distilled water until it became neutral. After treatment with $MgSO_4$ to remove water, the solution was filtered and CH_2Cl_2 was removed by rotary evaporation. Further drying was done under high vacuum at room temperature. A yield of approximately 85% was obtained. GPC and NMR confirmed the complete conversion of the hydroxyl to acrylate endgroups.

The synthesis of the polyDXL bis-macromonomers has been described in detail [22] and is based on the cationic ring-opening polymerization of 1,3-dioxolane in the presence of the transfer agent, methylene bis(oxyethylacrylate),

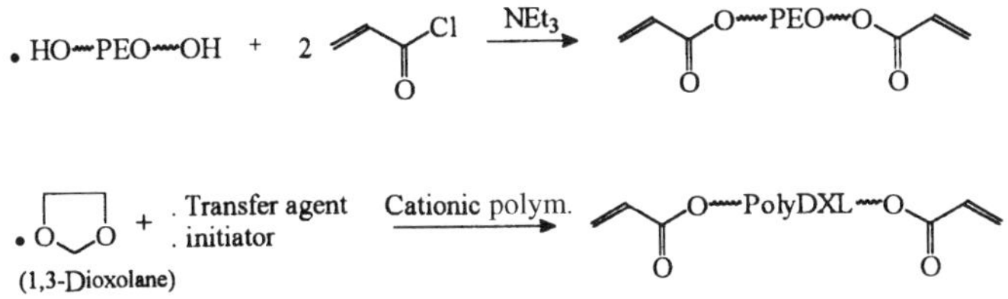

Figure 22.2 Synthesis scheme of the bis-macromonomers from PEO and polyDXL (acrylate endgroups)

that contains two acrylate groups. The molecular weight is governed by the ratio of reacted monomer to transfer agent and is well controllable in the range of 1000–10 000 g/mol. The synthetic procedures for the preparation of both bis-macromonomers are schematically presented in Figure 22.2.

SYNTHESIS OF SEGMENTED NETWORKS

Certain amounts of the vinyl monomer (MMA, MA or BA) and of the bis-macromonomer (BM), depending on the desired weight ratio, are mixed vigorously with 0.2 mol% (relative to vinyl monomer) Perkadox-16 (thermal initiator) or TMBPO (UV-initiator) at 50 °C for 3 min. For the synthesis of the networks with a high content of BM, a minimal amount of toluene is added to obtain a homogeneous and less viscous reaction mixture. The viscous solution was degassed for a few seconds before transferring it, by means of a syringe, between two glass plates that are kept at the desired distance (ultimate film thickness) by a silicone rubber spacer. Before use, the glass plates were treated with H_2SO_4 (95%) and with a solution (10%) of trimethylsilyl chloride in toluene to facilitate the recovery of the membrane after preparation. The glass plate mould containing the solution was kept in an oven for 80 min at 60 °C, 3 h at 80 °C and 12 h at 110 °C (for thermal initiators) or for 10–20 min under strong UV-light (365 nm) with an intensity of 25 mW/cm^2 (for UV-initiators). All films were subjected to a heat treatment for 12 h in a vacuum oven at 100 °C. A piece of each film was treated in a soxhlet apparatus with boiling ethanol for 12 h. Only films with an extractable fraction of less than 5 wt% were retained for further characterization. The difference between the degradation temperature of both components allowed us to determine the network composition by thermogravimetric analysis [19]. The networks will be coded as illustrated in the following example: *net*-poly(DXL$_{4000}$(30)-*co*-MMA(70)). In this network, 30% polyDXL bis-macromonomer ($M_n = 4000$) has been copolymerized with 70% MMA. The formation of the segmented networks is schematically depicted in Figure 22.3.

MEASUREMENTS

Values of E', E'' and tan δ were measured in bending mode by dynamic mechanical analysis (DMA) on a Polymer Laboratories PL-MKII apparatus. The experiments

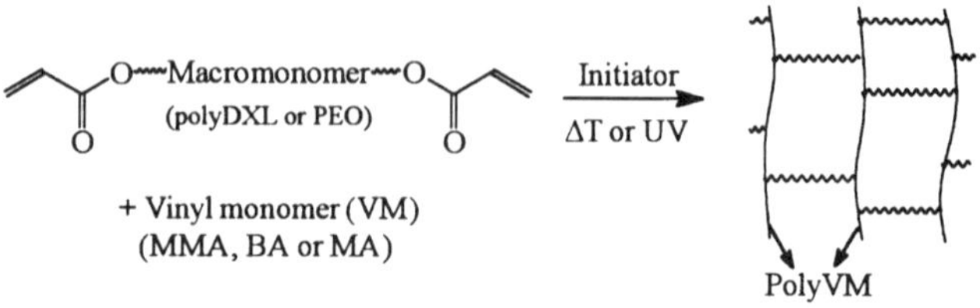

Figure 22.3 Reaction scheme for the synthesis of segmented networks

were performed on rectangular films (thickness 1 mm) at a heating rate of
2 °C/min (frequency 1 Hz). Thermogravimetric analysis was performed on
a Polymer Laboratories PL-TG1000 under nitrogen atmosphere. Differential
scanning calorimetry (DSC) curves were recorded on a DSC Perkin Elmer 7
equipped with a TAC 7/DX thermal analysis controller. After melting the samples
at 100 °C in the DSC apparatus, crystallization was performed at 0 °C for 15 min.
Then, the samples were quenched (-90 °C) and heated at a heating rate of
10 °C/min. The apparent enthalpies of melting were derived from the area of
the endothermic peaks. The degree of crystallinity (χ_c) of the PEO-fraction in
the sample was calculated from the following equation:

$$\chi_c(\text{sample}) = \frac{\Delta H^*_{(\text{PEO})}}{\Delta H^o_{(\text{PEO})}}$$

where $\Delta H^o_{(\text{PEO})}$ is the heat of melting per gram of 100% crystalline PEO
(205 J/g [23]) and where $\Delta H^*_{(\text{PEO})}$ is the apparent enthalpy of melting per gram
of PEO.

^{1}H NMR spectra were recorded in CDCl$_3$ on a Brüker AC360 FT-NMR apparatus. The molecular weight of the bis-macromonomers was determined by gel
permeation chromatography using a 60 cm 10^3 Å column from Tokyo Soda
Manufacturing Co and a Melz RI (LCD212) detector with chloroform as an
eluent at a flow rate of 1.0 ml/min (polystyrene standards). For the swelling
experiments, a piece of dried and weighed network was immersed in a solvent
(water or acetone) at 25 °C and then weighed periodically until constant weight
was reached. The degree of swelling Q was calculated from

$$Q = \frac{W_e - W_0}{W_0} \cdot \frac{\rho_n}{\rho_s}$$

where W_0 and W_e, respectively, denote the weight of the dry and swollen
network. ρ_n and ρ_s are the densities of the network and of the solvent respectively. The introduction of a density term ρ_n/ρ_s makes it possible to compare the
degrees of swelling in different solvents.

RESULTS AND DISCUSSION

DYNAMIC MECHANICAL PROPERTIES OF SEGMENTED NETWORKS

As can be seen in Table 22.1, vinyl polymers with both high and low T_g values have been used for the construction of the networks. The difference in solubility parameters (δ) between the bis-macromonomers (BM) and the vinyl polymers gives an indication of the intrinsic miscibility of the corresponding polymer blends.

Other research groups have concluded that blends of PEO and PMMA are mostly miscible in the molten state but undergo phase separation at temperatures below the melting point of PEO. This heterogeneity results in the crystallization of PEO in the PEO-rich domains for blends containing more than 20% PEO [23–26]. Recently, we reported similar results for polyDXL/PMMA blends [19]. For the segmented networks, in which the two components are kept together by covalent bonds, more miscible morphologies are expected due to a forced compatibility. The comparison of the tan δ–T curves of a blend and of a segmented network, with a ratio of polyDXL/PMMA equal to 50/50 w/w, clearly illustrates the different morphologies (Figure 22.4).

While curve I (segmented network) only shows one broad transition situated between the T_g values of the homo-polymers, curve II (polymer blend) shows three transitions situated at $-50\,^\circ$C (T_g of polyDXL-rich phase), 50–60$\,^\circ$C (melting point of polyDXL) and 110$\,^\circ$C (T_g of polyMMA-rich phase).

However, the ultimate phase morphology of the network structures also depends to a large extent on the nature of the vinyl polymer. Not only the miscibility of this vinyl polymer but also the copolymerization behaviour of the vinyl monomer with the BM are of great importance. The influence of the copolymerization reactivity, in terms of the reactivity ratios (r_1 and r_2), on the morphology of the networks is difficult to evaluate and is not further investigated in this paper. Because the specific features in the polymerization system of macromonomers influence the copolymerization ratio, it cannot be predicted by

Table 22.1 Glass transition temperatures and solubility parameters of the homo-polymers

Polymer		$T_g(^\circ\mathrm{C})^{(a)}$	$\delta(\mathrm{Mpa}^{1/2})^{(a)}$
Poly(1,3-dioxolane)	(PDXL)	-45	$20.9^{(b)}$
Poly(ethylene oxide)	(PEO)	-65	20.2
Poly(methyl methacrylate)	(PMMA)	105	19.4
Poly(methyl acrylate)	(PMA)	10	20.7
Poly(butyl acrylate)	(PBA)	-54	18.6

[a] The data are the average values reported in *Polymer Handbook*, 3rd edition, J. Brandrup and E.H. Immergut (eds), J. Wiley & Sons, NY, 1989.
[b] R. Alamo, J.G. Fatou and A. Bello, *Polym. J.*, **15**, 491 (1983).

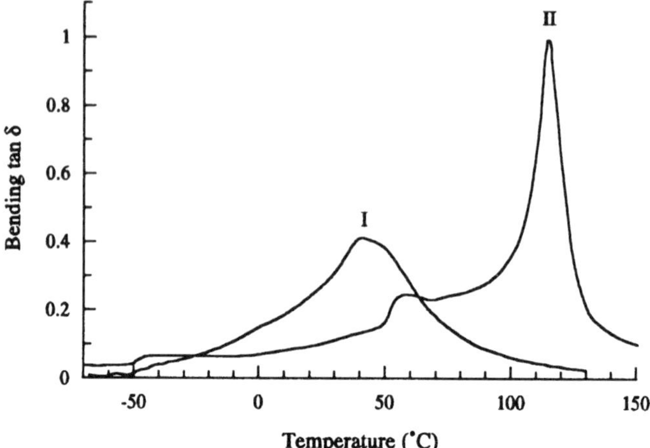

Figure 22.4 Comparison between the tan δ–T curve of a segmented network (curve I) and of a blend (curve II) of polyDXL and polyMMA (50/50 w/w) (redrawn from ref. [18])

the intrinsic reactivity of the polymerizable endgroup alone. Several parameters such as molecular weight, solvent quality and steric character of the substituents on the double bond of the endgroup should be considered [27,28].

The influence of the choice of the vinyl polymer on the morphology becomes clear from Figure 22.5 which shows the DMA-curves of three segmented networks in which the BM of polyDXL ($M_w = 4000$) was combined with PBA, PMA and PMMA respectively (in the same ratio 50/50 w/w).

The tan δ–T curves of the three networks show a single (curve 2), one broad (curve 3) or two transitions (curve 1). The polyMA- (curve 2) and the polyMMA-containing (curve 3) segmented networks show a single transition between the T_g of the homo-polymers. The difference in the width of the transitions could be explained by the small difference in solubility parameters (δ) between polyDXL and polyMA (0.2 Mpa$^{1/2}$) and the somewhat larger difference between polyDXL and polyMMA (1.5 Mpa$^{1/2}$) [3]. The temperature at the maximum value of tan δ for both networks is not only governed by the ratio of both components but also by the crosslink density as a consequence of reduced segmental mobility. Both networks are transparent, non-crystalline materials. Poly(DXL-*co*-BA) networks, on the other hand, are opaque, semi-crystalline materials. The DMA curve (curve 1) shows a superposition of two transitions at low temperatures corresponding to the T_g values of both components, together with a melting transition of crystalline polyDXL segments at 25 °C. The phase separation process can be attributed to the large difference between the δ-values of the homo-polymers (2.3 Mpa$^{1/2}$) and also to the low T_g of polyBA which offers the polyDXL segments enough flexibility to fold the polymer chains into regular crystal lattices at their crystallization temperature.

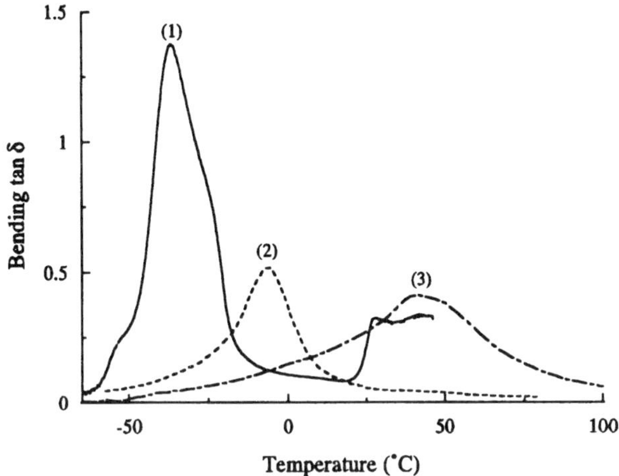

Figure 22.5 Tan δ versus temperature for three segmented networks with different nature of the vinyl polymer: *net*-poly(DXL-*co*-BA) (1); *net*-poly(DXL-*co*-MA) (2); *net*-poly(DXL-*co*-MMA) (3) (w/w 50/50; M_n(polyDXL) = 4000)

Another important parameter that determines the morphology of the segmented networks is the crosslink density. Both the ratio of the components and the molecular weight of the BM fix this density. Higher fractions of the BM decrease the average distance between the junction points along the vinyl polymer chains, while lower molecular weights of the BM bring the vinyl polymer chains closer together. The influence of the crosslink density on the dynamic mechanical properties of the materials is illustrated in Figure 22.6–22.8.

Figures 22.6 and 22.7 respectively show the tan δ–T curves of poly(DXL-*co*-MMA) and poly(EO-*co*-MMA) networks for different ratios of the components. In both cases, the maximum of the tan δ transition of the networks shifts to lower temperatures with increasing fractions of the BM, in spite of the increasing crosslink density. This is ascribed to the low T_g values of polyDXL and PEO, which lower the overall T_g of the materials when more of the corresponding BM is incorporated in the network. It can also be observed in Figure 22.6 that the transition broadens continuously for increasing polyDXL fractions, resulting in a plateau-like transition over a wide temperature range (from -50 to $+50\,^{\circ}$C) if 85% of polyDXL is incorporated (curve 1). For networks with a high fraction of the polyDXL segments (curves 1–3), the absolute values of the tan δ maxima decrease continuously due to the increasing fraction of crystalline polyDXL segments that does not contribute to the T_g transition. This has been confirmed by DSC analysis.

The most convenient method to study the effect of the crosslink density on the dynamic mechanical properties is to change the molecular weight of the BM while keeping the ratio of both components constant. In Figure 22.8, the

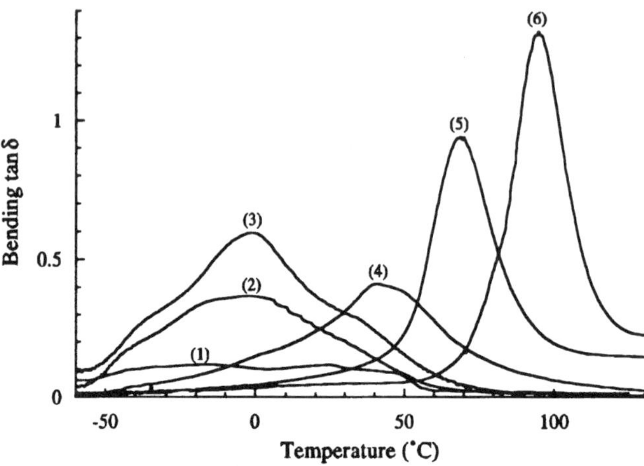

Figure 22.6 Tan δ versus temperature of *net*-poly(DXL$_{4000}$-*co*-MMA) with the ratio (w/w) of polyDXL/PMMA equal to 85/15 (1), 70/30 (2), 65/35 (3), 50/50 (4), 40/60 (5) and 20/80 (6)

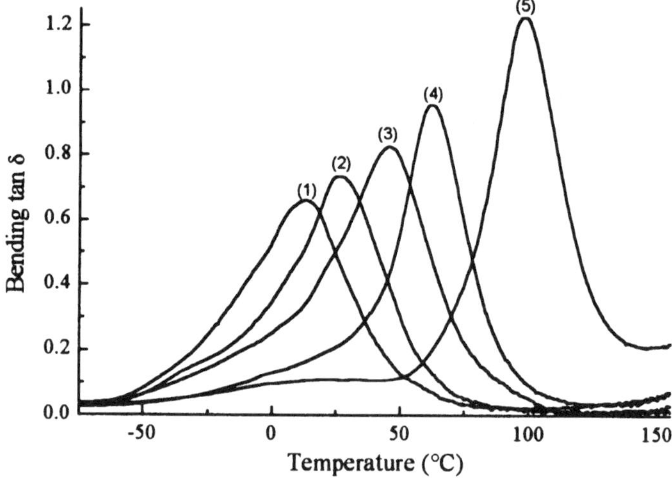

Figure 22.7 Tan δ versus temperature of *net*-poly(EO$_{4000}$-*co*-MMA) with the ratio (w/w) of PEO/PMMA equal to 60/40 (1), 55/45 (2), 45/55 (3), 40/60 (4) and 20/80 (5)

tan δ–T curves of three poly(EO-*co*-MMA) networks with a ratio of 40/60 w/w are given for which the molecular weight of the BM is varied between 1500 and 6500. The temperature, which corresponds to the maximum of the tan δ transition, only slightly moves to lower temperatures with decreasing crosslink density (curves 1 to 3). However, the transitions broaden and the tan δ maximum

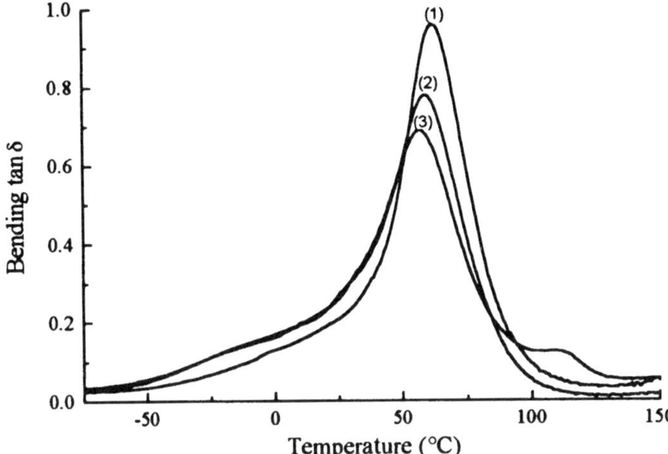

Figure 22.8 Tan δ versus temperature of *net*-poly(EO-*co*-MMA) (40/60 w/w) with the molecular weight of the bis-macromonomer of PEO equal to 1500 (1), 4000 (2) and 6500 (3)

decreases continuously. At the lowest crosslink density (curve 3), the transition appearing at 110 °C is due to the formation of a PMMA-rich phase and can be attributed to incomplete mixing of the system.

SWELLING PROPERTIES AND AMPHIPHILIC CHARACTER OF SEGMENTED NETWORKS

The investigated segmented networks are macromolecular substances containing covalent bonded segments of opposite philicity, i.e. hydrophilic and hydrophobic. It is well known that the amphiphilic character of copolymer structures consisting of incompatible blocks causes unique properties in bulk, as well as in selective solvents, due to so-called microphase separated morphologies [29,30]. From Figures 22.6–22.8, it could be concluded that the morphologies of the segmented networks, consisting of a hydrophilic polymer with low T_g and of a hydrophobic polymer with high T_g, can be varied to a large extent. When changing the ratio of both components (Figures 22.6 and 22.7), not only the crosslink density but also the hydrophilic/hydrophobic balance is influenced. The latter is clearly evidenced in Figure 22.9, which shows the swelling behaviour in water and in acetone of poly(EO_{1500}-*co*-MMA) networks with different ratios of the components.

For a given solvent, the degree of swelling depends on the fraction of the component that has a good affinity with the solvent. Cold water is a good solvent for PEO and a non-solvent for PMMA while acetone is a good solvent for PMMA but a poor solvent for PEO. While networks with a large amount of PEO (e.g.

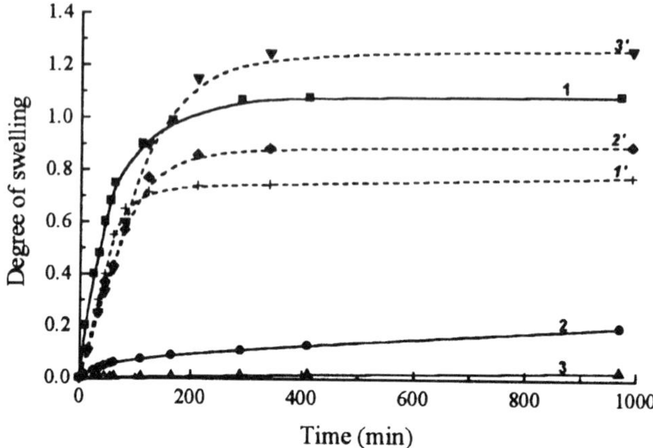

Figure 22.9 Degree of swelling at room temperature versus time for *net*-poly(EO$_{1500}$-*co*-MMA) in water (——) and in acetone (----) with ratios of PEO/PMMA equal to 80/20 (1,1'), 40/60 (2,2') and 20/80 (3,3')

80/20 w/w) swell considerably in water, the same networks swell much less in acetone compared to the networks with a high PMMA fraction (e.g. 20/80 w/w).

By plotting the equilibrium degree of swelling as a function of the fraction of PEO in the networks, a qualitative interpretation of the compatibility in the networks can be given.

If heterogeneous materials are swollen in solvents for one component, high degrees of swelling are expected only if this component forms the matrix of the material. Below a certain fraction of this component, the equilibrium swelling will show an abrupt decrease. On the contrary, in the case of homogeneous materials, the degree of swelling is expected to increase continuously with increasing fraction of one of the components. In the poly(EO-*co*-MMA) networks with three different molecular weights for PEO between the crosslinks, the degree of swelling at equilibrium in water increases fairly continuously over the whole range (curves A to C in Figure 22.10). The higher degrees of swelling for increasing molecular weights of PEO at a same PEO/PMMA ratio (curves A–C) are ascribed to the lowering of the degree of crosslinking of the networks.

On the other hand, curve D obtained for the *net*-poly(EO$_{1500}$-*co*-BA) exhibits an S-shaped form. This reveals the heterogeneity of the polyBA-containing networks, even at low molecular weights of PEO, and shows approximately the composition at which phase inversion occurs. These experiments confirm the conclusions drawn in the preceding chapter.

The swelling studies demonstrated that amphiphilic networks exhibiting various swelling characteristics can be obtained by the selection of the network components.

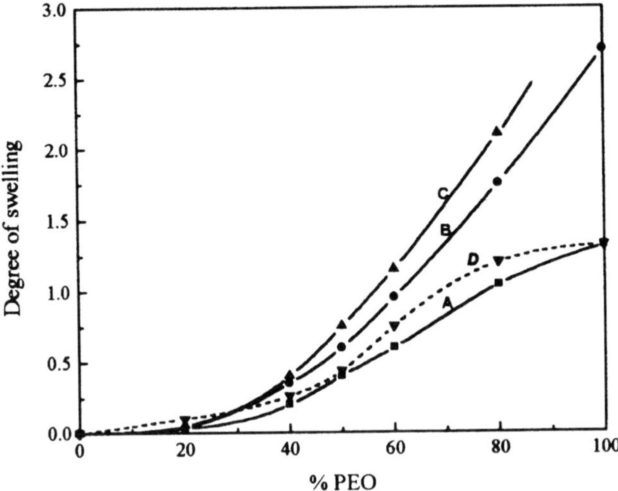

Figure 22.10 Degree of swelling in water (25 °C) as a function of the weight percent PEO in *net*-poly(EO$_{1500}$-*co*-MMA) (A), *net*-poly(EO$_{4000}$-*co*-MMA) (B), *net*-poly(EO$_{6500}$-*co*-MMA) (C) and *net*-poly(EO$_{1500}$-*co*-BA) (D)

CRYSTALLINITY OF SEGMENTED NETWORKS

As outlined in the introduction, PEO is a semi-crystalline polymer whose physical properties, such as conductivity after complexation with metal salts, depend on the degree of crystallinity. One of the purposes of this work was to investigate in how far the crystallinity behaviour of PEO could be changed by the incorporation of this polymer in the segmented networks. In Table 22.2, the degrees of crystallinity of segmented networks *net*-poly(EO-*co*-MMA), in which the molecular weight of PEO is varied between 1500 and 6500, are given.

The values in Table 22.2 indicate that the introduction of crosslinks (L → N) considerably decreases the crystallinity, especially in the case of networks with PEO$_{1500}$. The incorporation of only 20% of PMMA in the *net*-poly(EO$_{1500}$-*co*-MMA) prevents most of the crystallization. For higher molecular weights of PEO, the introduction of 50% of PMMA (PEO$_{4000}$) or even 60% of PMMA (PEO$_{6000}$) is necessary to avoid crystallization. For comparison, the melting heats of some polymer blends which have been prepared in the same way as the segmented networks (using PEG instead of the BM) are also given. Even at high PMMA fractions, the crystallization of PEO is only slightly decreased. Therefore, the incorporation of PEO in segmented networks may offer an interesting approach to the development of PEO-based conductive polymers at ambient temperature.

Table 22.2 Degrees of crystallinity (χ_c) of PEO-fraction in segmented networks and in some corresponding polymer blends (L = linear PEO; N = PEO-network)

Segmented network	Ratio PEO/PMMA	χ_c (network) (%)	χ_c (blend) (%)
PEO$_{1500}$/PMMA	100/0 (L)	76	
	100/0 (N)	12	
	80/20	0.4	
	60/40	0	
	50/50	0	44
PEO$_{4000}$/PMMA	100/0 (L)	85	
	100/0 (N)	49	
	80/20	31	
	60/40	13	
	50/50	0	47
PEO$_{6500}$/PMMA	100/0 (L)	86	
	80/20	47	
	60/40	21	
	50/50	1	
	40/60	0	56

CONCLUSION

This chapter is devoted to the synthesis and characterization of segmented networks, in which a bis-macromonomer of PEO or polyDXL acts as a hydrophilic crosslinker of hydrophobic polyMMA or polyBA chains. The investigation of the dynamic mechanical properties, swelling properties and crystallization behaviour revealed that the introduction of covalent bonds between the two components can cause a compatibilization of two incompatible polymers due to the restricted chain mobility. The final morphology and the amphiphilic character of the network structures can be varied over a broad range and are determined by the nature and the ratio of both components and by the molecular weight of the bis-macromonomer. Incorporation of PEO in such copolymer networks leads to a remarkable decrease in its crystallinity. For networks in which the PEO bis-macromonomer has a molecular weight of 1500, 20% of polyMMA was already sufficient to prevent crystallization.

ACKNOWLEDGMENTS

F. Du Prez thanks the Fonds voor Wetenschappelijk Onderzoek-Vlaanderen (FWO) for the financial support of postdoctoral research. D. Christova acknowledges the Belgian Office for Scientific, Technical and Cultural affairs (DWTC) for the research fellowships.

REFERENCES

1. D.R. Paul and L.H. Sperling, 'Multicomponent polymer materials', *Adv. in Chem. Series*, **211**, ACS, Washington DC, 1986.
2. D. Klempner and K.C. Frisch, *Polymer Alloys II: Blends, Blocks, Grafts, and Interpenetrating Networks*, Plenum press, New York, 1980.
3. F.E. Du Prez and E.J. Goethals, *Macromol. Chem. Phys.*, **196**, 903 (1995).
4. L.H. Sperling, *Interpenetrating Polymer Networks and Related Materials*, Plenum Press, New York, 1981.
5. D. Klempner and K.C. Frisch, *Advances in Interpenetrating Polymer Networks*, Technomic, Vols I–IV, Lancaster PA, 1989–1994.
6. Y.S. Lipatov and L.M. Sergeeva, *Interpenetrating Polymeric Networks*, Naukova Dumka, Kiev, 1979.
7. C.H. Bamford, G.C. Eastmond and D. Whittle, *Polymer*, **10**, 771, 885 (1969).
8. J. Liu, W. Liu, H. Zhou, C. Hou and S. Ni, *Polymer*, **32**, 1361 (1991).
9. P.S. Chang and M.A. Buese, *J. Am. Chem. Soc.*, **115**, 11475 (1993).
10. W. Meier, *Macromolecules*, **31**, 2212 (1998).
11. Y. Tezuka, T. Shida, T. Shiomi, K. Imai and E.J. Goethals, *Macromolecules*, **26**, 575 (1993).
12. T. Shiomi, K. Okada, Y. Tezuka, H. Kazama and K. Imai, *Makromol. Chem.*, **194**, 3405 (1993).
13. B. Ivan, J.P. Kennedy and P.W. Mackey, in *Polymer Drugs and Delivery Systems* (R.L. Dunn and R.M. Ottenbrite, eds), ACS Symposium Book Series 469, Washington DC, 1991, pp. 194 and 203.
14. M. Weber and R. Stadler, *Polymer*, **29**, 1071 (1988).
15. R.R. De Clercq and E.J. Goethals, *Macromolecules*, **25**, 1109 (1992).
16. F.E. Du Prez and E.J. Goethals, in *Ionic Polymerizations and Related Processes* (J.E. Puskas, Ed), Kluwer Academic Publishers, Netherlands, 1999, pp. 75–98.
17. C.B. Tsvetanov, R. Stamenova, D. Dotcheva, M. Doytcheva N. Belcheva and J. Smid *Macromol. Symp.*, **128**, 165 (1998).
18. F.E. Du Prez, P. Tan and E.J. Goethals, *Polym. Adv. Technol.*, **7**, 1 (1996).
19. F.E. Du Prez, E.J. Goethals, R. Schué, H. Qariouh and F. Schué, *Polym. Int.*, **46**, 117 (1998).
20. A. Hooper and J.M. North, *Solid State Ionics*, **9/10**, 1161 (1983).
21. C.S. Harris, D.S. Shriver and M.A. Ratner, *Macromolecules*, **19**, 987 (1986).
22. E.J. Goethals, R.R. De Clercq, H.C. De Clercq and P.J. Hartmann, *Makromol. Chem., Macromol. Symp.*, **47**, 151 (1991).
23. E. Martuscelli, C. Silvestre, M.L. Addonizio and L. Amelino, *Macromol. Chem.*, **187**, 1557 (1986).
24. N. Parizel, F. Lauprêtre and L. Monnerie, *Polymer*, **38**, 3719 (1997).
25. S.A. Liberman, A. Des Gomes and E.M. Macchi, *J. Polym. Sci.: Part A: Polym. Chem.*, **22**, 2809 (1984).
26. C. Silvestre, S. Cimmino, E. Martuscelli, F.E. Karasz and W.J. MacKnight, *Polymer*, **28**, 1190 (1987).
27. M.K. Mishra, *Macromolecular Design: Concept and Practice*, Polymer Frontiers International, New York, 1994.
28. K. Ito, *Prog. Polym. Sci.*, **23**, 581 (1998).
29. A. Noshay and J.E. McGrath, *Block Copolymers: Overview and Critical Survey*, Academic Press, Orlando, 1977.
30. R.A. Brown, A.J. Masters, C. Price and X.F. Yuan, in *Comprehensive Polymer Science*, (C. Booth and C. Price, eds), Pergamon Press, 1989, Vol. 2, p. 155.

23

Side Reactions during Polycyclotrimerization of Cyanates and their Influence on Network Structure and Properties

MONIKA BAUER[1], CHRISTOPH UHLIG[1], JÖRG BAUER[1], STEVEN HARRIS[2] and DAVID DIXON[2]

[1]Fraunhofer Institute for Reliability and Microintegration, Kantstr. 55, 14513 Teltow, Germany

[2]British Aerospace, Sowerby Research Centre, PO Box 5, Filton, Bristol BS12 7QW, UK

Wiley Polymer Networks Group Review Series Vol. 2. Edited by B.T. Stokke and A. Elgsaeter
© 1999 John Wiley & Sons Ltd

ABSTRACT

For a cyanate ester resin it is shown that there is the possibility of a deviating network formation mechanism resulting in structural elements other than the triazine ring which we name herein 'irregular structural elements'. Chemical structure of these irregular structural elements and the reaction mechanism leading to their formation is derived from chemical analysis. The relationship between content of these irregular structures in the cured TMDCBF and physical properties is discussed on a semi-quantitative basis.

INTRODUCTION

The influence of the network structure of highly crosslinked polymer networks on their thermophysical and mechanical properties has been studied by several workers (e.g. [1–6]). Crosslink density has been varied by using monomers of different chain lengths, different functionality or by curing networks to different conversions of functional groups. In many cases, where structure–property relationships of thermosetting resins are reported, assumptions on the network structure formed are derived from previous investigations of chemistry and structural formation or from literature. It is only rarely reported whether for the cured resin specimen subjected to physical and mechanical characterization also structural analysis has been carried out.

For a cyanate ester resin (tetramethyldicyanate of bisphenol F (TMDCBF) AroCy® M10 from Ciba Geigy) this work shows that it may be necessary to carry out both chemical structural analysis and characterization of physical and mechanical properties using *identical* resin samples. At least in the case reported herein structure–property relationships could only be understood on the basis of this approach.

CHEMISTRY AND NETWORK STRUCTURE

Aromatic cyanato groups undergo a cyclotrimerization reaction to form aroxy-substituted triazine rings (Scheme 23.1). In this way, three-dimensional polycyanurate networks can be synthesized, if one uses di-or polyfunctional cyanates as monomers. The present knowledge of chemistry, processing and properties of polycyanurate materials is summarized in a monograph [7].

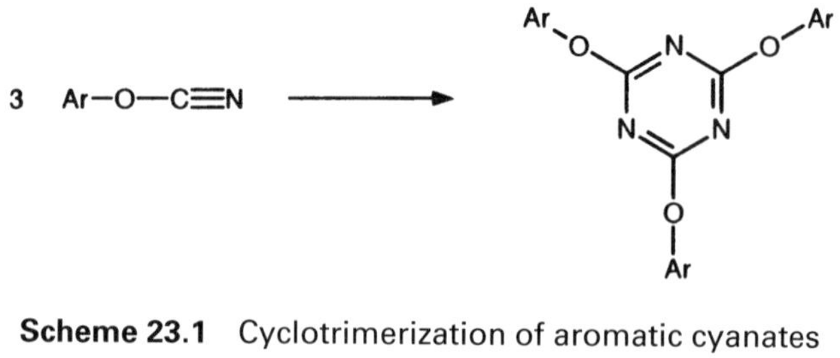

$$3 \quad Ar{-}O{-}C{\equiv}N \quad \longrightarrow$$

Scheme 23.1 Cyclotrimerization of aromatic cyanates

The peculiarity of the OCN polycyclotrimerization of dicyanates lies in the fact that at full conversion all functional groups are incorporated in trifunctional branching points. Therefore, cyanurate rings and aromatic units R of the monomer are the only structural elements of a fully cured network (Scheme 23.2).

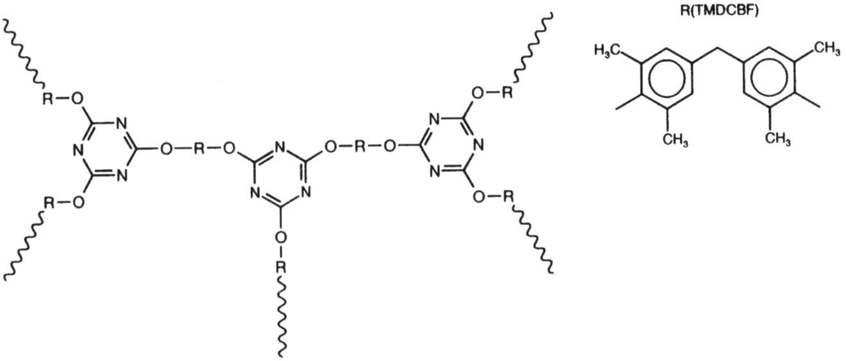

Scheme 23.2 Structural elements of a fully cured polycyanurate network based on TMDCBF

Amazingly, we found bands at 2215 cm^{-1} and 1638 cm^{-1} of additional structural elements in the IR spectra of several cured TMDCBF samples (see Figure 23. 1). According to the intensities of these two bands, which are apparently coupled, we could clearly distinguish two groups of cured TMDCBF polycyanurates:

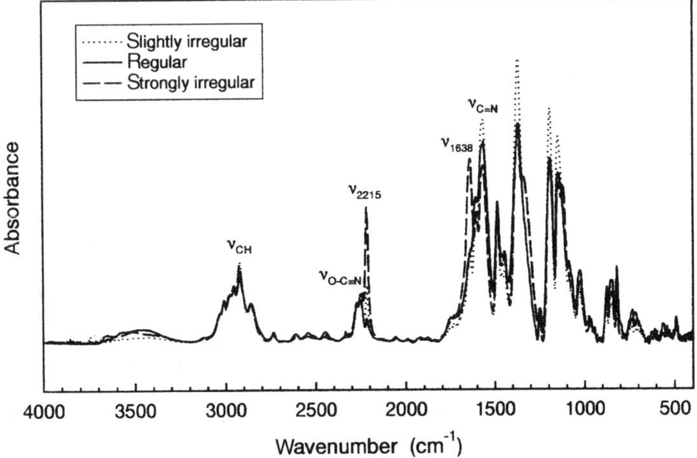

Figure 23.1 IR spectra (normalized to the areas under the CH-bands) of TMDCBF polycyanurates

1. 'regular' structures with no or only very little absorption at 2215 cm^{-1} and 1638 cm^{-1};
2. 'strongly irregular' structures with high absorption at 2215 cm^{-1} and 1638 cm^{-1}.

A third group of 'slightly irregular' structures shows an intermediate behaviour between low but significantly detectable and relatively high absorption at 2215 cm^{-1} and 1638 cm^{-1}.

The dependence of the normalized intensity of the 2215 cm^{-1} band on conversion of OCN groups (determined from the IR spectra as described in [8]) confirms the chosen classification of the TMDCBF polycyanurates into three groups. The data plotted in Figure 23.2 show a pronounced difference in the behaviour of the samples. For the regular structures the low intensity decreases monotonically with increasing conversion. A small part of this absorbance at 2215 cm^{-1} is caused by the split OCN-band of TMDCBF itself, and the remaining intensity comes from very small amounts of the additional structures also in regular polycyanurates (see SEC-analysis below).

On the other hand, the intensity at 2215 cm^{-1} for strongly irregular structures remains at a nearly constant high level independently of conversion. The slightly irregular structures show an intermediate behaviour: a part of the intensity is caused by irregular structures and remains constant, whereas the other part belongs to OCN groups and decreases with increasing conversion.

Sol fractions were extracted from the three specimens, the IR spectra of which are shown in Figure 23.1, to compare the amounts of irregular structures of the soluble parts with those of the corresponding networks. Sol fractions between

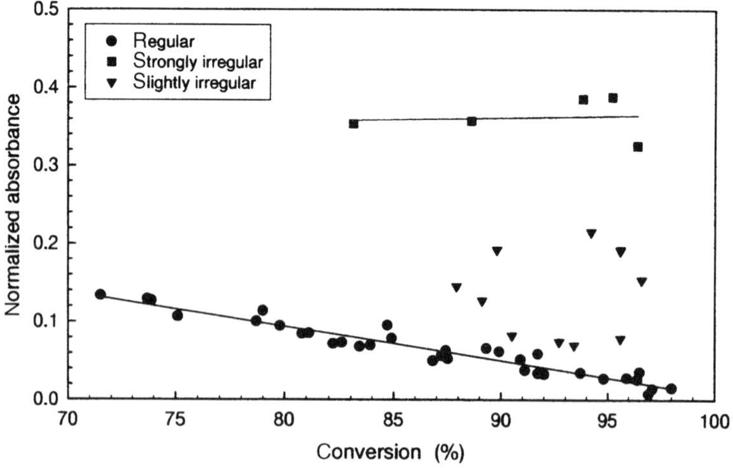

Figure 23.2 IR absorbance intensities (normalized to the areas under the CH-bands) versus OCN conversion of TMDCBF polycyanurates

0.8 and 2.5 wt % were found. The IR spectra of both the sol and the remaining network of strongly irregular specimens show high intensities at 2215 cm^{-1} and 1638 cm^{-1}. Also the spectra of sol and gel for the regular and slightly irregular samples are nearly identical. The only difference of all three types lies in the higher relative amount of unreacted OCN groups of the sol fractions compared to that of the corresponding gels as can be expected.

The sol fractions were analyzed by size-exclusion chromatography as well. Typical chromatograms of sol fractions as well as of a regular TMDCBF prepolymer are plotted in Figure 23.3. All sol fractions as well as the prepolymer contain the 'regular' structures: unreacted TMDCBF (1) and its cyclotrimerization products trimer (3), pentamer (5) and higher oligomers with an odd degree of polymerization.

On the other hand, additional substances can be detected in the sol fractions of the irregular samples, they are designated by 2'–6'. It was proved by additional MALDI-TOF measurements that these molecules have molar masses, which are exactly many times the mass of the monomer TMDCBF, and form their own oligomeric series with both even and odd degrees of polymerization. However, the lower elution volumes of the 'odd molecules' lead to the conclusion that their hydrodynamic volumes are slightly larger than those of the corresponding 'regular' trimerization products.

The band at 2215 cm^{-1} arises from a CN triple bond, which is not directly linked to an aliphatic group as well as an oxygen atom. Such a nitrile group is fairly stable under the curing schedule applied, so that it remains unreacted also in polycyanurates with the highest accessible conversions. We propose a

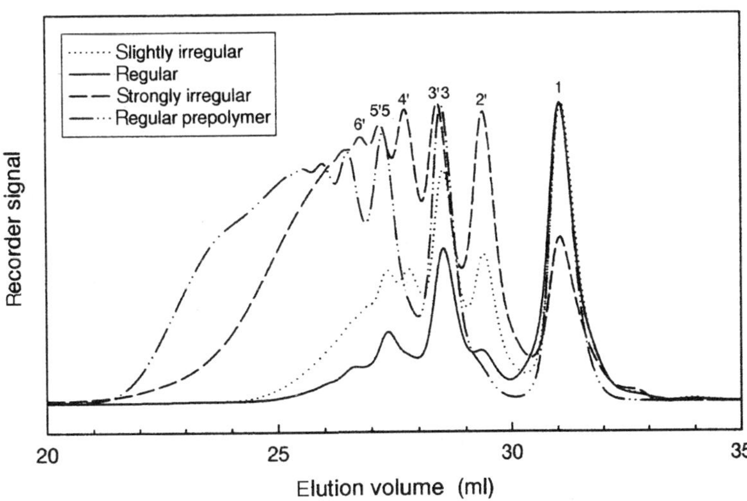

Figure 23.3 Chromatograms of sol fractions of the three types of TMDCBF polycyanurates and of a TMDCBF prepolymer

rearrangement of the cyanato group followed by an addition reaction, on the basis of which the simultaneous absorbance in the regions around 2215 cm^{-1} and 1638 cm^{-1} and the existence of odd and even oligomers can be explained (Scheme 23.3).

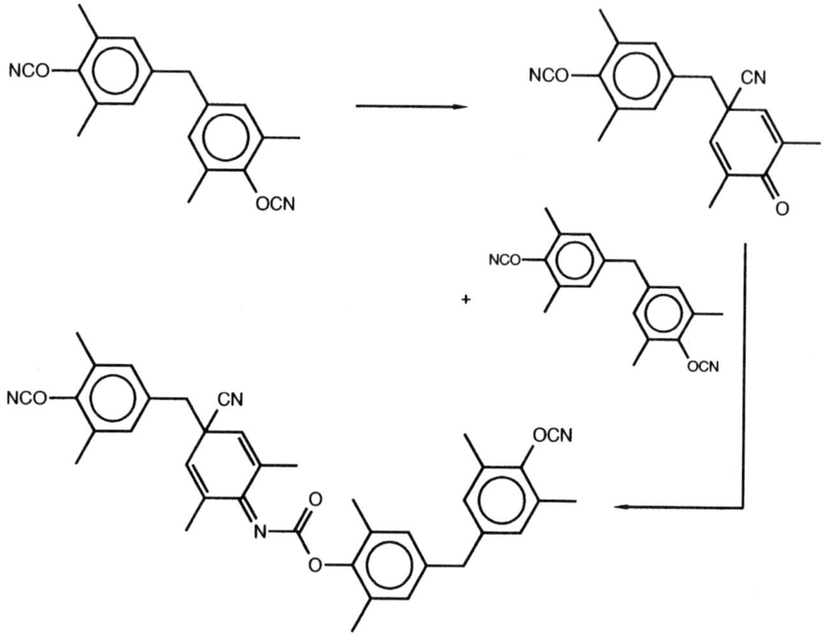

Scheme 23.3   Reaction forming the irregular structures

Summarizing the knowledge on the nature of the chemical structure of TMDCBF-based polycyanurate networks, the following facts can be stated:

- Under certain circumstances a side reaction, which plays only a minor role during 'regular' polycyclotrimerization, becomes important. Both the intensity of the 2215 cm^{-1} peak in the IR-spectrum and the amount of certain oligomers in the sol fraction, detected by SEC, are measures of the extent of this side reaction.
- Once the irregular structures are formed, their amount remains constant in the high-conversion region.
- During 'irregular' curing of dicyanates a part of the cyanate groups does not react to form cyanurate structures, but links the monomers together through difunctional bridges.
- As a consequence of this chemical structure, the branching density of an 'irregular' polycyanurate is lower than that of a regular one with the same conversion of OCN groups.

STRUCTURE–PROPERTY RELATIONSHIPS

DYNAMIC MECHANICAL ANALYSIS

In Figure 23.4 representative dynamic mechanical analysis (DMA) measurements are shown for regular networks from TMDCBF (with different conversions) and for a strongly irregular network. It can be seen that there are distinct differences in the dynamic mechanical behaviour of regular and irregular networks.

Glassy state Only one transition is found in the glassy region at about $-115\,^{\circ}$C for the regular networks (denominated herein as the γ-transition). Between this low temperature relaxation and the glass transition there is no further relaxation. Both the presence of the γ-transition and the absence of any other distinguished secondary transition at higher temperatures are common features of all regular type polycyanurates from TMDCBF. Conversion does not affect the temperature of the maximum of the γ-relaxation (at 1 Hz) but its intensity: Both the intensity of the γ-relaxation and the level of the loss factor between γ- and α-transition increase with conversion (this is possibly an effect of the increase in free volume when approaching the maximum conversion and consequently the maximum glass transition temperature [9]).

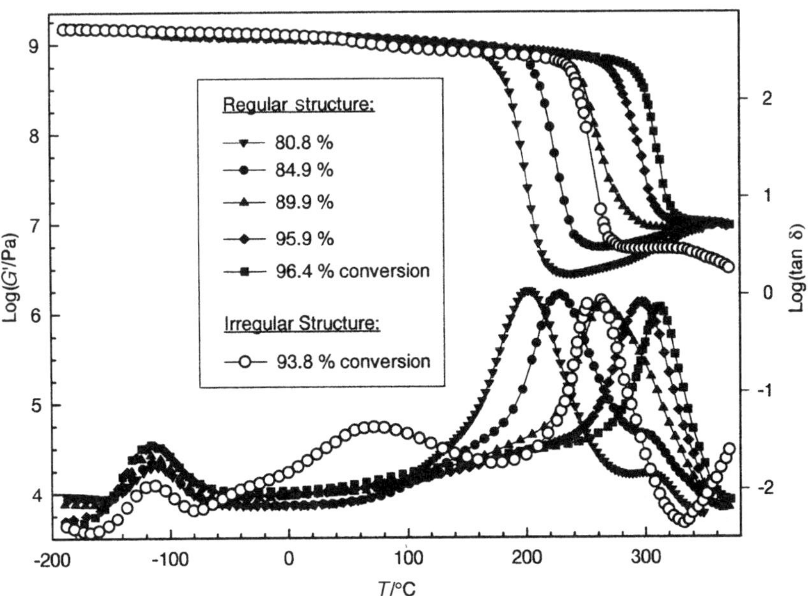

Figure 23.4 Storage shear modulus G' and loss factor tan δ (at 1 Hz) of regular and irregular TMDCBF networks

For strongly irregular networks an additional very broad transition is found between $-20\,^\circ$C and $+120\,^\circ$C. This transition is a unique feature of irregular networks. Therefore it will be denoted herein as 'β^*-transition'.

Glass transition and rubbery region above T_g Consistent thermomechanical behaviour within and above the glass transition is found for all specimen with a regular structure. Due to the low reactivity of TMDCBF, further crosslinking starts only well above the glass transition when conversion is much lower than 90% (T_g well below 250 °C). For conversions in the range of 90% up to full conversion further crosslinking reaction commences within the glass transition region because of the higher reactivity at higher temperatures. The minimum rubbery modulus increases with increasing conversion and thus increasing glass transition temperature as a consequence of the higher crosslink density. Further curing reaction during the DMA run increases the rubbery modulus of the incompletely cured networks and thus all modulus curves end up at the same level in the rubbery region before thermal decomposition starts at about 360 °C. The maximum T_g, i.e. the highest by any curing schedule attainable T_g, is in the range of 285 °C–295 °C (maximum in tan δ). For the strongly irregular type there is a lower maximum T_g ($T_{g\,max} \cong 245\,^\circ$C) and also a lower minimum rubbery modulus. Thermal decomposition (as indicated by a decrease in the rubbery modulus) commences at lower temperatures compared to regular networks.

Relationship between T_g *and conversion* It is interesting to further analyse quantitatively the correlation between the FTIR and the DMA results for identical castings. In Figure 23.5 T_g is plotted versus conversion for all investigated

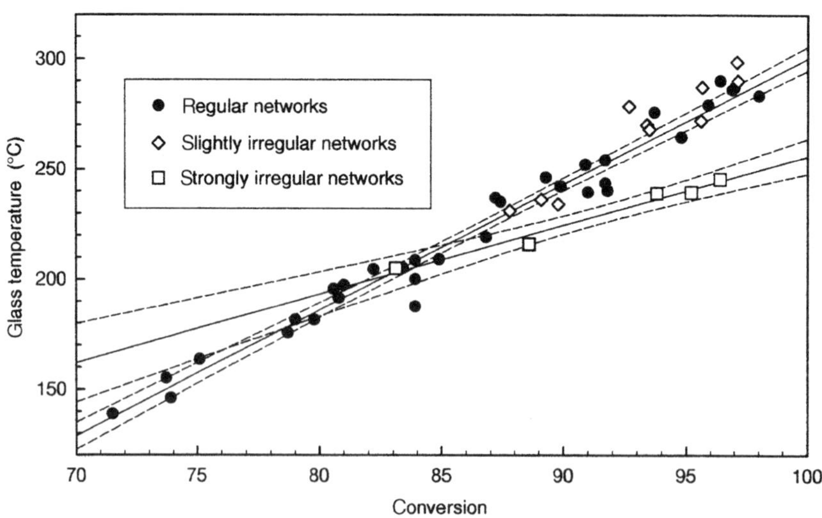

Figure 23.5 Glass transition temperature versus OCN conversion of TMDCBF networks

TMDCBF castings. (T_g is taken as peak maximum of tan δ. An error may result for incompletely cured samples if the tan δ peak corresponding to further curing reaction very much overlaps the glass transition peak. However, for TMDCBF curing reaction is comparatively slow also at high temperatures, so T_g can be determined for all conversions to a good accuracy.) Two different linear dependencies between T_g and conversion can be distinguished: a lower slope is found only for networks with a strongly irregular structure (normalized intensity of the 2215 cm^{-1} band > 0.3). All other networks ranging from regular to irregular networks with a lower content of irregularities (normalized intensity of the 2215 cm^{-1} band < 0.2) fall on the same line.

Relationship between intensity of the β^*-transition and content of irregular structures In Figure 23.6 the peak height of the β^*-transition is plotted versus the normalized intensity of the 2215 cm^{-1} band. It was observed that both the temperature and the height of the peak maximum of the β^*-transition vary slightly with OCN conversion, so only networks with equal conversion can be compared. Therefore, the data have been grouped rather arbitrarily into three groups (see Figure 23.6). For the regular networks (networks with a normalized intensity of the 2215 cm^{-1} band < 0.05) which do not show the β^*-transition the tan δ value at 50 °C was taken. Within these three arbitrarily selected groups a good correlation is found between the peak height of the β^*-transition and the intensity of the 2215 cm^{-1} band. For a low content of irregular structures (i.e. for values of the normalized intensity of the 2215 cm^{-1} band > 0.2) there is a near linear

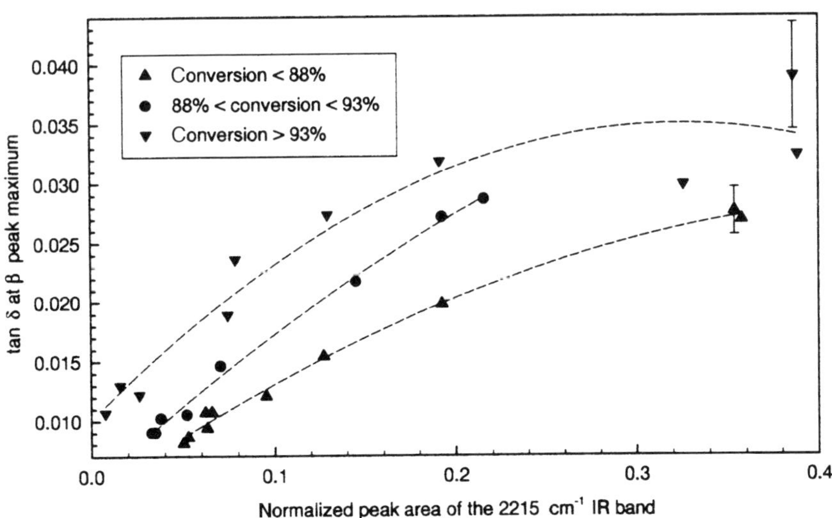

Figure 23.6 Loss factor tan δ at peak maximum of the β^*-relaxation versus normalized intensity of the 2215 cm^{-1} band

relationship between the β^*-transition peak height and the content of irregular structures. For higher contents (normalized intensity of the 2215 cm^{-1} band > 0.22) it seems that further increases in the 2215 cm^{-1} band intensity do not lead to a proportional increase of the β^*-transition. Furthermore it was observed that there is a macroscopic variation of dynamic mechanical properties within these castings: DMA measurements of rods which where machined out of the same casting from different positions showed that there are local differences in the intensity of the β^*-transition. These variations are indicated as error bars in Figure 23.6. However, even if the upper bar of these variations is considered it can be seen that the relationship between the β^*-transition intensity and the 2215 cm^{-1} band intensity is not linear — for higher 2215 cm^{-1} band intensities the curves level off.

Influence of irregular structures on thermal stability To summarize the above discussion, a different sensitivity to the concentration of irregular structures was found for the dependencies of glass transition and glassy state viscoelastic properties: whereas already for low contents of irregular structures (normalized intensity of the 2215 cm^{-1} band < 0.2) a pronounced influence on the intensity of the β^*-transition is seen, only for very high contents of irregular structures (normalized intensity of the 2215 cm^{-1} band > 0.22) a different T_g-conversion dependence with a lower slope is found. If this lower slope were interpreted as an effect of the lower crosslink density caused by the difunctional nature of the irregular structure (which is also seen by the lower rubbery modulus compared to regular networks) it would be difficult to explain why there is no proportional influence of the content of irregular structures on T_g as is observed for the behaviour in the glassy state (i.e. for the intensity of the β^*-transition). A possible explanation for this seemingly different influence on glassy state and glass transition properties may be derived from the influence of the content of irregular structures on thermal stability. Thermogravimetric analysis shows that the onset temperature for thermal decomposition is significantly lower (by 37 K) for the strongly irregular network. Therefore, one possible explanation for this observed 'critical transition' in the T_g-conversion relationship at a certain content of irregular structures can be derived from the hypothesis of a competing thermal decomposition (and/or thermally induced structural rearrangement) and crosslinking (trimerization) reaction. It can be supposed that only for high contents of irregular structures already at $T \cong 240\,^\circ$C the rate of such decomposition/rearrangement processes is higher than the rate of the trimerization reaction (i.e. the 'regular network formation mechanism'), so that the network structure is changed by such processes when T_g is approaching 240 °C in a way that no higher T_g can be reached by further thermal treatment. The fact that for lower contents of irregular structures no difference was found in the T_g-conversion relationship compared with the regular structure may then be explained by a lower rate of such network degradation processes due to the lower content of irregular structures, so that up to

290 °C trimerization is faster than decomposition and the maximum T_g of the regular network can be reached.

FRACTURE TOUGHNESS

Fracture toughness of networks from TMDCBF with both regular and strongly irregular structures is shown in Figure 23.7 as a function of T_g. It can be seen that there is a different relationship between T_g and fracture toughness for the irregular network structures; networks with a strongly irregular structure obviously produce a material with a higher toughness at the same T_g.

WATER DIFFUSION

Water uptake at saturation and diffusion coefficients have been determined for both a regular and strongly irregular sample of similar conversion (96% and 94%). Experimental and theoretical details of the procedure are described in Ref. [10]. In both cases deviations from Fickian behaviour have been found in that the water uptake is higher at longer times than predicted by a Fickian law. These deviations are more pronounced for 70 °C, so this may be an effect of a beginning chemical reaction in the network. The saturation water uptake for the strongly irregular network is 1.6 times as high as that for the regular network (an extrapolated uptake at saturation of 1.5% (35 °C) and 2.5% (70 °C) was determined for the regular type and 2.4% (35 °C) and 4.0% (70 °C) for the irregular type). On the other hand, diffusion coefficients have been found to be much higher for the regular network (3.8 (35 °C) and $5.8 \times 10^{-11} \mathrm{m^2 s^{-1}}$

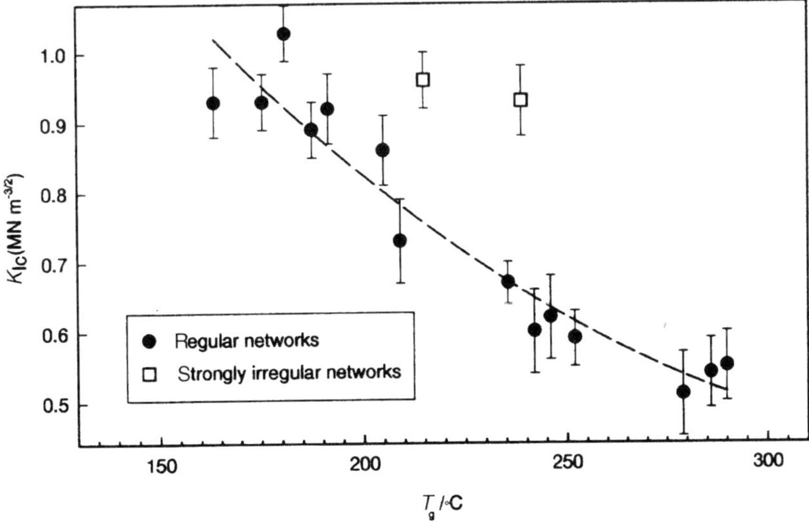

Figure 23.7 Critical stress intensity factor versus T_g for regular and irregular TMDCBF networks

the regular structure; 1.6 (35 °C) and 2.1×10^{-11} m^2s^{-1} (70 °C) for the strongly irregular structure), so there seems to be an inverse relationship between diffusion coefficient and saturation water uptake. With the structural elements shown in Scheme 23.3 more polar structures are contained in the irregular network than in the regular network structure (consisting of triazine ring links only). Thus, the regular network is assumed to be more hydrophobic than the irregular network. A highly hydrophobic structure provides a low interaction potential with the water molecules, so water is not very much hindered by interaction with the polymer and may go through the network more quickly. On the other hand, because of the more hydrophobic character of the regular network the saturation level remains lower than for the irregular structure. However, the real situation is surely more complex, so in the above argument free volume and molecular mobility of the networks were not considered.

FACTORS INFLUENCING THE FORMATION OF IRREGULAR STRUCTURES

It was omitted in the above discussion of structure–property relationships to discuss the factors controlling the formation of irregular structures. In fact, a lot of work has been done to identify these factors. However, our understanding is still rather incomplete. The following present state of knowledge may be summarized.

When casting and curing TMDCBF *without catalyst*, the formation of a regular network structure is the rule, i.e. it is near 100% reproducible. However, in rare cases, which could not have been reproduced so far, a very high content of irregular structures (with a normalized intensity of the 2215 cm^{-1} band $\cong 0.3$) was found. In total, we produced three castings with such a strongly irregular structure (compared to a total of more than 30 castings with a regular structure). All attempts to produce further strongly irregular samples by varying parameters of the process that are likely to be subjected to variations have been unsuccessful so far.

On the other hand, catalysts generally lead to a reproducible content of irregular structures. However, the content of (reproducibly formed) irregular structures is very different depending on the catalyst used. Three catalysts were investigated so far: the content of irregular structures ranges from only very slightly irregular (normalized intensity of the 2215 cm^{-1} band $\cong 0.08$) for catalyst 1 to a medium content of irregular structures (normalized intensity of the 2215 cm^{-1} band $\cong 0.2$) for catalyst 3.

CONCLUDING REMARKS

Structural analysis led to a conclusive assumption on the chemical structure of the irregular structural elements and by analysing the relationship between

IR-band peak intensities and physical properties a better understanding of the influence of these irregular structures on properties could be achieved. The next step is to extend these structure–property investigations to a more quantitative level. This requires derivation from IR-band intensities of the relative concentration of these irregular structural elements in the network, which is a very ambitious task for organic chemistry and chemical analysis.

However, there is another important aspect: The observed structural variations seem to be of a stochastic nature or triggered by only small variations in process parameters. Recent results in our lab indicate that there are also examples for other cyanate ester resins where a great sensitivity of structural formation to process parameters is observed. It may then be further argued that the variations which led to the reported structural changes are at any rate relatively small compared to the difference between any industrial process such as prepregging and laminating and the process of manufacturing neat resin castings. Therefore, when transforming such neat resin structure–property investigations into industrial processes and products, care about the sensitivity of structural formation to process parameters should be even greater.

AKNOWLEDGEMENT

Financial support of the European Commission for this work is gratefully acknowledged (Projects BREU-5104 and BE-96-3947).

REFERENCES

1. J.D. LeMay and F.N. Kelley, *Adv. Polym. Sci.*, **78**, 115 (1986).
2. M. Fischer, *Adv. Polym. Sci.*, **100**, 313 (1992).
3. Georjon, O., Schwach, G., Gerard, J.F., Galy, J., Albrand, M. and Dolmazon, R. *Polym. Eng. Sci.*, **37**(10), 1606 (1997).
4. Georjon, O. and Galy, J., *Polymer*, **39**(2), 339 (1998).
5. Georjon, O. and Galy, J., *J. Appl. Polym. Sci.*, **65**(12), 2471 (1997).
6. Georjon, O. and Galy, J., *Polym. Mater. Sci. Eng.*, **71**, 754 (1995).
7. I. Hamerton (ed.), *Chemistry and Technology of Cyanate Ester Resins*, Blackie Academical Press, Glasgow, 1994.
8. M. Bauer and J. Bauer, *Polym. Mater. Sci. Eng.*, **71**, 58 (1995)
9. J.P. Armistead and A. Snow, *ACS San Francisco PMSE*, 457 (1992).
10. T. Chang and D.A. Dillard, Report CASS/ESM-96-3.

PART 4

Biopolymer Networks and Gels and their Models

24

Physics of Solutions and Networks of Semiflexible Macromolecules and the Control of Cell Function

ERWIN FREY,[1,2] KLAUS KROY[2,3] and JAN WILHELM[2]

[1]Physics Department, Harvard University, Cambridge, MA 02138, USA
[2]Institut für Theoretische Physik, Physik-Department der Technischen Universität München, D-85747 Garching, Germany
[3]ESPCI, Physique Thermique, F-75231 Paris, France

Wiley Polymer Networks Group Review Series Vol. 2. Edited by B.T. Stokke and A. Elgsaeter
© 1999 John Wiley & Sons Ltd

ABSTRACT

Living cells are soft bodies of a characteristic form, but endowed with a capacity for a steady turnover of their structures. Both of these material properties, i.e. recovery of the shape after an external stress has been imposed and dynamic structural reorganization, are essential for many cellular phenomena. Examples are mechanical properties of tissue, cell motility, cell growth and division, and active intracellular transport. Numerous experiments *in vivo* and *in vitro* have shown that the structural element responsible for the extraordinary mechanical and dynamical properties of eukaryotic cells is the cytoskeleton, a three-dimensional assembly of protein fibers such as actin filaments and microtubules. In addition to those biopolymers various proteins with structural and regulatory functions have a major influence on the mechanical properties. At the relevant length scales (a few μm at most) the building blocks of these biomaterials are very different from conventional polymeric material. In contrast to flexible polymers the persistence length is of the same order of magnitude as their total contour length or even larger. This implies that the physics of such a system is determined by a subtle interplay between energetic and entropic contributions. We review our present understanding of the physics of biopolymers using concepts from macromolecular and statistical physics complemented by computer simulation. These systems open up a new field of soft condensed matter research, which to date is only poorly understood but has a great potential for interesting new physical phenomena.

INTRODUCTION

Most of the concepts used to understand the viscoelastic properties of chemical and physical gels of *flexible* polymers require the persistence length l_{p} [1] to be significantly smaller than other characteristic scales such as the filament length L, the distance between crosslinks or the width of reptation tubes. This condition no longer holds for networks of *semiflexible* polymers. One prominent family of such polymers consists of cytoskeletal biopolymers like F-actin, intermediate filaments and microtubules. An impression of the typical conformations of the filaments and the relative magnitude of the various length scales can be gained upon inspection of the following electron micrograph of a semidilute (0.4 mg/ml) actin solution. The most striking features of these networks are the enormous length and relatively elongated structures of the constituent biopolymers. Actin filaments have a diameter of 7 nm [2] and can reach lengths up to 30–100 μm *in vitro* [3], and at least several μm *in vivo*. The persistence length is approximately 17 μm [4,5,6], quite large compared to typical distances between neighboring filaments, which is in the range of a few tenths of a μm. This combination of length scales allows biopolymers to form networks at very low volume fraction, so that solutions of less than 0.1% volume fraction of polymer are still strongly entangled. Thus only a small amount of material needs to be produced by the cell in order to generate a sufficiently strong network. This fact is not only of considerable biological relevance but also facilitates interpretation of dynamic light scattering experiments and their relation to theory [7,8].

There are several motivations [9–12] for studying the viscoelasticity of cytoskeletal networks [13–22]:

(I) From a polymer physics perspective they are interesting because their behavior is expected to be determined by principles and mechanisms different from those established for flexible networks. In fact much of the physics behind the viscoelasticity of semiflexible polymer networks is only being explored recently [23–30].

(II) From an experimental point of view, semiflexible polymer networks are interesting because several semiflexible polymers have persistence lengths on the order of several µm or even mm. Thus techniques such as optical microscopy of single fluorescence labeled filaments or attached colloidal probes can be used to study the behavior of the network at the *single polymer level* [18,19,22,31]. For flexible polymers this has been possible only in simulations (see, e.g., [32]).

(III) Finally, the viscoelastic properties and regulation of semiflexible polymer networks both inside cells and in the extracellular matrix are of significant importance for the mechanical stability and properties of biological tissue, for cell locomotion, adhesion and force generation [9–11].

THE CYTOSKELETON: STRUCTURE AND BIOLOGICAL ROLE

Living cells need both the ability to maintain their shape when exposed to shear stresses exerted by their active contractile machinery or by fluid flow in blood vessels and the ability to reorganize their shape and internal architecture as is the case in cell migration and mitosis. The structure responsible for the mechanical and dynamic properties of the cell is the *cytoskeleton*, a rigid yet flexible and *dynamic network of proteins* of varying length and stiffness. Most cells contain three types of protein filaments comprising *actin, tubulin* and *intermediate filament, proteins* such as vimentin. These, as well as the plasma-membrane associated filaments make up the cytoskeleton [33]. There is also a range of *accessory proteins* for each of the cytoskeletal filaments which allow for *control* of nearly all mechanically relevant properties of the protein filaments [34,35]. Let us just mention one example, gelsolin, which is frequently used in rheological experiments. It caps the ends of a growing actin filament and can thus be used to regulate the average filament length in actin solutions. There are many other proteins with different tasks ranging from initiating and terminating polymerization over introducing crosslinks and forming lateral arrays of filaments to even changing the stiffness of the filaments.

An example of the biological role of the cytoskeleton is the migration of an amoeba. Its motion is initiated by adhesion driven spreading of the cell membrane on the substrate followed by gelation of actin in the advancing lobe (pseudopodium). The cycle is completed by retraction of the rear end and a gel–sol transition or fiber formation in the advancing front [11]. This process is of course quite complex and there is a subtle interplay between regulatory mechanisms and the material properties of the cytoskeleton. But, understanding the basic physical principles determining the viscoelasticity of actin networks

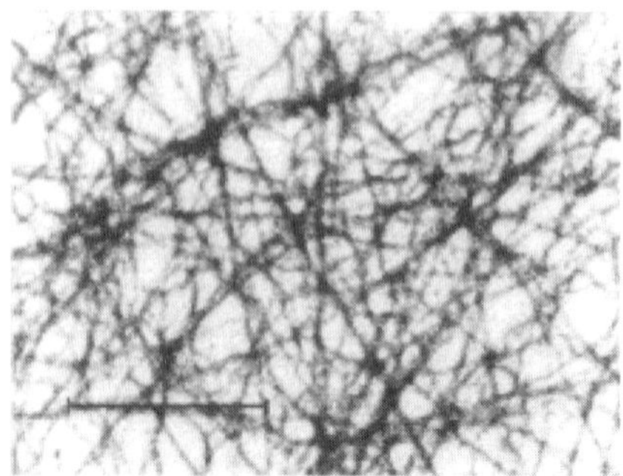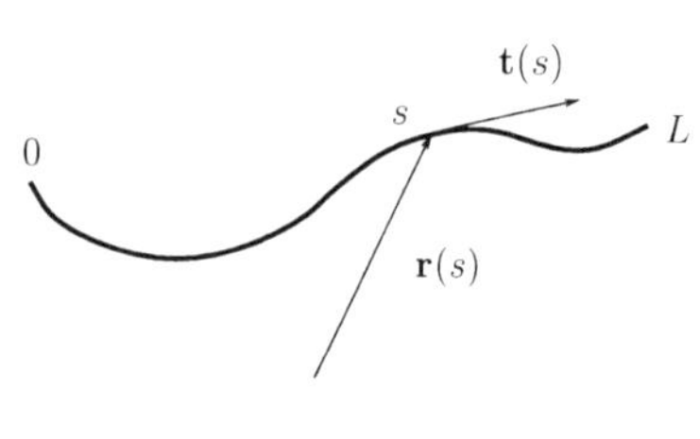

Figure 24.1   Left: Electron micrograph of a 0.4 mg/ml actin solution. The bar indicates 1 µm. Right: Sketch of the wormlike chain as a space curve **r**(s)

is certainly a prerequisite in understanding such a biological process. In the following we will address the following questions: (i) Can we even understand the physics of a one-component system such as a purified F-actin solution (depicted in Figure 24.1)? (ii) Can we identify the basic physical principles underlying the observed viscoelastic behavior? (iii) How is it different from the physics of long flexible coils?

SINGLE CHAIN PROPERTIES

The model usually adopted for a theoretical description of semiflexible chains is the *wormlike chain model* [1,36]. Here one describes the filament as a smooth inextensible line **r**(s) of total length L parameterized in terms of the arc length s. The statistical properties are determined by an effective free energy functional (the 'Hamiltonian')

$$\mathcal{H}(\{\mathbf{r}(s)\}) = \frac{\kappa}{2} \int_0^L ds \left(\frac{\partial^2 \mathbf{r}(s)}{\partial s^2} \right)^2 \tag{24.1}$$

which measures the total elastic energy of a particular conformation by the integral over the square of the local curvature weighted by the *bending modulus* κ. The inextensibilty of the chain is expressed by the local constraint, $|\mathbf{t}(s)| = 1$, on the tangent vector $\mathbf{t}(s) = \partial \mathbf{r}/\partial s$. We will see that this constraint is indeed essential for a correct description of the static as well as the dynamic properties of semiflexible polymers. Due to the mathematical complications resulting from the inextensibility only a few of the statistical properties of the wormlike chain can be extracted analytically, the best known being the exponential decay of the tangent–tangent correlation function $\langle \mathbf{t}(s)\mathbf{t}(s') \rangle = \exp(-|s - s'|/l_p)$ with the persistence length $l_p = \kappa/k_B T$, the mean-square end-to-end distance [1] and the radius of gyration [37],

$$\mathcal{R}^2 := \langle [\mathbf{r}(L) - \mathbf{r}(0)]^2 \rangle = L^2 f_D(L/l_p),$$

$$\mathcal{R}_g^2 = l_p^2 (f_D(L/l_p) - 1 + L/3l_p),$$

where we have introduced the Debye function $f_D(x) := 2(e^{-x} - 1 + x)/x^2$.

FORCE–EXTENSION RELATION

One of the most obvious differences between flexible and semiflexible polymers is their response to external forces (see Figure 24.2(left)). In the flexible case the response is isotropic and proportional to $1/k_\mathrm{B}T$, i.e., the Hookian force coefficient is proportional to the temperature. When the persistence length is of the same order of magnitude as the contour length, the response becomes increasingly *anisotropic*.

Then the linear response of the chain depends on the orientation of the force with respect to the tangent vector at the clamped end. Transverse forces give rise to ordinary mechanical-bending of the filaments and the *transverse spring coefficient* is proportional to κ. The linear response for longitudinal forces is due to the presence of thermal undulations, which tilt parts of the polymer contour with respect to the force direction. The effective *longitudinal spring coefficient* turns out to be proportional to κ^2/T indicating the breakdown of linear response at low temperatures ($T \to 0$) or very stiff filaments ($l_p \to \infty$). This is a consequence of the Euler buckling instability. Note also that for the special boundary conditions of a grafted chain (as depicted in Figure 24.2(left)) the linear response

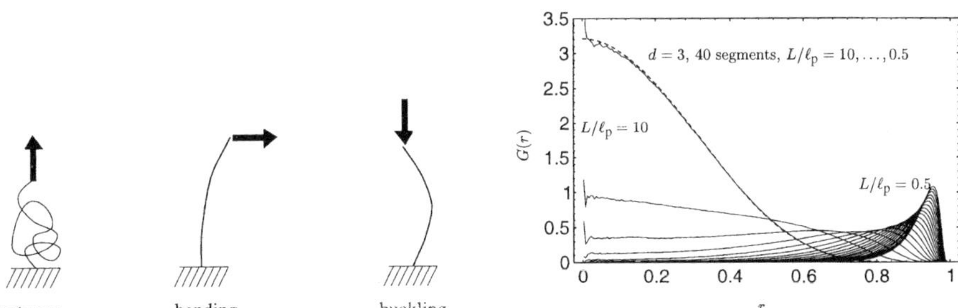

Figure 24.2 Left: A flexible chain's linear response is purely entropic and isotropic. The elastic response of a stiff rod is extremely anisotropic. Transverse forces (bending) lead to a purely mechanical response, whereas the longitudinal response (buckling) is characterized by a spring constant inversely proportional to the temperature. Right: Numerical results for the end-to-end distribution function of a (discretized) wormlike chain in $d = 3$ dimensional space taken from Ref. [38]. With increasing stiffness there is a pronounced crossover from a Gaussian shape to a form with the weight shifting towards full stretching. The dashed line indicates the Daniels approximation [39]

of the chain can even be worked out exactly for arbitrary stiffness [27]; these calculations use the fact that the conformational statistics of the wormlike chain is equivalent to the diffusion on the unit sphere [36].

RADIAL DISTRIBUTION FUNCTION

The function which contains the most comprehensive information about the statistical properties of the chain is the *probability distribution of the end-to-end vector* $G(\mathbf{r}; L) = \langle \delta(\mathbf{r} - \mathbf{R}) \rangle$. For a *freely jointed phantom chain* this function is known exactly [40]. As for any model with short-ranged interactions it converges quickly to a Gaussian distribution $G_0(\mathbf{r}; L) \sim \exp(-3r^2/4l_\mathrm{p}L)$ for increasing number of segments. For chains that are at least some $10\, l_\mathrm{p}$ long the Gaussian can serve as an excellent approximation to $G(\mathbf{r}; L)$ for many purposes. For the freely jointed chain and also for the so-called *freely rotating chain*, the persistence length l_p is independent of temperature because its microscopic origin lies in steric constraints rather than in the bending stiffness of the backbone.

For a Gaussian chain, the separation by a given distance r of any two segments with preferred mean-square distance $2l_\mathrm{p}s$ is punished by the free energy cost

$$F(\mathbf{r}) = -k_\mathrm{B}T \ln G_0(\mathbf{r}; s) = \mathrm{const.} + 3k_\mathrm{B}T\mathbf{r}^2/4l_\mathrm{p}s$$

quadratic in the end-to-end distance. Due to the Euler instability this is very different for semiflexible chains. The characteristic feature of the physics of beam buckling is that the energy E_cl of a straight rod is an almost *linear* function of its end-to-end distance R, $E_\mathrm{cl} \approx f_c \cdot (L - R)$. Here $f_c = \kappa\pi^2/L^2$ is the critical force for the onset of the Euler instability. Neglecting fluctuations around the classical contour this would lead to an end-to-end distribution function with maximum weight at $R = L$, $G(\mathbf{r}; L) \propto \exp[-f_c \cdot (L - r)/k_\mathrm{B}T]$. Note that with such an approach we completely ignore entropic effects which are the only contributions in case of the freely jointed chain, discussed above. In order to correct for this omission we have to multiply the above Boltzmann weight by the relative number of allowed conformations. This becomes most obvious for a completely stretched chain, where up to global rotations only one possible configuration exists and consequently the end-to-end distribution function has to vanish. These qualitative arguments lead to the shape of the distribution function shown in Figure 24.2(right). The actual form of the end-to-end distribution function can be obtained within a quantitative analysis [38] of the wormlike chain Hamiltonian.

DYNAMIC LIGHT SCATTERING

A useful experimental technique for investigating the short-term dynamics of semiflexible polymers is dynamic light scattering (DLS). In DLS experiments one directly observes the dynamic structure factor $g(\mathbf{k}, t)$. We focus on the ideal case of a dilute or semidilute solution of semiflexible polymers, where the scattering wavelength is much smaller than the mesh size. We also assume a separation of length scales, $a \ll \lambda \leqslant l_\mathrm{p}, L$, i.e., the scattering wavelength λ is large compared to the monomer size a but small compared to the characteristic mesoscopic scale defined by L and l_p. As a consequence the contributions to the time decay of $g(\mathbf{k}, t)$ from center of mass and rotational degrees of freedom of the chain are strongly suppressed as compared to contributions from bending undulations. Moreover, for this case [41–43,7,30] the structure factor can be written as $\exp(-k^2 r_\perp^2(t)/4)$ with the local mean square displacement $r_\perp^2(t) \sim t^{3/4}$:

$$g(\mathbf{k}, t) \propto \exp[-(\gamma_k t)^{3/4}] \tag{24.2}$$

where $\gamma_k \sim k^{8/3}/\zeta_\perp l_\mathrm{p}^{1/3}$ [7,8]. Such a stretched exponential behavior has been approved experimentally with very high accuracy for F-actin solutions [44]. However, a more careful analysis reveals that it cannot hold for very short times. For times shorter than $\zeta_\perp/\kappa k^4$ the bending forces can be considered weak and the contour obeys the fast wiggling motion imposed by hydrodynamic fluctuations. As a consequence the initial decay of the structure factor is of the form $g(\mathbf{k}, t) \propto \exp(-\gamma_k^{(0)} t)$ with [7]

$$\gamma_k^{(0)} = \frac{2k_\mathrm{B}T}{3\pi\zeta_\perp}k^3 = \frac{k_\mathrm{B}T}{6\pi^2\eta}k^3 \ln(e^{5/6}/ka) \tag{24.3}$$

For polymers, which are not quite as stiff as actin, e.g. for so-called intermediate filaments, this initial decay regime is readily observed in light scattering experiments. A very convincing confirmation of these theoretical results has recently been found in fibrin systems [47] (see Figure 24.3). Analyzing the data by equation (24.3) allows one to estimate the friction coefficient $\zeta_\perp$ entering the Langevin equation or, equivalently, the thickness a of these filaments.

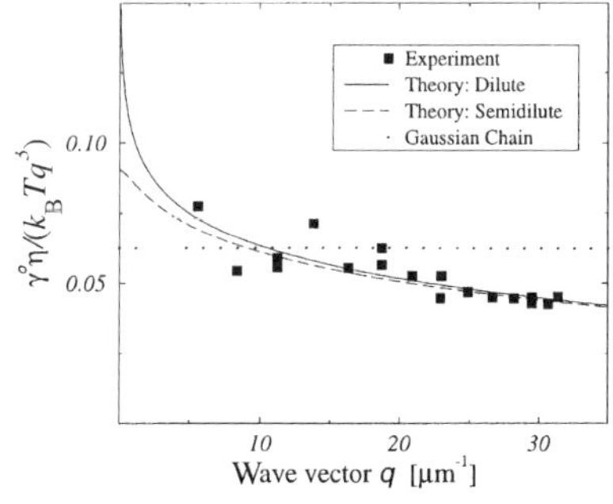

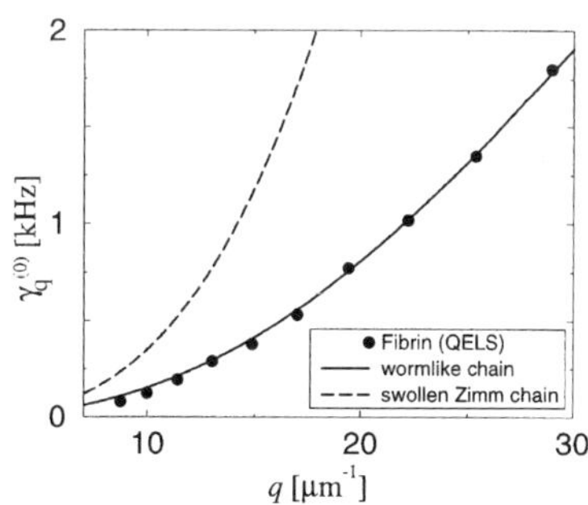

Figure 24.3 Left: The q-dependence to the initial decay rate compared to DLS data on actin [45]. Right: Comparison of the classical result for a swollen Zimm chain [46] with equation (24.3) and quasi-elastic light scattering experiments with the semiflexiable biopolymer fibrin [47]. The data were most kindly provide by G. Arcovito (see also this volume)

COLLECTIVE PROPERTIES

In conventional polymer systems made up of long flexible chain molecules the viscoelastic response is *entropic* in origin over a wide range of frequencies [46]. For semiflexible polymers a complete understanding of the viscoelastic response is complicated by several factors. First of all, there are several ways by which forces can be transmitted in a network. This can either happen by steric (or solvent-mediated) interactions between the filaments or by viscous couplings between the filaments and the solution. It is a priori not at all obvious which if any of these couplings will dominate. In the case of flexible polymers it is generally believed that macroscopic stresses are transmitted in such a way that these transformations stay affine locally, i.e. that the end-to-end distance of a single filament follows the macroscopic shear deformation [46]. Implicit in this hypothesis is the assumption that there is a very strong viscous coupling between polymers and solution and that interpolymer forces can be neglected. As a consequence most of the viscoelastic properties are modeled by a single-filament picture. The applicability of such a single-filament theory to semiflexible polymer networks may be seriously questioned. Second, single filaments are *anisotropic elastic elements* showing quite different response for forces perpendicular or parallel to its mean contour. Therefore one has to ask what kind of deformation of the actin filament is the dominant one and whether due to the anisotropy of the building blocks of the network macroscopically affine deformations stay affine locally. In the following we will address some of the issues raised.

PLATEAU MODULUS FOR ENTANGLED SOLUTIONS

If solutions of semiflexible polymers are sufficiently dense and are probed at sufficiently short time scales (typically in the range of 10^{-2} Hz to 1 Hz) they will exhibit a so-called 'rubber plateau'. The existence of such a plateau is in general traced back to a *time scale separation* between the internal dynamics and the center of mass motion of the polymers. But, even by anticipating a separation of time scales and neglecting the center of mass motion the calculation of the plateau modulus is still a complicated statistical mechanics problem. One has to answer the question how in a disordered network macroscopic stresses and strains are transmitted to individual filaments. However, little is known about these matters and all present theoretical approaches use a single chain picture where very different assumptions are made on the effect of the topological constraints on the conformation of a single filament.

In what might be called the *affine model* the 'phantom model' [48] is adopted to semiflexible polymer systems [25]. It is assumed that upon deforming the network macroscopically the path of a semiflexible polymer between two entanglement points is straightened out or shortened in an affine way with the sample. The macroscopic modulus is then calculated from the free energy cost associated with the resulting change in the end-to-end distance. Since in a solution forces between neighboring polymers can only be transmitted transverse to the polymer axis and there is no restoring force for sliding of one filament past another, it is however hard to imagine that entanglements are able to support longitudinal stresses in filaments. The modulus predicted in the affine model should scale as $G^0 \propto c^{11/5}$ and leads to absolute values of the order of 10 Pa; such high values are at odds with the low values observed in recent experiments on F-actin solutions [21,49]. It was therefore argued [12] that such models are more appropriate for crosslinked networks, where they would predict a plateau value $G^0 \simeq k_B T l_p^2 / \xi_m^5$. But, even in such chemical networks with crosslinks present it is a priori not obvious that local deformations on the scale of a single filament are actually affine and that longitudinal stresses in the filaments are the dominant contribution to the plateau modulus (see also Section 3.3).

Recent theoretical and experimental studies [26,21] based on pioneering work from the 1980's [50,51,52] suggest a different view. Here one considers the free energy cost of suppressed transverse fluctuations of the polymers that comes about by an *affine deformation of the tube* diameter. According to Odijk [51] the mean distance between collisions of a tagged polymer with its surrounding tube with diameter d is given by $L_e \simeq l_p^{1/3} d^{2/3}$. Since each of these collisions reduces the conformation space it costs free energy of the order of $k_B T$ the total free energy of $\nu = c/L$ polymers per unit volume becomes $F \simeq \nu k_B T L / L_e$. To be able to compare these results to experiments one needs to know how the tube diameter d depends on the concentration of the solution or equivalently on the mesh size $\xi_m := \sqrt{3/\nu L}$. In other words we have to determine the average thickness d of a bend cylindrical tube in a random array of polymers as depicted in Figure 24.4(left).

The contour and thickness of the tube will be determined by a competition between bending energy favoring a thin straight tube and entropy favoring a curved thick tube. These competing effects define a characteristic length scale

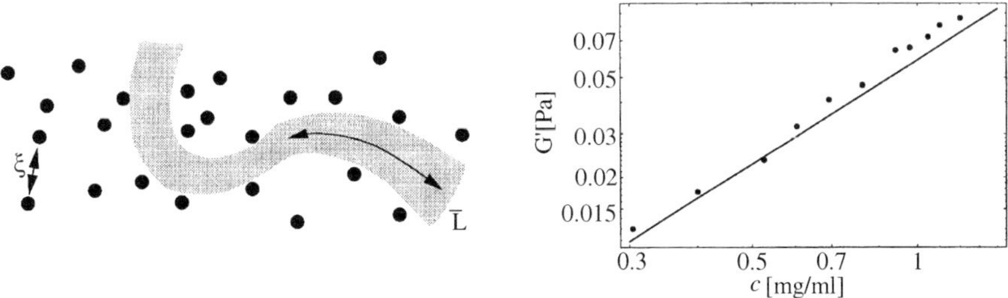

Figure 24.4   Left: A semiflexible polymer can trade bending energy for a wider tube. The configuration of the constraining polymers (dots) is the same as in the upper figure. Right: Comparison of the predicted shear modulus [53] to experiment [21]. The total length L and persistence length l_p were set to $l_p = 17\mu$m and $L = 16\mu$m, respectively [21]

which again turns out to be the deflection length L_e. For length scales below L_e the tube will be almost straight and we can estimate its thickness as follows. Upon restricting the orientations of the polymers to be parallel to the coordinate axes the density of intersection points (black dots in Figure 24.4) (left) will be $1/\xi^2$. Hence for a tube of length L_e the line density of these intersection points projected to a line perpendicular to the tube increases as L_e/ξ_m^2 which implies that the tube diameter decreases with increasing tube length as $d \simeq \xi_m^2/L_e$. Hence one finds $L_e = (\xi_m^2 l_p^{1/2})^{2/5}$ leading to the following form of the free energy and hence the plateau modulus

$$G^0 \simeq F \simeq k_B T l_p^{-1/5} c^{7/5} \tag{24.4}$$

The above scaling law is included as a limiting case in a more detailed analysis concerned with the calculation of the absolute value of the plateau modulus [53].

Recent experiments seem to favor the above tube picture, where the plateau modulus is thought to arise from free energy costs associated with deformed tubes due to macroscopic stresses. Figure 24.4(right) shows the results of a recent measurement of the concentration dependence of the plateau modulus in F-actin solutions [49] which confirms the scaling prediction $G^0 \propto c^{7/5}$ quite unambiguously.

VISCOELASTICITY AND HIGH FREQUENCY BEHAVIOR

At frequencies above the 'rubber plateau' (i.e. above 1 Hz for a typical F-actin solution) a power-law increase of the storage and loss modulus with frequency, $G'(\omega) \propto G''(\omega) \propto \omega^{3/4}$, has been observed [54–56]. It is tempting to speculate that it is somehow tightly connected with the anomalous subdiffusive behavior of the segment dynamics of a single filament. But in view of the actual micro-rheological experiments, where one observes the mean-square displacement of a bead of diameter larger than the mesh-size and hence couples to a large number of filaments, it is not obvious how this comes about. A thorough understanding would need to explore the nature of the crossover from local dynamics dominated by filament undulations to the collective dynamics of the network and the solvent.

At present there are two different theoretical approaches based on different assumptions on the nature of the dominant excitations of the individual filaments generated by the beads embedded in the network. In one class of theoretical models one simply takes over the 'phantom model' approach to the high frequency behavior [57,58]. It is assumed that under an applied shear deformation the filaments undergo affine deformations on a length scale of order L_e, implying longitudinal stresses on single filaments. In the high frequency regime this leads to

$$G^*(\omega) = \tfrac{1}{15} \nu (k_B T)^{1/4} l_p^{5/4} (i\omega\zeta_\perp)^{3/4} \tag{24.5}$$

independently of the entanglement length L_e. A complementary theoretical approach starts from the picture of an idealized micro-rheological experiment, where a point force $\mathbf{f}_s = \mathbf{f}\delta(s)$ is applied to the polymers by optical or magnetic tweezers techniques. This would require beads much smaller than the meshsize which are tightly connected with the filaments. Here one also finds [59] $G' = (\sqrt{2} - 1)G'' \propto \omega^{3/4}$ but with a prefactor which is now mainly due to transverse instead of longitudinal modes; it differs from equation (24.5) by a factor of order ξ_m/l_p, i.e. it is much smaller. This would contradict to the above model, where longitudinal deformations dominate, leading to a modulus (per polymer) which not only depends on the intrinsic properties of the filaments but through the meshsize also depends on the density of the polymer solution. Which one of these theoretical models is more favorable is not clear at present. It may well be that the actual physical mechanism is different from both. There is certainly a tremendous need for more detailed experimental studies which not only measure the power-law dependence of the modulus but also determine the concentration dependence of the prefactor.

EFFECT OF CROSSLINKING

For a crosslinked network of semiflexible polymers bending *and* compressing forces can be transmitted to the filaments. Both for networks where the mesh size is very small compared to the persistence length so that the longitudinal elastic response of the polymers is dominated by their Young's modulus and for networks with larger mesh widths where thermal undulations are crucial in understanding the elastic response of single filaments [25], compression is a much stiffer mode of deformation than bending. Unless highly ordered network geometries are assumed, it is not clear which of the two modes will dominate the elastic response. Different assumptions on the real or effective network geometry can favor either the bending modes as in [28] or the compressional modes as in [25], leading to substantially different predictions for the modulus.

We used a two-dimensional toy model to investigate which type of deformation mode is dominant in a disordered crosslinked network. Sticks of length L were placed randomly on the plane and crosslinked at every intersection with another stick. Crosslinks were inextensible. Sticks were assigned a Young's modulus E and a diameter r resulting in force constants $k_{\mathrm{comp}} = \pi r^2 E$ for compression and $k_{\mathrm{bend}} = 3\pi r^4 E/3L^2 = 3\kappa/L^2$ for bending the rod with one end clamped. Units were chosen such that $L = 1$ and $\kappa = 1$. The model was subjected to periodic boundary conditions, strained and the linear elastic response calculated by the method of finite elements. While this is a purely mechanical model it captures the essential features of two very different force constants and disorder. Entropic contributions from fluctuations of the crosslink positions are not expected to be significant for dense networks.

Which of the two modes dominates the elastic behavior was determined by keeping k_{bend} fixed and varying k_{comp}. We observe that beyond a certain point the modulus ceases to depend on k_{comp}, indicating that the elasticity is dominated by bending modes for slender rods. While these two-dimensional results are

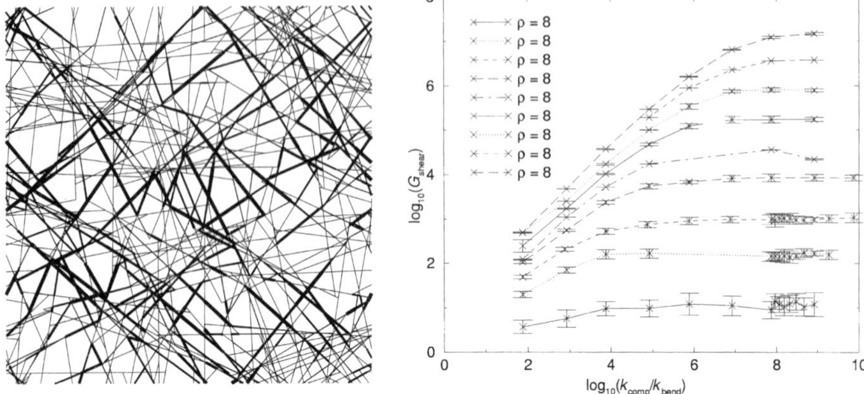

Figure 24.5 Left: Network of sticks for $\rho = 50$, $L = 2$ and $\alpha = 0.01$. The colorcode indicates the load distribution with energy decreasing from red to blue. Right: Dependence of the shear modulus on the ratio K_{comp}/K_{bend} for networks with $L = 15$ [60]

certainly not straightforwardly applicable to three-dimensional networks we will nevertheless try to get a feeling for the scales involved. Network densities can be compared roughly by using the average distance L_c between intersections as a measure: A cytoskeletal network might have $L_c \approx 0.1$ μm with typical filament lengths of 2 μm corresponding to a two-dimensional density of $\rho \approx 20$ (number of rods per area L^2) and an aspect ratio of $\alpha \approx 0.002$ resp. $k_{comp}/k_{bend} \approx 10^{-5}$. Comparison with Figure 24.5 shows that this would just place the network in the bending dominated regime. This might, however, be different for different scales or if more order were present in the network than assumed here. For a more detailed analysis of the random stick model see Ref. [60].

SUMMARY AND FUTURE PERSPECTIVES

We have seen that the cytoskeleton is a composite biomaterial with a wide variety of interesting viscoelastic properties. In particular F-actin solutions and networks provide a model system for a polymeric liquid composed of semiflexible polymers which is accessible to a complementary set of experimental techniques ranging from direct imaging techniques over dynamic light scattering to classical rheological methods. From these studies it has become quite obvious that semiflexible polymer networks require new theorectical models different from conventional theories for rubber elasticity. The nature of the entanglement in solutions of filaments is very different from flexible coils. In a frequency window where an elastic plateau is observed a tube picture where the modulus results from the free energy costs associated with the tube deformations seems to be sufficient to explain the observed concentration dependence of the plateau modulus and even its absolute value.

Outside the rubber plateau in the high frequency as well as the low-frequency regime the situation is less clear. Micro-rheology and dynamic light scattering experiments allow us to access the short-term dynamics of the filaments within a network. Here a theoretical model which describes the combined dynamics of network and solvent in this regime is still lacking. At present there are two quite different approaches which either start from a continuum medium approximation or from a single-filament picture. Obviously both are just limiting cases and a molecular theory needs to explain how starting from the single-filament dynamics including interactions with the solvent and the neighboring filaments leads at some length and time scale to collective behavior, which might be described in terms of some continuum model.

Another very important question is concerned with the effect of chemical crosslinks on the mechanical properties of semiflexible polymer networks. This is of prime interest for both cell biology and for polymer science. In cell biology one would like to know how the material properties (e.g. elastic modulus, time scales for structural rearrangement and stress propagation) change as a function of the network architecture and the mechanical and dynamic properties of the crosslinks. From the perspective of polymer science it connects cytoskeletal elasticity with the very active fields of transport in random media and elastic percolation. In Section 3.3 we have presented a numerical study using a two-dimensional toy model. One can certainly not expect that such a simplified model leads to quantitative results, but we think that some of its main features carry over to the more complicated situation of a three-dimensional network. Future research may concentrate on extending these studies to three-dimensional networks and study how distribution of crosslinks and different network architectures affect the elastic modulus.

ACKNOWLEDGMENT

This work has been supported by the Deutsche Forschungsgemeinschaft through a Heisenberg Fellowship (No. Fr 850/3) and through SFB 266 and SFB 413.

REFERENCES

1. O. Kratky and G. Porod, *Rec. Trav. Chim.*, **68**, 1106 (1949).
2. K. Holmes, D. Popp, W. Gebhard and W. Kabsch, *Nature*, **347**, 44 (1990).
3. S. Burlacu, P. Janmey and J. Borejdo, *Am. J. Physiol.*, **C5**, 69 (1992).
4. A. Ott, M. Magnasco, A. Simon and A. Libchaber, *Phys. Rev. E*, **48**, R1642 (1993).
5. F. Gittes, B. Mickey, J. Nettleton and J. Howard, *Journal of Cell Biology*, **120**, 923 (1993).
6. H. Isambert P. Venier, C. Maggs, Fattoum, R. Kassab, D. Pantaloni and M. Carlier. *J. Biol. Chem.*, **270**, 11437 (1995).

7. K. Kroy and E. Frey, *Phys. Rev.*, E **55**, 3092 (1997).
8. K. Kroy and E. Frey, in *Applications of Dynamic Light Scattering*, (W. Brown, ed.) Gordon and Breach, New York, 1998.
9. P.A. Janmey, *Curr. Op. Cell. Biol.*, **2**, 4 (1991).
10. K. Luby-Phelps, *Curr. Op. Cell. Biol.*, **6**, 3 (1994).
11. E. Sackmann, *Macromol. Chem. Phys.*, **195**, 7 (1994).
12. F. MacKintosh and P.A. Janmey, *Curr. Op. Cell. Biol.*, **2**, 350 (1997).
13. P.A. Janmey, S.O. Hvidt, J. Peetermans, J. Lamb, J.D. Ferry and T.P. Stossel, *Biochemistry*, **27**, 8218 (1988).
14. O. Müller, H.E. Gaub, M. Bärmann and E. Sackmann, *Macromol.*, **24**, 3111 (1991).
15. P.A. Janmey, U. Euteneuer, P. Traub and M. Schliwa, *J. Cell. Biol.*, **113**, 155 (1991).
16. T.D. Pollard, I. Goldberg and W.H. Schwarz, *J. Biol. Chem.*, **267**, 20339 (1992).
17. J. Newman, K.S. Zaner, K.L. Schick, L.C. Gershman, L.A. Selden, H.J. Kinosian, J.L. Travis and J.E. Estes, *Biophys. J.*, **64**, 1559 (1993).
18. P.A. Janmey, S. Hvidt, J. Käs, D. Lerche, A. Maggs, E. Sackmann, M. Schliwa, and T.P. Stossel. *J. Biol. Chem.*, **269**, 32503 (1994).
19. J. Käs, H. Strey, J.X. Tang, D. Finger, R. Ezzell, E. Sackmann and P.A. Janmey, *Biophys. J.*, **70**, 609 (1996).
20. M. Tempel, G. Isenberg and E. Sackmann, *Phys. Rev.*, E **54**, 1802 (1996).
21. B. Hinner, M. Tempel, E. Sackmann, K. Kroy and E. Frey, *Phys. Rev. Lett.*, **81**, 2614 (1998).
22. A. Caspi, M. Elbaum, R. Granek, A. Lachish, and D. Zbaida, *Phys. Rev. Lett.*, **80**, 1106 (1998).
23. S.M. Aharoni and S.F. Edwards, *Rigid Polymer Networks*, Vol. 118 of *Advances in Polymer Science*, Springer, Berlin, 1994.
24. J.L. Jones and R.C. Ball, *Macromol.*, **24**, 6369 (1991).
25. F. MacKintosh, J. Käs, and P. Janmey, *Phys. Rev. Lett.*, **75**, 4425 (1995).
26. H. Isambert and A.C. Maggs, *Macromol.*, **29**, 1036 (1996).
27. K. Kroy and E. Frey, *Phys. Rev. Lett.*, **77**, 306 (1996).
28. R.L. Satcher, Jr. and C.F. Dewey, Jr., *Biophys. J.*, **71**, 109 (1996).
29. E. Frey, K. Kroy, J. Wilhelm, and E. Sackmann, in *Dynamical Networks in Physics and Biology* (G. Forgacs and D. Beysens, eds.) Springer Verlag, Berlin, 1998.
30. R. Granek, *J. Phys. II France*, **7**, 1761 (1997).
31. J. Käs, H. Strey, and E. Sackmann, *Nature*, **368**, 226 (1994).
32. *Monte Carlo and Molecular Dynamics Simulations in Polymer Science* (K. Binder, ed.) Oxford University Press, Oxford, 1995.
33. M. Schliwa, *The Cytoskeleton: An Introductory Survey*, Springer Verlag, Berlin, 1985.
34. J. Hartwig and D. Kwiatkowski, *Curr. Opinion Cell Biol.*, **3**, 87 (1991).
35. J. Olmsted, *Annu. Rev. Cell Biol.*, **2**, 421 (1986).
36. N. Saitô, K. Takahashi and Y. Yunoki, *J. Phys. Soc. Jap.*, **22**, 219 (1967).
37. H. Benoit and P.M. Doty, *J. Chem. Phys.*, **87**, 958 (1953).
38. J. Wilhelm and E. Frey, Phys. Rev. Lett., **77**, 2581 (1996).
39. H.E. Daniels, *Proc. Roy. Soc. Edinburgh*, **63A**, 290 (1952).
40. H. Yamakawa, *Modern Theory of Polymer Solutions*, Harper & Row, New York, 1971.
41. E. Frey and D.R. Nelson, *J. Phys. I France*, **1**, 1715 (1991).
42. E. Farge and A.C. Maggs, Macromol., **26**, 5041 (1993).
43. L. Harnau, R.G. Winkler and P. Reineker, J. Chem. Phys., **140**, 6355 (1996).
44. R. Götter, K. Kroy, E. Frey, M. Bärmann and E. Sackmann, *Macromol.*, **29**, 30 (1996).
45. C.F. Schmidt, Ph. D. thesis, Technische Universität München, 1988.
46. M. Doi and S.F. Edwards, *The Theory of Polymer Dynamics*, Clarendon Press, Oxford, 1986.

47. G. Arcovito F.A. Bassi, M. Despirito, E. Distasio, and M. Sabetta, *Biophysical Chemistry*, **67**, 287 (1997).
48. L.R.G. Treloar, *The Physics of Rubber Elasticity*, Clarendon Press, Oxford, 1975.
49. B. Hinner, and E. Sackmann, unpublished.
50. W. Helfrich and W. Harbich, *Chem. Scr.*, **25**, 32 (1985).
51. T. Odijk, *Macromol.*, **19**, 2313 (1986).
52. A.N. Semenov, *J. Chem. Soc. Faraday Trans.*, **86**, 317 (1986).
53. J. Wilhelm and E. Frey, *Eur. Phys. J.*, B (1998), submitted.
54. F. Amblard A.C. Maggs, B. Yurke, A.N. Pargellis, and S. Leibler, *Phys. Rev. Lett.*, **77**, 4470 (1996).
55. F. Gittes, B. Schnurr, P.D. Olmsted, F.C. MacKintosh and C.F. Schmidt, *Phys. Rev. Lett.*, **79**, 3286 (1997).
56. B. Schnurr, F. Gittes, F.C. MacKintosh, and C.F. Schmidt, *Macromol.*, **30**, 7781 (1997).
57. F. Gittes and F.C. MacKintosh, Phys. Rev. E, **58**, R1241 (1998).
58. D. Morse, *Phys. Rev. E*, **58**, R1237 (1998).
59. K. Kroy and E. Frey, submitted for publication.
60. J. Wilhelm and E. Frey, in preparation.

25

Physical Network Formation under Shear

MADELEINE DJABOUROV, ISABELLE CAPRON, STÉPHANE COSTEUX and MOUSSA KANÉ

Laboratoire de Physique et Mécanique des Milieux Hétérogènes CNRS UMR 7636, ESPCI, 10 Rue Vauquelin 75231 Paris Cedex 5, France

Wiley Polymer Networks Group Review Series Vol. 2. Edited by B.T. Stokke and A. Elgsaeter
© 1999 John Wiley & Sons Ltd

ABSTRACT

Solutions undergoing a process of aggregation or gelation under flowing conditions suffer a profound modification of their natural way of structuring: under flow, a competition arises between the formation of clusters by physical interactions and their disruption under the action of shear forces. A brief survey of the literature reveals that the influence of shear on the structure of complex fluids has received increased attention during the last decade and various systems have been considered: polymer blends, block copolymers, colloidal suspensions, surfactant solutions, emulsions,.... Many industrial areas are affected by these problems. For example, in oil production, crystallization of heavy paraffins leads to gelation of crude oils cooled in quiescent conditions. However, when cooling and shearing are applied simultaneously, a fluid suspension is obtained. Other types of systems are phase separated polymer solutions submitted to shearing.

A review is presented and illustrated by recent experimental results on the effect of shear on physical gelation (for gelatin gels), on crystallization (for paraffinic crude oils and model solutions), on flexible (poly(ethylene oxide)) or rigid (xanthan) polymers in semi-dilute solutions and finally on two phase fluids (phase separated mixtures of alginate and caseinate in aqueous solution). The mechanisms controlling the size of inclusions (polymer clusters, droplets, dispersions,...) or the orientation of molecules during these processes are underlined and the consequences upon the rheological properties are suggested.

INTRODUCTION

Shearing plays a major role in industrial processes, where fluids are mixed, pumped, transported along pipes, while they are submitted to specific thermal treatments. In many cases, the thermal treatments induce physical phase transitions in the media which unavoidably take place under shear. Typical situations are: crystallization of small solutes in supersaturated solutions, physical gelation in polymer solutions, aggregation of colloidal particles in suspensions, which may be an irreversible process. Mixtures of polymers in solution or melts may also phase separate, leading to emulsion type structures. In some cases one observes phase separation and gelation at the same time under shear. The thermodynamic phase transitions or the spontaneous evolution of non-equilibrium states occur in either a well controlled process, and thus can be considered as a necessary step for elaborating special textures (polymer materials, paints, food,...) or may be an undesired phenomenon. For instance, in petroleum production, wax crystallization causes major difficulties in transport and storage of the fluids, in particular after a shut down in production, when strong gelation blocks the pipes. In processing, it is well known that shear plays a crucial role in controlling the microstructure and consequently the rheology of the systems. In laboratory work, one can use conventional rheometers and imagine different protocols in order to simulate the situations encountered in industry, with the advantage of imposing well-controlled conditions, for either the flow, the thermal treatment or any other parameter relevant to the process. Shear flows are the most common ones, but elongational flows may also be considered. Consequently, rheological experiments can serve both to induce disturbances to the phase transition and to explore the properties of the medium after perturbation. In order to develop reliable models for the mechanisms of structural modifications induced by shear, it is important to combine various methods, rheology being one of them and molecular spectroscopy, microscopy, scattering techniques,... being complementary ones. In the literature, one can find many different systems which have been investigated under shear. Some are interesting strictly from the theoretical point of view (binary liquid mixtures near the critical point [1,2]) and some should be closer to application problems: polymer blends [3], surfactant solutions [4,5], emulsions [6]. From the historical point of view, it is interesting to find a pioneer work on aggregation of globular proteins after thermal denaturation by using rheooptical techniques (von Muralt and Edsall [7], 1930, Edsall and Mehl [8], 1940, Bailey [9] 1948, Maruyama [10], 1959, Noda and Ebashi [11], 1960 and a series of papers starting in 1949 by M. Joly and co-workers [12]). In the polymer field, birefringence experiments have been first reported in rubber-like solids (Philippoff [13,14] and co-workers, in 1938 and 1956). Later on, a rheological equation has been adapted for flexible polymers in non-crosslinked states by extension of the statistical theory of rubber elasticity to polymer liquids. Experimental evidence for flow-induced birefringence in visco-elastic polymer systems is extensively reported in the review of Janeschitz-Kriegl [15]. Modern developments in the field of rheooptical instrumentation can be found in the recent book of G. Fuller with numerous examples on complex fluids [16].

The aim of this paper is to briefly recall the basic mechanisms which are at the origin of the shearing effects in complex fluids, trying to classify these complex fluids into suspensions of rigid spherical particles, solutions of elongated rigid or flexible molecules (polymers), aggregating colloids (flocculation) and emulsions (non miscible fluids). In most of the examples quoted above, shear induces a decrease of the viscosity (shear-thinning behaviour). There are, however, examples where shear thickening occurs (increase of the viscosity under shear); they will be mentioned as well. In the second part of the paper, our recent experimental work is presented, which illustrates some of the ideas presented above for various systems: physical gelation under shear for gelatin gels, crystallization under shear for paraffinic crude oils or model solutions, alignment of flexible or rigid polymers in semi-dilute solutions (rheooptical experiments), and eventually phase separated solutions of binary polymer mixtures (alginate + caseinate) under shear. We conclude with an outlook to future work.

ORIGIN OF SHEARING EFFECTS

The different cases which are presented belong either to the colloidal or to the polymer field, but the basic concepts are the same.

CROWDING EFFECTS IN SUSPENSIONS

The description that we adopt was initially proposed by Krieger and Dougherty [17], for the basic ideas. The theory has been reviewed [18], and

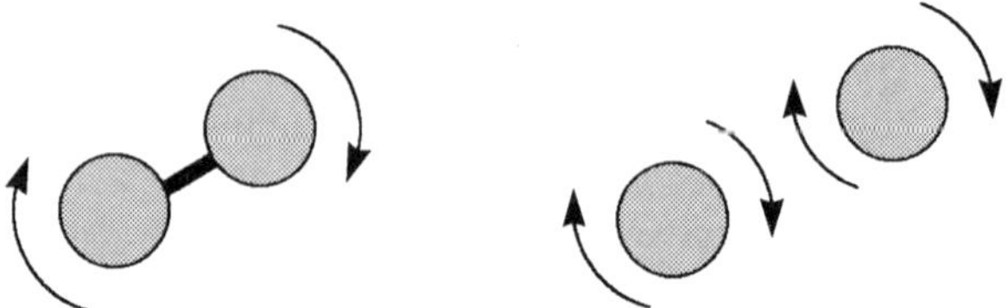

Figure 25.1 Pairs of particles at (a) low and (b) high shear rates

expanded more recently by several authors [19,20]. It is established that suspensions of colloidal rigid particles which exhibit Brownian motions, behave under flow as shear thinning fluids. The explanation for this effect was first given by Krieger and Dougherty. In their model, shear thinning arises from concentration fluctuations and crowding effects in the suspension. Indeed, due to Brownian movements, instant pairs of particles form and break spontaneously. In presence of a low shear rate, the pairs or doublets of particles tend to rotate as dumbbells. At high shear rates the dumbbells are progressively dissociated and the particles rotate independently as singlets, as shown schematically in Figure 25.1.

The number of singlets (P_1) and of doublets (P_2), can be represented, in the absence of shear, by a formal equation similar to a chemical reaction:

$$2P_1 \underset{k_b}{\overset{k_f}{\rightleftharpoons}} P_2$$

where k_f is the spontaneous rate of formation and k_b rate of breakage of the doublets. In the presence of shear, there is a second mechanism of dissociation of the doublets at a rate k_s which is a function of the shear rate:

$$P_2 \xrightarrow{k_s} 2P_1$$

Krieger and Dougherty assumed a linear relation between the viscosity of the suspension and the fraction of doublets and singlets. They calculated the equilibrium distribution of singlets and doublets under shear, and finally derived, with additional simplifications, the equation relating the viscosity of the suspension to the rate constants k_s and k_b. Let η_0 be the viscosity of the suspension at low shear rate, $\dot{\gamma} \to 0$, and η_∞ be the viscosity at high shear rates, $\dot{\gamma} \to \infty$,

$$\frac{\eta - \eta_\infty}{\eta_0 - \eta_\infty} = \left(1 + \frac{k_S}{k_b}\right)^{-1} \tag{25.1}$$

The reaction constant k_b is equal to the inverse of the life time t_b of the doublet, in absence of shear $t_b = 1/k_b$. For Brownian motions, the life time t_b can be approximated by the time for diffusion of the particle along a distance comparable to its own size:

$$t_b \approx \frac{\overline{x^2}}{D_{tr}} \approx \frac{a^2}{D_{tr}} \tag{25.2}$$

where D_{tr} is the translational diffusion coefficient of the particle which is given by the Stokes–Einstein relation:

$$D_{tr} = \frac{kT}{6\pi\eta a} \tag{25.3}$$

with a the particle radius, η the viscosity of the suspending medium (solvent for dilute solutions), k the Boltzmann constant, T the temperature. Under shear, the mean lifetime t_s of the doublet $t_s = 1/k_s$ is a function of the shear rate. If the suspension flows at a shear rate $\dot{\gamma}$, the dumbbell rotates at an angular velocity $\dot{\gamma}/2$. Further on, one may estimate that if the dumbbell has rotated a certain angle α, then the two particles dissociate. Thus:

$$t_s = \frac{\alpha}{\dot{\gamma}/2} \tag{25.4}$$

The ratio of the characteristic times t_b and t_s can be derived from (25.2)–(25.4) as a function of the shear rate

$$\frac{t_b}{t_s} = \frac{\beta a^3 \eta \dot{\gamma}}{kT} \tag{25.5}$$

β being a numerical factor. At this stage different important assumptions should be made concerning the effective viscosity of the suspending medium. If the viscosity is the macroscopic viscosity, $\eta\dot{\gamma} = \tau$, τ is the macroscopic stress and equation (25.5) can be rewritten by introducing a characteristic stress τ_c which depends on the size of the individual particles a and on temperature T:

$$\frac{t_b}{t_s} = \frac{\tau}{\tau_c}$$

with
$$\tau_c = kT/(\beta a^3). \tag{25.6}$$

The characteristic shear stress τ_c controls the appearance of shear thinning of the medium. For this shear stress the viscosity equals $(\eta_0 - \eta_\infty)/2$. The viscosity is a function of the shear stress which in the Krieger–Dougherty model is written:

$$\frac{\eta - \eta_\infty}{\eta_0 - \eta_\infty} = \frac{1}{1 + \tau/\tau_c} \tag{25.7}$$

Alternatively, if one makes the assumption that the viscosity of the surrounding medium is the solvent viscosity, then one observes a shear rate dependence for the ratio:

$$\frac{t_b}{t_s} = \frac{\dot{\gamma}}{\dot{\gamma}_c}$$

with
$$\dot{\gamma}_c = kT/(\beta a^3 \eta_s) \tag{25.8}$$

The characteristic shear rate $\dot{\gamma}_c$ is the inverse of the time for Brownian motion of the particle of size a diffusing on a mean square distance of the order of its squared radius (with $\beta = 6\pi$ and the approximation of equation (25.2)). The viscosity is then related to the ratio $\dot{\gamma}/\dot{\gamma}_c$, of the actual shear rate of the flow to the characteristic shear rate of the suspension. This dimensionless parameter is the reduced shear rate $\dot{\gamma}_r$:

$$\dot{\gamma}_r = \dot{\gamma}/\dot{\gamma}_c \tag{25.9}$$

In some cases it appears as the reciprocal of the Péclet number, which is precisely

$$\mathrm{Pe} = \frac{6\pi\eta a^3 (kT)}{(\dot{\gamma})^{-1}} = \frac{\text{characteristic time of Brownian motion}}{\text{characteristic time of the flow}} \tag{25.10}$$

The characteristic shear rate $\dot{\gamma}_c$ corresponds to the Péclet number $\mathrm{Pe} \approx 1$ (with $\beta = 6\pi$). For $\mathrm{Pe} \ll 1$, the suspension has a viscosity η_0, at large $\mathrm{Pe} \gg 1$, the viscosity reaches the lowest limit η_∞. The viscosity of the suspension thus depends on the shear rate through the relation:

$$\frac{\eta - \eta_\infty}{\eta_0 - \eta_\infty} = \frac{1}{1 + \dot{\gamma}/\dot{\gamma}_c} = \frac{1}{1 + \dot{\gamma}_r} \tag{25.11}$$

This relation determines the steady-state viscosity: the flow curves are reversible upon increasing or decreasing the shear rate (equation (25.11)) or the shear stress (equation (25.7)).

An alternative derivation of this equation is due to Cross [21] which established the well-known equation which is widely used in polymer flows:

$$\frac{\eta - \eta_\infty}{\eta_0 - \eta_\infty} = \frac{1}{1 + (\dot{\gamma}/\dot{\gamma}_c)^m} = \frac{1}{1 + (\dot{\gamma}_r)^m} \tag{25.12}$$

The exponent m ($0 < m < 1$) is related to polydispersity or it appears as a phenomenological or an adjustable parameter.

ORIENTATION OF ROD-LIKE MOLECULES

Rigid anisotropic molecules are easily oriented by external fields, such as electric, magnetic or hydrodynamic fields. At rest, anisotropic rigid molecules are in a random orientational state due to Brownian motions. Under a shear flow, the velocity gradient tends to orient them in the direction of the stream lines, whereas Brownian motion acts against orientation. Under shear, molecules spend a greater part of time at a preferred orientation with respect to the stream line [22] (Figure 25.2).

The degree of orientation is characterized by the ratio of the shear rate $\dot{\gamma}$ to the rotational diffusion coefficient of the rods, D_{rot}^0. At high shear rates (assuming that we are still in the laminar regime) the molecules tend to align in the flow

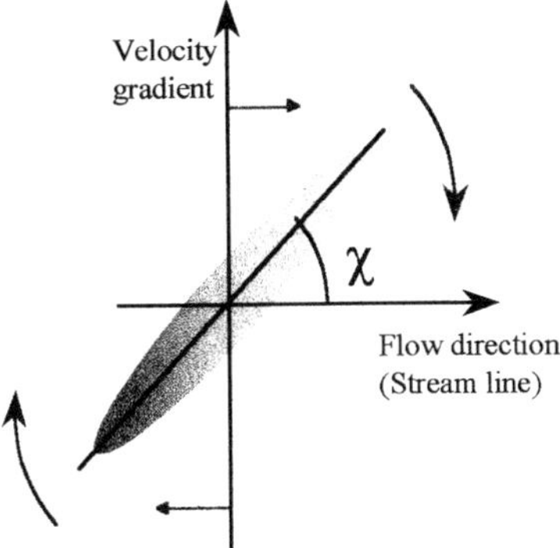

Figure 25.2 Preferred molecular orientation

making a small angle with the stream direction. At small shear rates, the orientation approaches 45° of the stream lines (random orientation). For rigid rod-like molecules, in the limit of dilute solutions, the rotational diffusion coefficient is given by [23]

$$D_{rot}^{0} = \frac{kT \ln(L/d)}{3\pi\eta_s L^3} \qquad (25.13)$$

where L is the length of the rod, d its diameter, η_s the viscosity of the solvent. The relation is valid for $L/d \gg 1$. The coefficient D_{rot}^{0} has the dimension of the inverse of time (s^{-1}) and thus is comparable to the characteristic shear rate defined in the preceding section, for rigid spherical particles. Indeed, the rotational diffusion coefficient for a sphere is $D_{rot} = kT/8\pi\eta a^3$ which can be compared to equation (25.8) giving $\dot{\gamma}_c$ with $\beta = 8\pi$. Thus these two descriptions are comparable, differing only by the value of the numerical factor β. The orientation of the molecules in the flow creates an anisotropy of the index of refraction due to the difference of electronic polarizabilities of the molecules between their axial and transversal directions. The difference of indices is measured in these two perpendicular directions. Flow birefringence appears, which can be detected experimentally. This effect can be used as a direct means for measuring the rotational diffusion coefficients and for deriving molecular parameters. The orientation angle, called the extinction angle χ, versus the shear rate is very sensitive to polydispersity effects, as can be seen from the dependence of D_{rot}^{0} on the length L.

Different concentration regimes are known for solutions of rod like molecules: in the dilute regime, the average distance between the molecules is much larger than their length L (case outlined above). Due to their large hydrodynamic volumes (as compared to flexible polymers), the rods tend to overlap, which severely restricts their possibility of rotation at relatively low concentrations. According to the Doi and Edwards theory [24], the semi-dilute regime for rigid rod like molecules is reached when the number concentration of the molecules ν, is such as:

$$1/L^3 \ll \nu \ll 1/(L^2 d)$$

the lower limit restricts their rotational movements, the higher limit is the transition to an anisotropic liquid state (nematic phase) ($L^2 d$ is the excluded volume of one rod). In the semi-dilute range of concentrations, Doi and Edwards have predicted the change of the rotational diffusion coefficient with the concentration. They propose the following equation:

$$D_{\text{rot}} = \kappa(\nu L^3)^{-2} D_{\text{rot}}^0 \tag{25.14}$$

where κ is a non-specified numerical constant. D_{rot} becomes very much smaller than D_{rot}^0. The rheological consequences are quite important: while in the dilute

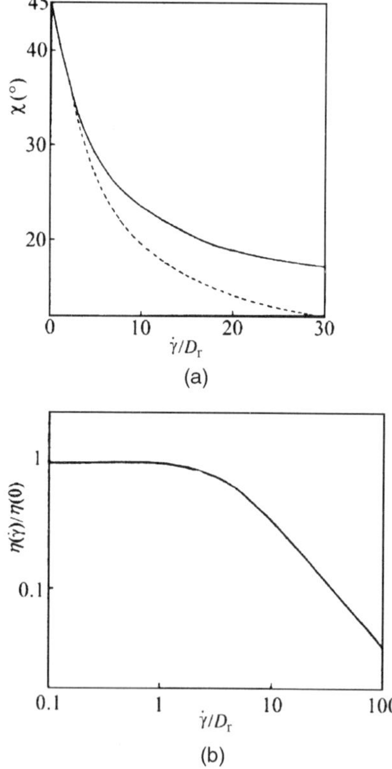

Figure 25.3 (a) Decrease of the extinction angle with the reduced shear rate for rod like molecules, according to Doi and Edwards. Continuous line: semi-dilute solutions, broken line: dilute solutions (after ref. [24]). (b) Shear thinning in semi-dilute solutions of rod like molecules (after ref. [24])

regime, the viscosity due to the rods is always small as compared to the solvent (the correction is of the order of ($\nu L^3 \ll 1$), in the semi-dilute range ($\nu L^3 \gg 1$), the low shear viscosity is much larger and shear thinning effects become very important. Doi and Edwards have constructed a molecular model for the rheological properties of these solutions. In this model, both the orientation of the rods under shear and the viscosity of the solution decrease with a characteristic shear rate $\dot{\gamma}_c = D_{rot}$. The decrease of the extinction angle χ and the decrease of the reduced viscosity $\eta(\dot{\gamma})/\eta(0)$ with the reduced shear rate

$$\dot{\gamma}_r = \dot{\gamma}/D_{rot} \tag{25.15}$$

for a semi-dilute solution are represented in Figure 25.3(a,b). These curves are numerical solutions of the equations proposed by Doi and Edwards. The dotted line is the dilute regime. When collective effects are included (continuous line Figure 25.3(a)) the decrease of the extinction angle with the shear rate is shifted to the right : the χ angle decreases less steeply than in dilute solutions. At high shear rates flow instabilities are predicted (slipping at the walls or turbulent flows above $\dot{\gamma}_r \sim 20$). Jain and Cohen [25] have extended the theory to a wider range of shear rates and also derived the dynamic spectrum of the solutions.

DEFORMATION AND ORIENTATION OF FLEXIBLE POLYMERS

Shear thinning is also observed in semi-dilute solutions of flexible polymers. The theoretical modelling of this effect is delicate. In dilute solutions, the shear rate dependence of the viscosity is ascribed to molecular orientation and deformationd [26].

Various molecular theories of Kuhn, Rouse and Zimm yield to an expression of the rotational diffusion coefficients as the reciprocal of the longest molecular relaxation time $(t_{relax})^{-1}$ of the chain:

$$t_{relax} = \frac{[\eta]\eta_s M}{kN_A T} \tag{25.16}$$

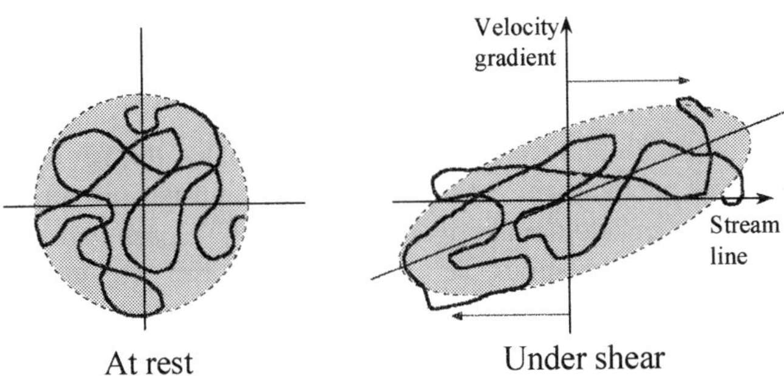

Figure 25.4 A flexible polymer coil at rest and under flow

where M is the molar mass, $[\eta]$ the intrinsic viscosity, η_s the solvent viscosity, N_A the Avogadro number. The characteristic shear rate for the solution under flow is given by the rotational diffusion coefficient of the molecules. Trying to compare equation (25.16) with the previous ones, let us assume that the chain has a radius of gyration a and is non-permeable to the flow of the solvent (Zimm model), then the intrinsic viscosity varies like $N_A a^3/M$ and the longest molecular relaxation time can be written as $t_{\text{relax}} \approx a^3 \eta s/kT$ which is close to $(\dot{\gamma}_c)^{-1}$ in equation (25.8). Orientation of the molecules is accompanied by deformation of the coils whose shape changes from sphere to ellipsoid. The orientation angle χ between the longest axis of the ellipsoid and the flow velocity direction varies with the reduced shear rate:

$$\dot{\gamma}_r = \dot{\gamma} t_{\text{relax}} \tag{25.17}$$

according to the relation

$$\chi = \frac{1}{2} \arctan \frac{C}{\dot{\gamma}_r} \tag{25.18}$$

with the numerical factor C being: $C = 1$, Kuhn model; $C = 2.5$ Rouse model; $C = 4.88$ Zimm model [26]. The mean square radius of gyration of the chains under flow is a quadratic function of the reduced shear rate. These predictions apply only at very low polymer concentrations. Concerning the viscosity, deformation and orientation operate in opposite ways: deformation of the coils increases flow resistance, whereas orientation decreases the friction and thus the viscosity. If these two effects compensate, viscosity should be independent of the shear rate. Whenever orientation effects predominate, shear thinning is observed.

For concentrated polymer solutions and melts, the predictions are more complicated. A large number of experiments have shown the cooperativity in orientational relaxation for melts and rubber networks. A nematic interaction parameter has thus been introduced to account for this effect. The orientation distribution of polymer segments between entanglements has been treated in a similar way as for crosslinked systems [27].

CONTROL OF GROWTH FOR FRACTAL CLUSTERS

Initially dispersed colloidal suspensions flocculate when the Brownian particles undergo attractive interactions. Tenuous networks or clusters are formed to which the shear may produce additional effects: (i) the elastic clusters are deformed and break and (ii) the shear modifies the rate on encounter of the clusters.

According to Wessel and Ball [28] the force F experienced by a cluster of size R in presence of a stress τ is

$$F \sim R^2 \tau$$

Supposing we have a dilute suspension of clusters of size R, the bending moment (torque) Γ applied to the tips of the cluster, in a solvent of viscosity η_s, is

$$\Gamma \sim RF \sim R^3 \tau \sim R^3 \eta_s \dot{\gamma} \tag{25.19}$$

There is a critical value for the torque, Γ_c, that the cluster can support without breaking, which depends on the type of attractive interactions between the particles (electrostatic, van der Waals, bridging between individual particles...). This torque limits the maximum size R_{max} that a cluster can grow in presence of a shear rate $\dot{\gamma}$. From equation (25.19):

$$R_{max} \sim (\Gamma_c)^{1/3} (\eta_s \dot{\gamma})^{-1/3} \tag{25.20}$$

This equation predicts a power law behaviour with an exponent $-1/3$ for the maximum size of the clusters versus the shear rate. Let us assume that the clusters have a fractal structure (Figure 25.5).

Let Φ be the volume fraction of the clusters in the suspension, ψ the volume fraction of primary particles of size a, for a fractal dimension D_f the relation between the two volume fractions is

$$\Phi = \psi \left(\frac{R}{a} \right)^{3-D_f} \tag{25.21}$$

In the limit of dilute suspensions, the relative viscosity varies linearly with the volume fraction of the clusters Φ:

$$\frac{\eta - \eta_s}{\eta_s} \sim \Phi \sim \psi \left[\frac{\eta_s \dot{\gamma} a^3}{\Gamma_{max}} \right]^{-(3-D_f)/3} \tag{25.22}$$

The model predicts a shear thinning behaviour due to the change in size of the clusters (change of the volume fraction of clusters) with the shear rate, for a given volume fraction Ψ of primary particles. The viscosity decreases with the shear rate with a power law with an exponent $-(3 - D_f)/3$. The Péclet number

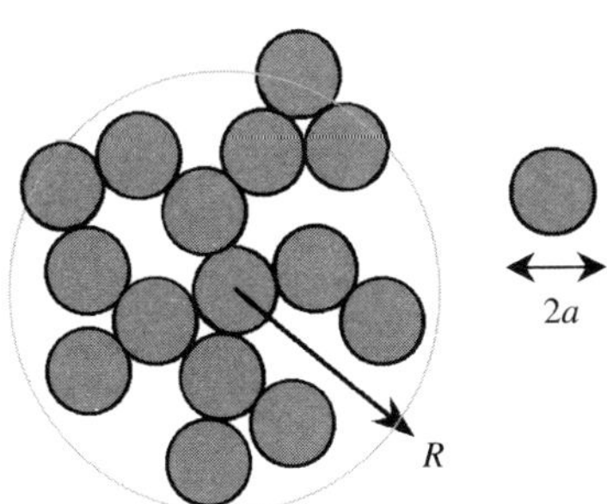

Figure 25.5   Cluster structure

of these suspensions, as in the Krieger–Dougherty model, can be evaluated by the ratio of the diffusion time of the cluster over a distance comparable to its size, $t_b \sim \eta_s R^3 / kT$ to the reciprocal shear rate of the flow, $t_s \sim (\dot{\gamma})^{-1}$, which then gives

$$\mathrm{Pe} \sim \frac{\eta_s \dot{\gamma} R^3}{kT} \sim \frac{\Gamma}{kT} \gg 1$$

Because the attractive energy is much larger than kT, in general, to ensure the cohesion of the aggregates, the Péclet number is always large, the solution is shear thinning to all shear rates. This means also that the Brownian rate of collision $(t_b)^{-1}$ between clusters is much smaller than the shear induced rate of collision, $\dot{\gamma}$. The contribution of the shear flow to the encounter of clusters dominates over the Brownian motion, in the steady state.

BREAK-UP OF DROPLETS UNDER SHEAR IN EMULSION-TYPE SYSTEMS

In phase separated systems, emulsions, non-miscible blends of polymers,... the microstructure is controlled by hydrodynamic forces. The microstructure is composed of droplets of a dispersed phase into a continuous phase. The typical size of the droplets is between 1 and 100 µm. The microstructure may appear spontaneously during phase separation or result from emulsification processes in industry, when two immiscible liquids are mixed to obtain a dispersion of one liquid in the other. Droplet break-up and coalescence occurs during industrial processing. These non-stable systems should eventually separate in two phases by sedimentation. However, emulsions or phase separated systems may remain stable during long enough periods of time so as to be characterized by their microstructure (droplet distribution) and their linear dynamic properties.

A liquid droplet dispersed in another immiscible liquid has a spherical shape at rest. Under a steady flow of the continuous phase the droplet will deform until it reaches a steady shape or break up occurs, depending on the so-called capillary number C_a, which is given by

$$C_a = \frac{\eta_{\mathrm{cont}} \dot{\gamma} R}{\sigma} \tag{25.23}$$

where σ is the interfacial tension between the two liquids, R the radius of the droplet and η_{cont} the viscosity of the continuous phase. Let p be the ratio of viscosities of the inclusions (dispersed phase), η_{incl}, versus the continuous phase, η_{cont}:

$$p = \frac{\eta_{\mathrm{incl}}}{\eta_{\mathrm{cont}}} \tag{25.24}$$

The critical capillary number corresponds to a dimensionless shear rate at which the droplet can no longer assume a steady shape: the interfacial tension is no more able to balance the forces induced by the flow. It has been established from

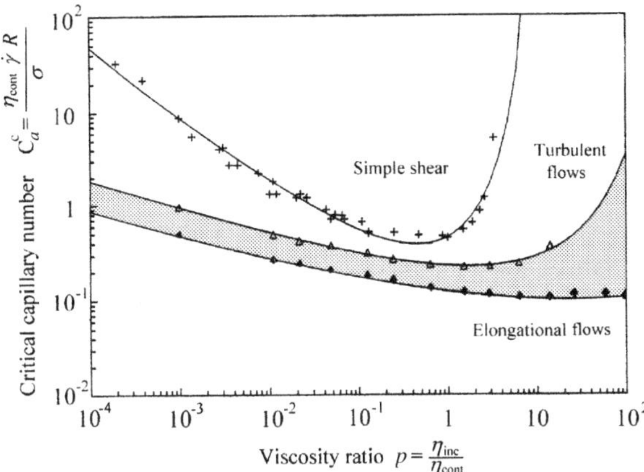

Figure 25.6 Critical capillary number C_a^c versus viscosity ratio for droplet break-up. Simple shear, turbulent and elongational flows are represented (adapted from refs. [29,30])

experimental, theoretical and numerical approaches on single droplets, that there is relation between the critical capillary number C_a^c and the ratio of viscosities p, which is reproduced in Figure 25.6. Simple shear and elongational flows are represented in this figure [29,30]. According to this diagram, for a given ratio of the viscosities, the critical capillary number of the emulsion is fixed. If the interfacial tension σ and the viscosity of the continuous phase η_{cont} are known, then the product $\dot{\gamma}R$ is fixed. This correlation thus can be used to predict the maximum droplet size in simple shear flows. This size decreases inversely to the shear rate. When a droplet is broken it gives rise to several smaller droplets. When the ratio p exceeds a value of 4, which means that the inclusion are four times as viscous as the continuous phase, the simple shear cannot break the droplets, only elongational and turbulent flows can do it.

In real emulsions, several effects are to be taken into account: (i) the continuous and the dispersed phase are in general non-Newtonian liquids; (ii) in presence of surfactants, which are used to stabilize emulsions against coalescence and to decrease the interfacial tensions, the interfacial tension cannot be considered as constant: it depends on the shear deformation and variation of area; (iii) interactions between droplets have to be taken into account (volume fraction of droplets) which may influence the local shear which is 'seen' by a droplet. Thus the field of research is rather wide. The theoretical modelling which has been recently developed by Palierne [31] is devoted to the interpretation of the visco-elastic properties of emulsions with interfacial tension, possibly in presence of surfactants, in the linear dynamic regime. The theoretical predictions should allow the determination of the distribution of radii to the droplets when the interfacial tension is known. So, by varying the shear rate applied to the emulsion, before the measurement of the dynamic spectrum, one may find out the influence of the particular shear on the distribution of the droplets, in the steady state. The higher shear rates should generate the smaller sizes of inclusions.

In order to complete this short review, we have to mention the shear thickening effects. From a macroscopic point of view, what is observed is an abrupt increase of the viscosity at a certain critical shear rate, the viscosity reaches a maximum, and then decreases again at higher shear rates. Although there are numerous examples in the literature with various systems, there seems to be a general mechanism for the shear thickening which is the flow-induced complex formation. For colloidal suspensions in presence of adsorbed flexible polymer, which do not have very strong affinities for the surface of the particle and when the coils have a size comparable to the size of the particles [32], one flexible coil can bind reversibly two particles. The fraction of loops that are formed at the surface of the particle is important as compared to those which are in direct contact with the surface (trains). The adsorption–desorption of the polymer occurs by Brownian motion. The onset of the shear thickening corresponds to a shear rate which becomes comparable to the relaxation time of the suspension. When this particular shear rate is reached, the polymer coils bridging the particles are extended by the flow, then are desorbed. In this model the shear thickening is related to the existence of a three-dimensional network. Some other situations deal with concentrated (phase volume around 0.5) suspensions of non-aggregating solid particles where shear thickening corresponds to a transition from a two-dimensional layering, brought about by the initial shear thinning of the suspensions, which is then disrupted above a critical shear stress [33].

EXPERIMENTAL DETAILS

The analysis presented under 'Origin of shearing effects' is now illustrated by different examples in the polymer and collodial field. The rheological measurements were performed with Carrimed CSL^2 100 and AR 1000 rheometers (TA Instruments) working with a cone and plate geometry (diameter 4 cm, angle 2°). The rheooptical experiments were performed with a home-built instrument adapted on a Haake RS 100 constant stress rheometer, with a coaxial cylindrical geometry with a gap of 1 mm, and a height of 57 mm (optical path). The He–Ne laser beam crosses the gap between cylinders parallel to the axis. The top and the bottom lids of the cylinders are quartz windows which have no stay birefringence. The amplitude and phase of the detected signal give access to the macroscopic birefringence and to the average orientation angle.

The samples use in these various experiments are gelatin from SKW Biosystems, whose molecular characteristics have been reported elsewhere [34], poly(ethylene oxide) (PEO) from Polysciences with a molecular weight $M_w = 5 \times 10^6$ g/mole, poly(styrene sulphonate) (PSS) from National Starch and Chemicals ($M_w = 10^6$ g/mole), xanthan from SKW Biosystems. The samples

(PEO, PSS, xanthan) used for the rheooptical experiments have been filtered under Millipore 1 μm before used. All the samples were dissolved into demineralized water. Other aqueous phase separated solutions were prepared with Na-alginate from Kelco and Na-caseinate from DMV. Finally, a non-aqueous suspension is reported: a crude oil sample containing a few percent of paraffinic long chains (between C_{20} and C_{30}) which was provided by Elf EP.

RESULTS

The basic concepts recalled under 'Origin of shearing effects' are now illustrated with various (complex) examples taken from the colloidal or the polymeric field: gelation under shear for a physical gel (gelatin gel), crystallization of supersaturated solutions (paraffinic crude oils) under shear, Orientation effects in semi-dilute aqueous polymer solutions and finally phase separated ternary solutions (two polymers in aqueous solutions) submitted to shear. Most of the results presented are rheological experiments; some of them include a rheooptical investigation. It is important, as has been underlined in the introduction, that one can use different techniques to corroborate the interpretation of the rheological data. Flow birefringence experiments like those presented in this paper require fully transparent fluids and thus cannot be used for all types of solutions.

GELATION UNDER SHEAR FOR A PHYSICAL GEL

The problem that we considered in this experiment is the phenomenon of gelation for a physical polymeric gel, which is well known for its applications in the food or the photographic industries, that is gelatin gel. Gelation is obtained by cooling solutions below $30\,^\circ$C, inducing the coil to helix transition of the gelatin chains. As triple helices are formed (collagen type structure), stabilized by hydrogen bonds, the conformational transition leads to gelation by physically crosslinking the chains. The helix–coil transition exhibits a time dependence of the helix amount in two steps: first a rapid increase by nucleation of helices, followed by a slow increase, which proceeds indefinitely, through perfection and growth of the helical sequences. The helix renaturation is never complete and reaches approximately 50% of the residues. The question that we addressed was how application of a well-controlled shear would disturb the formation of the gel network. Under flow, a competition arises between formation of the clusters by chain-to-chain crosslinking and their disruption by shear forces. Rheological experiments [34,35] were performed for the same temperature history (cooling and keeping the sample at a fixed temperature) by imposing different protocols: either a permanent shear stress or a permanent shear rate. In order to characterize the structure of the fluid at any stage of the process, we added dynamic measurements (liner regime) and instant flow curves (non linear regime) during very brief interruptions of the process. In order to avoid memory effects provoked by the measurements in the non-linear regime, a new sample was taken for successive instant measurements. The time evolution of the flow curves (given for instance by the viscosity versus shear rate plots in double logarithmic scales) was very indicative of the state of the fluid. Figure 25.7 one can see the beginning of the process where the instant rheograms are shown successively at times t_1, t_2, t_3, ..., for a solution evolving under a constant shear stress. One clearly notices that the flow curves are shifted to the left, towards the low shear rates.

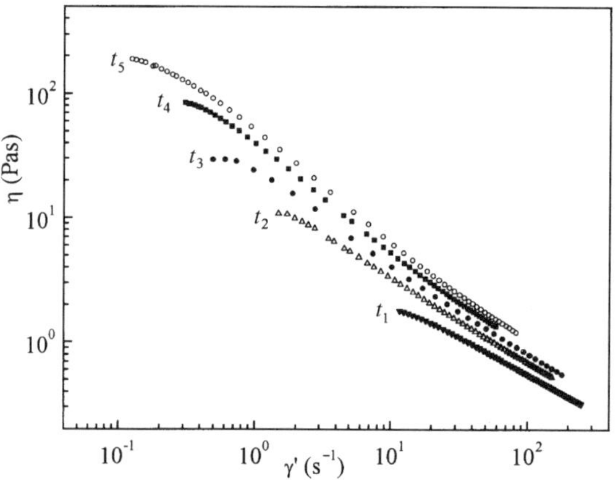

Figure 25.7   Flow curves during gelation of a gelatin solution of concentration 6.5% at a temperature of 26 °C. The different measurements are made at times t_1, t_2, ... during the kinetics of gelation under a permanent shear stress

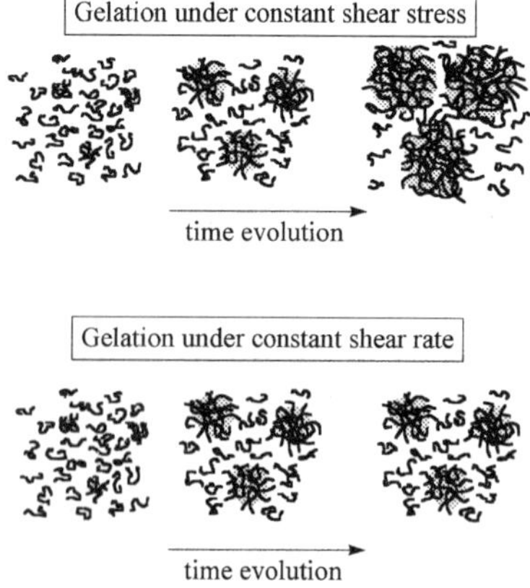

Figure 25.8   Schematic representation of the time evolution of the gelatin microgels during gelation at constant temperature, under imposed shear stress or imposed shear rate

Recalling the Cross equation (25.12), one can fit these curves and derive the characteristic shear rate $\dot{\gamma}_c$ of the solution and the exponent m. Because for these curves $\eta_0 \gg \eta_\infty$ the characteristic shear rates are easily identified as the shear rate for which $\eta \cong \eta_0/2$. $\dot{\gamma}_c$ decreases by several decades in the course of gelation. After some time, the sol–gel transition takes place at a particular moment which is detected by [34]: (i) an abrupt increase of the viscosity, (ii) a steep increase of $G' \gg G''$, (iii) the appearance of a yield stress on the flow curves. The time when the transition occurred was very sensitive to the level of the stress, the higher this level, the later the transition occurred. The behaviour was totally different for the gelation under a fixed shear rate. A critical shear rate $\dot{\gamma}^*$ was determined experimentally: well above this shear rate there was no gelation, the solution remained in a liquid state; at a value close to $\dot{\gamma}^*$, the flow was difficult to control, below this limit no perturbation was seen and gelation occurred very quickly. The whole rheological investigation allowed the proposal of a schematic evolution for the cluster formation under shear. This is shown in Figure 25.8: from the position of the characteristic shear rate in the instant rheograms, using the Péclet number (equation (25.10)), we derived the size of the clusters of chains (called microgels) in the course of gelation. Under a fixed shear stress, the size of the microgels increased with time, until percolation was observed. A particulate gel was built under shear, by contrast with a homogeneous network. Under an imposed shear rate, the size remained constant and did not allow the formation of a gel. Under a shear of $100 \ \mathrm{s}^{-1}$ no gelation was observed after 15 h for a gelatin solution of 6.5% concentration which normally gels in 15 min! This schematic representation of the process will be examined in the near future by using rheooptical techniques, as described below. Microscopic details such as orientation of the chains of helix amount under shear should be derived from these new measurements.

CRYSTALLIZATION UNDER SHEAR

We deal here with a practical problem encountered in oil production. Crude oils which contain around 20% of n-paraffins $>C_{10}$ exhibit complex flow behaviours which strongly depend on the thermal and flow conditions [36]. When the oil is cooled below a certain temperature (around $30\,^\circ\mathrm{C}$ for instance) the longest paraffinic chains become insoluble and start to crystallize. These are called waxy crude oils. The flow characteristics for these oils are time and history dependent.

When the oil is cooled in quiescent conditions, a strong gel is formed. Such a transition is shown in Figure 25.9. Within a few degrees the elastic modulus of the gel increases over six decades (up to $10^6\mathrm{Pa}$). The gel is brittle and supports only small deformations ($<10^{-3}$). The moduli G' and G'' are frequency independent. Although the moduli are very high, the amount of crystalline material is rather low (a few percent) when the gel appears. However, when the oil is cooled under flow, as usually in oil production, only a moderate increase of the viscosity is seen, such as in Figure 25.10. The oil has been cooled under a fixed shear rate,

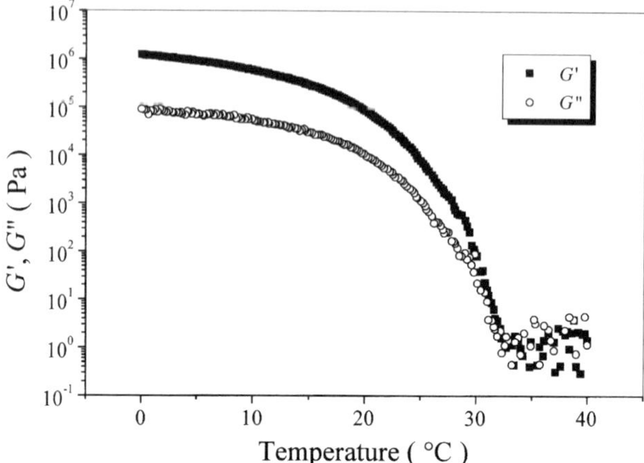

Figure 25.9 Gelation of a crude oil containing a small amount of paraffins. The shear moduli G' and G'' were measured at the frequency of 1 Hz during cooling at a rate of 0.5 °C/min

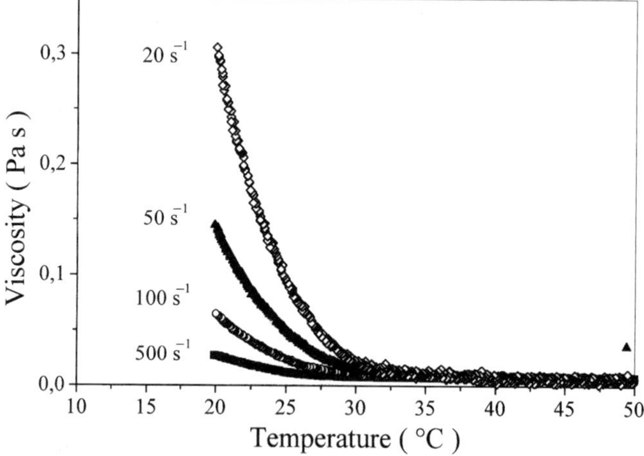

Figure 25.10 The apparent viscosity of the crude oil is measured as a function of temperature during cooling under fixed shear rates, indicated on the figure

indicated on the figure. The apparent viscosity is plotted versus temperature. One can see that it varies strongly with the shear rate, for a given range of temperatures and a fixed cooling rate (0.5 °C/min). The higher viscosities are reached for the lowest shear rates, but the values are quite low (0.30 Pa s, for instance). Time dependent effects are also seen when the fluids are kept over long periods of time at the same temperature and under an imposed shear rate, usually a decrease of

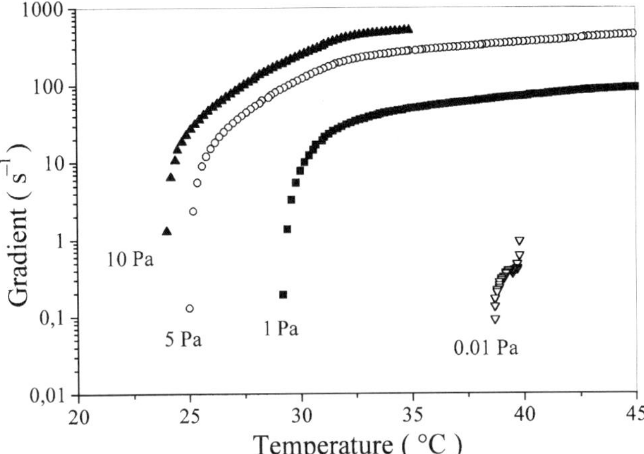

Figure 25.11 The velocity gradient versus temperature for the crude oil cooled under imposed shear stresses indicated on the figure. The gelation transition occurs at different temperatures depending on the stress

viscosity. When the process of cooling takes place under a fixed shear stress, there is a gelation transition which clearly takes place, which can be compared to gelatin gelation under shear. The results are shown in Figure 25.11. The Shear rate decreases strongly as the oil is cooled: the apparent gelation transition occurs at different temperatures when the cooling rate is fixed and even seems to take place at a higher temperature than expected, under a stress of 0.01 Pa. Although these are preliminary results the process of crystallisation under shear can be interpreted by following the ideas of formation of clusters with a fractal structure. The shear rate controls the size of the clusters, the shear thinning effect should be due to variation of the volume fraction of the cluster, for a given volume fraction of primary particles (Ψ in equation (25.21)). The latter may be assimilated to the crystalline content for each temperature. The clusters which were broken by shear do not reaggregate. In parallel to the rheological data microscopic investigations of the structure are under way in order to characterize the architecture of the network and the size and structure of the clusters under shear. The crystallization of model paraffinic solutions exhibit a similar behaviour and are also currently investigated.

FLOW BIREFRINGENCE OF SEMIDILUTE POLYMER SOLUTIONS

The experimental device which has been recently built in our laboratory with the aim of investigating complex processes, such as gelation under shear, for optically transparent solutions, has been only under the testing procedure so far and the first results that we report in this paper concern the behaviour of semi-dilute solutions of flexible and rigid macromolecules under shear [36]. The main parameters that

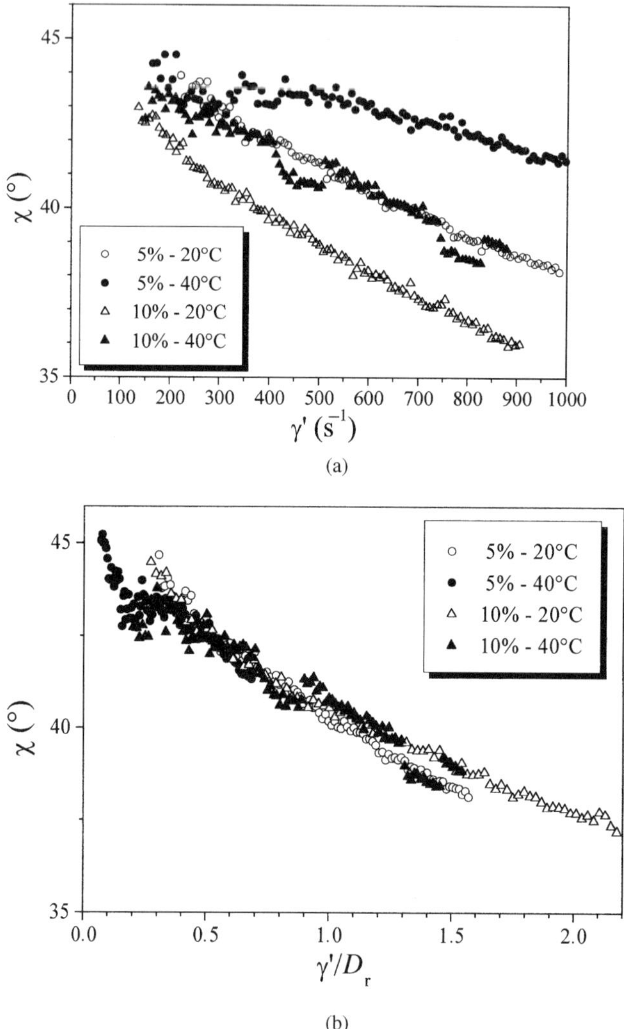

Figure 25.12   (a) The extinction angle versus the shear rate for PSS solutions at two concentrations (5% and 10%) and two temperatures (20 °C and 40 °C) (b) Extinction angle versus reduced shear rates for the PSS solutions

we derived are the macroscopic birefringence and the average orientation angle χ for these solutions. The flow curves were measured either simultaneously to the optical measurements, in the Couette device or independently on the Carrimed CSL^2 with the cone and plate geometry, in order to get confirmation. The results for the PSS samples are shown in Figure 25.12(a). The extinction angle decreases from 45° to 37° when the shear rate increases from 0 to 10^3 s^{-1} for PSS at two different concentrations, 5% and 10%. This orientation is relatively weak, while the shear rate covers the whole accessible range in currently used instruments. According to the theory, the following relation determines a rotational diffusion coefficient of the macromolecules:

$$\lim_{\dot{\gamma}\to 0} \frac{d\chi}{d\dot{\gamma}} = -\frac{1}{12 D_{\text{rot}}} \tag{25.25}$$

Where D_{rot} is given by equation (25.14) for rigid rods, and for random coils by

$$D_{\text{rot}} = \frac{C}{6}\,(\tau_{\text{relax}})^{-1}$$

Thus, if the slope at origin for $\chi(\dot{\gamma})$ is derived from the graph, for different concentrations and temperatures, then one can normalize the shear rate for the different plots. This is shown in Figure 25.12(b) where the decrease of the extinction angle from different experiments is shown with this scaling assumption. One can see that indeed there is a good superposition of the various curves. The stress optical rule is also obeyed with this sample:

$$\Delta n \sin 2\chi = 2 C_{\text{so}} \tau \tag{25.26}$$

where C_{so} is the stress-optical coefficient and Δn the birefringence.

We found for $C_{\text{so}} \sim 10^{-8}$ Pa^{-1} which is a very low value showing that our experimental device was able to detect a very low birefringence. For PEO solutions the results were rather different. The stress optical coefficient is even lower (10^{-9} Pa^{-1}). The decrease of the orientation angle with the shear rate (Figure 25.13(a)) is much steeper, for very low polymer concentrations (0.5 and 1%) the χ angle decreasing abruptly from 45° to 25° from 0 to 30 s^{-1}, followed by a more progressive decrease. The reduced shear rate is shown in Figure 25.13(b): one can see a scaling up to $\dot{\gamma}_r \approx 400$. The last example is a xanthan sample [38]: the extinction angles for five concentrations versus normalized shear rate $\dot{\gamma}_r$ are plotted in Figure 25.14(a) and the flow curves in Figure 25.14(b). Xanthan gum displays an extensive shear thinning due to large orientation of the semi-rigid rods, the viscosity being close to the viscosity of water for high shear rates. Nonetheless the extinction angles level off around 20°. The flow curves cannot be superposed using the normalized shear rate, in disagreement with the Doi–Edwards theory. The presence of cooperative effects may be at the origin of this discrepancy. The flow curves for PSS can be normalized with the reduced shear rate, showing that the slight shear-thinning behaviour can be related to the weak orientation of the molecules. This unexpected behaviour for entangled polyelectrolyte solutions, especially at such high concentrations, may be explained if the chains were in a collapsed state due to the presence of salt in this commercial sample. For PEO, the normalization of the flow curves (data not shown) is valid for low shear rates, in the range where χ decreases rapidly, and a discrepancy appears for higher shear rates: this could be explained by a two-step process in

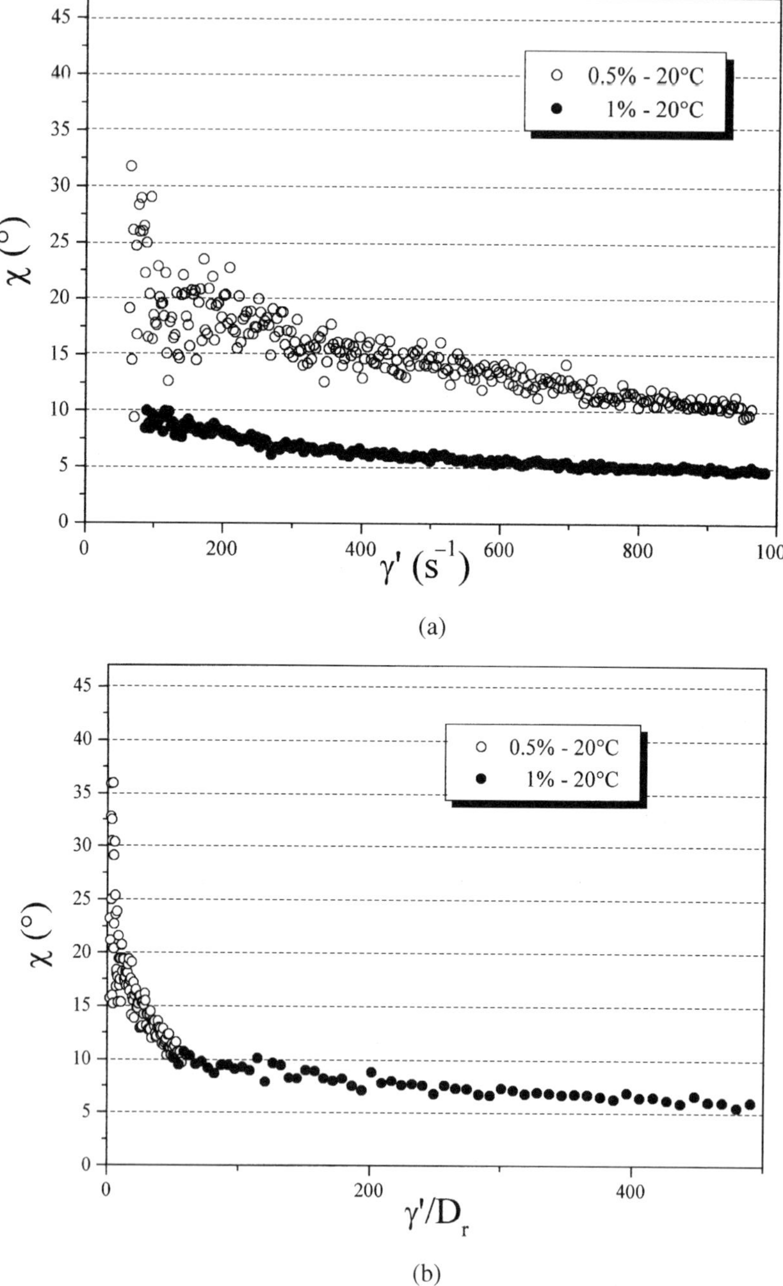

(a)

(b)

Figure 25.13   (a) The extinction angle versus the shear rate for PEO solutions of concentrations 0.5 and 1% at 20 °C (b) Extinction angle versus reduced shear rates for the PEO solutions

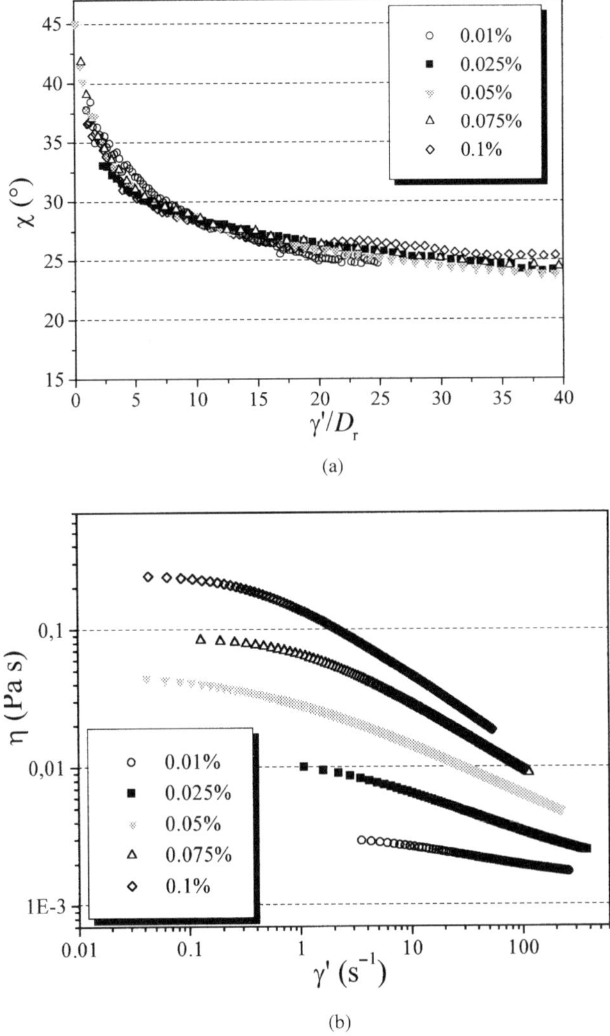

Figure 25.14 (a) Extinction angle versus reduced shear rates for Xanthan solutions of various concentrations indicated on the figure, at room temperature. (b) Viscosity versus shear rate for the Xanthan solutions at the same concentrations, at room temperature

the flow behaviour of this very flexible polymer, with an orientation of the chains at low shear rates and a disentanglement for larger $\dot{\gamma}$.

These preliminary experiments illustrate the effect of shear upon the orientation of flexible or rigid molecules. The relation between the orientation of the molecules and the shear-thinning of the solutions is not straightforward. Also entanglement effects should be taken explicitly into account. This is a large area for experimental and theoretical investigations.

PHASE SEPARATED SOLUTIONS UNDER SHEAR

The last case that we examine is a phase separated aqueous solution containing alginate and caseinate. Only the general features of these experiments are reported. An initial mixture in the two phase region of the solution was prepared. Then the solution was centrifuged and the two phases collected, one being a caseinate-rich phase and the other an alginate-rich phase, both phases containing the two biopolymers. Various amounts of each phase were then mixed together varying the volume fraction of one phase from 0 to 100%. Phase separated solutions of alginate and caseinate [39] show microstructures which are close to emulsions, droplets of various sizes scaling in the range 10–30 µm appear. By mixing the emulsion under various shear rates, it is, in principle, possible to change the microstructure by controlling the maximum size of the droplets in agreement with the ideas presented under 'Origin of shearing effects'. Rheological experiments in the two phase region are difficult to carry on. One requires a certain time stability for the system, in order to investigate the dynamic spectrum. One also assumes that the size of the droplets small enough that the conventional rheometers can be used. When these requirements are fulfilled the spectrum of the emulsion can be performed. Knowing the volume fraction of each phase and its dynamic spectrum, the interpretation of the dynamic spectrum

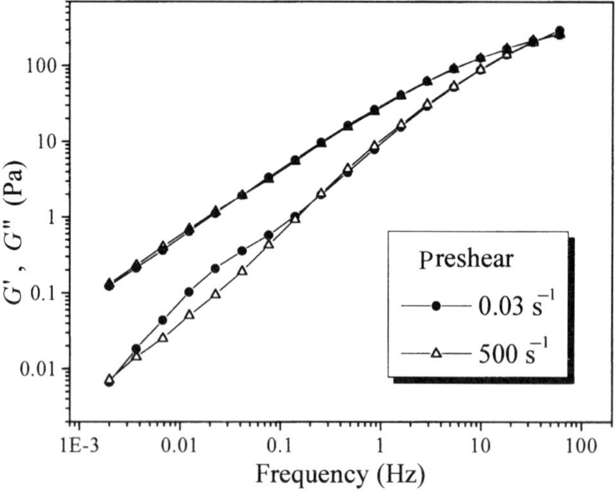

Figure 25.15 Dynamic spectra for an alginate–caseinate solution in the two phase region. The solution has been submitted to two different shear rates (0.03 and 500 s⁻¹) before measurement of the dynamic spectra. The continuous lines are guides for the eye

of the emulsion can be performed according to the Palierne theory [31]. The fitting parameter for this theory in its simplified version is the ratio σ/R of the interfacial tension σ to the radius R of the (monodisperse) droplets. By submitting the phase separated solutions to various shear rates before measuring the spectrum, one can explore possible changes of the size of the droplets. For instance in Figure 25.15, we illustrate this effect. A phase separated solution containing $\Phi = 10\%$ of the caseinate rich phase, which is the less viscous one, was submitted to intensive shearing at either 0.03 s^{-1} or 500 s^{-1} then the dynamic spectrum was performed. It appears in Figure 25.15 that at the low frequency limit of the spectrum, the storage modulus G' exhibits a visible change, while the loss modulus G'' is insensitive to the preshear. We observed similar effects for the symmetric composition of $\Phi = 10\%$ of the alginate-rich phase. The linear viscoelastic behaviour of two phase polymer blends in the melt has been characterized by several authors [40–42] and it has been shown that the storage modulus G' has a higher value at low frequencies and that the terminal relaxation times of the blends of viscoelastic liquids are generally higher than the terminal (longest) relaxation times of the individual components. The relaxation time of the emulsion is attributed to the relaxation of shape of the droplets. Changes in size of the droplets induce a visible shift of the characteristic frequencies. The full interpretation of the data requires the knowledge of the interfacial tension between the phases which can be measured independently. A direct observation of the droplets in the emulsion should help to validate the existing models. When the two phase volumes become comparable ($\Phi = 50\%$) the predictions are less clear. Shear thinning effects are expected to modify the correlation between the critical capillary number and the ratio of viscosities (Figure 25.6).

CONCLUSION AND OUTLOOK

This review intends to underline the effects of shear on the molecular conformation or the microstructure of complex fluids containing polymers and colloids. Although the basic mechanisms have been identified and classified, processes involving aggregation, physical gelation, phase separation,... involve a juxtaposition of these mechanisms and thus are more difficult to analyse. Combining rheological and structural techniques one should be able to better understand such systems. Models relating the local structure and the flow properties of these solutions can then be tested and improved. This area of research seems very promising and exciting for the future.

REFERENCES

1. A. Onuki and K. Kawasaki, *Ann. Phys.*, NY, **121**, 456 (1979).
2. D. Beysens, M. Gbadamassi and B. Moncef-Bouanz, *Phys. Rev. A*, **28**, 2491 (1983).

3. A.I. Nakatatni, H. Kim, Y. Takahashi, Y. Masushita, A. Takano, B.J. Bauer and C.C. Man, *J. Chem. Phys.*, **93**, 795 (1990).
4. J.F. Berret, D.C. Roux and G. Porte, *J. Phys. II France*, **4**, 1261 (1994).
5. J.P. Decruppe, R. Cressely, R. Makhloufi and E. Cappelaere, *Colloid & Polymer Sci.*, **273**, 346 (1994).
6. T.G. Masson and J. Bibette, *Phys. Rev. Lett.*, **77**, 3481 (1996).
7. A.L. Von Muralt and J.T. Edsall, *J. Biol. Chem.*, **89**, 315 (1930).
8. J.T. Edsall and J.W. Mehl, *J. Biol. Chem.*, **113**, 409 (1940).
9. K. Bailey, *Biochem. J.*, **43**, 271 (1948).
10. K. Maruyama, *Sci. Pap. Coll. Gen. Educ. Univ. Tokyo*, **9**, 147 (1959).
11. H. Noda and S. Ebashi, *Biochem. Biophys. Acta*, **41**, 386 (1960).
12. M. Joly and E. Barbu, *Bull. Sté. Chim. Biol.*, **31**, 1642 (1949).
13. W. Buchheim and W. Philippoff, *Naturwissenschaften*, **26**, 694 (1938).
14. W. Philippoff, *Nature*, **178**, 811 (1956).
15. H. Janeschitz-Kriegl, *Adv. Polym. Sci.*, **170** (1969).
16. G.G. Fuller, *Optical Rheometry of Complex Fluids*, Oxford University Press, New York (1995).
17. I.M. Krieger and T.J. Dougherty, *Trans. Soc. Rheol.*, **III**, 137 (1959).
18. I.M. Krieger, Adv. *Coll. Inter. Sci.*, **3**, 111 (1972).
19. W.P.B. Russel, D.A. Saville, and W.R. Schowalter, *Colloidal Dispersions* Cambridge University Press, Cambridge, (1986).
20. D. Quemada, *Rheol. Acta*, **17**, 632 (1978).
21. M.M. Cross, *J. Colloid. Sci.*, **20**, 417 (1965).
22. R. Cerf and H.A. Scheraga, *Chem. Revs.*, **51**, 185 (1952).
23. M. Doi and S.F. Edwards, *J. Chem. Soc., Farad. Trans. II*, **74**, 560 (1978).
24. M. Doi and S.F. Edwards, *J. Chem. Soc., Farad. Trans. II*, **74**, 918 (1978).
25. S. Jain and C. Cohen, *Macromolecules*, **14**, 759 (1981).
26. A. Link, M. Zisenis, B. Prötzl and J. Springer, *Makrom. Chem. Makrom. Symp.*, **61**, 358 (1992).
27. M. Doi, D. Pearson, J. Kornfield and G. Fuller, *Macromolecules*, **22**, 1488 (1989).
28. R. Wessel and R.C. Ball, *Phys. Rev. A*, **46**, R3008 (1992).
29. B.J. Bentley and L.G. Leal, *J. Fluid Mech.*, **167**, 241 (1986).
30. H.P. Grace, *Chem. Eng. Commun.*, **14**, 225 (1982).
31. J.F. Palierne, *Rheol. Acta*, **29**, 204 (1990).
32. Y. Otsubo, *Adv. Colloid and Interf. Sci.*, **53**, 1 (1994).
33. H.A. Barnes, J.F. Hutton and K. Walters, *An Introduction to Rheology*, Elsevier Sci. Publ. B.V. (1989).
34. W. de Carvalho and M. Djabourov, *Rheol. Acta*, **36**, 591 (1997).
35. W. de Carvalho and M. Djabourov, The Wiley Polymer Networks Group Review Series (K. te Nijenhuis and W.J. Mijs, eds), Vol. 1, Chap. 7, Wiley Chichester.
36. O. Lorge, M. Djabourov and F. Brucy, *Rev. Inst. Fran. Petrole*, **52**, 235 (1997).
37. S. Costeux, M. Djabourov and J.C. Charmet, *Proceedings of the 5th European Rheology Conference, Progress and Trends in Rheology*, (I. Emri, ed.), Darmstadt, Steinkopff, 459, (1998).
38. I. Capron, G. Brigand and G. Muller, *Polymer*, **38**, 5289 (1997).
39. J.C.G. Blonk, J. van Eendenburg, M.M.G. Koning, P.C.M. Weisenborn and C. Winkel, *Carbohyd. Polym.*, **28**, 287 (1995).
40. P. Scholtz, D. Froelich and R. Muller, *J. Rheol.*, **33**, 481 (1989).
41. D. Graebling, D. Froelich and R. Muller, *J. Rheol.*, **33**, 1283 (1989).
42. D. Graebling and R. Muller, *J. Rheol.*, **34**, 193 (1990); I. Vinckier, P. Moldanaers and J. Mewis, *J. Rheol.*, **40**, 613 (1996).

26

Viscoelastic Properties of Some Biopolymer Systems in Relation with their Nano-structure

M.A.V. AXELOS[1], D. RENARD, C. CHEVILLARD[2] and J. LEFEBVRE[1]

[1]INRA-Laboratoire de Physico-Chimie des Macromolécules, Rue de la Géraudière, BP 71627, 44316 Nantes cedex 03, France

[2]University of Massachusetts, Chem. Eng. Dept., 159 Goessmann laboratory, Amherst, MA, 01003 USA

Wiley Polymer Networks Group Review Series Vol. 2. Edited by B.T. Stokke and A. Elgsaeter
© 1999 John Wiley & Sons Ltd

ABSTRACT

We will present some experimental results obtained using neutron and X-ray scattering techniques on a few systems representatives of different classes of biopolymers and we will compare these structural data to the linear viscoelastic response or flow behaviour. The first example, on methylcellulose gelation with and without surfactant, will illustrate the direct consequence of a change in the polymer organization on the rheological properties. The second example, on pectin gelation in presence of calcium or copper cations, will illustrate that no clear evidence of a difference was found in the nanostructure of both systems while large differences in their mechanical properties were observed. The third example, on bovine serum albumin thermal gelation, will illustrate that differences in the nano-structure do not necessarily lead to large changes in the viscoelastic responses.

INTRODUCTION

The understanding of the viscoelastic properties of materials from their micro-scopic characteristics is a fundamental question in rheology, but always a difficult one. In polymeric systems, despite their complexity, some universal rheological behaviours have been outlined, showing that macroscopic properties depend only on a few molecular parameters [1]. Molecular theories of viscoelasticity have been developed which allow prediction of various dynamical properties of polymer solutions or gels [2,3]. However, many discrepancies between models and experiments remain and the incidence of structural details and interactions on mechanical behaviour of real polymers is not completely grasped. This is particu-larly true in the case of biopolymer solutions or gels where the ideas of completely flexible chains or permanent cross-link points are clearly inappropriate [4]. The monomer is no longer the length scale which allows a statistical description of the polymer. Larger length scales are needed to characterize biopolymer systems due to the intrinsic rigidity of the backbone, and to the formation of extended junction zones involving a large number of individual chains, or of dense aggregates at an intermediate scale, which is involved in biopolymer gelation.

In this respect, neutron and X-ray scattering methods provide a powerful tool for structural studies because of the rather wide range of distances they probe [5]. In the following, some experimental results obtained on three different biopolymer systems using these techniques will be presented and compared to their linear viscoelastic response or flow behaviour. The systems studied are (i) the thermogelation of methylcellulose, (ii) pectin gelation and phase separa-tion induced by addition of divalent cations and (iii) the thermal aggregation and gelation of bovine serum albumin.

EXPERIMENTAL

MATERIAL

The methylcellulose sample used: Methocel A4C, was from Dow Chemical Company. The average methoxyl content is around 30%, corresponding to a degree of substitution of 1.8. The intrinsic viscosity measured in water at 25 °C was 4.5 dl/g. Prior to use, the commercial sample was dialysed extensively against distilled water and then freeze-dried. Methylcellulose aqueous solutions were prepared at 4 °C under stirring for 12 h, then centrifuged at 4 °C; the supernatant was filtered on 0.45 µm porosity membrane, and degassed under vacuum [6].

The source material of pectin samples was a high methoxyl citrus pectin provided by Copenhagen Pectin Factory. Other degrees of esterification (DE) were obtained by controlled acid de-esterification [7]. Pectin aqueous solutions were obtained by gentle stirring in distilled water overnight, at room temperature; they were filtered through a 0.8 µm membrane.

The bovine serum albumin (BSA) sample used was from ICN, with a purity of 98–99%. The BSA sample was defatted by cold pentane before use to eliminate bound fatty acid, which affects its thermal denaturation. Solutions were prepared by dissolving BSA in pure water or D_2O (isoelectric conditions), or in NaCl solutions in water or D_2O with subsequent adjustment of the pH to 7, and then were carefully degassed [8].

SMALL ANGLE SCATTERING MEASUREMENTS

Small angle X-ray scattering (SAXS) measurements were performed using the synchrotron radiation of the DCI storage ring at LURE (Université d'Orsay, France). The collected data, on beam D24, covered the scattering vector q range from 0.06 to 2.5 nm^{-1} ($q = (4\pi \sin \theta/2)/\lambda$, where θ is the scattering angle and λ the wavelength).

Small angle neutron scattering (SANS) measurements were performed on the PACE and PAXY spectrometers of Laboratoire Léon Brillouin (Saclay, France) and on the D22 spectrometer of the Laue–Langevin Institute (Grenoble, France). Using different combinations of wavelength and sample-to-detector distance, the q range spanned was 2.7×10^{-2}–2 nm^{-1}. For SANS measurements water was replaced by D_2O.

RHEOLOGICAL MEASUREMENTS

The rheological studies were carried out with two Carri-Med rotational stress-controlled instruments: CSL 50 and CSL 100, fitted with a cone and plate device, and with a Contraves Low Shear 40 instrument working in the Couette geometry. Whatever the system studied, the gels were formed in situ in the measuring device using a specific procedure [6,8,9]. After each heat treatment both dynamic shear storage G' and loss moduli G'' were measured at a given frequency, until equilibrium values were reached. In the case of BSA gels, the systems were heated from 20 °C to 80 °C in 20 min following a linear temperature ramp and quenched to 20 °C in the same way at the end of the heating period. When the gels were formed in pure water (isoelectric gels), the moduli reached a plateau after 4 h heating: the length of the 80 °C temperature plateau was fixed to 4 h accordingly; at pH 7, the moduli continued to increase after 17 h of heating, so that the 80 °C temperature plateau was limited to 2 h, whatever the BSA concentration or the ionic strength, for convenience reasons.

RESULTS AND DISCUSSION

THERMOGELATION OF METHYLCELLULOSE

Methylcellulose is a water-soluble hydrophobically modified cellulose derivative largely used in industry to obtain thermothickening effect [10,11]. Commercial methylcellulose samples have a heterogeneous distribution of the CH_3 groups along their backbone resulting in a complex block copolymer structure [12]. Hydrophobic interactions between the highly derivatised zones of the polymer chains cause polymer–polymer association induced by increasing temperature. Upon heating, gel formation competes with phase separation [13]. As shown in Figure 26.1, the concave shape of the cloud point curve indicates that methylcellulose exhibits a lower critical solution temperature (LCST) [6]. Inside the upper domain of the phase diagram turbid gels were obtained, and the system was found to evolve to a complete phase separation all the more easily as the temperature is high and the concentration is low. Below the binodal, the phase diagram is complex with a clear gel phase in the low concentration range and viscous solutions at higher concentrations.

The evolution of the mechanical spectra recorded at the end of the kinetics for different temperatures and for polymer concentrations less than 20 g/l allows determination first of the sol–gel transition, then the onset of phase separation. As shown in Figure 26.2, where G' measured at 0.005 Hz is plotted versus T,

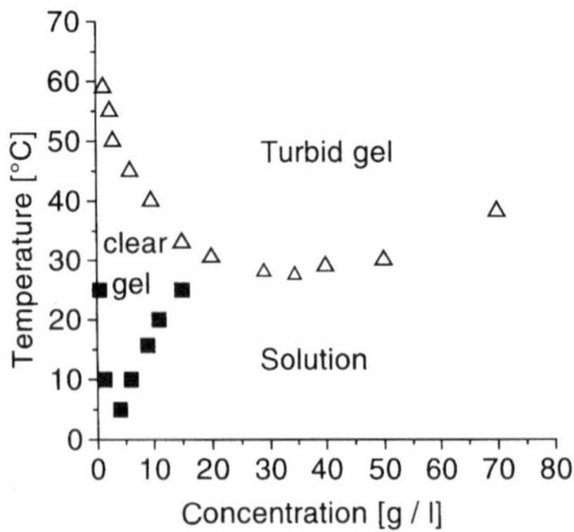

Figure 26.1 The low concentration region of the phase diagram of methylcellulose in water including the binodal curve (△), and the sol–gel transition line (■)

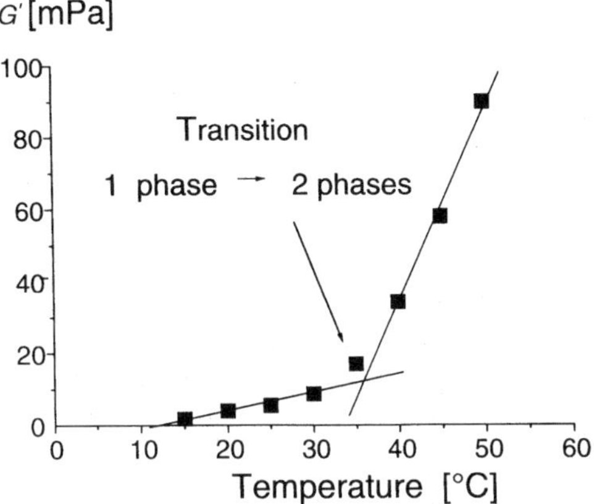

Figure 26.2 Evolution of G' (measured at 5×10^{-3} Hz) with temperature for the methylcellulose–water system at 9 g/l concentration, showing the transition between the one-phase and the two-phase region of the phase diagram

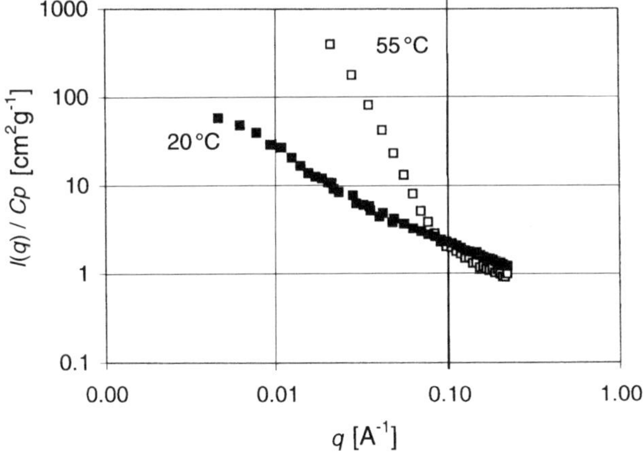

Figure 26.3 Small-angle neutron scattering spectrum of methylcellulose (9 g/l) in water at 20 °C (■) and at 55 °C (□)

the turbid gel domain displays a larger dependence on temperature than the clear gel phase. The large increase of G' was found to be closely linked with the development of turbidity and thus related to the important structural change induced by the phase separation.

SANS experiments have been carried out at a given polymer concentration (9 g/l) and different temperatures in order to follow the structural evolution through the different domains of the phase diagram. No structural change is detected in the 5–30 °C interval. The intensity curve at 20 °C (Figure 26.3) is indicative of a semiflexible polymer in good solvent: $I(q) \sim q^{-1.7}$ for $q <$ 0.02 Å^{-1}. Above the binodal, at 55 °C, the large increase in the scattered intensity in the intermediate q range together with the $I(q) \sim q^{-4}$ dependence indicate the formation of dense domains; at higher q no change is observed. The lower q range remains to be explored to determine the size and the structural organization of these dense domains.

Addition of an anionic surfactant such as sodium dodecyl sulphate (SDS) leads to complete disappearance of the clear gel phase (as shown in Figure 26.4, $G'(\omega)$ drops when 8 mM of SDS is added), and to an increase of the cloud point temperature. SANS experiments have been carried out at 55 °C with an increasing amount of SDS, between one and ten times the critical micellar concentration of the surfactant. In Figure 26.5, the decrease of the intensity upon addition of the surfactant and the appearance of a maximum on the scattering curve at high enough SDS concentration may be interpretated as a consequence of the progressive disappearence of the dense domains and of the appearance of a correlation distance due to electrostatic repulsions between SDS micelles bound to the polymer chain. This system looks very similar to the poly(ethylene oxide)/SDS one [14], with the difference that here SDS micelles may not be equally distributed on the macromolecules but may interact with the hydrophobic zone of the methylcellulose preventing polymer–polymer associations.

For the methylcellulose system, the structural observations made around the cloud point temperature and inside the turbid gel domain with and without surfactant, were found to be directly linked to the change observed in the rheological data. Despite the large discrepancy between the time scale explored through rheological measurements which corresponds to a macroscopic scale, and the

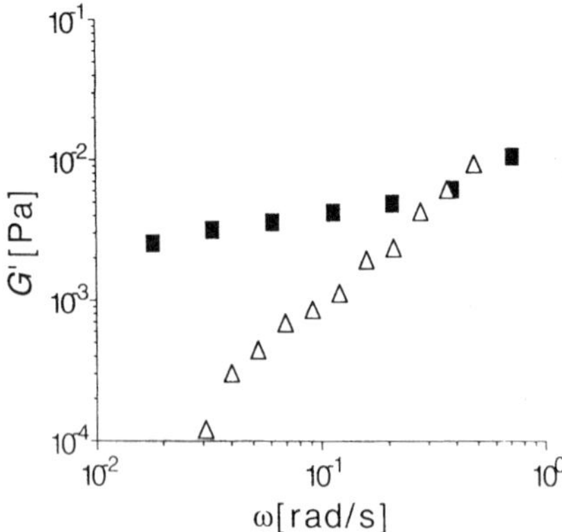

Figure 26.4 Decrease of G' upon addition of sodium dodecyl sulfate (SDS) to a methylcellulose–water system in the clear gel region of the phase diagram. Methylcellulose (9 g/l)-water at 25 °C: ■; the same system after addition of SDS (8 mM): △

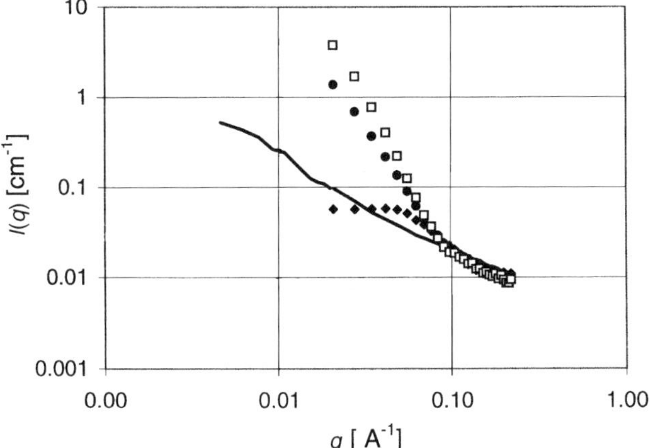

Figure 26.5 Changes in the neutron scattering spectrum induced by the addition of increasing amounts of surfactant to a 9 g/L methylcellulose–water system at 55 °C: □ without surfactant, ● with 2.5 g/L SDS, ♦ with 22.5 g/L SDS. The spectrum at 20 °C in absence of SDS (——) is shown for comparison

nanometric scale investigated through radiation scattering, the good correlation observed between these two types of measurements may be ascribed to the phase separation phenomenon which intrinsically affects all the structural scales with large fluctuations. On the other hand, the sol–gel transition inside the one-phase domain, which results only in the formation of some discrete cross-links between the macromolecular chains, does not lead to detectable structural changes.

PECTINS GELATION AND PHASE SEPARATION INDUCED BY ADDITION OF DIVALENT CATIONS

Pectins are anionic polysaccharides extracted from the plant cell wall. They consist of a linear backbone of a randomly $\alpha(1-4)$ linked D-galacturonic acid units and their methyl esters [15]. Changing the degree of esterification allows variation of the charge density of the polymer. Much attention has been paid to the pectin–calcium system because of its applications in the food industry as a gelling agent [16]. More generally, pectins interact strongly with cations, as is expected for a polycarboxylate polymer in the presence of oppositely charged electrolytes [17]. The effect of addition of a divalent cation to aqueous solutions of pectins is best illustrated by the phase diagram with the added cation concentration and the pectin concentration as variables, as shown on Figure 26.6. Two transitions can be detected: a phase separation always occurs when the cation concentration increases, dividing the diagram into two main regions; in addition, in the lower, one-phase region, a sol–gel transition can occur. The position of these two transitions depends strongly on the charge density of the polymer and the nature of the cation, as illustrated in Figure 26.6, allowing easy determination

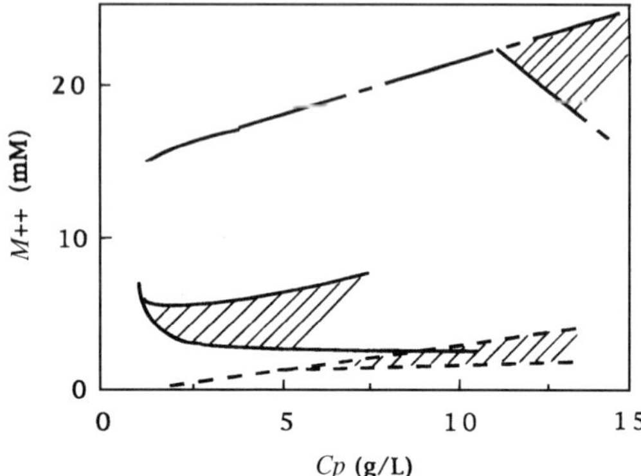

Figure 26.6 Phase diagrams of a low methoxyl pectin sample at pH 7 in NaCl 0.1M in presence of: CuCl₂, (----), CaCl₂ (——) and MnCl₂ (— - —), showing the 'gel' domain (hatched zone)

of a scale of affinity: Cu > Ca ≫ Mn. In general when the cation–polymer interactions are very strong, precipitation is observed at low cation concentrations, in stoichiometric proportion with respect to the number of charged groups of the polymer; this is the case of pectin with copper. When the cation–polymer interactions are weaker, larger numbers of cations are required to get the phase separation as the degree of esterification increases for a given number of COO^- groups. This behaviour, observed in the case of pectin in presence of calcium, indicates that the charge distribution, and not only the total number of charges, must be taken into account to explain the pectin–calcium interactions [18].

The rheological data of Figure 26.7 indicate that in presence of copper, highly viscous solutions are generally obtained under the demixing line, instead of a real gel with a clear solid-like behaviour as with calcium or manganese.

SAXS measurements were performed on the same pectin sample in the presence of copper or calcium. Assuming that the scattering is mainly due to the electronic contrast between the cation and the polymer rather than that between the polymer and the solvent, the measured intensity allows determination of the structure of the cation-rich zones. The average size of the cross-section radius of these structures is obtained from the slope of the Guinier plot Ln $qI(q)$ versus q. As shown in Figure 26.8, the slopes are identical for the two cation/polymer systems at the same cation and polymer concentrations. The radius thus measured is about 7 Å compared to around 3 Å for an isolated polymer chain. Nevertheless, in presence of 3.4×10^{-3} M of divalent cations, the calcium/pectin system is a gel while the copper/pectin is a precipitate. Thus for two completely different physical states and rheological properties, the local structure was found to be very similar. Further experiments will be undertaken using small angle X-ray scattering, but also other techniques like cryo-electron microscopy in order to probe the micron structural scale at which larger differences are to be expected.

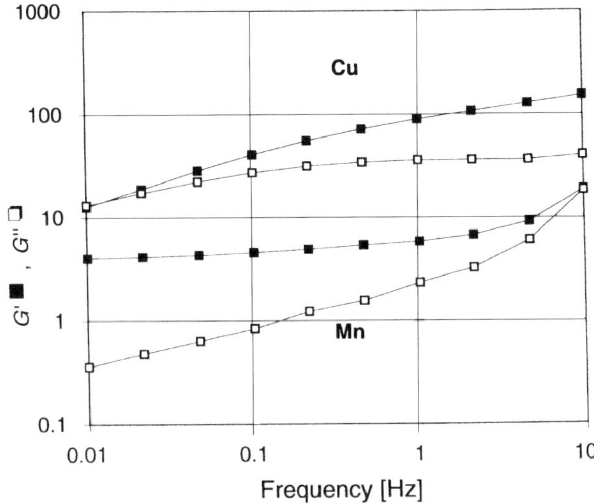

Figure 26.7 Mechanical spectra of a low methoxyl pectin sample in the presence of copper (upper curves) and manganese (lower curves). In both cases, the system is in the gel domain

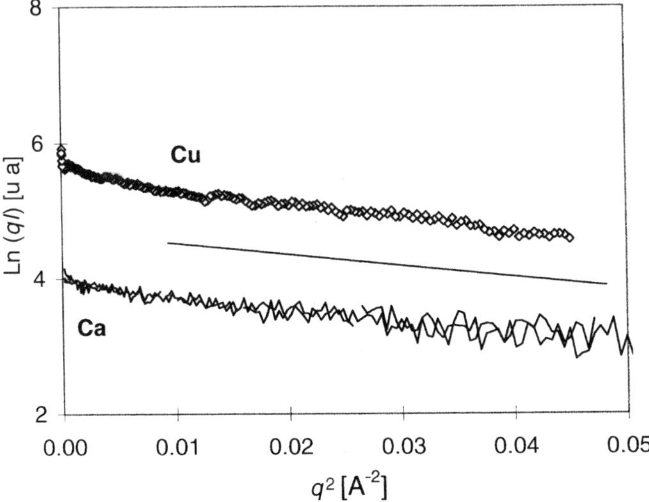

Figure 26.8 Guinier plot of the X-ray scattered intensity of a 6 g/l pectin solution in 0.1 M NaCl in the presence of 3.4×10^{-3} M calcium or copper, showing the formation of rod-like particles in the system

THE THERMAL AGGREGATION AND GELATION OF BOVINE SERUM ALBUMIN

Bovine serum albumin is a typical globular protein. Globular proteins are zwitterionic macromolecules which can be considered in their native state as compact quasi-spherical or slightly ellipsoidal small particles with diameters in the $1-10$ nm range (~ 6 nm for BSA), with a complex tightly packed inner organization specific to each protein. When their solutions are heated above a certain temperature, they undergo a conformational change ('denaturation') resulting in the loss of their native structure; but it has been shown that they keep in most cases a corpuscular character, with a moderate increase in diameter, generally of the order of 20% [19]. The temperature-induced conformational change brings hydrophobic amino-acid residues which were buried into the native structure to the contact of the aqueous solvent; as a consequence, it triggers an *irreversible aggregation process* of the denatured protein molecules, which results eventually in the formation of a gel if the concentration of the protein is high enough. Heat-induced aggregation and gelation of globular proteins are by far less studied and less understood than the initiating conformational change.

Of course, the structure and properties of such gels, which are of the colloidal type, differ completely from those of the gels examined above. On the other hand, they depend stringently, as can be expected, on the pH and the ionic strength in the solvent, which control the balance between repulsive and attractive interactions between the denatured protein molecules. Visual inspection and electron microscopy observations have shown that in conditions of strong electrostatic repulsion the aggregation process is rather linear and clear, homogeneous, fine-stranded gels are formed, whereas opaque gels with coarse lumpy structure are obtained near the isoelectric point of the protein, i.e. when the net charge on the macromolecule is very small [8].

Comparison of Figures 26.9(a,b) shows that the structures of the two types of BSA gels is also strikingly different on the $5-200$ nm length scale probed by SANS measurements. In the case of isoelectric BSA gels (zero protein net charge), the scattered intensity follows Porod's law ($I(q) \propto q^{-4}$) up to $q \sim 0.3$ nm^{-1}. This means that on the $20 < 2\pi/q < 200$ nm size range the gel is made of compact clusters with a smooth surface. A more thorough analysis of the results suggests that these aggregates are formed of closely packed BSA molecules and that their size is of the order of the micrometer. Such a dense liquid-like structure of the clusters results probably from reordering of the molecules within the aggregates, as has been proposed in the case of some flocculated colloidal systems showing a similar scattering behaviour, but very different in nature [20, 21].

Serum albumin gels formed at pH 7 (the pH value at which the BSA molecule bears a fairly large negative net charge) display scattering spectra which are more complex (Figure 26.9(b)). In the absence of added salt, a correlation peak is observed with its maximum situated at $q_{\mathrm{m}} = 0.14$ nm^{-1}, which could reflect

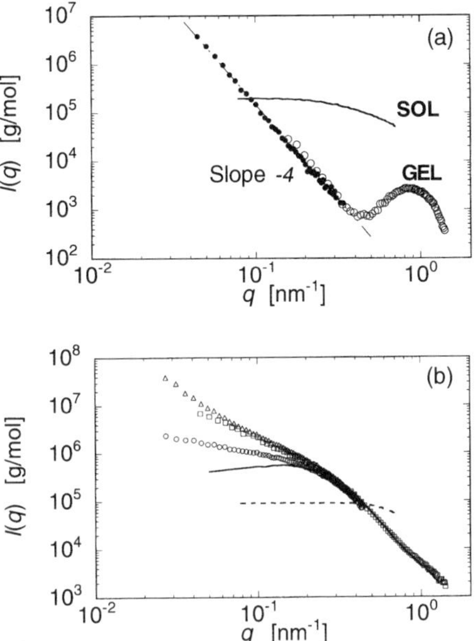

Figure 26.9 SANS spectra of bovine serum albumin (BSA) heat-set gels. Gels are formed in the measuring cells by heating BSA solutions in D_2O (protein concentration: 30 mg/ml) during 4 h (isoelectric gels) or 2 h (pH 7 gels), and then quenched at 20 °C. SANS measurements are performed at 20 °C.
(a) Isoelectric gel (pure D_2O) (symbols), compared to the initial solution (continuous line).
(b) Gels at pH 7. Continuous line: no salt added; circles: 0.02 M NaCl; squares: 0.05 M NaCl; triangles: 0.08 M NaCl. The interrupted line shows the spectrum of the initial solution containing 0.05 M NaCl

the existence of a preferential distance ($2\pi/q = 45$ nm) between charged aggregates in the absence of screening of double layer repulsions [22]. When the ionic strength increases, this correlation peak vanishes for a salt concentration as low as 0.02 M NaCl, and the scattered intensity at low q values ($q < \sim0.13$ nm^{-1}) follows a power law in q with an exponent which increases from 0.7 in 0.02 M NaCl to 1.75 in 0.05 M NaCl. These results can be considered tentatively as indicative of fractal structures in BSA aggregates formed at pH 7 in the presence of salt. The fractal aggregates generated in 0.02 M NaCl are quite linear (apparent fractal dimension of 0.7); as expected, the structure becomes more branched and compact (apparent fractal dimension of 1.75) when the salt concentration is increased to 0.05 M. In 0.08 M NaCl, two linear regimes of the scattered intensity appear within the experimental window, with slope values of 2.3 ($q < 0.08$ nm^{-1}) and 1.75 ($0.08 < q < 0.2$nm^{-1}), respectively, suggesting a crossover between two fractal regimes. It is interesting to note that, whereas the position of the lower internal cut-off for the fractal regime varies somewhat (from $q^{-1} = 6.5$ nm to $q^{-1} = 5$ nm) when salt concentration increases from 0.02 M to 0.08 M, the corresponding scattered intensity keeps the same value of 660 000 g/mol, equivalent to about 10 BSA molecules; besides, for $q > 0.3$ nm^{-1} the scattering curves obtained at the three salt concentrations superimpose. These features suggest that the building blocks of the aggregates at pH 7 are not the BSA molecule, but an elementary aggregate comprising about 10 BSA molecules, the size of which does not depend on the ionic strength. An analogous situation has been described for heat-set β-lactoglobulin aggregation at pH 7.

Whatever the interpretation of SANS results, they demonstrate that the structure of BSA heat-set gels formed at pH 7 differs completely from that obtained at the isoelectric pH in absence of added salt. But the rheological behaviours of the two types of gels show great similarity. This is illustrated first on Figure 26.10, which compares the viscoelastic response over the $10^{-3}-10$ Hz frequency range of gels formed in different solvent conditions at about the same protein concentration. The long-term rheological behaviour, investigated with the creep and recovery test, is in all cases that of a viscoelastic liquid. Figure 26.11 shows the remarkable proximity between isoelectric gels and pH 7–0.02 M NaCl ones regarding the dependence on protein concentration of the steady-state compliance J_e° and viscosity η. Paradoxically, although ionic strength is expected to screen off electrostatic repulsions between charged protein molecules, quantitative differences in J_e° and η between the two types of gels increase when NaCl concentration is increased in pH 7 gels; the creep behaviour remains nevertheless of the same type whatever the solvent conditions.

Obviously, the length scale of the structural characteristics underlying the viscoelastic behaviour of BSA heat-set gels accessible to our rheological measurements is larger than the length scale explored by our SANS measurements.

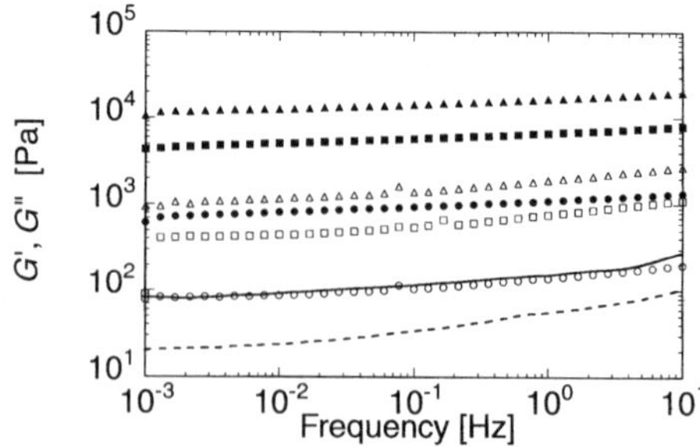

Figure 26.10 Mechanical spectra of 80 mg/ml BSA heat-set gels. Gels are formed in the measuring device of the rheometer as indicated for Figure 26.9. Measurements are performed at 20 °C with 3% strain amplitude. Isoelectric gel (pure water): G' (———), G'' (----). Gels at pH 7: G' (filled symbols), G'' (open symbols); circles: 0.02 M NaCl; squares: 0.05 M NaCl; triangles: 0.08 M NaCl

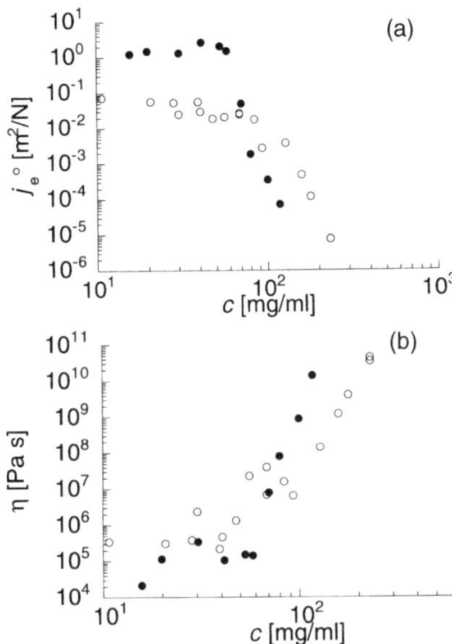

Figure 26.11 Variation with protein concentration of the steady-state compliance J_e° (a) and viscosity η (b) of BSA heat-set gels: isoelectric gels (open circles); pH 7, 0.02 M NaCl gels (filled circles). Gels are formed in the measuring device of the rheometer (see Figure 26.9 for the conditions); creep and recovery measurements are performed at 20 °C

CONCLUSIONS

These three examples have been chosen to illustrate the three different cases it is possible to find when considering the relation between rheological properties and the structure at a nanometric scale:

(1) large changes in the local structure concomitant with remarkable rheological signatures, as for the thermogelation of methylcelulose;
(2) no clear difference in the nano-structure with large differences in mechanical properties, as for pectins in presence of calcium or copper;
(3) at the opposite, large differences in the nano-structure with almost similar viscoelastic responses, as for the thermal gelation of bovine serum albumin.

The first case is that of systems in which the different levels of structure are strongly interlinked; any change at some local scale affects the organization at larger length scales and ultimately the macroscopic structure. In the two other cases, the nano-structure does not determine in a univocal or straightforward way the macrostructure of the system; even the two levels of organization are practically independent, because the global structure of the system is a more complex and hierarchic one.

By way of conclusion, we will point out that one of the main difficulties when studying the relation between structure and mechanical properties is to find the relevant level of structural organization: this is not easy to determine beforehand and often implies a comparative approach as illustrated in this paper. In addition, we have generally to investigate rather concentrated systems, with the consequence that application of some structural methods becomes severely restricted, such as that of conventional light scattering techniques for turbid systems (associative polymers or globular protein gels for instance).

REFERENCES

1. P.G. de Gennes, *Scaling Concepts in Polymer Physics*, Cornell University Press, Ithaca, 1979.
2. M. Doi and S.F. Edwards, *The Theory of Polymer Dynamics*, Clarendon Press, Oxford, 1986.
3. J.E. Mark, in *Physical Properties of Polymers*, (J.E. Mark, A. Eisenberg, W.W. Graessley, L. Mandelkern, E.T. Samulski, J.L. Koenig and G.D. Wignall, eds), American Chemical Society, Washington, DC, 1993, p. 3.
4. W. Burchard, *Brit. Polymer J.*, **17**, 154 (1985).
5. J. Bastide and S.J. Candau, in *The Physical Properties of Polymeric Gels*, (J.P. Cohen Addad, ed.), Wiley, Chichester, 1996, p. 144.
6. C. Chevillard and M.A.V. Axelos, *Colloid. Polym. Sci.*, **275**, 537 (1997).
7. C. Garnier, M.A.V. Axelos and J.-F. Thibault, *Carbohydr. Res.*, **240**, 219 (1993).
8. J. Lefebvre, D. Renard and A.C. Sanchez Gimeno, *Rheol. Acta*, **37**, 345 (1998).
9. M.A.V. Axelos and M. Kolb, *Phys. Rev. Lett.*, **64**, 1457 (1990).
10. E. Heymann, *Trans. Farad. Soc.*, **31**, 846 (1935).
11. N. Sarkar, *J. Appl. Polym. Sci.*, **24**, 1073 (1979).
12. M. Hirrien, C. Chevillard, J. Desbrieres, M.A.V. Axelos and M. Rinaudo, *Polymer*, **39**, 6251 (1998).
13. N. Sarkar and L.C. Walker, *Carbohydr. Polym.*, **27**, 177 (1995).
14. B. Cabanne and R. Duplessix, *J. Phys. France*, **48**, 651 (1987).
15. A.G.J. Voragen, W. Pilnik, J.-F. Thibault, M.A.V. Axelos and C.M.G.C. Renard, in *Food Polysaccharides* (A.M. Stephen and I. Dea, eds), Marcel Dekker, 1995, p. 287.
16. C. Rolin and J. de Vries, in *Food gels* (P. Harris, ed.), Elsevier Applied Science, London, 1990, p. 401.
17. M.A.V. Axelos, M.M. Mestdagh and J. François, *Macromolecules*, **27**, 6594 (1994).
18. C. Garnier, M.A.V. Axelos and J.-F. Thibault, *Carbohydr. Res.*, **240**, 219 (1993).
19. J. Lefebvre and P. Relkin, in *Surface activity of proteins* (S. Magdassi, ed.), Marcel Dekker, New York, 1996, p. 181.
20. K. Wong, P. Lixon, F. Lafuma, P. Lindner, O. Aguerre Charriol and B. Cabane, *J. Colloid Interface Sci.*, **153**, 55 (1992).
21. M. L. Broide, Y. Garrabos and D. Beysens, *Phys. Rev. E*, **47**, 3768 (1993).
22. D. Renard, M.A.V. Axelos, F. Boué and J. Lefebvre, *Biopolymers*, **39**, 149 (1996).
23. P. Aymard, J.C. Gimel, T. Nicolai and D. Durand, *J. Chim. Phys.*, **93**, 987 (1996).

27

Network Characterization of Some Polysaccharides Observed by Small Angle X-ray Scattering

YOSHIAKI YUGUCHI, MITSURU MIMURA, HIROSHI URAKAWA and KANJI KAJIWARA

Faculty of Engineering and Design, Kyoto Institute of Technology, Kyoto, Sakyo-ku, Matsugasaki, 606-8585 Japan

ABSTRACT

Two examples are shown to analyze the small-angle X-ray scattering profile from the gelling system (especially in the case of physical gels) with appropriate models in order to explore the potential of the scattering method.

In the first example, the classic Flory–Stockmayer model of gelation was adapted to evaluate quantitatively the gelation process of methylhydroxypropyl cellulose aqueous solution by heating. Here the micro phase separation is viewed as the network formation of the basic domains constituted of fringed micelles.

The supra-structure formation in potassium ι-carrageenate gel was analyzed as a step-by step association of double helices, and the intermolecular interaction was evaluated in terms of the range of interaction from the SAXS profiles in the second example.

INTRODUCTION

Many polysaccharides form gel in aqueous solution, but the gelation mechanism is not unique. Gelation in general has been described in terms of a multi-functional reaction in the classic Flory–Stockmayer (FS) model [1]. Although many models are proposed to improve our understanding of the gelation mechanism, the FS model has an advantage in its simplicity to specify the gelling system explicitly with two parameters (i.e., the functionality f and the fraction of reacted functional groups α) and define a gel point uniquely. The model is also applied to calculate the scattering profile from the gelling system within the framework of the cascade process [2]. Here the scattered intensity from the FS model diverges by gelation toward the zero scattering angle due to infinite connectivity in the gel.

In reality, the scattering at a finite concentration from the gelling polysaccharide aqueous solutions exhibits a characteristic profile depending on the system, and does not always diverge at zero angle because of intersegmental interaction. The scattering in this case is considered to be mainly from the associated or crosslinked domains, which are more densely packed and electromagnetically distinguished in contrast to solvent and network chains moving freely. Hydration also causes the change of the local density, and the apparent scattering behavior in a higher scattering vector range does not necessarily reflects the local structure of dissolved molecular chains.

Thus the problem in analyzing the scattering profile from the gelling system involves two aspects. One is to identify what is observed in the scattering behavior, and another is to characterize its structure with an appropriate model. However, there is no criterion available to identify the scattering units. Even when we could identify the scattering units, the scattering profile might reveal an artifact by the variation of density contrasts caused by the local structure formation and/or hydration. We are in consequence obliged to analyze the scattering results in terms of a certain model, which could account for the observed scattering profile, and to deduce the network structure of gels indirectly. In most cases of polysaccharide gels (physical gels), the model consists of heterogeneous domains randomly distributed in the system, and those domains are considered as forming junction zones. The domains might be constituted of fringed micelles or microcrystallines. The formation of the domains varies with the species of polysaccharides with respect to the size and structure, and in consequence the resulted gels differ in their optical and mechanical properties.

The aim of the present paper is to generalize the scattering theory from the gelling system from the viewpoints of the classic FS model [1] and the broken rod model [3]. The idea behind this generalization relies on the assumption that the formation of the domains takes place simultaneously in various places. Here the spatial arrangement of the domains could be specified by the existing theories of gelation including the FS theory and the percolation process. That is, the inhomogeneity in the physical gel is assumed to be due to multiple stages of structure formation in the process of gelation. Here each stage of structure formation may follow a specific mechanism including an order–disorder transition, crystallization, and random branching. The following examples will demonstrate how those specific mechanisms are incorporated in the model for physical gelation in order to explain the scattering behavior from the aqueous solutions of a cellulose derivative and ι-carrageenate, which undergo the sol–gel transition by a different mechanism.

BASIC FORMULATION

The cascade theory for random f-functional polycondensation [4] is extended to calculate the scattering intensity as a function of the magnitude of the scattering vector q for the gelling system [2,5]. Since the scattering intensity $I(q)$ is given for the system where scattering units are dispersed homogeneously as [6]

$$I(q) = \sum_{i=1}^{n} \sum_{j=1}^{n} A_i A_j \cdot \frac{\sin(q r_{ij})}{q r_{ij}} \tag{27.1}$$

the scattering intensity from random f-functional polycondensates is calculated as

$$I(q) = A^2(q)(1 + \alpha\phi)/[1 - (f - 1)\alpha\phi] \tag{27.2}$$

$$\phi \equiv \exp(-b^2 q^2/6) \tag{27.3}$$

Here α is the fraction of reacted functional groups, and b^2 the mean-square distance between two nearest units. The magnitude of the scattering vector is defined in terms of the wavelength of an incident beam λ and the scattering angle θ as

$$q = (4\pi/\lambda)\sin(\theta/2) \tag{27.4}$$

Equation (27.2) involves three parameters α, f and b. However, in reality $(f - 1)\alpha \leqslant 1$ up to the gel point and $\phi < 1$, so that $\alpha\phi \ll 1$ when $f \gg 1$. In these conditions, equation (27.2) is determined by two parameters b and $(f - 1)\alpha$.

A_i denotes the scattering amplitude of the ith scattering unit, r_{ij} the distance between the ith and the jth unit. Equation (27.2) implies that the scattering amplitude is assumed to be common for all scattering units distributed randomly in the system, so that $A^2(q)$ corresponds to the particle scattering factor of the scattering unit. $A^2(q) = 1$ for a point-like scattering unit as originally derived [2].

In the case that spherical particles associate and form gel, the gelation process will be described in terms of the FS model by introducing the form factor of a sphere in equation (27.1) as

$$A_i(q) = g_i \cdot \left[\frac{3(\sin qR_i - qR_i \cos qR_i)}{(qR_i)^3} \right] \tag{27.5}$$

where R_i and g_i denote the radius and the scattering factor of the ith sphere. An example is shown in Figure 27.1 for the association of spherical particles, which are represented by a rigid sphere of a uniform radius 5 Å.

It is often found that the scattering units consist of the domain specified by the generalized Ornstein–Zernike density correlation function $(\xi/r)^{3-D} \exp(-r/\xi)$ [7]. Then $A^2(q)$ will be given approximately by the scattering function for particle associates in an appropriate range of q as [2,8,9]

$$A^2(q) = \frac{1}{[1 + (D+1)\xi^2 q^2/3]^{D/2}} \tag{27.6}$$

Here D denotes the fractal dimension specifying the geometrical arrangement of scattering units in the domain.

The broken rod model was introduced to fit the scattering profile from the associates and swollen gels composed of cylindrical domains with cross-sectional heterogeneity [3]. The scattering in this model is due to the rods immersed in solvent, where the rods are broken at random and their cross-section varies. The scattering intensity from the broken rod model is given as

$$q^2 I(q) \approx \sum \pi q w_i M_{Li} \cdot \frac{4J_1^2(qr_{ci})}{(qr_{ci})^2} + \text{const.} \tag{27.7}$$

J_1 denotes the first-order Bessel function of the first kind and w_i is the weight fraction of the ith component specified by the cross-sectional radius r_{ci} and the linear mass M_{Li}. The constant term in equation (27.7) takes into account the inter-rod correlation [10] given in terms of the number of rod ends, the number of kinks and the number of crossings with other rods. Since the model is free from a specific mode of connectivity, equation (27.7) can be applied to either gel or sol systems composed of rod-like associates. In other words, no information emerges on the network structure from the analysis according to equation (27.7).

The Debye formula equation (27.1) can also be applied to calculate the scattering intensity directly from a molecular model. A molecular model is generated by, for example, the Monte Carlo method, and the scattering intensity is calculated from equation (27.1) as an ensemble average over generated molecules [11]. Here a molecular model is composed of the atoms represented by the spheres with the corresponding van der Waals radius. That is, the scattering amplitude

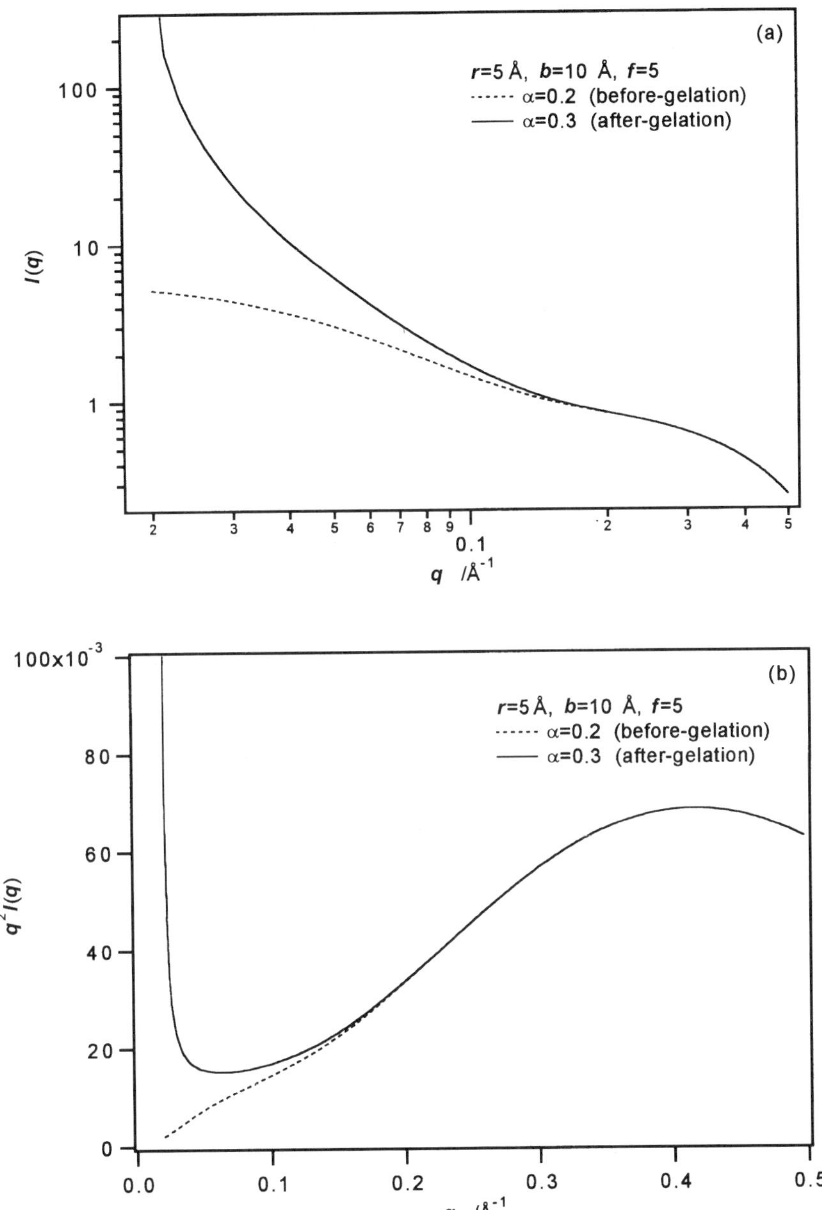

Figure 27.1 Scattering from associated spheres of a uniform radius 5 Å, calculated according to equations (27.2) and (27.6). The scattering profiles are shown in the double logarithmic plots (a) and the Kratky plots (b). Broken and solid lines denote the situation before gelation, i.e., $(f-1)\alpha = 0.8$, and after gelation, i.e., $(f-1)\alpha = 1.2$, respectively

in equation (27.1) is equivalent to the form factor for a sphere equation (27.5), where the radius and the scattering factor of a sphere are replaced with the van der Waals radius and the atomic scattering factor of the ith atom. The sum in equation (27.1) should be taken over all pairs of atoms in a molecule.

APPLICATION OF BASIC FORMALISM

METHYLHYDROXYPROPYL CELLULOSE (MHPC)

Methylhydroxypropyl cellulose (MHPC) in aqueous solution (the chemical structure is shown in Figure 27.2) undergoes a thermoreversible gelation on heating, and its gelation mechanism was investigated by various methods including light scattering [12] and small angle X-ray scattering (SAXS) [13]. The DSC analysis suggests that the gelation proceeds in two thermodynamic steps [13]. Here the gelation is probably caused by the fringed micelle formation and the subsequent micro-phase separation, resulting in the loss of the internal mobility of chain segments. The fringed micelles are present even in the very dilute regime [12,14]. Some chains may not be incorporated partially or totally in micelles, and constitute dangling chains or free chains, respectively. Upon heating, the fringed micelles grow in size and number due to association, and the dangling or free chains could be incorporated further in other fringed micelles to form a network. The loss of the mobility of dangling chains by coupling is observed by the appearance of the maximum in the slow relaxation time in the rheological and dynamic light scattering experiments in the same system [15].

The sharp upturn of the SAXS profile was observed at $q \to 0$ in the Kratky plots from the gelling MHPC aqueous solution ($C_p = 2$wt%) as shown in Figure 27.3. Here MHPC ($M_w = 4.89 \times 10^5$) possesses methyl and hydroxypropyl substituents of the same amount and its total degree of substitution is estimated as 1.80 [12]. Two alternative models will be presented to account for the gelation of MHPC in aqueous solution. In the previous analysis, the

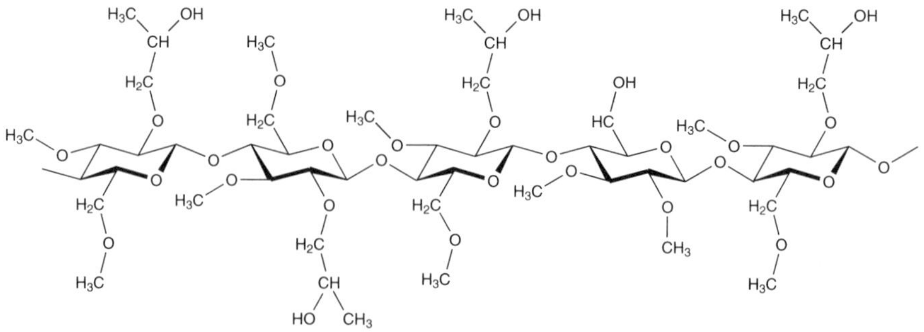

Figure 27.2 Chemical structure of methylhydroxypropyl cellulose (MHPC)

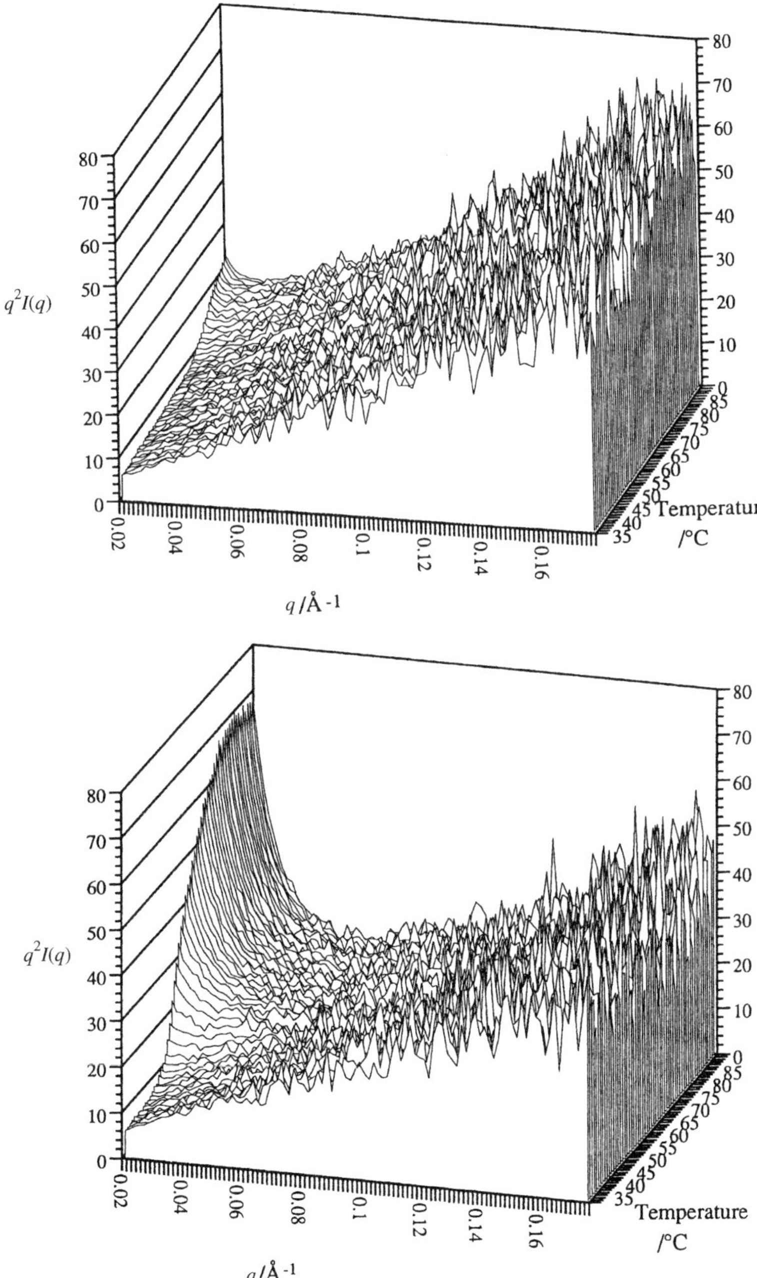

Figure 27.3 Scattering profiles (the Kratky plots) from the MHPC aqueous solution (C_p = 2wt%) undergoing sol–gel transition, observed in the process of heating from 30 °C to 90 °C at a rate of 0.5 °C/min (above) and gel–sol transition, in cooling from 90 °C to 30 °C at a rate of 0.5 °C/min (below)

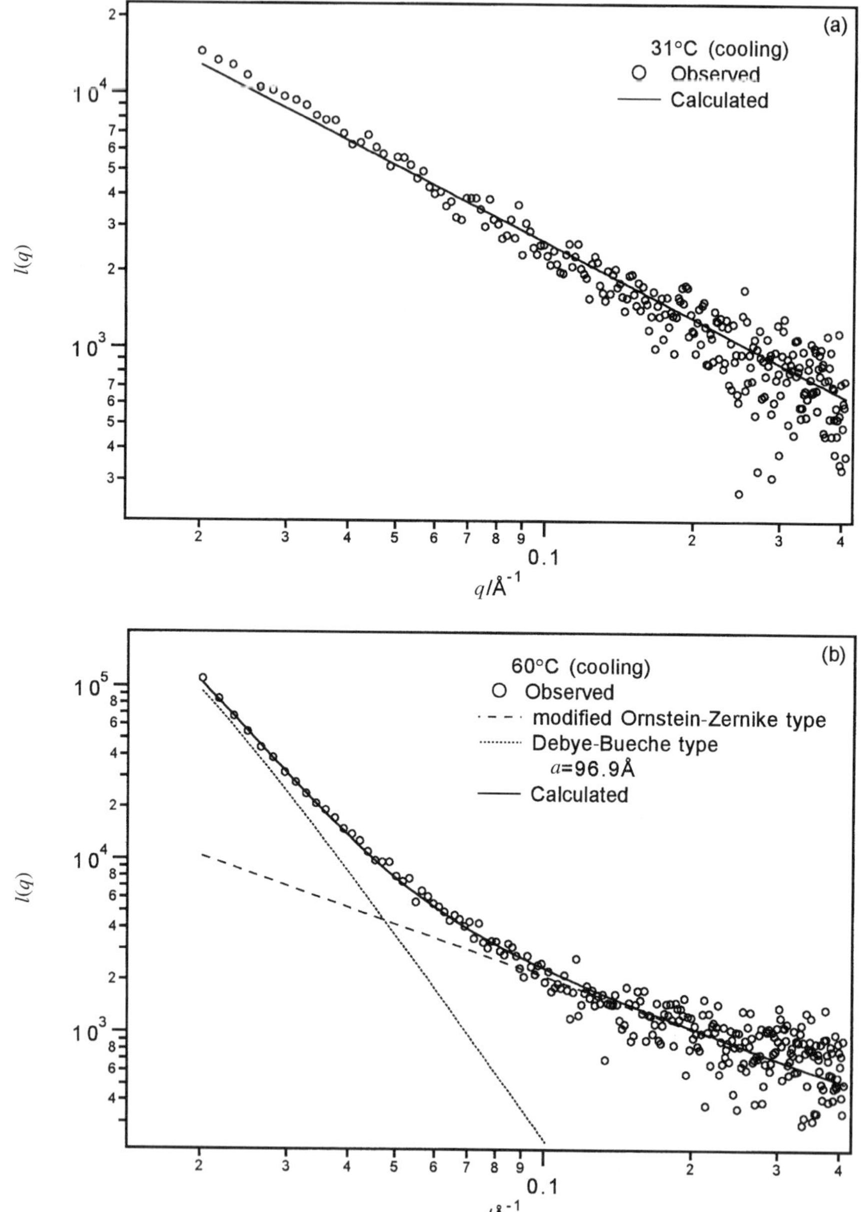

Figure 27.4 Observed SAXS profiles from the MHPC aqueous solution (C_p = 2wt%) decomposed into two terms of the modified Ornstein–Zernike type and the Debye–Bueche type. Two examples are shown for sol and gel

gelation of MHPC was assumed to take place by the formation of larger domains characterized by the Debye–Bueche type random aggregation [16]. The total scattering profile is given by the sum of two scattering functions corresponding to each domain (see Figure 27.4). Here larger domains (specified by the Debye–Bueche correlation length a) were found to start growing above the gel point, while smaller domains (specified by the modified Ornstein–Zernike correlation length ξ) remain invariant [17] as shown in Figure 27.5. Here the correlation function for a smaller domain was given by equation (27.6) with $D = 1$ in order to fit better to the observed scattering behavior at higher q region [17]. Although the model describes the gelation of MHPC reasonably well in its scattering behavior, a question remains why two distinctive domains should be needed for gelation. The basic domains are thought to be composed of fringed micelles, and one can argue that the random association of the basic domains yields larger domains. Phase separation might cause the change of the density correlation function from the Ornstein–Zernike type to the Debye–Bueche type, although the size of random associates (specified by the correlation length a) is too small in comparison with the condensed phase.

An alternative model for the gelation of MHPC in aqueous solution is based on the classic FS model where the units are composed of domains specified by a generalized Ornstein–Zernike density correlation function $(\xi/r)^{3-D} \exp(-r/\xi)$. The domains, probably constituted of a fringed micelle, grow by heating, and are linked with each other in a tree-like fashion described by the FS process. The gelation scheme is sketched in Figure 27.6. Equations (27.2) and (27.6) are then applied to fit the scattering profile from the gelling system of MHPC with four parameters $D, b, (f - 1)\alpha$ and ξ. A sol–gel transition is marked around $55\,^\circ\mathrm{C}$ where $(f - 1)\alpha$ exceeds unity, as expected from the FS model. The fitting examples are shown in Figure 27.7 for the MHPC aqueous solution at $45\,^\circ\mathrm{C}$ (sol), $55\,^\circ\mathrm{C}$ (near a gel point) and $80\,^\circ\mathrm{C}$ (gel). The evaluated parameters $(D, b, (f - 1)\alpha$ and $\xi)$ from the fittings are summarized in Figure 27.8 as the functions of temperature. When a gel point is reached, then $(f - 1)\alpha$ and the inter-domain distance become almost invariant, but the correlation length ξ starts increasing its value from 70 Å. The apparent fractal dimension D increases monotonously by cooling. Although no physical significance could be attributed to the evaluated fractal dimensions being less than unity, the model describes the association of the fringed micelles as the FS process. The apparent fractal dimension less than unity may be attributed to the inhomogeneous scattering contrast along MHPC chains with respect to water, some parts of which could be transparent electromagnetically due to the change of the electron density distribution caused by the substituents. Here no larger domains are required to constitute junction zones. The infinitely connected domains form a network and will be phase-separated. Most of the domains are still finitely connected (at most two or three) and remain in sol phase.

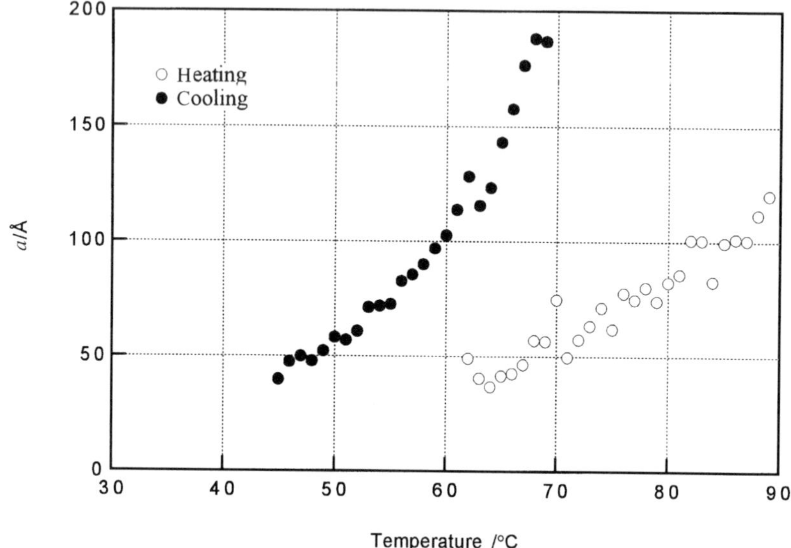

Figure 27.5 Change of the size *a* of larger domains in the process of heating and cooling. ξ (the correlation length of a modified Ornstein–Zernike type) is fixed to 500 Å in the present analysis, but *a* is insensitive to the choice of the value for ξ

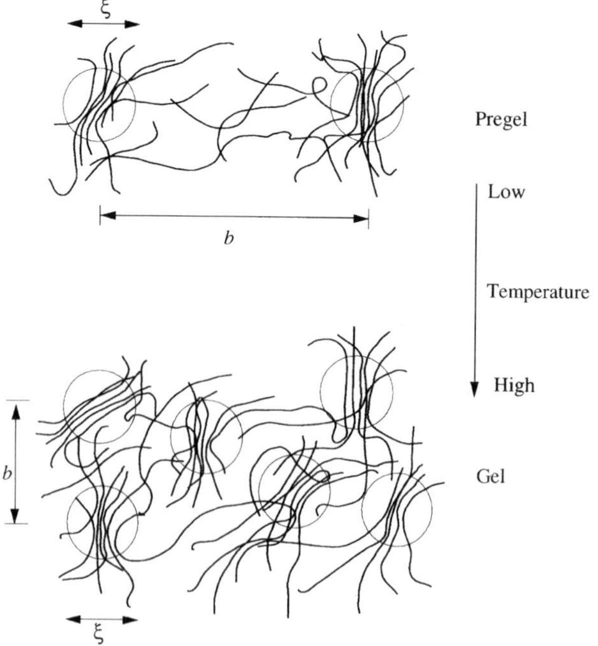

Figure 27.6 Gelation scheme of fringed micelles

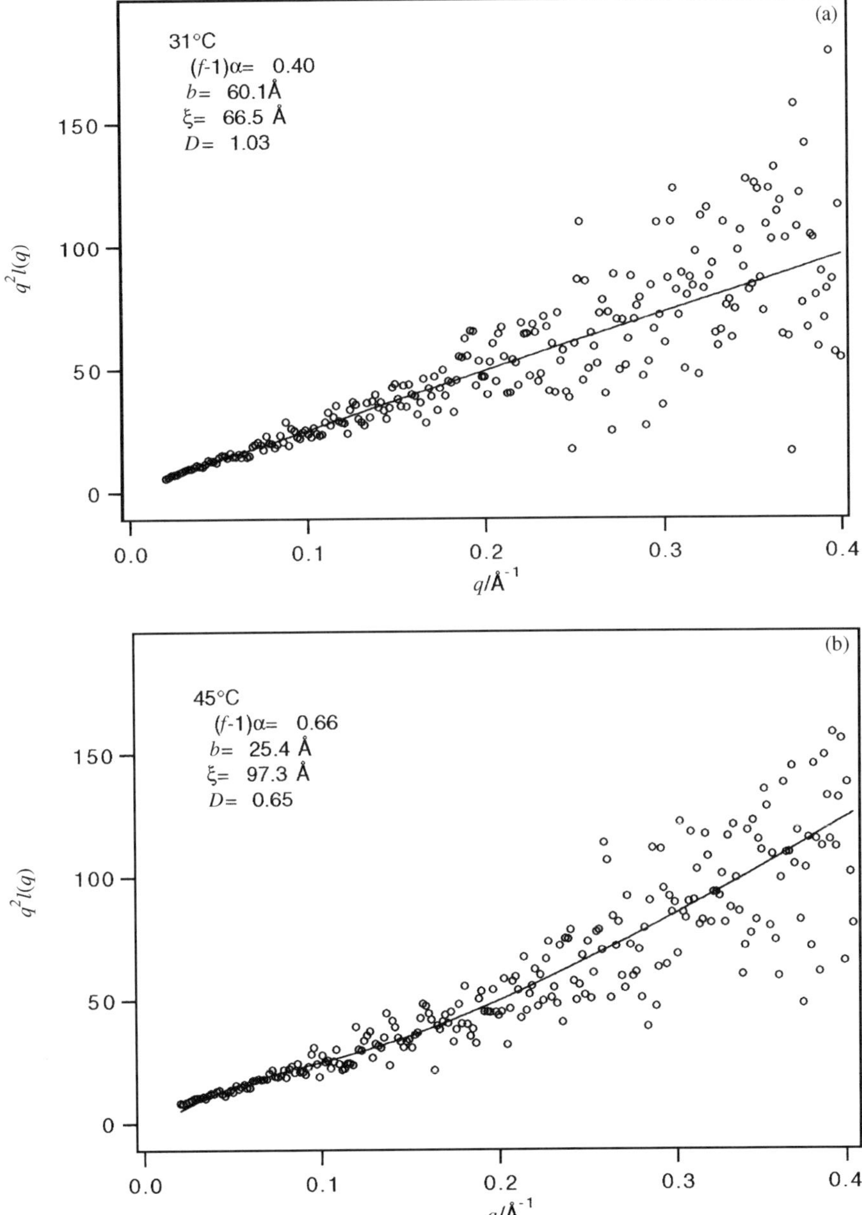

Figure 27.7 Observed and calculated SAXS profiles of MHPC aqueous solution ($C_p = 2$wt%). The calculated profiles are due to equation (27.2) and (27.6). Here the gelation is regarded as a classic Flory–Stockmayer process

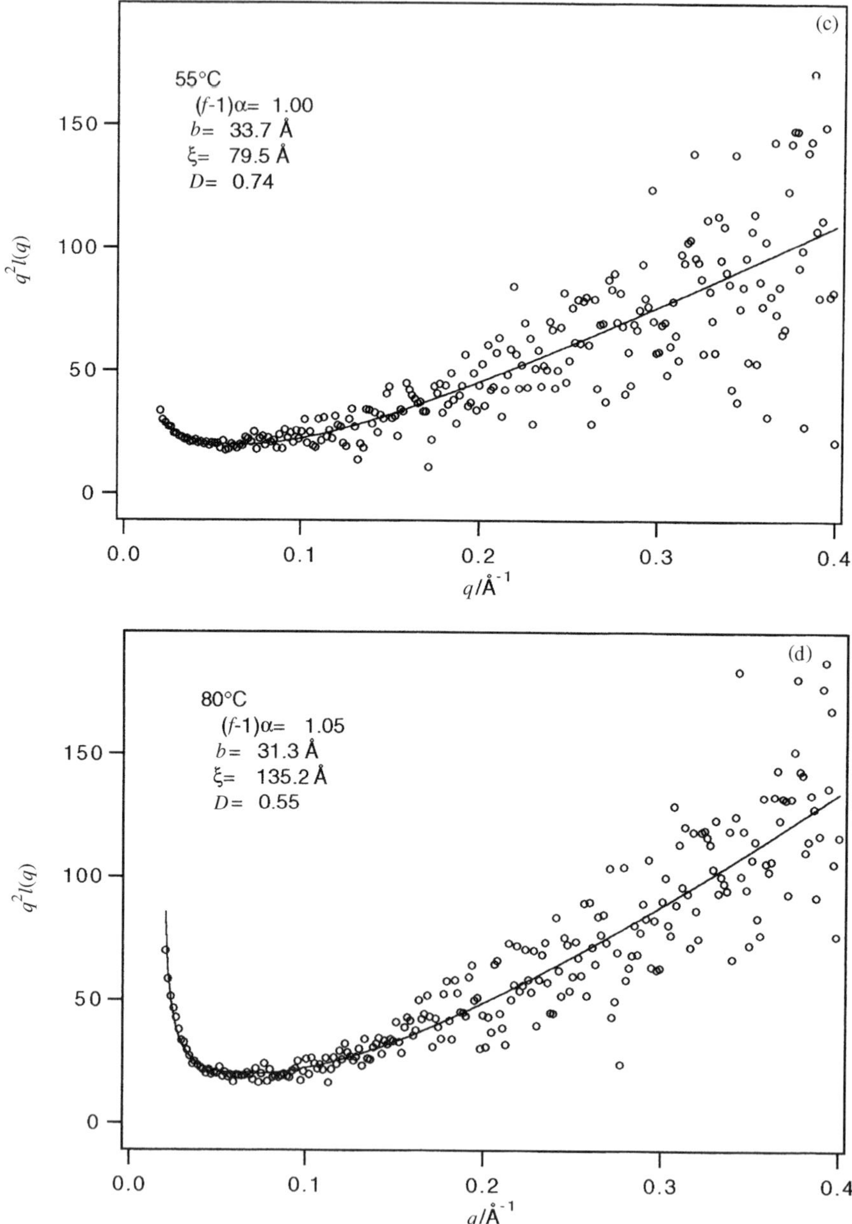

Figure 27.7 (*continued*)

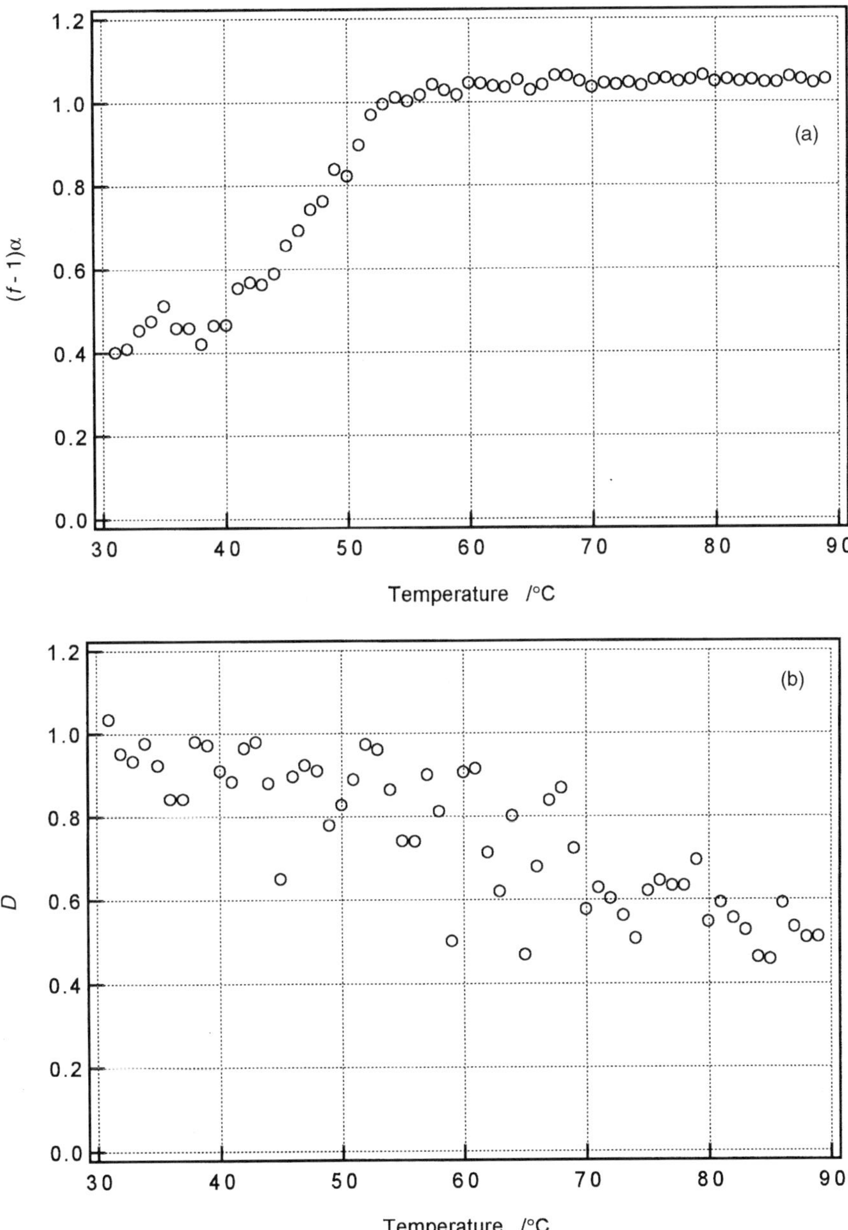

Figure 27.8 Evaluated parameters by fitting as a function of temperature

Y. YUGUCHI *ET AL.*

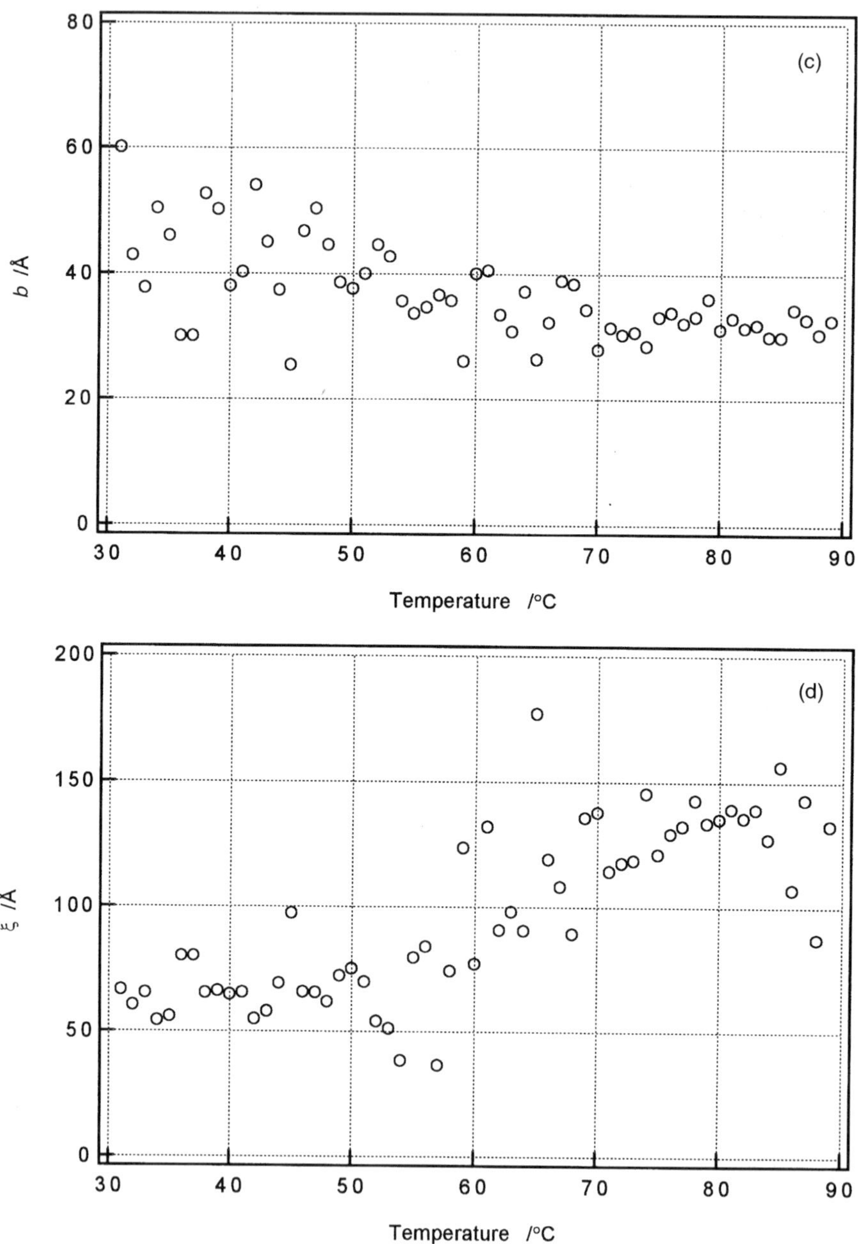

Figure 27.8 (*continued*)

ι-CARRAGEENATE

ι-Carrageenate is a family of a sulfated galactan extracted from red seaweed (*Rhodophyceae*) and is composed of a disaccharide repeat unit of 1,3-linked β-D-galactose-4-sulfate and 1,4-linked α-3,6-D-anhydro-galactose-2-sulfate. Commercial ι-carrageenate (Sigma Lot108F0046, Ca^{2+}4.7%, K^{+}2.7% and Na^{+}1.5%) was purified by dialysis in 10 mM EDTA and deionized water to remove metal salts and other contaminants. Then the sample was percolated through an ion exchange resin (IR-120). The resulted acidic carrageenan was neutralized with KOH immediately and freeze-dried to obtain potassium ι-carrageenate. Potassium ι-carrageenate forms gel in aqueous solution by lowering the temperature, but its scattering profile shows no sharp upturn at $q \to 0$ as expected from the classic model of gelation. It is thought that the double-helix formation promotes gelation by associating side by side as in the case of gellan aqueous solutions [18]. This gelation process can be visualized by the series of molecular models for ι-carrageenate ordered structure. The molecular models for a single helix, a double helix and associated double helices are constructed from the available crystallographic data [19] as shown in Figure 27.9. Here the domain of associated double helices are considered to constitute the junction zone, and the observed SAXS is dominated by those junction zones and does not reflect the network structure of gel.

The scattering profile from ι-carrageenate aqueous solution was analyzed in terms of a broken rod model equation (27.8). Since the first term of the right side of equation (27.7) represents the contribution of rod-like parts, it is replaced with the scattering factors directly calculated from the molecular models as

$$q^{2}I(q) \approx \sum w_{i}M_{Li} \cdot \Theta_{i}(q) + \text{const.} \tag{27.8}$$

where the subscript i denotes a single coil component, a double-stranded helix component or associated double-stranded helix components, and the constant term accounts for the spatial correlation of cylindrical parts being random. The scattering profiles are calculated in Figure 27.9 from the molecular models of a single coil component, a double-stranded helix component and an associated double-stranded helix component.

Since ι-carrageenate possesses sulfate groups, the electrostatic interaction between the groups causes an interference effect in the scattering. The interference effect is taken into account by the interference term $S(q)$, and the apparent scattering intensity $I_{\text{app}}(q)$ is given [20] approximately by the product of the scattering intensity from an isolated particle $I(q)$ and the interference term $S(q)$ as

$$I_{\text{app}} \approx I(q) \cdot S(q) \tag{27.9}$$

$S(q)$ is given in the case of rigid sphere interaction as [20]

$$S(q) \approx \frac{1}{1 + 8(v_0/v_1)\Phi(2qR)} \tag{27.10}$$

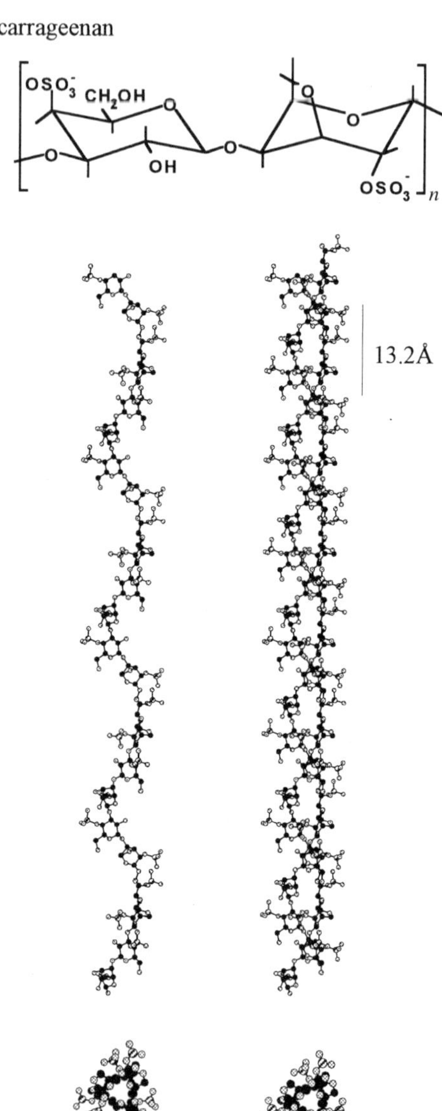

Figure 27.9 Molecular models for a single coil, a double helix and associated double helices of ι-carrageenan (the chemical structure is also illustrated at the top of the figure). The corresponding scattering profiles are calculated according to equation (27.1)

Packing model of ι-carrageenan double helix

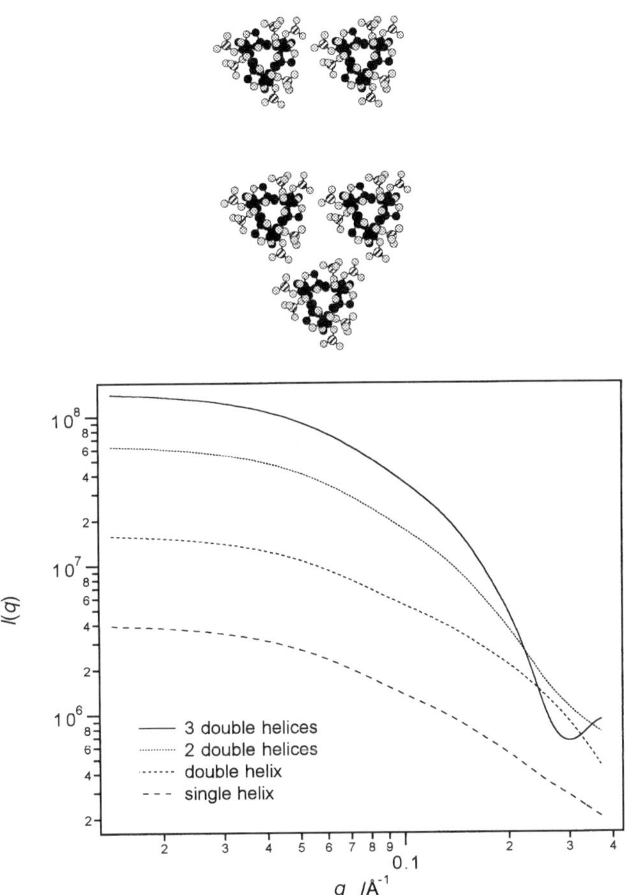

Figure 27.9 (*continued*)

where Φ denotes the potential function for a rigid sphere and is given by the form factor of a sphere equation (27.5) with g_i being unity. R is the range of interaction, and (v_0/v_1) corresponds to the volume fraction of the interacting spheres. Equation (27.9) is valid only when the interaction is isotropic and spherically symmetric. This condition will be fulfilled for solutions and gels, where the interacting units are thermally agitated and move almost freely. The thermal motion of interacting units in solutions and gels is so fast as to smear the interference. The effect of the thermal motion can be taken into account by multiplying the Debye–Waller factor $\exp(-\sigma^2 q^2)$ to the potential function Φ in equation (27.10), where σ corresponds to the range of thermal fluctuation.

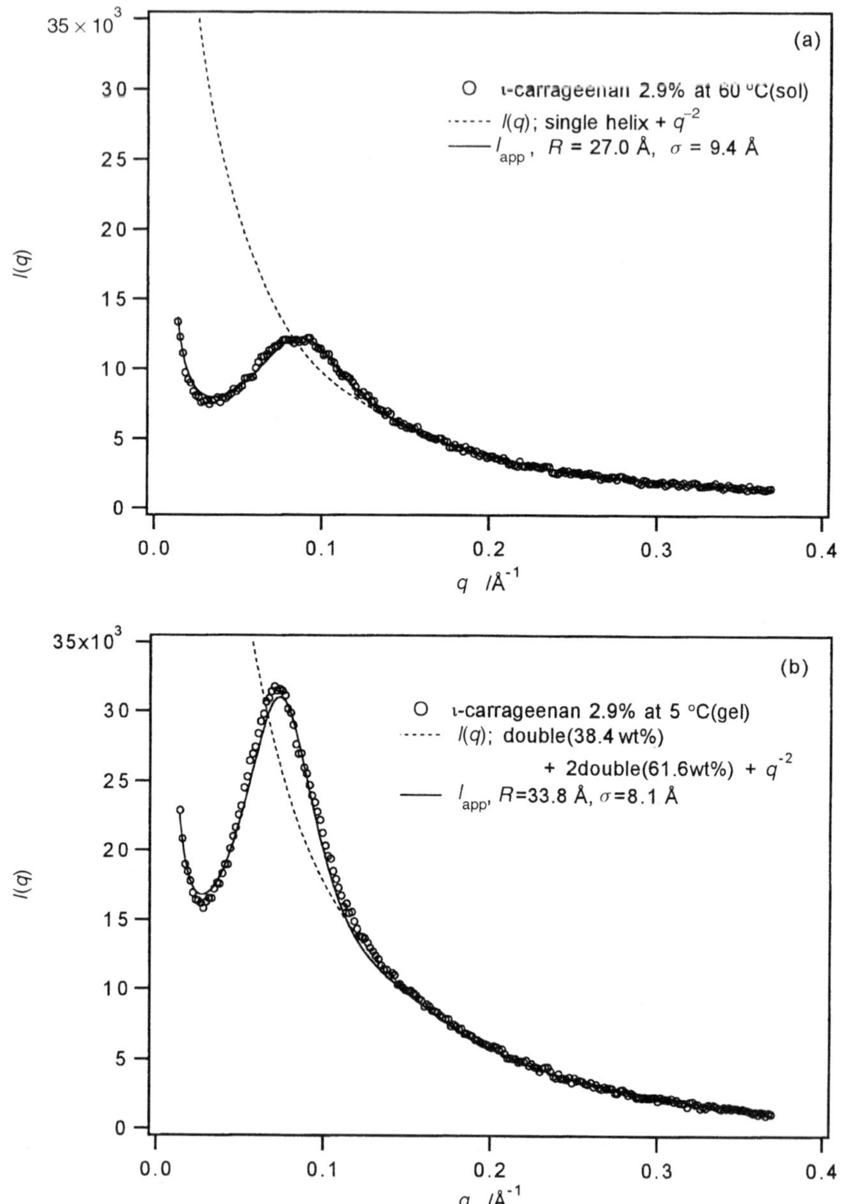

Figure 27.10 Observed and calculated SAXS profiles from potassium ι-carrageenate in aqueous solution ($C_p = 2.9$wt.%). The scattering profiles are calculated from a modified broken rod model equation (27.8) without the interference effect (dotted lines) and with the interference effect (solid lines). (a) At 60 °C (sol). A single helix was a sole component in equation (27.8). (b) At 5 °C (gel). Two components(a double helix + two associated double helices) are considered in equation (27.8)

Table 27.1 Broken rod components, and the range of interaction and thermal motion in the aqueous solution of potassium ι-carrageenate at two temperatures

Temperature	Components	Range of Interaction R (Å)	Range of thermal fluctuation σ (Å)
60 °C (sol)	single helix (100 %)	27.0 Å	9.4
5 °C (gel)	double helix (38.4 wt%) + two associated double helices (61.6 wt%)	33.8 Å	8.1

The final results of fitting the observed SAXS profiles are shown in Figure 27.10, and the evaluated parameters are summarized in Table 27.1. Potassium ι-carrageenate assumes a broken single helical conformation at 60 °C (sol), while it forms double helices by lowering the temperature and the side-by-side association of two double helices causes gelation. The range of interaction expands from 27 Å to 34 Å and the interference peak becomes more pronounced by gelation, indicating that the interacting units will be localized and concentrated due to the formation of double helices and their subsequent association. The range of thermal motion decreases slightly in gel as could be expected.

CONCLUSION

Two examples are shown to demonstrate how the molecular models are incorporated in the analysis of the SAXS profiles from the gelling system.

The classic Flory–Stockmayer model of gelation is proved to be useful in evaluating the gelation process in a quantitative way. Here the micro phase separation is viewed as the network formation of the basic domains constituted of fringed micelles, microcrystallines or other ordered/disordered structures.

Once an appropriate molecular model is constructed, the supra-structure formation can be analyzed as a step-by-step association of basic units and the intermolecular interaction be evaluated from the SAXS profiles. Gelation in this case is observed as the supra-structure formation, and it is often found that the infinite connectivity of gel is not reflected in the observed scattering profile because of the high contrast of formed supra-structure. A strong inter-domain interaction also contributes to yield an apparent finite connectivity [21].

ACKNOWLEDGMENTS

The SAXS measurements were performed under the approval of the Photon Factory Advisory Committee (Proposal Nos. 91-217 and 94G291). K.K. is indebted to Professor W. Burchard and Dr L. Schulz for fruitful discussion, and to DAAD and JSPS for financial support. We would like to thank Dr R. Dönges, Clariant AG, Wiesbaden, Germany, for providing MHPC samples. Y. Y. acknowledges the financial support of JSPS Research Fellowships for Young Scienctists.

REFERENCES

1. See, for example, P.J. Flory, *Principles of Polymer Chemistry*, Cornell University Press, Ithaca, NY, 1953.
2. K. Kajiwara, S. Kohjiya, M. Shibayama, H. Urakawa, in *Polymer Gels*, (D. De Rossi, K. Kajiwara, Y. Osada and A. Yamauchi, eds), Plenum, 1991.
3. J.-M. Guenet, *Thermoreversle Gelation of Polymers and Biopolymers*, Academic Press, London, 1992.
4. M. Gordon, *Proc. Roy. Soc. (London)*, **A268**, 240 (1961).
5. K. Kajiwara, W. Burchard, M. Gordon, *Brit. Polym. J.*, **2**, 110 (1970).
6. P. Debye, in *Light Scatteng from Dilute Polymer Solutions*, (D. McIntyre and F. Gornick, eds), Gordon & Breech, New York, 1964.
7. P.-G. de Gennes, *Scaling Concepts in Polymer Physics*, Cornell University Press, Ithaca, NY, 1979.
8. T. Freltoft, J.K. Kjems, S.K. Sinha, *Phys. Rev.*, **B33**, 269 (1986).
9. M. Shibayama, H. Kurokawa, S. Nomura, M. Muthkumar, R.S. Stein, S. Roy, *Polymer*, **33**, 2883 (1992).
10. V. Luzzati, H. Benoit, *J. Appl. Crystallogr.*, **14**, 297 (1961).
11. M. Mimura, H. Urakawa, K. Kajiwara, S. Kitamura, K. Takeo, *Macromol. Symp.*, **99**, 43 (1995).
12. L. Schulz, W. Burchard, *Das Papier*, **47**, 1 (1993).
13. Y. Yuguchi, H. Urakawa, S. Kitamura, S. Ohno, K. Kajiwara, *Food Hydrocolloid.*, **9**, 173 (1995).
14. W. Burchard, *Adv. Colloid Interface Sci.*, **64**, 45 (1996).
15. L. Schulz, W. Burchard, R. Dönges, in *Cellulose Derivatives*, (ed. by T.J. Heinz, W.G. Glasser), *ASC Symp. Series*, **688**, ACS, 1998.
16. P. Debye, A.M. Bueche, *J. Appl. Phys.*, **20**, 518 (1949).
17. Y. Yuguchi, M. Mimura, H. Urakawa, K. Kajiwara, M. Shirakawa, K. Yamatoya, S. Kitamura, *Proceedings of the International Workshop on Green Polymers*, pp.306, 4–8 November, 1996, Bandung-Bogor, Indonesia.
18. Y. Yuguchi, H. Urakawa, K. Kajiwara, *Macromol. Symp.*, **120**, 77 (1997).
19. R.P. Millane, R. Chandrasekaran, S. Arnott, I.C.M. Dea, *Carbohydr. Res.*, **182**, 1 (1988).
20. A. Guinier, G. Fournet, *Small-Angle Scattering of X-Rays*, Wiley, New York, 1955.
21. W. Burchard, P. Lang, L. Schulz, T. Coviello, *Macromol. Symp.*, **58**, 21 (1992).

28

Structure and Rheology of Gelatin Gels

PAULA M. GILSENAN and SIMON B. ROSS-MURPHY
Biopolymers Group, Division of Life Sciences, King's College London,
Franklin-Wilkins Building, 150 Stamford Street, London SEI 8WA, UK

ABSTRACT

This article is essentially a review of past work, including structural and rheological studies on gelatin gels and progress made in the last decade or so. Now many aspects of gelatin gelation are unchallenged. For example it seems clear that the helical junction zones do not have a large cross-sectional radius of gyration, and both the gel modulus and the absolute optical rotation appear to increase slowly, but without limit, even when plotted on a log time axis. Other topics of current interest include the nature of gelatin gels at very long times (creep *vs.* dynamic measurements), and the future of bovine gelatin

Wiley Polymer Networks Group Review Series Vol. 2. Edited by B.T. Stokke and A. Elgsaeter
© 1999 John Wiley & Sons Ltd

and the development of novel sources following the bovine spongiform encephalopathy (BSE) crisis. This paper also introduces some work by the present authors on gelatins from fish sources and draws comparisons with results from the more traditional bovine gelatins.

INTRODUCTION

The structural and mechanical behaviour of gels from the polypeptide gelatin have been very widely studied in the past [1,2]. We summarize below the major topics in gelatin research, with the view that the vast majority of workers in the field now accept these principles, even though some were previously regarded as controversial. We also present preliminary data on the concentration dependence of melt behaviour and of gel modulus for two fish gelatin samples, compared with that for three 'normal' bovine samples.

MOLECULAR STRUCTURE OF GELATIN SOLS AND GELS

Gelatin is a proteinaceous material derived by hydrolytic degradation of collagen, the principal protein component of white fibrous connective tissue; the fundamental molecular unit is the tropocollagen rod. The precise amino acid content and sequence varies from one source to another, but always consists of large amounts of proline, hydroxyproline and glycine. The proline content is particularly important, as it tends to promote formation of the polyproline II helix, which ultimately determines the form of the tropocollagen trimer.

Gelatins normally dissolve in warm water ($> \sim 40\,^{\circ}\mathrm{C}$) and above this temperature the polypeptide exists essentially as isolated, flexible, and usually lightly cross-linked chains. On re-cooling, transparent gels are formed (provided the concentration is greater than some critical concentration, C_{o}, typically 0.4 to 1.0%). These are now generally accepted to contain extended physical cross-links or 'junction zones' formed by a partial reversion to 'ordered' triple helical collagen-like sequences, separated along the chain contour by peptide residues in the 'disordered' conformation. Figure 28.1 illustrates this simple picture.

The main evidence for this reversion has involved optical rotation (OR) measurements. The sign and magnitude of rotation of plane polarized light transmitted through a solution can be correlated, at least semi-empirically, with the torsion angle specifying the relative orientation of adjacent peptide residues [3]. Measurements of the specific rotation of tropocollagen, and of 'denatured' gelatin at high temperatures can be made readily. The measurement of specific rotation of cooled gelatin sols (and gels) then enables an estimate of the relative amount of helix, i.e. the proportion of peptide residues in the triple helical state. Work by Djabourov has studied the OR of gelatin in some depth [4].

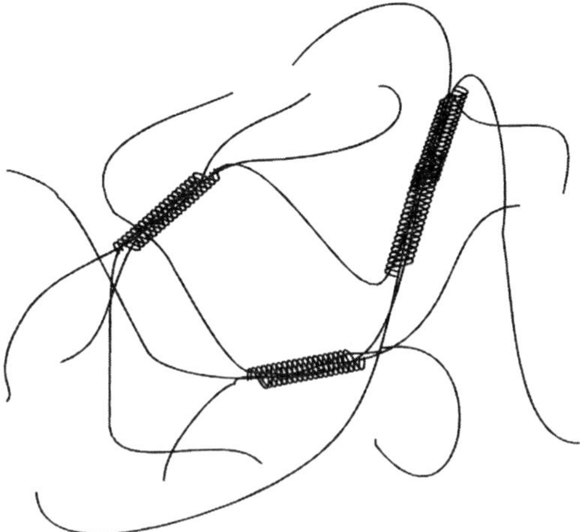

Figure 28.1 Structure of a gelatin gel (idealized) showing the triple helical junction zones

PHYSICAL GELATION OF GELATIN

The following points summarize the current views on the physical gelation of gelatin, reflecting many years of work and countless primary publications. We note that some of these points are very obvious, while others are still mildly controversial.

(1) *Gelatin physical gels form on cooling.* A coil to helix transition occurs when a warm solution of gelatin is cooled to below a critical gelation temperature T_C, ($\sim$35 °C for mammalian sourced gelatin).

(2) *This coil–helix transition is a thermodynamic phase transition.* The absolute tendency is for coils to form helices below the transition temperature.

(3) *The precise form of the helix is still controversial.* It was originally thought that cross-links are formed by intermolecular triple helices. However, this is inconsistent with work on optical rotation kinetics, where second or lower order behaviour is seen. This has been explained by a bimolecular hairpin form [5] which is supported by some spectroscopic evidence [6].

(4) *Gelation only occurs above a certain concentration, C_o, the critical gelation concentration.*

(5) *C_o is dependent on the chain molecular weight.* This is reflected in the Bloom strength parameter; high Bloom $=$ lower C_o $=$ stronger gels when $C > C_o$.

(6) *C_o depends on the difference between actual gelling temperature T and T_C.* The relationship between temperature and gelation characteristics is complex [7]. If a gelatin solution is overcooled the rate of initiation of helices is increased, but the rate of propagation is slowed. This has been

described in terms of Avrami-type crystallization kinetics, although this is an over-simplification. Better gels result when the temperature is cycled and the time at each temperature is controlled. Intramolecular helices are then formed at concentrations below C_o.

(7) *At a given temperature gelation occurs at a critical degree of helix conversion.* Djabourov and Papon [8] used OR to show that at the gel point $\sim\frac{1}{15}$ of the residues are in the helical conformation independent of the temperature. Durand *et al.* [9] also found the same generality and in addition established a linear relation between the proportion of residues in the helical conformation, h_C and reciprocal concentration. These results are consistent with a competition between inter- and intramolecular helix formation.

(8) *Gelation is a kinetic process which never goes to completion.* Long term measurements of the gel modulus and the absolute OR [4] increase slowly but without limit, even when plotted on a log time axis. This suggests a substantial amount of conformational flexibility after gelation, so that the proportion of peptide units involved in junction zones increases even in a solid gel.

(9) *It was believed that the second stage process was akin to crystallization.* It is now widely accepted that inter-chain aggregation, while common in other biopolymer gelling systems (agarose, carrageenan) [10,11] is of minor significance for gelatin. If such a maturation process did occur there would be no further helix growth. However, OR and other evidence clearly suggests the contrary.

(10) *The critical concentration, C_o, is indirectly related to the chain overlap concentration, C^*.* The latter $\approx 1/[\eta]$, whereas C_o is not a basic chain length property, since it depends upon prior treatment. Thus two samples subjected to different cooling regimes will show a different C_o but the same C^*

(11) *All relevant gel mechanical properties can be related back to C_o.* The final modulus can be related directly to C/C_o. For example the shear modulus, G of a gelatin gel $\propto C^2$ is true [10] only where $C/C_o \geqslant 5$. At lower concentrations the exponent x in the expression $G \propto C^x$ will be >2 and will tend to infinity as $C \to C_o$. Similarly the gelation time (at a given temperature, T) varies inversely with C/C_o and depends on the difference between T and T_C.

CREEP AND THE LONG TERM FLOW BEHAVIOUR OF GELATIN GELS

Small controlled strain oscillatory measurements have shown that the storage modulus $G'(\omega)$ of gelatin gels is almost constant down to frequencies lower than 10^{-2} rad s^{-1}, and $G''(\omega)$ is lower than, but largely parallel to $G'(\omega)$ as expected for a gel network [10,11]. For highly swollen gels (w/w concentration of gelatin < say 10%) there is often a pronounced minimum in the mechanical spectrum, centred around 1 rad s^{-1} (0.16 Hz), which is assumed to be associated with relaxation processes occurring over much longer time scales. Over typical oscillatory frequencies there is no indication of 'terminal flow', at least reasonably below the gel melting temperature. Gelatin gels are therefore much more solid than typical entangled melts of linear chains, where motion of large regions of chain (e.g. reptation) leads to substantial reduction of stress at equivalent time scales. At the same time these oscillatory experiments leave open the question of the exact behaviour at very long times. In particular is there a true equilibrium modulus? In other words are gelatin gels viscoelastic solids, or merely very high viscosity fluids, like window glass?

A series of measurements by Higgs and one of the present authors [12] has been useful in exploring this particular question. Shear creep experiments were performed on a high molecular weight alkaline processed ossein gelatin. The crucial observation was that there was an apparent flow at long times, and with concentration dependent viscosities in the range $10^8 - 10^{11}$ Pa s, with an approximately linear dependence on concentration (actually $\eta \propto C^{1.1}$). This observation seems to be contradicted by the flatness of the mechanical spectrum and is commented upon by te Nijenhuis, in his review on the rheology of thermoreversible gels [11].

Although initially sceptical of such creep measurements, he has used his own oscillatory data to furnish an estimate of the very low frequency behaviour of both G' and G''. The (commonly observed for gelatin) minimum in G'' reflects a maximum at lower frequencies which, in turn, prefaces a terminal flow regime. This is illustrated in Figure 28.2, derived from his article. This strongly suggests

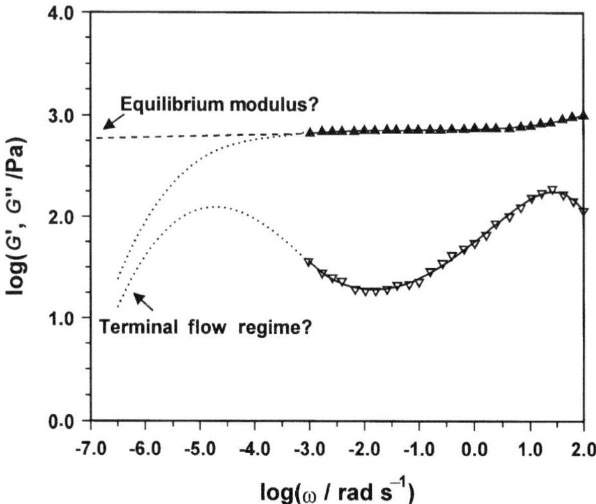

Figure 28.2 G' (filled) and G'' (open symbols) for a gelatin sample as measured by te Nijenhuis [7]; dashed lines represent one assumption about the behaviour at lower frequencies after te Nijenhuis [11]

that there is no equilibrium modulus in a physical gelatin gel, and the system is, in fact, more akin to a glass. In the absence of microbial proteolysis or water evaporation, Figure 28.2 suggests flow will tend to occur on a time scale of several years (10^8 s is $\sim$3.2 years). Of course other bounds are calculable, but this view reopens the question as to whether gelatin is a viscoelastic solid or a viscoelastic liquid.

GELATINS FROM ALTERNATIVE SOURCES

Most of the commercial gelatins available are derived from mammalian, and particularly bovine, sources. From the food industry viewpoint, recent UK problems with BSE have proved problematic, and there is a preference expressed for a move away from bovine gelatins. These generally have a melting point of 35–40 °C. Another degree of flexibility in design of gelatin products for the food industries would arise if gelatins with a rather lower melting temperature could be obtained. Some work has been done on fish gelatins, showing quite varied results. One of the main findings, however, is that fish gelatins have low gelation and melting temperatures but relatively high solution viscosities [13]. One possible commercial usage of these gelatins would be in industrial applications where high solution viscosity without gel formation is required.

It has been known for a long time that the actual melting temperature of gelatin is directly related to the shrinkage temperature of the source collagen, and it is well appreciated that this depends upon the primary sequence of the collagen which, in turn, depends upon its source [14]. Collagen is an unusual protein, because it has a high number of glycines, as well as the imino acids proline and hydroxyproline. Both proline and hydroxyproline stabilize the helical form of collagen (gelatin), so gelatins produced from collagens with a lower proportion of the imino acids have a lower melting and gelling temperature. Indeed Veis' book [14] illustrates how the collagen shrinkage temperature (closely related to the melting temperature of the resultant gelatin) can be correlated with imino acid percentage. This is typically around 24% for mammals, and 16–18% for most fish species. The stability of the collagen helix results mainly from the pyrrolidine rings of proline and hydroxyproline rather than from hydrogen bonding through the hydroxy groups [15], although this does also contribute to the intermolecular stability of the helix. Consequently the variation in shrinkage temperature is related to the total number of imino acid residues rather than hydroxyproline residues alone. Cold water fish, e.g. cod, have the lowest hydroxyproline content. This is offset, however, by an increase in the serine content and to a lesser extent threonine and hydroxylysine, so that the hydroxyl content for fish gelatins (and for vertebrate collagens in general) is almost constant [15]. Gelatins extracted from collagens from reptilian and amphibian sources often have intermediate melting points, whereas that from the polar ice fish is still lower.

EXPERIMENTAL

Here we briefly present some recent data from a current and continuing study on fish gelatins. Although results are preliminary they are still considered appropriate. Gelation and melting of a series of gelatins has been investigated by rheological measurements under low-amplitude oscillatory shear. Three samples of bovine gelatin were supplied by SBI (SKW, Biosystems), France and two fish gelatins (cod, 2747) and tilapia (a tropical fresh water fish, 7056) by Croda Colloids, Luton, England. Appropriate % w/w concentrations, C, were made up in deionized water. Samples were heated to well above the gel melting temperature and then cooled at a constant rate while monitoring the gelation process using the viscoelastic parameters G' and G'' until the gel was well formed. After some times and some further measurements, the sample was reheated to melt the gel, again monitoring these parameters. Measurements were carried out using a Rheometric Scientific Fluids Spectrometer RFSII, parallel plate geometry and a circulating fluid jacket system using protocols described previously [16] (Rheometric Scientific, Surrey, UK). The initial cooling regime employed was usually from $30\,^\circ$C to $8\,^\circ$C at $1\,^\circ$C min^{-1}. For the cod gelatin sample, cooling was carried out to $1\,^\circ$C. Frequency sweeps and subsequent gel melting profiles were determined as described previously.

RESULTS

Figure 28.3 illustrates the concentration dependence of gel modulus for one sample of the bovine gelatin and for the cod and tilapia gelatins. Data has been plotted as log (G') measured at small strain and a frequency of 10 rad s^{-1}.

Figure 28.4 shows the corresponding melting temperatures (interpolated from traces of G' vs temperature on heating at $1\,^\circ$C min^{-1}); these have been plotted in Eldridge–Ferry (EF) coordinates [10,17] of log (C) versus $1/T$ in kelvin. We note that neither of the plots employed are standard, so more detail is given below.

DISCUSSION

Figure 28.3 illustrates a preliminary analysis of the modulus vs concentration behaviour for the three samples. As might be expected from the structural considerations above, the bovine gelatin sample has the lowest critical concentration, $\sim0.6\%$, with the tilapia a little higher, and the cod up to an order of magnitude higher still (although this would ultimately depend also on chain length) The expected high concentration $G' \propto C^2$ asymptote is also illustrated, i.e. at high polymer concentrations (say $>5C_\mathrm{o}$) where C_o is the minimum gelation concentration, G' is proportional to C^x where x is ~2–2.5, and at lower concentrations, as $C \to C_\mathrm{o}$, x becomes significantly greater. In general, C_o itself varies

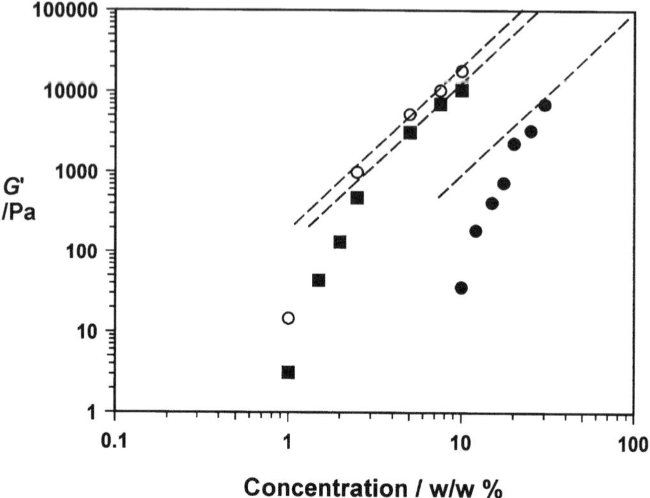

Figure 28.3 Log G' versus concentration for a typical bovine gelatin sample (open circles), and for two fish gelatin samples from tilapia (filled squares) and cod (filled circles). Dotted lines represent the $G' \sim C^2$ high concentration asymptote

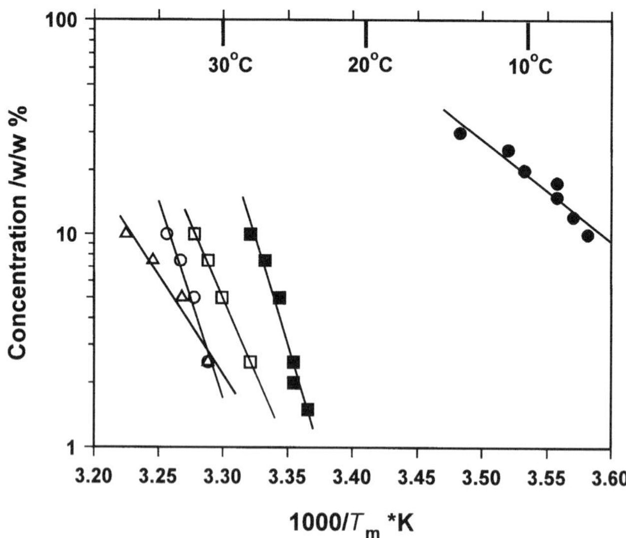

Figure 28.4 Eldridge–Ferry plot for three bovine (open symbols) and fish gelatin samples; tilapia (filled squares) and cod (filled circles). The corresponding temperatures in °C are given on the top axis. Molecular weights are 168.5, 145.7, 133.3, 89.6 and 60 kg/mol respectively

from <0.05% in the case of some microbial polysaccharides to >10% for more particulate gels, but the shape of the scaled log (G) versus log (C/C_o) curve remains almost the same. Details of the form of these curves, and what can be deduced from them, are given in refs. [10] and [11].

Eldridge and Ferry [17] showed that the concentrations (C) and weight average molecular weight (M) dependence of the gel melting temperature for gelatin gels could be described by a simple model. They found that a plot of ln (C) vs $1/T_m$ (gel melting temperature in kelvin) was almost linear, as was ln (M) vs $1/T_m$. This is a van't Hoff form, reflecting a single van't Hoff enthalpy. However, many data for gelatin show curvature, presumably reflecting finite chain effects and more a complex mechanism than the simple two cross-linking sites $\leftrightarrow$ one cross-link would imply [17]. Nevertheless it illustrates effectively the range of temperatures over which different gelatins can melt.

According to Eldridge and Ferry, a linear plot of ln (C) vs $1/T_m$ has a slope relating to the melting enthalpy ΔH_m^o. The actual slope depends on the molecular weight and on T_m. Figure 28.4 illustrates the variation in slope and intercept, reflecting the differences in melting temperature and molecular weight of the gelatin samples. The lower the melting temperature, the smaller the helical structures present and consequently the lower the values of $-\Delta H_m^o$. Thus the slope of the EF plot becomes less negative as the temperature decreases, as can be seen for the cod sample. A similar pattern can be seen for the variation in molecular weight. The lower the molecular weight, the greater the number of cross-links per unit volume needed to form a gel. Hence low molecular weight gels melt at a lower temperature than high molecular weight ones, which is again evident in Figure 28.4.

Our future work will concentrate on extending the number of measured samples, fitting data to more detailed models, and trying to understand the relationship between the simple Eldridge–Ferry model enthalpy and other more realistic extensions. The blending of different samples (say cod and bovine) also has some attractive and intriguing possibilities. For example, will these produce a broad melting profile, or a single sharp melting peak at some intermediate temperature? Concomitant differential scanning calorimetry (DSC) measurements may also be of interest here.

CONCLUSIONS

Many aspects of gelatin gelation have been well documented and are now widely accepted. One of the major advances in elucidating the gelation process was the observation that gelation occurs at a critical degree of helix conversion. This has enabled gelatin gelation to be discussed in a similar manner to conventional models of covalent network formation.

The exploitation of novel gelatin sources is certainly a major topic for the future, whether viewed from an applications standpoint, or because it will help us to make progress in the understanding of the structure-property relationships of conventional gelatins. The data illustrated here suggests that in many applications there is no real alternative to gelatin, and improved materials require a still greater understanding of the factors controlling gelation properties.

ACKNOWLEDGMENTS

We are grateful to Dr D.S. Field of Croda Colloids, Luton, England for the samples of the fish gelatins, and Dr G. Takerkart of SKW Biosystems, France for the bovine gelatin samples. Part of this studied was funded by the European Commission under contract FAIR.CT97.3055.

REFERENCES

1. Djabourov, M., *Contemp. Phys.*, **29**, 273 (1988).
2. Ross-Murphy, S.B., *Polymer*, **33**, 2622 (1992).
3. Morris, E.R., *Physical Methods for the Study of Food Biopolymers*, (S.B. Ross-Murphy, ed.), Blackie Academic and Professional, Glaogow, 1994, p. 15.
4. Djabourov, M., Maquet, J., Theveneau, H., Leblond, J. and Papon, P., *Brit. Polym. J.*, **17**, 164 (1985).
5. Busnel, J.-P., Morris, E.R. and Ross-Murphy, S.B., *Int. J. Macromol*, **11**, 119 (1989).
6. Prystupa, D. and Donald, A., *Polymer Gels and Networks*, **4**, 87 (1996).
7. te Nijenhuis, K., *Colloid Polym. Sci.*, **259**, 522 (1981).
8. Djabourov, M. and Papon, P., *Polymer*, **24**, 537 (1983).
9. Durand, D., Emery, J.R. and Chatellier, J.Y., *Int. J. Biol., Macromol.*, **7**, 315 (1985).
10. Clark, A.H. and Ross-Murphy, S.B., *Adv. Polym. Sci.*, **85**, 57 (1987).
11. te Nijenhuis, K., *Adv. Polym. Sci.*, **130**, 1 (1997).
12. Higgs, P.G. and Ross-Murphy, S.B., *Int. J. Biol. Macromol.*, **12**, 233 (1990).
13. Leuenberger, B.H., *Food Hydrocolloids*, **5**, 353 (1991).
14. Veis, A., *The Macromolecular Chemistry of Gelatin*, Academic Press, London (1964).
15. Leach, A.A., *Biochem. J.*, **67**, 83 (1957).
16. Ross-Murphy, S.B. *Rheol. Acta*, **30**, 401 (1991).
17. Ferry J.D. and Eldridge, J.E. *J. Phys. Chem.*, **53**, 184 (1949).

29

Kinetical, Structural and Dynamical Properties of Fibrin Networks

G. ARCOVITO[1], M. DE SPIRITO[1], F. ANDREASI BASSI[1],
M. ROCCO[2], S. BERNOCCO[2], E. PAGANINI[3] and F. FERRI[4]

[1]Istituto di Fisica and Istituto Nazionale Fisica della Materia (INFM), Facoltà di Medicina e Chirurgia, Universita' Cattolica del Sacro Cuore, L.go F. Vito 1, 00168 Roma, Italy

[2]Gruppo di Biostrutture, Istituto Nazionale per la Ricerca sul Cancro (IST), Centro Biotecnologie Avanzate (CBA), L.go R. Benzi 10, 16132 Genova, Italy

[3]CISE, P.O. Box 12081, 20134 Milano, Italy

[4]Istituto di Scienze Matematiche, Fisiche e Chimiche and Istituto Nazionale Fisica della Materia (INFM), Università degli Studi di Milano a Como, via Lucini 3, 22100 Como, Italy

Wiley Polymer Networks Group Review Series Vol. 2. Edited by B.T. Stokke and A. Elgsaeter
© 1999 John Wiley & Sons Ltd

ABSTRACT

The structure of fibrin gels grown at room temperature from fibrinogen solution at several fibrinogen concentrations was investigated by means of elastic light scattering. By combining classical static light scattering (CSLS) and low angle elastic light scattering (LAELS) an overall wavevector range of more than two decades was spanned, from 3×10^2 to 3×10^5 cm^{-1}. The scattered intensity distribution $I(q)$ of all the gels was characterized by three different regimes, delimited for each gel by a pair of q values, q_1 and q_2, by which, for $q_1 < q < q_2 I(q)$ decays as a power law, $I(q) \propto q^{-Dm}$, typical of mass fractals, with fractal dimension $D_m = 1.2 \pm 0.05$, equal for all the samples. For $q = q_1, I(q)$ exhibits a maximum indicating a long range order in the gel structure, with an average mesh size $\xi_1 = 2\pi/q_1$. At larger values of q there is a crossover to the scattering from surface fractals and $I(q)$, for $q > q_2$, decays with a new power law characterized by an exponent of -4. The experimental investigation is completed by reporting classical static and dynamic light scattering (DLS) measurements, made up on 'fine' fibrin gel, formed in solution with a salt concentration of 0.5 M NaCl. Here, at higher q values, the crossover of the $I(q)$ to a surface fractal behavior does not appear, while a fractal dimension $D_m = 1.15 \pm 0.05$ is observed. By fitting the equation suggested by a recent theoretical approach in studying semiflexible polymer networks to the experimental data, a value of the average fibrin diameter $a = 30 \pm 2$ nm, for fibrinogen concentration $c_F = 1676$ nM, has been obtained, in good agreement with reported electron microscopy results.

INTRODUCTION

The fibrin fibers network, which forms a three-dimensional polymer gel, is the basic element of blood clot, the final event in a series of reactions that occurs physio- or pathologically at sites of lesions in the blood vessel wall. The knowledge of the physical properties of the fibrin gel is of fundamental importance to understand its role in the physio-pathological processes, and it is known [1–3] that these properties are determined, to a large extent, by the physical properties of the fibrin fibers, such as their diameter, degree of aggregation, stability, resistance to proteolysis, etc.

Much experimental work has been performed since the early 1950s on the fibrin gel structure, mainly using electron microscopy techniques, demonstrating the importance of fibrin fiber diameter in determining the physical properties of the fibrin gel (see for example Refs. [4–7] and also Ref. [3] for a review of more recent data). Fibrin gels have been classified in two limiting classes: 'coarse' and 'fine' gels. Coarse gels are large-pore gels made up of thick fibers while fine gels are narrow-pore gels made up of thin fibers. All the possible intermediate structures are allowed, just varying the physico-chemical parameters of the gelling solution such as salt concentration, ionic strength, pH values, etc. [1–3].

The aim of this paper is to present low angle elastic light scattering (LAELS) and classical static light scattering (CSLS) data on coarse fibrin gels, made from purified, and, as far as possible, undegraded, fibrinogen, and in the absence of chemical crosslinks. From these data the kinetics of formation as well as the structural properties of the gels, such as the fractal dimension, the gel pore size and the fiber thickness, have been obtained. The experimental investigation is completed by reporting classical static and dynamic light scattering (DLS) measurements made up on 'fine' fibrin gels.

THE FIBRIN NETWORK

Fibrin networks are gels formed by branched polymers grown in solutions of fibrinogen macromolecules activated by some specific enzyme. Fibrinogen is a centro-symmetric high molecular weight plasma glycoprotein (MW = 340 000) made up of two pairs each of three different polypeptide chains, $A\alpha$, $B\beta$, and γ. The macromolecule, rod-like in shape, $\sim$47 nm in length and $\sim$6 nm thick, basically consist of a central globular domains joined to two outer globular domains by two extended connectors [1–3, and references therein].

When the fibrinogen interacts with some specific enzyme, such as thrombin in the case of vertebrates, it becomes a reactive protein called monomeric fibrin (2α, 2β, 2γ). This activation occurs by cleaving two pairs of short peptides called fibrinopeptides A and B (FPA and FPB) from the central domain of the molecules, making the binding sites A and B exposed for forming bonds with the corresponding complementary sites a and b located on the distal domains on another molecule. The aA interactions are responsible for linear aggregation, while the bB govern the lateral growth of the fibrin fibers. This mode of action results in a polymerization mechanism where, after the cleavage of FPAs, the monomers first proceed to form two-stranded, half-staggered protofibrils. Only successively, because of the cleavage of FPBs, the protofibrils aggregate with each other to a variable extent, depending on the ionic strength of the solution, to form fiber bundles. The fibers branch at some stage, not yet accurately pinpointed, during this process, yielding a three-dimensional gel.

MATERIALS AND METHODS

SAMPLE PREPARATION AND CONTROL

The sample preparation techniques, as well as the sample integrity control procedures, have been described in detail in previous works [8,9]. In short, lyophilized, plasminogen free, human fibrinogen (Calbiochem, lot.B10707 (341576), USA) was used to prepare samples at different concentrations in freshly prepared buffer (50 mM TRIS, 0.1 M NaCl, 1 mM EDTA, pH 7.4), which were divided into aliquots and stored at $-20\,^{\circ}$C. In order to activate fibrinogen, lyophilized human thrombin from Sigma Aldrich (Milan, Italy, lot.104H9314) was reconstituted in the same buffer and used in a ratio 1/100 to the fibrinogen. It is worth while to emphasize that, in order to compare experimental results obtained by LAELS, CSLS and DLS measurements, carried out in different laboratories, extra attention was paid in defining common procedures for sample preparation and control.

LIGHT SCATTERING TECHNIQUES

The intensity of light elastically scattered from samples at different c_F was measured as a function of the wave vector $q = 4\pi n/\lambda \sin(\theta/2)$, where λ is the vacuum wavelength of the light, n is the refraction index of the medium, and θ is the scattering angle. Thus, by measuring the scattered intensity distribution within a wavevector range (q_{min}, q_{max}), it is possible to investigate the sample structure on length scales in the range $2\pi/q_{max} < \xi < 2\pi/q_{min}$. LAELS operated at scattering angles ranging from $\sim 0.1°$ to $\sim 10°$, covering a wavevector range from $\sim 3 \times 10^2$ to $\sim 3 \times 10^4$ cm^{-1}. A detailed description of the LAELS instrument used in this work can be found [10].

The CSLS technique operates at larger angles, typically from $\sim 15°$ to $\sim 150°$, corresponding to q values ranging between $\sim 3 \times 10^4$ and $\sim 3 \times 10^5$ cm^{-1}. The CSLS instrument used was an ALV-5000 system from ALV, Langen, Germany, equipped with a 4 W Argon laser (Innova 70, Coherent, FRG) operating with a power of 800 mW at 488 nm.

The same instrument was used to measure the normalized intensity–intensity autocorrelation function, $g_2(\mathbf{q}, t)$. The technical features of this instrument together with the measurement method are largely described [8,9]. As is well known [11], $g_2(\mathbf{q}, t)$ is related to the dynamic structure factor, $s(\mathbf{q}, t)$, of the scattering sample by the equation:

$$g_2(\mathbf{q}, t) = (1 + \beta |s(\mathbf{q}, t)|^2)$$

where $\beta \leqslant 1$ is the coherence factor depending on the geometry of the optical set-up.

RESULTS

STATIC LIGHT SCATTERING

Figure 29.1 shows the intensity distributions $I(q)$ scattered from two fibrin gels of concentration $c_F = 789$ and 190 nM, plotted versus the wavevector q. Measurements have been carried out by LAELS (circles) and CSLS (squares) techniques. For the $I(q)$ behavior of both gels, three different regimes, delimited by two wavevectors values q_1 and q_2, can be observed. For $q_1 < q < q_2$, $I(q)$ decays as a power law function, $I(q) = Aq^{-D_m}$, to which the data were fitted with A, D_m as floating parameters. Moreover, the data corresponding to CSLS were rescaled by a factor B, which was also a floating parameter of the fitting. Thus, in this q range, the two gels behave as mass fractals with the same value of the fractal dimension $D_m \sim 1.20 \pm 0.05$ consistent with structures formed by branched linear polymers. For $q > q_2$, there is a crossover to a behavior typical of surface fractals of dimension D_s where $I(q)$ decays as a power law $I(q) \approx q^{-6+D_s} = q^{-4}$, for $D_s \approx 2$ [12]. Finally, for $q = q_1$, $I(q)$ exhibits a maximum beyond which it decreases toward very small values. The peak indicates [13] the presence of

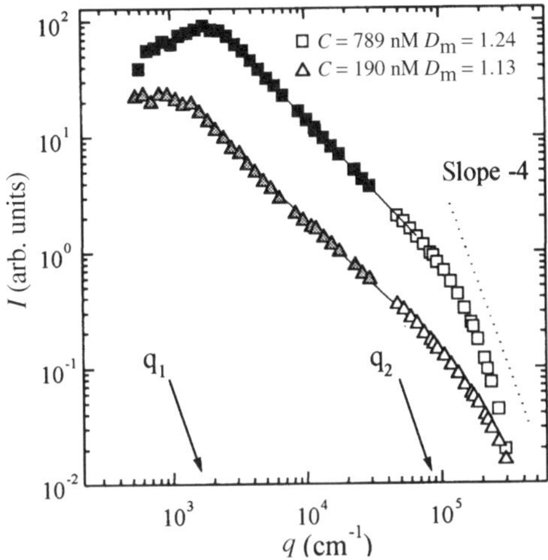

Figure 29.1 Log-log plot of the scattered intensity distribution $I(q)$ as a function of the wavevector q for two aged fibrin gels prepared at the fibrinogen concentration $c_F = 789$ nM and 190 nM. The data have been taken by using LAELS (filled symbols) and CSLS (open symbols) techniques. The arrows indicate the q values at which the three regimes described in the text take place. The solid line is the best fit of part of the data of the central region to a power law function characterized by the exponent D_m. The dotted line shows a slope of -4, indicating the expected intensity decay for surface fractals of smooth surface

a long-range order in the gel structure with average mesh sizes $\xi_1 = 2\pi/q_1 \sim 36$ and 68 μm, respectively.

The kinetics of the gelling process can be followed only by LAELS technique which allows the simultaneous detection of the scattered intensity distributions at all q. In Figure 29.2 the time evolution of $I(q)$ for a gel grown at $c_F = 2371$ nM is reported as a function of q for several time points before gelation. The time $t = 0$ refers to when the fibrinogen was activated by addition of the thrombin enzyme to the solution. As it can be clearly seen, a peak in $I(q)$ is formed very soon, within ~70 s. Its position, q_{peak}, varies somewhat during the first 100–150 s, but beyond this time its value does not change any more. Conversely, the amplitude of scattered intensity keeps increasing up to a saturation value, which is much higher than its initial value at 70 s. This result is better elucidated by the data reported in Figure 29.3, where $I(q = 10^4 \text{ cm}^{-1})$ is reported as a function of time t, during the gelling process. The figure shows two key features: (i) at a time of approximately 60–80 s the scattered intensity undergoes an abrupt change in its behavior and starts to increase very fast, (ii) not reaching its saturation level until much longer times. This result, together with that reported in Figure 29.2, suggests a mechanism of gel formation in which the scaffold of polymer network is formed very soon, within the first 100 s, whereas its further growth (with the

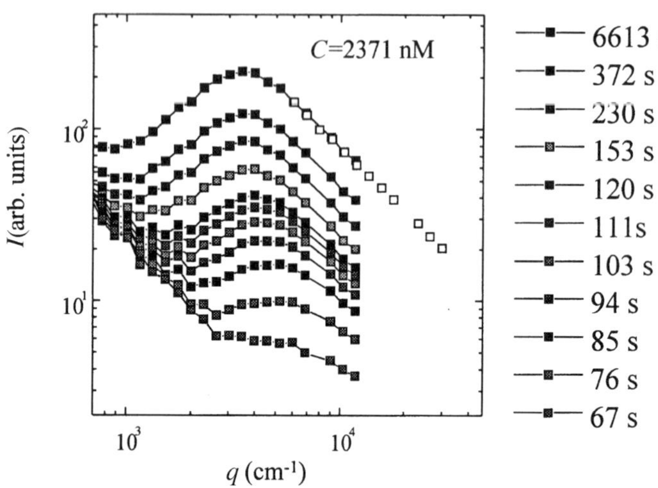

Figure 29.2 LAELS study of the kinetics of gelation for a fibrinogen solution leading to a gel similar to those of Figure 29.1. The fibrinogen concentration was 2371 nM. The different curves refer to $I(q)$ measured at different times after the addition of thrombin to the fibrinogen solution ($t = 0$ time). The curve at the time 6613 s corresponds to the aged gel. The data show that, after approximately 70 s, the angular distribution of $I(q)$ attains a steady-state shape with $\xi_1 = 2\pi/q_1 \sim 25\,\mu m$ and $D_m \sim 1.2$, whereas its amplitude increases with time

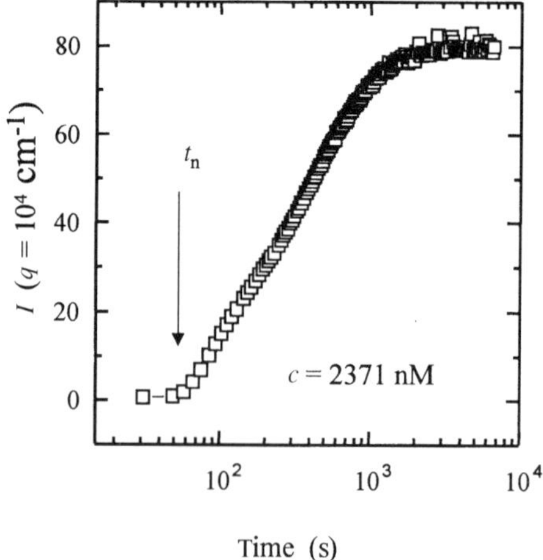

Figure 29.3 Behavior of the intensity amplitude scattered at $q = 10^4$ cm^{-1} as a function of time t for the same sample of Figure 29.2. The time $t_n = 70$ s is that at which the gel basic structure seems to be already formed. The aged gel is formed after approximately 3000 s

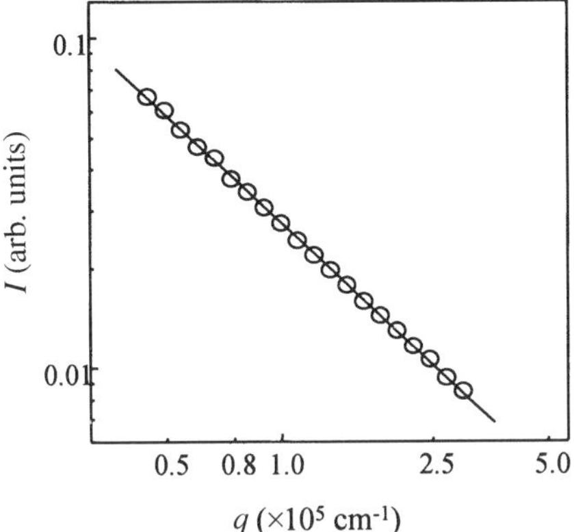

Figure 29.4   Log-log plot of the scattered intensity $I(q)$ versus q as obtained by CSLS measurements for an aged gel of $c_F = 3020$ nM, prepared in a solution with salt concentration (0.5 M NaCl)

corresponding increasing of the scattered intensity) may be attributed only to a process of fiber thickening.

Finally, to investigate the effect of the ionic strength of the gelling solution on the structure of the resulting gel, Figure 29.4 reports an example of the intensity distribution obtained with CSLS when a gel prepared with a salt concentration of 0.5 M NaCl (instead of 0.1 M NaCl) is formed. At this higher salt concentration only the mass fractal behaviour is observed, with $D_m \sim 1.15$. The crossover to a surface fractal behaviour moves towards higher wavevectors, outside the range of q accessible with CSLS. This indicates that, in this case, the gel fibers are very thin and their size cannot be estimated by means of elastic light scattering techniques.

DYNAMIC LIGHT SCATTERING

It is now accepted that fibrin networks, as well as all biopolymer networks [14], are formed by semiflexible polymers, that is by macromolecules with a stiffness intermediate between the two extreme cases of random coils and rigid rods. In a recent paper, some of us [8] reported DLS measurements which have been analyzed in the framework of a recent theoretical model [14,15] proposed to study semiflexible polymer networks restricted to the ideal case by which these systems are fulfilling the conditions $a < q^{-1} < \xi < l_p, L$, where a is the average diameter of fibrin fibers, ξ the gel mesh size, l_p the persistence length and L the contour length. By assuming that the internal configurational dynamics dominates over the center of mass and the rotational motions, and in the weakly bending rod limit, the dynamic structure factor, $s(\mathbf{q},t)$, is calculated to a good approximation with a stretched exponential at the intermediate times, and as a mono-exponential

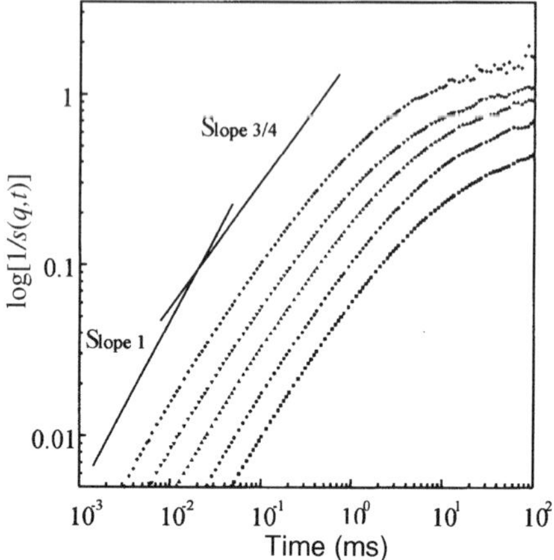

Figure 29.5 Log-log plot of log $[1/s(q, t)]$ of a fibrin network sample with fibrinogen concentration $c_f = 3020$ nM, as a function of the delay time t. The scattering wavevectors were, from left to right, $q = 1.00 \times 10^5$ cm^{-1}, 1.14×10^5 cm^{-1}, 1.69×10^5 cm^{-1}, 2.40×10^5 cm^{-1} and 2.89×10^5 cm^{-1}. The drawn lines have slope of 1 and of 3/4, respectively

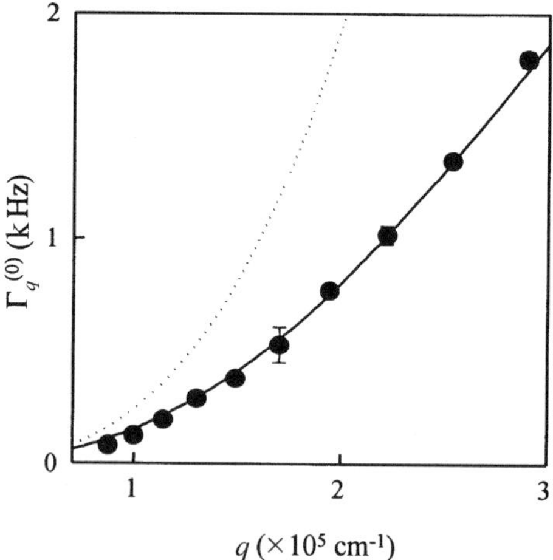

Figure 29.6 The initial decay rate $\Gamma_q^{(0)}$ of the dynamic structure factor of a sample at $c_f = 3020$ nM is plotted against the wavevector q. From the fit of equation (29.1) to the experimental data (full line) a value of the fiber mean diameter $a = 30 \pm 2$ nm has been found. The theoretical prediction for Gaussian chains (dashed line) is also reported for comparison

function for $t \to 0$, with a decay rate, $\Gamma_q^{(0)}$ given by the equation [14,15]:

$$\Gamma_q^{(0)} = \frac{k_B T}{6\pi^2 \eta} q^3 \left[\frac{5}{6} - \log qa \right] \tag{29.1}$$

where k_B is the Boltzman constant, T is the temperature and η the solution viscosity. The effective average hydrodynamic diameter a of semiflexible polymers forming the network can be obtained.

Figure 29.5 reports the log-log plot of the $\log[1/s(\boldsymbol{q},t)]$ versus the delay time t for the sample at $c_f = 3202$ nM. As can be seen, at the shortest time $s(\boldsymbol{q},t)$ is an exponential function of time, $s(\boldsymbol{q},t) \propto \exp(-\Gamma t)$, while at intermediate delay times $s(\boldsymbol{q},t)$ decays as a stretched exponential function, $s(\boldsymbol{q},t) \propto \exp[-(\Gamma t)^\beta]$, with $\beta = 3/4$. At longest times more and more deviation from that behavior can be observed. The initial slope of $s(\boldsymbol{q},t)$ has been analyzed by applying the cumulant method [11] to obtain $\Gamma_q^{(0)}$ values as a function of q, shown in Figure 29.6. As reported, Equation (29.1) fits well to the experimental data giving a value $a = 30 \pm 2$ nm for the average diameter of the fibrin fibers. In the same Figure, the clear deviation of the experimental data from the behavior predicted for Gaussian chains is also shown.

DISCUSSION AND CONCLUSIONS

In this report, it was shown that the kinetics, the structure and the viscoelastic properties of fibrin networks could be characterized by using static and dynamic light scattering techniques.

By static light scattering measurements, due to the wide range of length scale $(0.2 \ \mu m < \xi = 2\pi/q < 200 \ \mu m)$, it was possible to measure both the mass fractal $D_m \sim 1.2$ of the networks and the surface fractal dimension $D_s \sim 2$ of the fibrin fibers, which appears to be independent of the initial monomer concentration.

The gelling kinetics has been also investigated and it was shown that the scaffold of the gel structure seems to be formed soon at time t_n, then the structure ripens and in about 30–50 times t_n, depending on solution concentration c_F, a constant value of $I(q)$ is reached. During this time only a fiber thickening appears evident.

Finally, DLS data proved that fibrin networks are formed by branched semiflexible polymers. Moreover, the possibility to measure with DLS the fiber diameter, in a size range not accessible by CSLS, shows how DLS can be considered an effective complementary technique for CSLS.

As proved, mainly by electron microscopy measurements, the fibrin fibers are not formed by simple lateral aggregation of protofibrils, but by a twisting mechanism which limits the size of fiber diameter, as a consequence of the flexibility of fibrin monomers. Moreover, it was shown that the average fiber diameter decreases as the NaCl concentration of the solution increases, but, at the same time, the fiber diameter distribution narrows, so that more homogeneous and rigid networks are built up. Thus the average fibrin diameter becomes an important parameter, being closely linked to the elastic properties not only of the entire gel but also of the single fibrin monomer.

In agreement and extending over the electron microscopy data, the non-invasive light scattering techniques employed seem to suggest that the physico-chemical properties of the gelling solution determine the structure and the dynamics of fibrin gels, while the main effect of the initial fibrinogen concentration is to grow gels with thicker fibers and somewhat different mesh size. Further studies are in progress in our laboratories to investigate more in depth, by using various light scattering techniques, both the ionic strength- and monomer concentration-dependence of the structure and dynamics of fibrin gels.

REFERENCES

1. M.W. Mosesson, R.F. Doolittle (eds), *Molecular Biology of Fibrinogen and Fibrin* (New York, June 2–4, 1982), *Ann. NY Acad. Sci.*, Vol. 408, 1983.
2. R.F. Doolittle, *Ann. Rev. Biochem.*, **53**, 195 (1984).
3. Blomback B. *Thromb. Res.*, **83**, 1–75 (1996).
4. J.D. Ferry and P.R. Morrison, *J. Am. Chem. Soc.*, **63**, 388 (1947).
5. C.V.Z. Hawn and K.R. Porter, *J. Exp. Med.*, **86**, 285 (1947).
6. J.W. Weisel, *Biophys. J.*, **50**, 1079 (1986).
7. J.W. Weisel, C. Nagaswami and L. Makosky, *Prot. Natl. Acad. Sci., USA*, **84**, 8891 (1987).
8. M. De Spirito, G. Arcovito, F. Andreasi Bassi, M. Rocco, E. Paganini, M. Greco and F. Ferri, *Nuovo Cimento* **20D**, 2409 (1998).
9. G. Arcovito, F. Andreasi Bassi, M. De Spirito, E. Di Stasio and M. Sabetta, *Biophys. Chem.*, **67**, 287 (1997).
10. M. Carpineti, F. Ferri, M. Giglio, E. Paganini, and U. Perini, *Phys. Rev. A*, **42**, 7347 (1990).
11. B. Berne, R. Pecora, *Dynamic Light Scattering*, Wiley, New York (1976).
12. P.W Schmidt, *J. Appl. Cryst.*, **24**, 414 (1991).
13. M. Carpineti, M. Giglio, V. Degiorgio, *Phys. Rev. E*, **51**, 590 (1995).
14. K. Kroy and E. Frey, *Phys, Rev. E*, **55**, 3092 (1996).
15. K. Kroy and E. Frey, 'Dynamic light scattering from semiflexible polymers,' book chapter in *Scattering in Polymeric and Colloidal Systems*, (W. Brown and K. Mortensen, ed.) Gordon and Breach, New York (in press).

30

Anomalous Crosslink Density Dependence of Spatial Inhomogeneities in Polymer Gels

F. IKKAI and M. SHIBAYAMA

Department of Polymer Science and Engineering, Kyoto Institute of Technology, Matsugasaki, Sakyo-ku, Kyoto, 606-8585 Japan

ABSTRACT

Spatial inhomogeneities in polymer gels have been studied empirically and theoretically as a function of crosslink density (CD). Detailed discussions on the structure factor are given for gels with various sets of gel parameters, i.e., CD, network volume fraction (ϕ), degree of ionization (f), and Flory's χ parameter or temperature. Light scattering (LS) and small-angle neutron scattering (SANS) experiments were carried out on chemically crosslinked poly(N-isopropylacrylamide) - *co* - poly (acrylic acid) gels under the following conditions: CD = 0–20 wt%, $\phi = 0.078$, $f = 0$–0.091, and temperature = 20–55 °C. In LS, the CD dependence of the ensemble averaged scattered intensity ($\langle I(q)\rangle_E$) shows

Wiley Polymer Networks Group Review Series Vol. 2. Edited by B.T. Stokke and A. Elgsaeter

an inversion phenomenon, that is, $\langle I(q) \rangle_E$ increases with CD in a good solvent, but decreases with CD in a poor solvent. The SANS results show appearance of the inversion at $q \leqslant 0.03$ Å^{-1}, where q is the scattering vector. The Rabin–Panyukov theory for the structure factor of weakly charged polyelectrolyte gels in a poor solvent is employed to interpret the inversion. The theoretical prediction of the scattering function agrees well with the experimental results.

INTRODUCTION

The aim of this study is to elucidate the basis of spatial inhomogeneities in polymer gels, both empirically and theoretically, particularly focussing on the crosslink density (CD) dependence of the gel microstructure. Microstructure in polymer gels depends on three 'how's. (1) How was the gel prepared? This includes problems of the chemistry of gel formation. (2) How was the gel studied? This concerns the final conditions of gels, such as, temperature, pH, network volume fraction. (3) How did the gel reach its present state? This is just the problem of hysteresis. (3) is one of the most important topics of discussion and some papers have been published recently [1–3]. In this paper, all the gel samples were prepared in the same way, hence, we consider only (1) and (2).

Regarding (1), the degree of crosslinking at gel preparation is one of the most important parameters. Although the difference between a gel and the corresponding polymer solution at the same concentration has been discussed in many papers [4–8], the CD dependence of gel microstructures has been given only limited attention. As a result, it is believed that the microstructure of polymer gels becomes more inhomogeneous as CD increases due to the formation of structural inhomogeneities with an inhomogeneous distribution of crosslinks in the gel. For (2), in particular, theoretical treatments have been restricted to discussions with empirical conditions [5,9–12].

First of all in this paper, we show some experimental results obtained by light scattering (LS) [13,14] and small-angle neutron scattering (SANS) [15]. These indicate the existence of an anomolous CD dependence, an inversion of scattered intensity $(I(q))$, as well as the influence of surrounding conditions of gels on $I(q)$. Second, we demonstrate that our experimental results are reproduced very well by the Rabin–Panyukov (RP) theory [16]. Theoretical fitting to the LS as well as SANS functions and a simulation are carried out by using the gel parameters at preparation and under the conditions of the experiment. Third, we make some suggestions concerning the relationship between microstructure and the inversion.

MATERIALS AND METHODS

PREPARATION OF GEL SAMPLES

N-Isopropylacrylamide (NIPA) monomers were kindly supplied by Kohjin Chemical Co. Ltd and were purified by recrystallization from toluene solution by petroleum ether at $0\,°$C. Acrylic acid (AAc) of reagent grade was purchased from

Wako Chemical Co. and used without further purification. Mixtures of NIPA and AAc monomers with different NIPA/AAc ratios were dissolved in 40 mL of H_2O (for LS) or D_2O (for SANS) including 20 mg of ammonium persulfate (polymerization initiator). Two NIPA/AAc molar concentration ratios at preparation were chosen, i.e., NIPA/AAc = 668 mM/32 mM and 636 mM/ 64 mM. Each solution was divided into eight vessels containing equal amounts. Then, the given amounts of N, N'- methylenebisacrylamide (BIS) crosslinker were added to each solution in order that the final concentrations of BIS (C_{BIS}) were 0(CD = 0 mol%), 2(0.28), 4(0.57), 6(0.85), 8(1.1), 12(1.7), 16(2.2), and 24(3.3) mM. By adding 24 μl of N, N, N', N'-tetramethylethylenediamine (TEMED) accelerator to those solutions filtered with a 0.2 μm micropore filter, gelation was initiated in test tubes at 20 °C. It took about a day to complete gelation. Here, the pH of the gels was more than six, so most of AAc groups were dissociated. No noticeable sol fraction was detected by swelling the gels for CD $\geqslant$ 2 mM. For LS, the transparent gels in test tubes were employed without further treatment. For SANS, gels were passed through a 500 μm sieve to produce small pieces in order to enhance the rate of thermal equilibration of a gel upon a change of temperature. Then, without any further treatment, the sieved gels in D_2O were sealed in a brass cell with a pair of quartz windows about 2 mm apart.

LIGHT SCATTERING

LS experiments were performed using a laboratory-made dynamic light scattering instrument with a 10 mW He–Ne laser (wavelength, $\lambda = 632.8$ nm) coupled with a photon correlator (DLS-7, Otsuka Electric Co.). The details of the instrument were described in ref.[17]. A sample in a test tube (inner diameter, $d = 10$ mm) was placed in the silicone oil bath, the temperature of which was controlled to within ± 0.1 °C. The scattered intensity $I(q)$, where q is the magnitude of the scattering vector, was measured at a fixed scattering angle of 60°. This corresponds to $q \approx 1.31 \times 10^{-3}$ Å$^{-1}(= (4n\pi/\lambda)\sin(\theta/2))$, where n is the refractive index of water. The ensemble average of the scattered intensity ($\langle I(q)\rangle_E$), in units of counts per second (cps), was calculated by measuring $I(q)$ at 100 different sample positions chosen arbitrarily in the gel. The gel was kept in the bath for more than a day before each run in order to obtain thermal equilibrium at the given temperature.

SMALL-ANGLE NEUTRON SCATTERING

SANS experiments were carried out on the 8m-SANS facility (NG1) at the National Institute of Standards and Technology. A gel sample in a brass cell was irradiated by a neutron beam with a wavelength of 9 Å. The temperature in the cell was changed from 30 to 54 °C in steps of 3 °C. At least 30 min was allowed before each measurement in order to ensure thermal equilibrium. The absolute scattered intensity functions ($I(q)$; $0.01 \leqslant q \leqslant 0.09$ Å^{-1}) were obtained by the following procedure. The scattered neutrons were counted with a two-dimensional detector and were circularly averaged by taking account of the detector inhomogeneities, followed by correction for cell scattering, fast neutrons, and transmission. Then, $I(q)$ as absolute intensity was obtained by multiplying by the scaling factor determined with a silica standard.

THEORETICAL BACKGROUND

As proposed by the pioneering work of Bastide *et al.*, [18,19] there are two types of concentration fluctuations in polymer gels; liquid-like (thermal) fluctuations and solid-like (static) inhomogeneities. In the RP theory [16], the structure factor $(S(q))$, proportional to $I(q)$, for instantaneously crosslinked networks of Gaussian phantom chains with excluded volume, is given by the sum of the thermal correlator $G(q)$ and the static correlator $C(q)$ as follows:

$$S(q) = G(q) + C(q) \tag{30.1}$$

$$G(q) = \frac{\phi N g(q)}{1 + w(q)g(q)} \tag{30.2}$$

$$C(q) = \frac{\phi N}{[1 + w(q)g(q)]^2(1 + Q^2)^2}\left[6 + \frac{9}{w_0(q) - 1 + (\frac{1}{2})Q^2(\phi_0/\phi)^{2/3}\phi_0^{-1/4}}\right] \tag{30.3}$$

where a, N, ϕ, f, and $w(q)$ indicate the segment length (8.12 Å for NIPA polymer chains [20]), the average degree of polymerization between crosslinks, the network volume fraction, the degree of ionization, and the effective second virial coefficient, respectively. Note that the parameters with and without subscript 0 denote those in preparation and during the experiment, respectively. $Q(= aN^{1/2}q)$ is the dimensionless wavevector. The function $g(q)$ is given by

$$g(q) = \frac{1}{Q^2/2 + (4Q^2)^{-1} + 1} + \frac{2(\phi/\phi_0)^{2/3}\phi_0^{1/4}}{(1 + Q^2)^2} \tag{30.4}$$

Furthermore, the functions of $w(q)$ and $w_0(q)$ are expressed by

$$w(q) = (1 - 2\chi + \phi)\phi N + \frac{l_B f^2 \phi N^2}{Q^2 + l_B f \phi N} \tag{30.5}$$

$$w_0(q) = \phi_0^{5/4}N + \frac{l_B f_0^2 \phi_0^{5/4} N^2}{Q^2(\phi_0/\phi)^{2/3} + l_B f_0 \phi_0^{5/4} N} \tag{30.6}$$

χ is Flory's interaction parameter under the experimental conditions. l_B is the dimensionless Bjerrum length, given by $l_B = 4\pi L_B/a$, where L_B is the Bjerrum length (≈ 7 Å for aqueous solution at $25\,^\circ$C) [21].

RESULTS AND DISCUSSION

Figure 30.1 shows the plots of $\langle I(q)\rangle_E$ of (N-isopropylacrylamide/acrylic acid) (NIPA/AAc) copolymer gels with NIPA/AAc = 668 mM/32 mM at various temperatures as functions of the degree of polymerization between crosslinks (N). Note that N is inversely proportional to CD. We estimated the value of N from the amounts of reactants. $\langle I(q)\rangle_E$ was given by averaging the scattered intensity obtained by LS at 100 arbitrary positions in the gel. The inset shows the scattered intensity variation at various sample positions, i.e., the so-called speckle pattern, and the horizontal solid line indicates $\langle I(q)\rangle_E$. $\langle I(q)\rangle_E$ decreases with increasing N at low temperatures up to 40 °C, which means that $\langle I(q)\rangle_E$ increases with CD. Note that these NIPA/AAc gels have a hydrophobic nature. However, for temperatures $\geqslant 45$ °C, $\langle I(q)\rangle_E$ increases with N, for $N \geqslant 87.5$. We call this opposite type of N-dependence on $\langle I(q)\rangle_E$ an *inversion*. Such an anomalous phenomenon is strongly supported by the results of SANS over a wider range of $q(0 < q \leqslant 0.03 \text{ Å}^{-1})$. Note that the value of q for LS corresponds to $1.31 \times 10^{-3} \text{ Å}^{-1}$.

Figure 30.2 shows the CD dependence of $I(q)$ at various temperatures for NIPA/AAc (636 mM/64 mM) gels. In Figure 30.2, $I(q)$ has a maximum

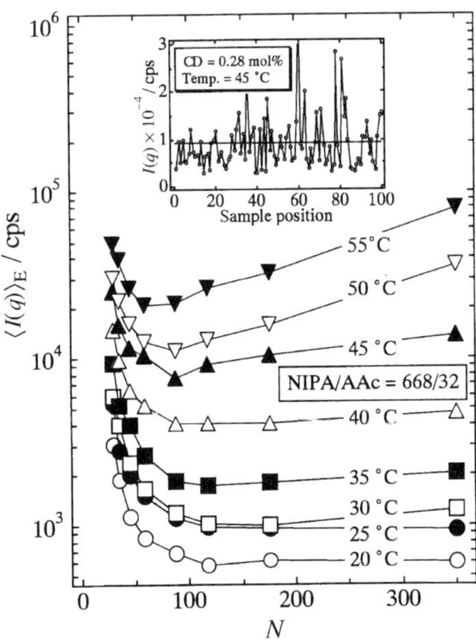

Figure 30.1 N dependence of $\langle I(q)\rangle_E$ of NIPA/AAc (= 668 mM/32 mM) copolymer gels. The inset shows the speckle patterns. The horizontal solid line indicates $\langle I(q)\rangle_E$

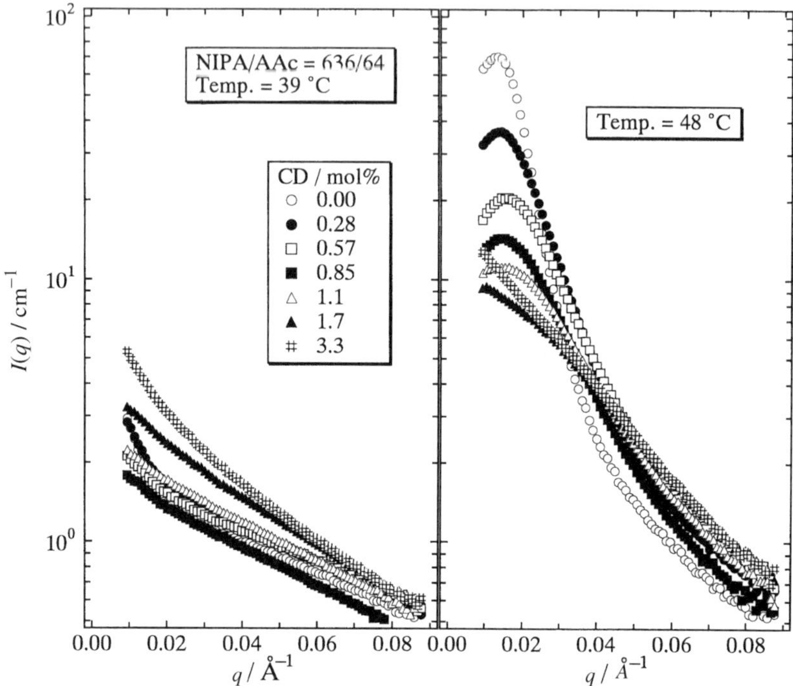

Figure 30.2   CD dependence of $I(q)$ at 39 and 48 °C for NIPA/AAc (= 636 mM/ 64 mM) gels

around $q = 0.02$ Å^{-1}, which appears at temperatures higher than 42 °C. It has been proposed that this peak results from microphase separation due to competition between hydrophobic (shrinking) and electrostatic (or Donnan potential) (swelling) interactions. Here, the appearance of the peak and the change of $I(q)$ are not related to the gels shrinking, because all of the gels at these temperatures were in swollen states. $I(q)$ at 39 °C increases with CD for CD > 0.57 mol%. On the other hand, at 48 °C, $I(q)$ decreases with increasing CD for $0 < q \leqslant 0.03$ Å^{-1}. The inversion phenomenon may be explained in terms of competition between two effects of crosslinking: (1) the introduction of inhomogeneities in a gel due to crosslinking and (2) the suppression of inhomogenizing movements by topological freezing. The former effect results in the increase of $I(q)$ with CD at lower temperatures and the latter in the decrease of $I(q)$ with increasing CD at higher temperatures.

Figure 30.3 shows a schematic interpretation of the phenomena in Figure 30.1. The solid lines and filled small circles in the insets denote the polymer network chains and the crosslinking points, respectively. The dashed vertical line, at $N = N_{\text{inv}}$, indicates the value of N at which inversion takes place. The insets

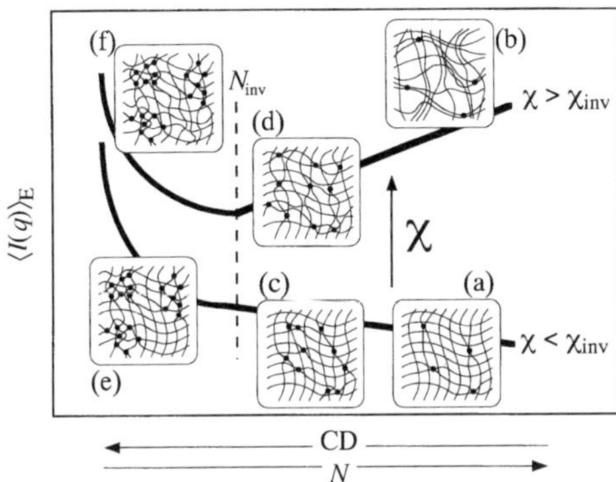

Figure 30.3 Network models on the basis of a schematic interpretation of the phenomenon in Figure 30.1

(a), (c), and (e) indicate networks at given values of N and a low temperature ($\chi < \chi_{\text{inv}}$). Here, χ_{inv} indicates the value of χ at which inversion occurs. The insets (b), (d), and (f) correspond to developments from (a), (c), and (e), respectively, with an increase of χ. By increasing N (decreasing CD) at $\chi < \chi_{\text{inv}}$, the inhomogeneity becomes less, resulting in a decrease in $\langle I(q)\rangle_{\text{E}}$. The increase in inhomogeneities at small N results from a structural disturbance due to the inhomogeneous distribution of the crosslinks frozen-in during the polymerization process. On the other hand, when χ is increased to $\chi > \chi_{\text{inv}}$ (a poor solvent regime) opposing interactions (repulsive and attractive) coexist in the gel. Then, it can be conjectured that the intragel structure is a disturbed one, with both expanded and contracted chain parts. The more mobile polymer chains (with smaller CD) shown as (b) can easily move to give this type of inhomogeneous structure. However, the presence of crosslinks prevents such an inhomogenizing movement. Thus, as CD is increased from (b) to (d), the gel becomes more homogeneous. This tendency may be strong in weakly charged polymer gels in a poor solvent. Note that, in the region of (f), the network inhomogeneity increases dramatically with CD.

Now, we focus on a numerical calculation which gives us more definite information. Figure 30.4 shows examples of scattering functions calculated using RP theory at different values of N for $\chi = 0.70$ (a) and 0.83 (b) and a set of known parameters, $\phi(= \phi_0) = 0.07$ and $f(= f_0) = 0.0457$. In addition, the insets indicate the q dependences of the two scattering correlators, $G(q)$ and $C(q)$. The theoretical functions reproduce well (1) the presence of a scattering peak at $q \approx 0.02$ Å^{-1}, which is found to be insensitive to N, and (2) the existence of the inversion phenomenon, depending on the value of χ. $S(q)$ increases with decreasing N at $\chi = 0.70$, i.e., $S(q)$ increases with increasing CD. On the other hand, at $\chi = 0.83$, $S(q)$ increases with increasing N, i.e., $S(q)$ increases with decreasing CD. Furthermore, the changes of $G(q)$ and $C(q)$ with N give some

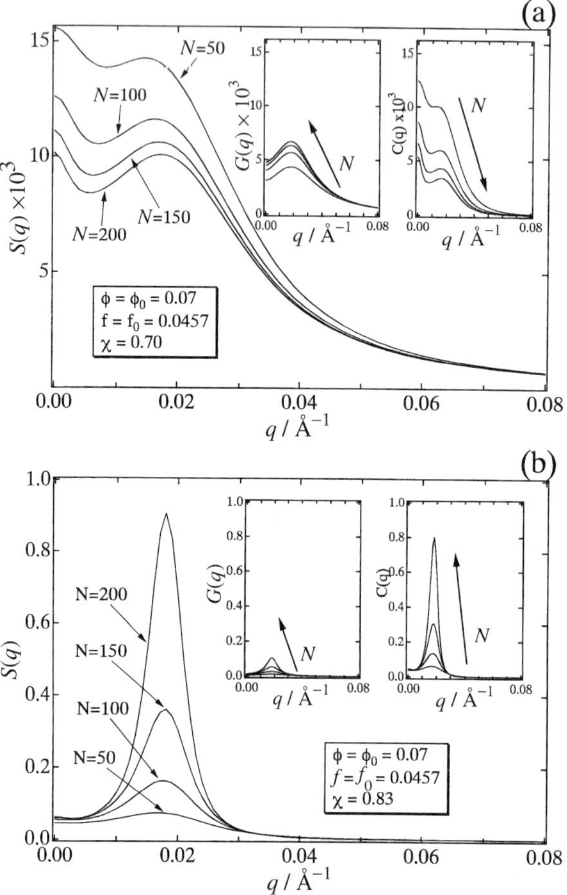

Figure 30.4 N dependence of the theoretical structure factor, $S(q)$, with $\phi = \phi_0 = 0.07$, $f = f_0 = 0.0457$, and $\chi = 0.70$ (a) and 0.83 (b). The insets show the q dependences of $G(q)$ and $C(q)$

important predictions, i.e., (1) the appearance of the inversion phenomenon is mainly determined by the contribution of the static correlator $C(q)$ and (2) the thermal correlator $G(q)$ always increases with N. This behavior helps understanding of the spatial inhomogeneities in gels. Figure 30.4 also indicates that an inversion in the N dependence of $S(q)$ occurs by changing χ from 0.70 to 0.83. Here, we define χ_{inv}, as the value of χ where $[\partial S(q)/\partial N]_q = 0$.

Figure 30.5 shows theoretical phase diagrams for the inversion at $q = 0.02$ Å^{-1}. The solid curves indicate χ_{inv}, where the inversion phenomenon takes place by varying ϕ at constant f. On the other hand, the dotted curves indicate the critical values of $\chi(\chi_c)$ where phase separation occurs. By imposing $S(q) \to \infty$

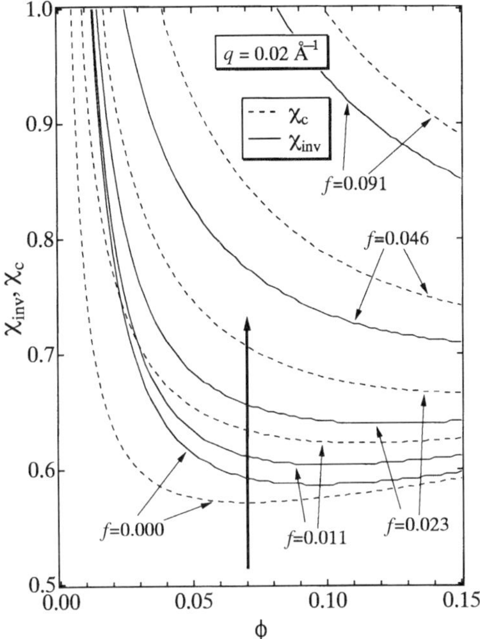

Figure 30.5 Theoretical phase diagrams for the phase separation and inversion calculated using RP theory

and $N \gg 1$, these curves were obtained from the following equation [14].

$$\chi_c = \frac{1}{2}\left(1 + \phi + \frac{f}{\phi}\right) \tag{30.7}$$

In the region below the inversion curves, the spatial inhomogeneity, i.e., $I(q)$, increases with increasing CD (decreasing N), resulting in behavior such as that in Figure 30.4 (a). In the narrow regions between the inversion curves and the phase separation curves, spatial inhomogeneity decreases with increasing CD, corresponding to behavior, such as that in Figure 30.4 (b). Here, the inversion curve (solid curve) is above the phase separation curve (dotted curve) only in the case of $f = 0$. Otherwise, the inversion curve is always below the phase separation curve for $f \neq 0$. Therefore, for $f \neq 0$, a system first encounters inversion and then enters a phase separated region with increasing χ, as shown by the arrow in the figure.

CONCLUSIONS

Spatial inhomogeneities in polymer gels have been investigated experimentally and theoretically. The microscopic inhomogeneities indicate an anomaly, i.e., an inversion in crosslinking density (CD) dependence of light scattering and small-angle neutron scattering intensities. At $\chi < \chi_{inv}$, the inhomogeneities in a gel increase with CD. At $\chi > \chi_{inv}$, on the other hand, the inhomogeneities decreased with increasing CD. The inversion occurs as a result of a competition between two effects of crosslinking; (1) the introduction of static inhomogeneities due to a spatial distribution of crosslinking points and (2) suppression of inhomogenizing moves to a phase segregated structure due to the restricting effect of crosslinks. The inversion phenomenon is reproduced very well using the Rabin–Panyukov (RP) theory.

ACKNOWLEDGMENTS

We acknowledge Professors Y. Rabin and Y. Shiwa for valuable discussions. F.I. is grateful for financial support from the Research Fellowship of the Japan Society for the Promotion of Science for Young Scientists. This work is partially supported by the Ministry of Education, Science Sports and Culture, Japan (grant-in-aid No. 08231245 and 09450362 to M.S.)

REFERENCES

1 M. Annaka and T. Tanaka, *Nature*, **355**, 430 (1992).
2 M. Annaka and T. Tanaka, Phase Transitions, **47**, 143 (1994).
3 H. Hirose and M. Shibayama, Macromolecules, **31**, 5336 (1998).
4 P.G. de Gennes, *Scaling Concepts in Polymer Physics*, Cornell University Press, Ithaca, NY, 1979.
5 A. Onuki, *Adv. Polym. Sci.*, **109**, 63 (1993).
6 J. Bastide and S.J. Candau, in *The Physical Properties of Polymer Gels* (J.P. Cohen Addad ed.) Wiley, New York, 1996, Chapter 9, p. 143.
7 E. Geissler, F. Horkay and A. Hecht, *Phys. Rev. Lett.*, **71**, 645 (1993).
8 S. Mallam, F. Horkay, A.M. Hecht, and E. Geissler, *Macromolecules*, **22**, 3356 (1989).
9 T. Tanaka, L.O. Hocker, and G.B. Benedek, *J. Chem. Phys.*, **59**, 5151 (1973).
10 A. Onuki, *J. Phys. II France*, **2**, 45 (1992).
11 V. Borue and I. Erukhimovich, *Macromolecules*, **21**, 3240 (1988).
12 S. Mallam, F. Horkay, A.M. Hecht, A.R. Rennie, and E. Geissler, *Macromolecules*, **24**, 543 (1991).
13 F. Ikkai and M. Shibayama, *Phys. Rev. E*, **56**, R51 (1997).
14 M. Shibayama, F. Ikkai, Y. Shiwa, and Y. Rabin, *J. Chem. Phys.*, **107**, 5227 (1997).
15 F. Ikkai, M. Shibayama and C.C. Han, *Macromolecules*, **31**, 3275 (1998).
16 Y. Rabin and S. Panyukov, *Macromolecules*, **30**, 301 (1997).
17 M. Shibayama, T. Takeuchi, and S. Nomura, *Macromolecules*, **27**, 5350 (1994).
18 J. Bastide and L. Leibler, *Macromolecules*, **21**, 2647 (1988).
19 J. Bastide, F. Boue and M. Buzier, in *Molecular Basis of Polymer Networks* (A. Baumgartner and C.E. Picot, eds) Springer-Verlag, Berlin, 1989, p. 48.
20 K. Kubota, S. Fujishige, and I. Ando, *Polymer J.*, **22**, 15 (1990).
21 M. Shibayama, T. Tanaka, and C.C. Han, *J. Chem. Phys.*, **97**, 6842 (1992).

PART 5

Biomedical Applications of Polymer Networks

31

Transtissue Photopolymerization of Poly(Vinyl Alcohol) Hydrogels

STEPHANIE BRYANT[1], PENNY MARTENS[1],
JENNIFER ELISSEEFF[2], MARK RANDOLPH[3],
ROBERT LANGER[2], and KRISTI S. ANSETH[1],[*]

[1]Department of Chemical Engineering, University of Colorado, Boulder, CO 80309 USA

[2]Harvard-MIT Division of Health Sciences and Technology and MIT Department of Chemical Engineering, Cambridge, MA 02139 USA

[3]Department of Surgery, Massachusetts General Hospital, Boston, MA 02114 USA

ABSTRACT

This work demonstrates the feasibility of a new noninvasive technique to place polymer implants subcutaneously. The technique uses light transmitted through the skin to

---

[*] To whom correspondence should be addressed

Wiley Polymer Networks Group Review Series Vol. 2. Edited by B.T. Stokke and A. Elgsaeter
© 1999 John Wiley & Sons Ltd

photopolymerize liquid macromer solutions that have been injected subcutaneously. In this study, photoinitiation conditions were identified for an ultraviolet initiating system that would allow photocrosslinking of an acrylated poly(vinyl alcohol) solution through skin while minimizing potential cell damage.

INTRODUCTION

Numerous medical applications could benefit from the ability to fabricate polymers *in vivo*. For example, as an alternative to mercury amalgam fillings, ceramic filled dimethacrylate monomers have been developed that can be photopolymerized with blue light to produce a tooth colored restoration *in situ* [1]. Bone cements, based on acrylate monomers (e.g., methyl methacrylate), are redox or thermally cured *in vivo* to form a rigid polymer to secure metallic orthopedic implants [2]. Hubbell *et al.* [3] presented some of the first work using *degradable* polymers that were photocured *in vivo* to prevent post-operative adhesions. The gels were formed from macromers of block copolymers of poly(ethylene glycol) and lactic acid (or glycolic acid) that were end-capped with acrylate groups [4]. The authors were very successful in developing methods to *in vivo* cure these polymers and to produce adhesion preventing barriers that degraded in a few days post-surgery.

The advantages of *in vivo* polymerization are many. First, because the initial materials are liquid solutions or moldable putties, the systems are easily placed in complex shapes (e.g., tooth caries) and subsequently reacted to form a polymer of the exact required dimensions. No additional shaping or modification of the implant is required. Second, the adhesion of the polymer to surrounding tissue is generally significantly improved because of intimate contact of the polymer with the tissue during formation and the mechanical interlocking that can result from surface microroughness. Third, the invasiveness of some surgical techniques can be minimized as liquid solutions are easily introduced through needle injections and can be photocured with fiber optic cables using arthroscopic techniques.

In addition to these advantages, however, *in vivo* polymerization also introduces many new challenges. Polymerization conditions for *in vivo* applications are quite adverse, including a narrow range of physiologically acceptable temperatures, requirement for nontoxic monomers and/or solvents, moist and oxygen-rich environments, and clinically suitable rates of polymerization. However, photopolymerization can overcome many of these limitations since the initiation does not require elevated temperatures as in thermal polymerization and the polymerization process is typically rapid (and can be controlled to occur over a time period of a few seconds to a couple of minutes) which allows the system to overcome oxygen inhibition and moisture effects. Thus, a main challenge to broadening the application of *in vivo* polymerization is the development of nontoxic monomers that are photopolymerizable and react to form polymers with desirable material properties for their proposed function.

Thus, the objective of this work was first to develop a new, photocrosslinkable macromer with properties that could be easily varied to suit the chosen application. In addition to synthesizing new materials, we also investigated a novel technique, called transtissue or transdermal polymerization, to polymerize these materials *in vivo* [5]. The technique involves injecting a liquid macromer solution through a tissue (e.g., skin) and subsequently exposing the exterior surface of the tissue to light to induce photocrosslinking. Compared to existing *in vivo* polymerization methods, transdermal polymerization allows one to place a polymer device subcutaneously without the need for any incision or surgical intervention. The potential applications for this process are numerous. For example, chemotherapy agents could be loaded in the polymer and locally delivered to skin tumors, or chondrocytes could be encapsulated in the polymer matrix and used to tissue engineer cartilage for the broad field of reconstructive plastic surgery.

EXPERIMENTAL

MATERIALS

Photocrosslinkable poly(vinyl alcohol) was prepared from functionalization of linear poly(vinyl alcohol) (PVA, Polysciences) with acrylate groups (see Figure 31.1). In this study, PVA ($M_n = 25\,000$) was dissolved in distilled deionized water at 80 °C and a solution concentration of 10 wt%. The solution was cooled to room temperature and the pendant hydroxy groups were converted to acrylates by addition of glycidyl acrylate (Aldrich), along with HCl to produce an acidic environment. The resulting solution was reacted for at least 12 h at room temperature. The final polymer solution was purified by precipitation in acetone, followed by filtration and drying. For the stoichiometric addition of glycidyl

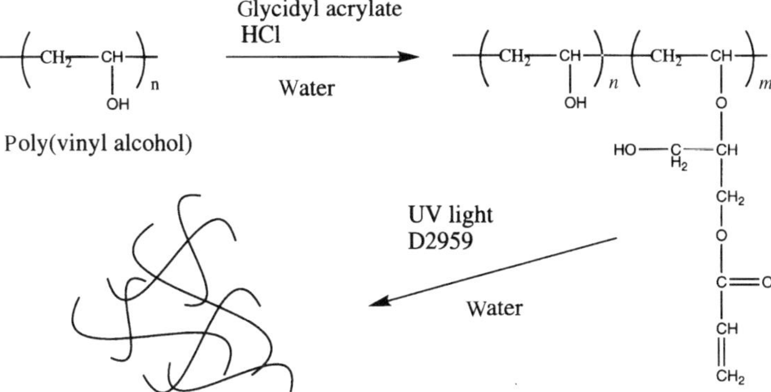

Figure 31.1 General synthesis of crosslinked hydrogels from acrylated poly(vinyl alcohol)

acrylate to produce 10% acrylation of the pendant PVA hydroxy groups, 4% acrylation resulted as characterized by ^{1}H NMR.

Crosslinked gels were prepared from the acrylated PVA by dissolving the polymer and a photoinitiator (Darocur 2959, D2959, Ciba-Geigy) in water at varying concentrations. The solutions were subsequently exposed to ultraviolet light (365 nm) at intensities ranging from 0.1 to 20 mW/cm^2.

METHODS

A differential scanning calorimeter equipped with a photoaccessory (Perkin Elmer, DSC/DPA-7) was used to characterize the general polymerization behavior of the macromers in solution. The DSC monitors heat flux, which is directly proportional to the rate of polymerization, while the integrated peak is related to the conversion of double bonds in the system.

Cytocompatibility studies of the photoinitiation process were performed using a fibroblast cell line NIH/3T3 (ATCC CRL-1658, Rockville, MD). Cells were plated on 12 well plates with 2 ml media/well and a seeding density of 2600 cells/cm^2. Dulbecco's Modified Eagle Medium supplemented with 1.5 g/l sodium bicarbonate, 1.0 mmol/l sodium pyruvate and 10% calf serum was used to culture the cells. The water soluble photoinitiator, Darocur 2959, was dissolved in media at varying concentrations. Studies were conducted on fibroblasts that were seeded and allowed to adhere and proliferate for at least two days. The initiator solution was added 2 ml/well and incubated for an additional two days. Cell viability and metabolism were determined using the MTT Assay (Sigma–Aldrich) by reading the absorbance at 560 nm using a UV-VIS spectrometer. The cell viability was recorded in terms of a relative survival in which the absorbance of a control containing cells and media only was used to normalized the absorbance of cells treated with initiator. The effects of ultraviolet light, initiator concentration, and photoinitiated radicals on the cell viability were examined.

RESULTS AND DISCUSSION

In evaluating the efficacy of transtissue or transdermal photopolymerizations as a noninvasive technique to implant polymers, several critical issues were examined and are reported here. First, a material needed to be identified or synthesized that could be photopolymerized *in vivo* (i.e. in direct contact with tissues using ultraviolet or visible light). To this extent, polymeric hydrogels were synthesized from a poly(vinyl alcohol) backbone (see Figure 31.1). PVA is advantageous because of its water solubility, biocompatibility, and ease with which it is cleared by the body [6,7]. Acrylate groups were chosen as the reactive functionality because they undergo rapid photopolymerization, and they have a long history of use in dentistry, orthopedics, and adhesion prevention barriers [1–3]. Since the macromers have several reactive groups per backbone chain (and the density of reactive groups is easily changed), the macromers crosslink when reacted to form a hydrogel. This simple macromer system was chosen as a model to examine the feasibility of transdermal photopolymerization, but the material can be easily modified to span a wide variety of properties. For example, degradable gels can be synthesized in a manner similar to Hubbell *et al.* [4] by grafting of poly(α-hydroxy acids) (e.g., poly(glycolic acid) (PGA) or poly(lactic acid) (PLA)) followed by end-capping the grafts with acrylates. Both PGA and PLA have a long-standing history in medical applications (e.g., sutures); their *in vivo* and *in vitro* degradation is well understood; and their degradation products are readily eliminated by the body.

In addition to rationally designing a photopolymerizable macromer, preliminary studies were also conducted to examine the cytotoxicity of the photoinitiation process. In particular, the effects of a water soluble, ultraviolet photoinitiator, D2959, on a fibroblast cell-line were investigated to identify reaction conditions (e.g., initiating wavelength and initiator concentration) with minimal adverse effects to cells. The purpose of this study was to examine initially the potential for the process to encapsulate cells since this ability is important to certain potential applications of transdermal photopolymerization (e.g., tissue engineering of cartilage). Figure 31.2 illustrates the effects of D2959 concentration on the cell survival relative to a control with no initiator. At D2959 concentrations less than 0.05 wt%, the cell viability and metabolism is similar to the control; however,

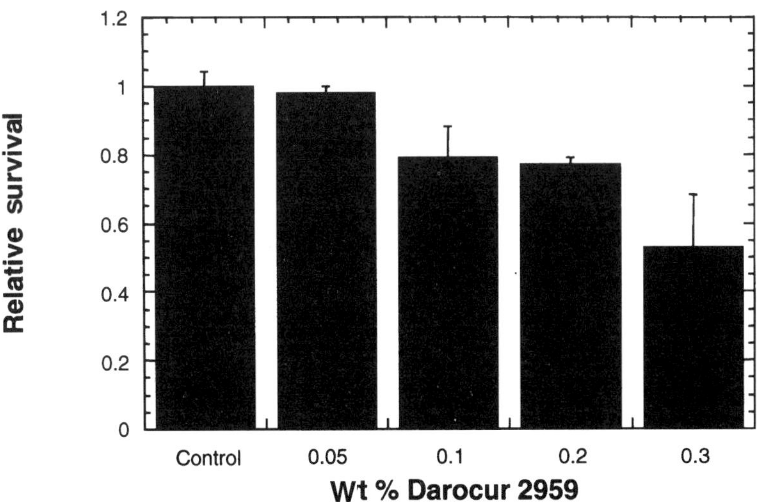

Figure 31.2 The effect of initiator concentration (Darocur 2959) on the cell viability and metabolism

the cell survival decreases to about 80% of the control at higher initiator concentrations of 0.1 and 0.2 wt%.

In addition to examining the cytocompatibility of the photoinitiator itself, the fibroblasts were also treated with 0.05 wt% initiator solution and exposed to ultraviolet light for different lengths of time. The effective intensity on the cells was 6 mW/cm^2. As measured by the MTT assay, this chosen light intensity was shown to have no significant effect on the cells when no initiator was present. Furthermore, since no reactive macromer is present in the media, this study should represent the highest dose of radicals, and potential radical damage, that the cells would experience during a photopolymerization, encapsulation process. After exposure, the fibroblasts were incubated for two days and cell viability was then determined. At exposure times of 1 min and 5 min, which are ranges of acceptable clinical curing times, the relative survival was 1.04±0.02 and 0.96±0.04, respectively. Thus, the metabolism and viability of cells treated with 0.05 wt% initiator and exposed to UV light for less than 5 min remain comparable to the control.

In addition to identifying ranges of photoinitiator concentrations and light intensities that can be used during the photopolymerization process when encapsulating cells, the attenuation of light through skin must also be characterized. In particular, is the intensity of light transmitted through the tissue high enough to initiate a photopolymerization? In Figure 31.3, the percentage of transmitted

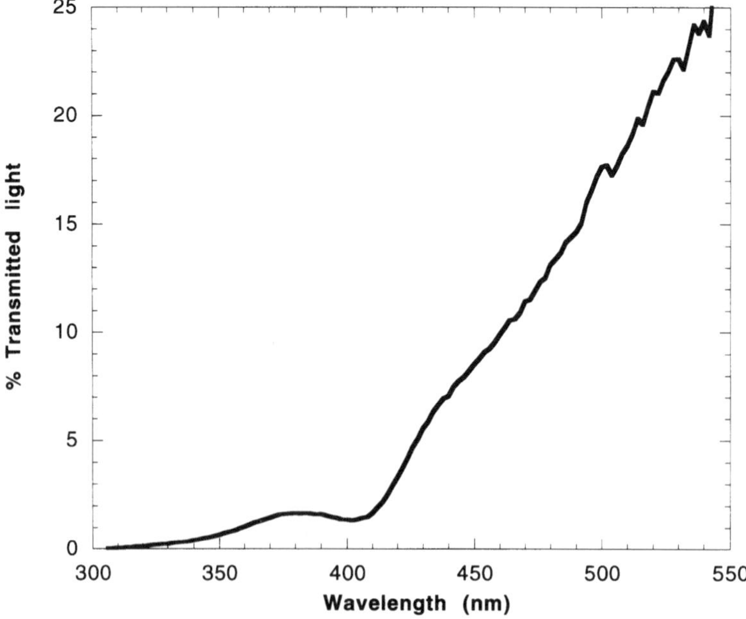

Figure 31.3 The percentage of light transmitted as a function of wavelength in human skin (thickness ~1.1 mm)

light is plotted as a function of wavelength of the light for human skin. The data was collected using an integrating sphere combined with a monochromater. The thickness of the skin sample was 1.1 mm. While the thickness of skin on the human body can vary widely depending on its location (e.g., several millimeters on the soles of feet to <1 mm on the tips of noses), 1.2 mm is an average thickness. As illustrated in Figure 31.3, as the wavelength of the light is increased, the % transmitted light increases dramatically. For our initial study, we examined 365 nm ultraviolet light, which is widely used in the photopolymerization industry. At 365 nm, approximately 1.3% of the light exposed to the exterior surface of the skin is transmitted through the skin to the subcutaneous space. Thus, if 20 mW/cm^2 is used as an upper limit intensity of ultraviolet light for the exterior surface of skin (i.e., the intensity of commercial tanning beds) [8], then a macromer injected underneath the skin could be photopolymerized with ∼0.26 mW/cm^2 of light. Current work is under way investigating visible light initiating systems, such as the blue light initiating system used in dentistry (470–90 nm) [1]. As shown in the tissue transmittance data, blue light is less attenuated by skin and the exterior surface of skin can be exposed to much higher intensities (e.g., hundreds of mW/cm^2).

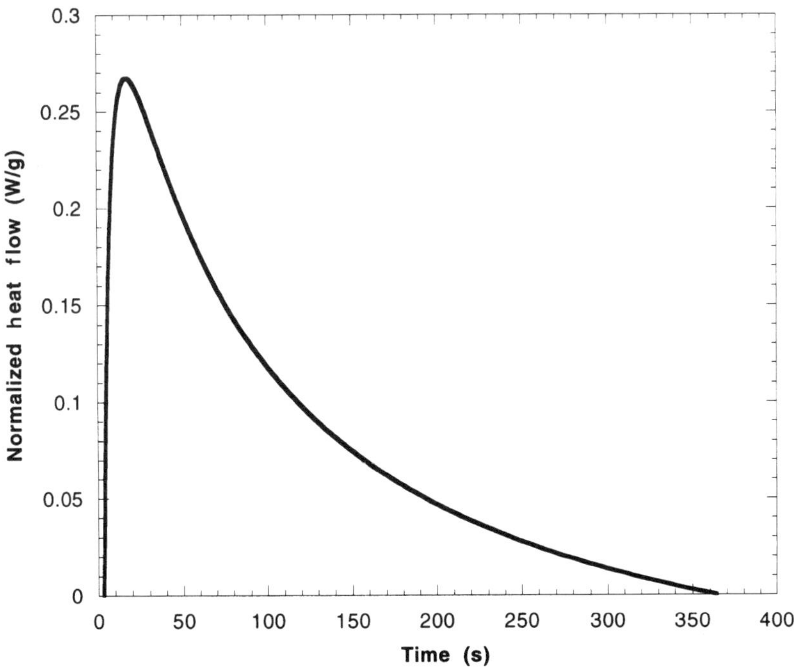

Figure 31.4 The normalized heat flow as a function of time for the photopolymerization of a 20 wt% solution of acrylated PVA in water at 37 °C with 0.3 mW/cm^2 of ultraviolet light and 0.05 wt% D2959

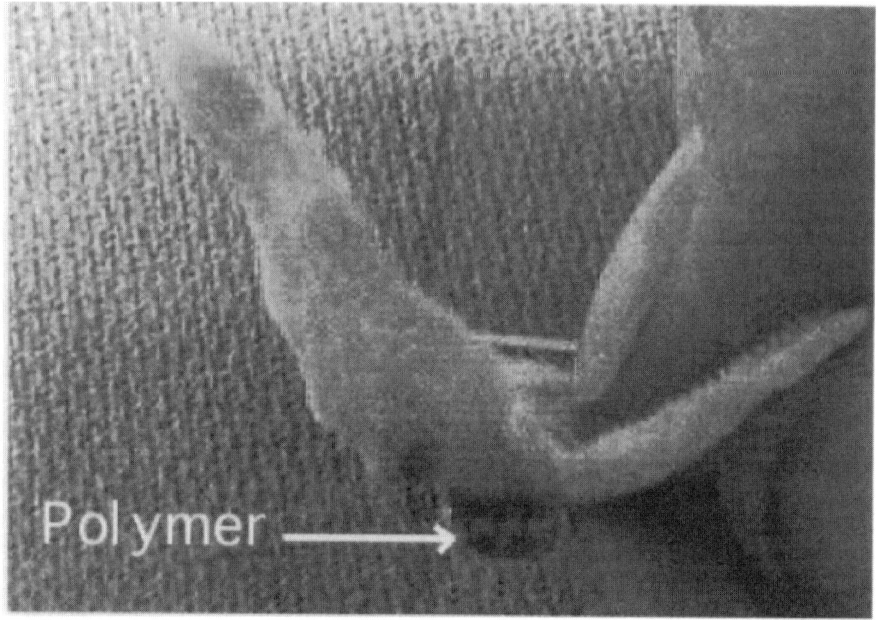

Figure 31.5 Picture of a PVA gel adhered to rat skin (thickness ~1 mm) through which it was photopolymerized

Differential scanning photocalorimetry was used to examine the feasibility of photocrosslinking the macromer transdermally on clinically suitable time scales at these low initiator concentrations and light intensities. Figure 31.4 shows the polymerization behavior of a 20 wt% acrylated PVA solution with 0.05 wt% D2959 when exposed to 0.3 mW/cm^2 of 365 nm ultraviolet light. The polymerization was monitored at 37 °C. In general, the polymerization behavior was typical of crosslinking polymerization [9] and the polymerization was essentially complete in 5–6 minutes. As a further demonstration of feasibility, Figure 31.5 shows a picture of a gel (containing a dye for visualization purposes) that was photopolymerized through rat skin (~1 mm thick). The macromer solution was injected underneath the skin using a needle and the system was photopolymerized by exposing the surface of the skin to ~4 mW/cm^2 of ultraviolet light for 10 min.

ACKNOWLEDGMENTS

The authors would like to thank the Dreyfus Foundation and the Packard Foundation for their support of this work through grants.

REFERENCES

1. K.S. Anseth, S.M. Newman and C.N. Bowman, *Adv. Polym. Sci.*, **122**, 177 (1995).
2. D.H. Kohn and P. Ducheyne, in *Materials Science and Technology*, Vol. 14, *Medical and Dental Materials* (R. Cahn, P. Haasen and E. Kramer, eds), p. 29, (1992).
3. J. Hill-West, S. Chowdhury, A. Sawhney, C. Pathak, R. Dunn and J. Hubbell, *Obstet. Gynecol.*, **83**, 59 (1994).
4. A.S. Sawhney, C.P. Pathak and J.A. Hubbell, *Macromolecules*, **26**, 581 (1993).
5. J. Elisseeff, K. Anseth, D. Sims, M. Mclntosh, W. Randolph and R. Langer, *Proc. Natl. Acad. Sci. USA*, **96**, 3104 (1999).
6. N.A. Peppas, *Hydrogels in Medicine and Pharmacy*, Vol. III, *Properties and Applications*, CRC Press, Boca Raton, 1987.
7. Y. Murakami, Y. Tabata and Y. Ikada, *Drug Delivery*, **3**, 231 (1996).
8. Svaldi Muggli, D.C.; M.S. Dissertation, *Chemical Engineering Department*, University of Colorado, 1997.
9. Kloosterboer, J.G. *Adv. Polym. Sci.*, **84**, 1 (1988).

32

Rheological Properties and Microstructure of Tissue Engineered Cartilage

M. STADING[1] and R. LANGER[2]

[1]Chalmers University of Technology/SIK, The Swedish Institute for Food and Biotechnology, P O Box 5401, S-402 29 Göteborg, Sweden

[2]Massachusetts Institute of Technology, Department of Chemical Engineering E25-342, Cambridge, MA 02139, USA

ABSTRACT

Tissue engineered cartilage can be used to treat injuries on the articular cartilage which covers the end surfaces of bones within diarthrodial joints and functions as a bearing material providing low friction and high resistance to wear.

Chondrocytes have earlier been shown to grow well on scaffolds of polyglycolic acid (PGA). They secrete the extracellular matrix (ECM) that fills the empty spaces in the polymer scaffold at the same time as the polymer degrades. Cartilaginous constructs were created by using bovine chondrocytes on synthetic, biodegradable scaffolds made of

Wiley Polymer Networks Group Review Series Vol. 2. Edited by B.T. Stokke and A. Elgsaeter
© 1999 John Wiley & Sons Ltd

fibrous polyglycolic acid. The constructs have previously been shown to resemble natural articular cartilage biochemically and histologically. The mechanical properties of articular cartilage mainly depend on the swollen ECM, which is a gel consisting of collagen fibres and proteoglycans in a fluid phase of water and electrolytes. The mechanical properties of the constructs and the build-up of the ECM were studied using dynamic, non-destructive measurements in shear. The shear modulus G^* during *in vitro* cultivation correlated well with both the collagen and glycose–amino–glycan content of the constructs but did not reach the same level as in natural cartilage. The difference in G^* between the constructs and nature cartilage was shown to depend on both the biochemical composition and the microstructure of the constructs.

INTRODUCTION

Despite current sophisticated medical technology, tissue failure remains a serious clinical problem. Although transplantation techniques are available, the shortage of donor tissue limits the number of transplantation performed. One possible solution is therefore tissue engineering which has been defined as 'the use of living cells, together with extracellular components, either natural or synthetic, in the development of implantable parts or devices for the restoration of function' [1]. It has been used to produce tissues *in vitro* such as liver, intestine, urothelial tissue, bone, tendon and cartilage [2]. The present paper focuses only on cartilage and, specifically, on articular cartilage.

Cartilage is connective tissue found in, e.g., the nose and the ears. Articular cartilage also covers the surface of bones in joints, where it functions as an almost frictionless bearing surface. Under healthy conditions, articular cartilage is almost completely wear-resistant throughout our lives. An injury to the articular cartilage by, e.g., trauma or arthritis is a serious condition since the capacity for self-repair is limited due to the avascular nature of the tissue [3]. The current treatment is therefore either transplantation or the use of prostheses. Transplants are limited by a shortage of accessible donor tissue; prostheses have limited lifetimes owing to wear, and there is also an increased risk of infections. Tissue engineering of cartilage is consequently an attractive alternative.

Our method of making tisue engineered cartilage [4] is shown in Figure 32.1. The cartilage producing cells, chondrocytes, are first isolated from healthy articular cartilage. The chondrocytes are then seeded on a three-dimensional polymer scaffold which has the shape of the desired missing cartilage. This cell–polymer construct is cultivated *in vitro* in tissue culture medium, where the chondrocytes produce the extracellular matrix (ECM) typical of articular cartilage. Meanwhile, the polymer scaffold degrades, leaving a piece of cartilage which is genetically identical to the donor tissue that can be re-transplanted to heal the donor injury.

A related method has been developed and tested clinically [5,6]. Healthy chondrocytes were obtained from an uninvolved area of an injured knee by arthroscopy. The chondrocytes were cultured in the laboratory for 2–3 weeks without a scaffold and were then injected back into the area of the defect. The

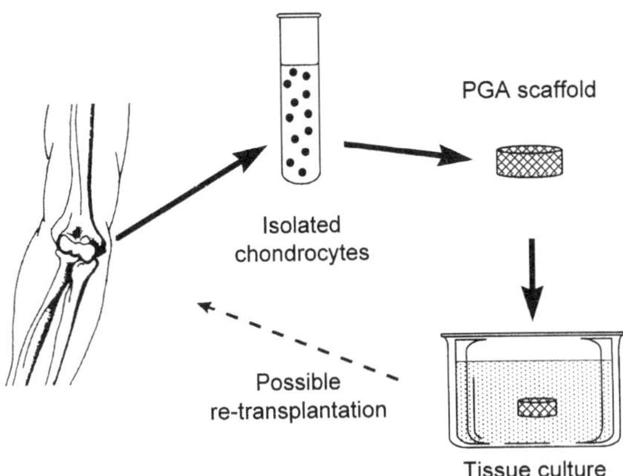

Figure 32.1   Tissue engineering of articular cartilage

defect was covered with a sutured periosteal flap from the tibia bone. A majority of the patients developed hyaline cartilage and showed good-to-excellent results after two years.

The scaffold is made of fibrous polyglycolic acid (PGA), which is approved for clinical use by e.g. the US Food and Drug Administration (FDA). PGA is used clinically in resorbable, surgical sutures. The scaffold used in the present study was formed from 12 μm thick PGA fibers into a lightly porous, nonwoven mesh using textile processing techniques [7]. The PGA mesh is a highly porous (97%) network with sufficient mechanical integrity to maintain its dimensions during cell seeding and *in vitro* cultivation. The use of a scaffold is important for several reasons: (1) the cells need a support to which they can adhere in order to proliferate and develop the correct extracellular matrix (ECM), (2) the shape of the scaffold determines the shape of the finished cartilage piece and (3) it gives mechanical stability to the construct before the ECM is developed. The PGA fibers degrade hydrolytically in tissue culture, ultimately to CO_2 and water, such that about 50% of the initial polymer mass remains after four weeks in culture and only 30% after eight weeks in culture [7].

The chondrocyte cells are seeded on the polymer scaffolds and immersed in tissue culture medium at well-mixed conditions in a variety of bioreactors. Mixed flasks were used in the present study. In one day, essentially all cells attach to the PGA fibers [8]. The cells proliferate along the fibers and excrete the ECM, as shown in Figure 32.2. The ECM consists mainly of collagen fibrils, proteoglycans and other glycoproteins which form a three-dimensional network. The collagen fibrils form a strong and stiff network which provides the main contribution to the shear and tensile modulus [9]. This network coexists with a weak proteoglycan network. The main feature of the charged proteoglycans is that they cause an osmotic pressure which contributes up to 50% to the total compressive modulus

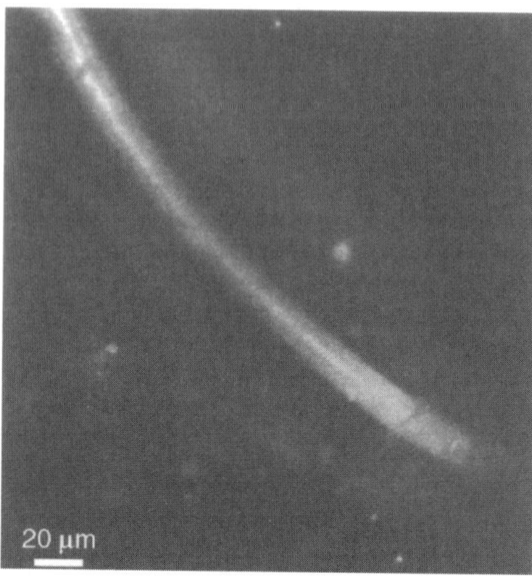

Figure 32.2   A confocal laser scanning micrograph of a cell-polymer construct cultivated for two weeks. The PGA fiber is covered with chondrocytes that have excreted collagen (white in the image). The bright spots in the collagen are chondrocytes that have migrated into the extracellular matrix

of cartilage [9,10]. The strength of the cell–polymer construct depends on the properties of the ECM, especially when the PGA fibers degrade. The strength of the construct is also crucial for the final clinical use, since the construct must be able to stand the stresses present in the joint where they are re-transplanted.

MECHANICAL PROPERTIES

Articular cartilage must provide low friction under high loads in the joints, and the mechanical properties are therefore crucial. The mechanical strength can be monitored both in compression and in shear, and both modes of loading are found under physiological conditions. The physiological loading is predominantly compressive, and the dominant mechanism causing the viscoelastic response comes from frictional drag caused by interstitial fluid flow [9]. The most common method for measuring the mechanical properties of cartilage is confined compression in combination with the biphasic theory [11,12]. In confined compression, a cylindrical sample is enclosed in a rigid confining ring, preventing radial expansion. The compressive load is applied via a porous filter, which allows fluid exudation from the tissue, see Figure 32.3(a). The purely axial loading simplifies the mathematical analysis by the biphasic theory. The theory considers cartilage as two immiscible phases: a solid network phase and an interstitial fluid phase. It is used to calculate a compressive and an aggregate modulus, as well as an apparent permeability.

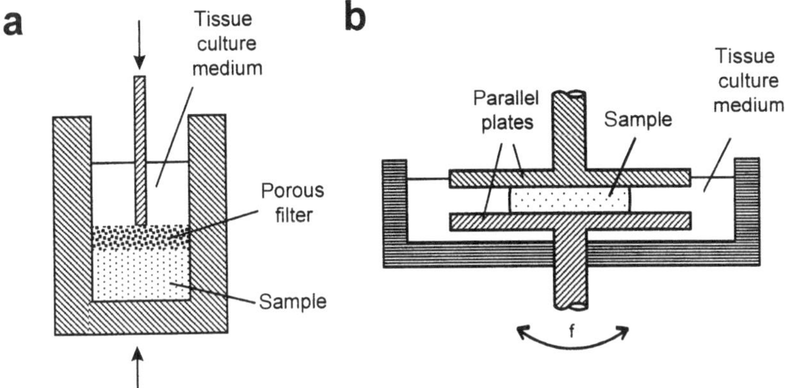

Figure 32.3 Measuring methods for viscoelastic properties of soft tissues: (a) confined compression and (b) shear between parallel plates

To be able to monitor the evolution of the ECM network, it is suitable to use shear since shear properties depend mainly on the strength of the network and invole only minimal fluid flow [13]. When studying the properties at very small strain, it is difficult to use confined compression on cultivated constructs because they are not perfectly cylindrical and therefore require large deformations to fill the confining ring. The imperfect geometry of the constructs also requires special precautions regarding the shear measurements. The samples in this study were in the shape of disks, 10 mm in diameter and 2 mm thick. The constructs had slightly convex surfaces, and the area in contact with the plates therefore had to be monitored. The compression of the construct was adjusted until most of the sample was in contact with the plate, and the sample was then compressed an additional 5% axially to ensure full contact. The modulus scales as the fourth power of the sample diameter, i.e. the contact diameter, which illustrates that the experimental technique is crucial for the measured magnitude of the modulus [13].

The measurements were performed by applying a sinusoidal, angular deflection of frequency, f, to the sample by the lower plate and measuring the resulting deflection of the upper plate, as shown in Figure 32.3(b). The measured signals were processed, and the complex shear modulus G^* was calculated using the deflections of the two plates, the known measuring geometry and instrument constants. G^* was divided into the storage modulus, G' and the loss modulus, G'' which are related through

$$G^* = G' + jG'' \qquad (32.1)$$

where $j^2 = -1$

The phase angle, δ, is

$$\tan \delta = G''/G' \qquad (32.2)$$

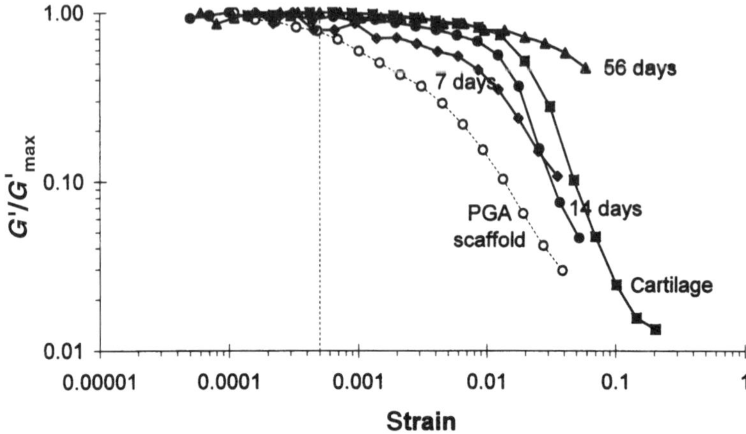

Figure 32.4 The storage modulus as a function of strain for natural cartilage (■), the PGA scaffold without cells (O) and constructs cultured 7 days (♦), 14 days (●) and 56 days (▲). The dashed line shows the strain used for the rest of the measurements

STRAIN DEPENDENCE

The possible occurrence of slip and the linear region of the samples were tested by performing a strain sweep at a constant frequency of 1 Hz on the samples, as shown in Figure 32.4. G' was normalized to the maximum G' for each sample because of the large difference in absolute G' between natural cartilage and the constructs. All samples first show a fairly constant G' below a strain of approximately 10^{-3}. There is then a slowly decreasing G' between a strain of 10^{-3} and 10^{-2}, after which a rapid drop in G' occurs at strains $>10^{-2}$.

The decrease in G' with increasing strain, shown in Figure 32.4, can be the results of either a change in the material, reversible or irreversible (fracture), or it may be slip at the sample-plate interface. It is probable that the first slow decrease is a reversible change and that the more rapid drop at higher strain is caused by slip. To ensure measurements within the linear region, a shear strain of 5×10^{-4} was chosen for the rest of the measurements.

EFFECT OF AXIAL COMPRESSION

Figure 32.5 shows the mechanical spectra for articular cartilage from the femoropatellar grooves of 2–3-week-old bovine calves. The magnitude of G' and the characteristics of the spectra change when the axial deformation increases. The first increase in modulus on compression was also reported in earlier work [14–16]. At axial deformations $>10\%$, the magnitude of G' decreases, and there is also a change to a more frequency-dependent behavior at frequencies lower than 0.1 Hz. Similar mechanical spectra were observed for inhomogeneous

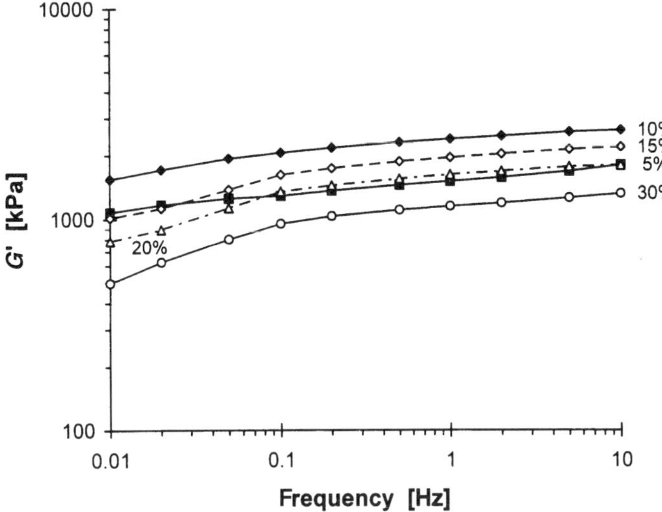

Figure 32.5 Mechanical spectra for natural cartilage at increasing axial deformation. The legends show the axial compression of the disk-shaped samples

protein gels [17,18]. This was explained by different relaxation times in areas with different densities. The high axial deformations of the cartilage may also induce changes in the microstructure, which in turn, affect the modulus. However, the axial deformation affects the shear modulus, and care must be taken in designing the protocol for the shear measurements. In this study, an axial deformation of 5% was chosen.

DEVELOPMENT OF MECHANICAL PROPERTIES DURING TISSUE CULTURE

The development of the extracellular matrix in the constructs was monitored by G' and δ. At time $t = 0$, the chondrocytes were seeded onto the PGA scaffold. The PGA fibers were then the only structure that could take up any mechanical stress. Over a period of two weeks, the PGA fibers degraded enough to lose their mechanical integrity, as shown in Figure 32.6. The PGA fibers degrade in several steps, and they lose their tensile strength as a result of e.g. cleavage of ester bonds, before losing mass [19]. It takes about four weeks in culture to degrade 50% of the initial PGA polymer mass [7]. After one week in tissue culture, the chondrocytes produced enough ECM to give a G' greater than in the PGA scaffold without chondrocytes. The G' of the constructs then increases almost linearly over time. The phase angle is fairly constant, about $\delta = 12°$ for the constructs during the whole time in culture.

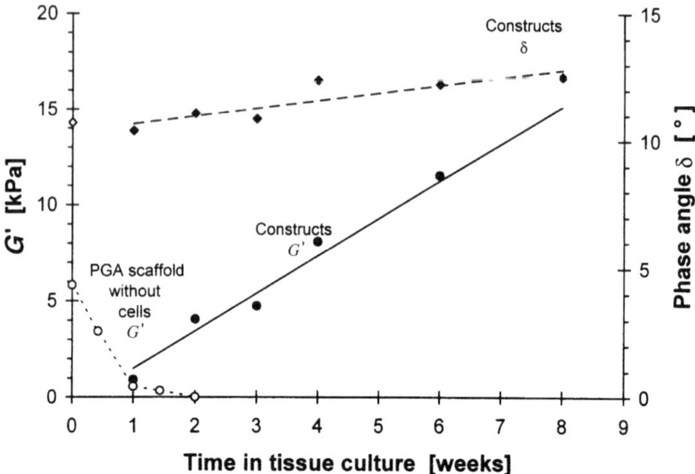

Figure 32.6 Storage modulus and phase angle at 1 Hz for cell-polymer constructs during tissue culture. G' in PGA scaffold without cells (----), G' in constructs (——), δ in constructs (---)

The final value of G' after eight weeks in culture under the reactor and media conditions given in this paper is 15 kPa. This is almost 100 times smaller than in natural cartilage (see Figure 32.5). The difference can be explained primarily by two factors: composition and microstructure. The concentration of both collagen and proteoglycans directly influences the modulus. The collagen fibrils are the component that mainly provides shear stiffness to the tissue [9]. While the proteoglycan also forms a network, this is much weaker than the collagen network. The proteoglycans are also important for the shear stiffness, however, because they cause an osmotic pressure that keeps the collagen network in an inflated and stretched state. The concentrations of collagen and glycosaminoglycans (GAG) were measured in the constructs as a function of time in tissue culture, which is shown in Figure 32.7. They increase with time up to final concentrations of 3.2% (wet wt) for collagen and 1.1% (wet wt) for GAG, which is about one-fifth of the concentrations found in natural cartilage [9].

The dependence of G' on concentrations of collagen and GAG can be extracted from the results shown in Figures 32.6 and 32.7. The storage modulus can be assumed to scale as a power law of the concentration for concentrations significantly higher than the critical gel concentration [20],

$$G' \propto c^a \tag{32.3}$$

By fitting the experimental results to equation (32.3), the exponent a was found to be $a = 2.1$ for collagen and $a = 1.6$ for GAG.

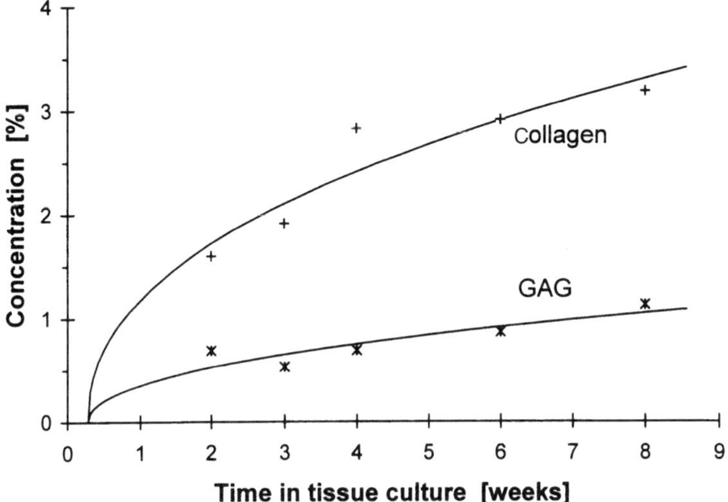

Figure 32.7 Concentration of collagen and glycoseaminoglycans (GAG) for cell–polymer constructs during tissue culture

The difference in modulus between natural cartilage and the constructs can partly be explained by the lower polymer concentrations, but a five-times lower concentration accounts only for a 25-times difference in G' assuming a quadratic modulus dependence. This means that the microstructure also has a major influence on the modulus. Histological sections and biochemical measurements of the constructs have shown that the structure is similar to natural cartilage, with an outer collagenous capsule and a high fraction of collagen type II in the interior [7,21]. Nutrients are only transported from the outside of the constructs, and the mass transfer becomes slower as the constructs become denser. This leads to limitations in thickness for the cultivation of constructs and probably also to weaker constructs. The lower modulus of the constructs is probably not a serious limitation to a number of clinical applications, since further chondrogenesis following *in vivo* implantation has been demonstrated [4]. The *in vivo* conditions are more favourable than those in tissue culture because of e.g. a correct balance of nutrients and growth factors. Much work is currently being done to improve the *in vitro* cultivation, e.g., by using bioreactors, growth factors, microgravity, and this may further improve the *in vitro* modulus [7,8,21,22].

ACKNOWLEDGMENTS

Gordana Vunjak-Novakovic and Lisa Freed are gratefully acknowledged for their help and support with tissue culture and analytical biochemistry. The authors also thank Gordana Vunjak-Novakovic for her review of the manuscript and Ms Elvy Jordansson for the

CLSM microscopy. Financial support to M.S. from The Swedish Council for Forestry and Agricultural Research and from The Swedish Institute is gratefully acknowledged.

REFERENCES

1. R.M. Neren, *Ann. Biomed. Eng.*, **19**, 529 (1991).
2. R. Langer, J.P. Vacanti, C.A. Vacanti, A. Atala, L.E. Freed and G. Vunjak-Novakovic, *Tissue Engineering*, **1**, 151 (1995).
3. L.E. Freed and G. Vunjak-Novakovic, in *The Biomedical Engineering Handbook*, (J.D. Bronzino, ed.), CRC Press, Boca Raton, 1995, p. 1.
4. L.E. Freed, J.C. Marquis, A. Nohria, J. Emmanual, A.G. Mikos and R. Langer, *J. Biomed. Mater. Res.*, **27**, 11 (1993).
5. M. Brittberg, A. Lindahl, A. Nilsson, C. Ohlsson, O. Isaksson and L. Peterson, *N. Engl. J. Med.*, **331**, 889 (1994).
6. M. Brittberg, A. Nilsson, A. Lindahl, C. Ohlsson and L. Peterson, *N. Engl. J. Med*, **326**, 270 (1996).
7. L.E. Freed, G. Vunjak-Novakovic, R.J. Biron, D.B. Eagles, D.C. Lesnoy, S.K. Barlow and R. Langer, *Bio/Technology*, **12**, 689 (1994).
8. G. Vunjak-Novakovic, B. Obradovic, I. Martin, P.M. Bursac, R. Langer and L. Freed, *Biotech. Progr.*, **14**, 193 (1998).
9. V.C. Mow, A. Ratcliffe and A.R. Poole, *Biomaterials*, **13**, 67 (1992).
10. W.M. Lai, J.S. Hou and V.C. Mow, in *Biomechanics of diarthrodial joints* (V.C. Mow, A. Ratcliffe and S.L.Y. Woo, eds), Springer, New York, 1990, p. 283.
11. E. Myers, W. Zhu and V.C. Mow, in *Collagen, Biochemistry and Biomechanics*, (M.E. Nimni, ed.), CRC Press, Boca Raton, 1988, p. 267.
12. V.C. Mow, J.S. Hou, J.M. Owens and A. Ratcliffe, in *Biomechanics of Diarthrodial Joints* (V.C. Mow, A. Ratcliffe and S.L.Y. Woo, eds), Springer, New York, 1990, p. 215.
13. M. Stading and R. Langer, *Tissue Engineering*, **5**, 241 (1999).
14. W.C. Hayes and A.J. Bodine, *J Biomech.*, **11**, 407 (1978).
15. V. Roth, J.M. Schoonbeck and V.C. Mow, *Transactions of the Annual Meeting of the Orthopaedic Research Society*, 150 (1982).
16. W. Zhu, V.C. Mow, T.J. Koob and D.R. Eyre, *J. Orthop. Res.*, **11**, 771 (1993).
17. M. Stading, M. Langton and A.-M. Hermansson, *Food Hydrocolloids*, **6**, 455 (1992).
18. M. Stading, M. Langton and A.-M. Hermansson, *Food Hydrocolloids*, **7**, 195 (1993).
19. R.L. Kronentahl, in *Polymers in medicine and surgery* (R.L. Kronentahl, ed.), Plenum Press, New York, 1975, p. 118.
20. A.H. Clark, in *Food Polymers, Gels and Colloids* (E. Dickinson, ed.), The Royal Society of Chemistry, Cambridge, 1991, p. 322.
21. L.E. Freed, A.P. Hollander, I. Martin, J.R. Barry, R. Langer and G. Vunjak-Novakovic, *Exp. Cell. Res.*, **240**, 58 (1998).
22. L.E. Freed, R. Langer, I. Martin, N.R. Pellis and G. Vunjak-Novakovic, *Proc. Natl. Acad. Sci. USA*, **94**, 13885 (1998).

33

Blocking of Microchannels in Teeth by Alginate Polyelectrolyte Ionotropic Hydrogels

LARS-ÅKE LINDÉN[1], NIE JUN[2], EWA ADAMCZAK[2], JAN F. RABEK[2] and ANDRZEJ WRZYSZCZYNSKI[3]

[1]Department of Dental Materials Science, Umeå University, S-901 87 Umeå Sweden

[2]Polymer Research Group, Department of Dental Biomaterials Science, Karolinska Institute, Box 4064, S-141 04 Huddinge (Stockholm), Sweden.

[3]Faculty of Chemical Technology and Engineering, University of Technology and Engineering, Seminaryjna 3, 85-326 Bydgoszcz, Poland

Wiley Polymer Networks Group Review Series Vol. 2. Edited by B.T. Stokke and A. Elgsaeter

© 1999 John Wiley & Sons Ltd

ABSTRACT

Microchannels in teeth can be blocked by alginate polyelectrolyte ionotropic hydrogels under clinical conditions with the purpose to reduce dentinal hypersensitivity and hinder bacterial invasion (caries). The ionotropic gelation of alginate with bivalent cation such as Ca^{++}, give products that are compatible with most bioactive materials. Alginate hydrogels have very high water swelling capacities (up to 4000%), and can serve as media for ion flow in micro-channels of teeth. Diffusion of alginate into the micro-channels depends on the molecular weight of the alginate and its solution viscosity. In order to decrease the molecular weight of the alginate, photodegradation was proposed. However, UV irradiation of sodium alginate in water solution causes partial decarboxylation and branching of alginate molecules, instead of the expected chain scission and ring opening photo-reactions. UV photoirradiation was found not be a useful method for decreasing the molecular weight of the alginate.

INTRODUCTION

Human teeth are penetrated by microchannels and in dentine the channels (tubules) seen in Figure 33.1, 1–3 µm diameter and 30 000–40 000 per mm², are radiating from the pulp throughout the dentine to the enamel or cementum–dentine junction. *In vivo* the microchannels are filled with a biohydrogel consisting of a fibrous protein and a fluid phase [1,2]. This native hydrogel has a network with a pore diameter of about 60 nm and is bound to the organic collagen matrix and to hydroxyapatite crystals in the walls of the dentine tubules [2]. In SEM, the tubules look empty due to dehydration and collapse of the native gel as seen in Figure 33.1(a,b). The microchannels play an important role during the development of the tooth and for dentinal sensitivity and for progression of dentine caries throughout life. Fluid movements within the microchannels [2,3], due to a positive or negative hydrostatic pressure, induce dentinal pain [4–6]. This fluid flow induces an activation of A–δ nerves in

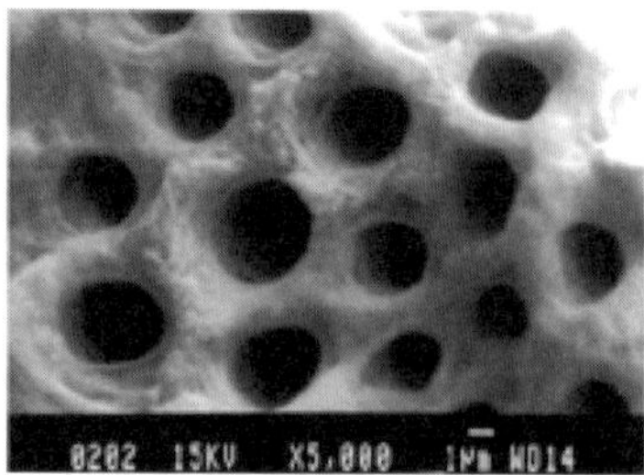
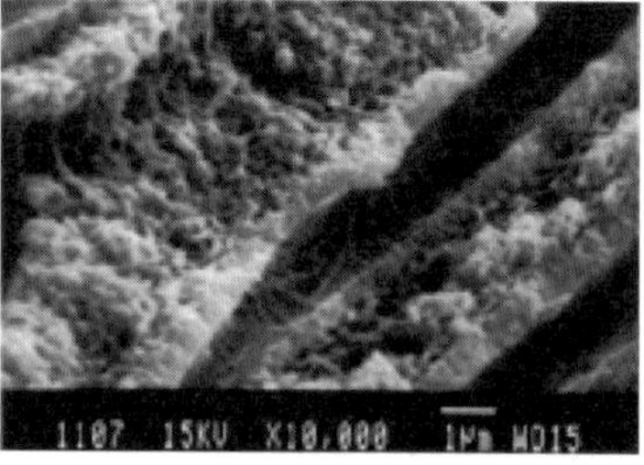

Figure 33.1 Human tooth structure: SEM microphotographs of dentinal tubules: (a) fractured perpendicular to their length and (b) fractured parallel to their length. (b) shows the residues of dehydrated biohydrogel consisting of protein fibres

the pulp, which causes a sensation of sharp and/or shooting pain [7]. Micro-organisms, such as *Streptococcus sanguis* and *mutants*, present in the oral cavity, ferment carbohydrates (sugars) to form lactic acid and result in a decreased pH. Acid environments, with a pH below 5.5, cause dissolution of the hydroxyapatite crystals in the hard walls of the microchannels [8,9]. This increases the hydraulic permeability of the channels and thus facilitates the penetration of micro-organisms. Diffusion of bacterial toxins causes inflammatory reactions in the pulp, which give symptoms such as dentinal hypersensitivity and pain. This is the first physiological signal of the caries process and a remainder for the patient to seek dentist attention.

The idea of blocking the microchannels in tooth dentine by polymeric materials, in order to decrease the fluid movements within the microchannels and protect against tooth decay, has been developed in our laboratory over the last five years [10–13]. In this paper we present results of using calcium alginate hydrogels (CaAH) to block microchannels in teeth.

MATERIALS AND METHODS

CHEMICALS

All chemicals were reagent grade and purchased from Aldrich; sodium alginate (from Macrocystis pyrifera, low viscosity of 2 wt-% approx. 250 cps at 25 °C) (NaA), poly(acrylic acid) (MW 2000) (PAA), $CaCl_2$, $Ca(NO_3)_2$, $Ca(H_2PO_4)_2$, and calcium gluconate ($[HOCH_2[CH(OH)_4]CO_2]Ca$), ethylenediaminetetraaceti-cacid (EDTA), sodium 4-cyclohexanebutyrate ($C_6H_{11}(CH_2)_3COONa$) and calcium 2-ethylhexanoate ($CH_3(CH_2)_3CH(C_2H_5)COO)_2Ca$. Artificial saliva (AS) (similar to that of human saliva but without proteins) was prepared as described earlier [14].

PREPARATION OF ALGINATE IONOTROPIC HYDROGELS

The calcium alginate hydrogel (CaAH) was prepared *in vitro* by mixing a 3 wt-% aqueous solution of NaA with a 3 wt-% calcium salt dissolved in water. After mixing these solutions, a colourless, transparent, insoluble hydrogel (CaAH) is precipitated.

The CaAH was very effective for the blocking of microchannels (tubules) in teeth. For the *in situ* blocking procedure, 200 μm thick cross sections of the human tooth dentine were cleaned from debris (smear layer) with a 5 wt-% EDTA solution and then exposed first to the 3-wt% calcium salt solution in water (or artificial saliva (AS)) for 5 minutes. After draining and cleaning with filter paper, the sections were then exposed for 5 minutes to a 3 wt-% solution of NaA in water (or in AS). The hydrodynamic effect of formation of hydrogel in the tubules was monitored by examining the reduction in hydraulic permeability of the dentine sections [11].

DESWELLING AND SWELLING CHARACTERIZATION

Deswelling and swelling of CaAH was measured by a common gravimetric method at 37 °C (human body temperature) and 70% relative humidity. The amount of water retained by a hydrogel was estimated from the following equation:

$$\text{Deswelling (or swelling)} = (W_t - W_o)/W_o \times 100(\%)$$

Where: W_t is the weight of a partially swollen CaAH after a given time t, and W_o is the weight of completely dried CaAH.

IR AND ATOMIC SPECTROSCOPY

IR spectra of NaA and CaAH were recorded with a FT-IR spectrophotometer (Perkin Elmer 1650). The ion content of Na^+ (589.0 nm) and Ca^{++} (422.7 nm) in NaA and CaAH (in water) was measured with an atomic absorption spectrometer (Varian Spectra 20ABQ), using sodium 4-cyclohexanebutyrate and calcium 2-ethylhexanoate as organo-metallic standards.

GEL PERMEATION CHROMATOGRAPHY

Molecular weight distribution curves (GPC) were measured with a GPC Spectrophysics Model GP-8810 and a PL Aquagel Column type 50.

IRRADIATION METHODS

Water solutions of NaA (3 wt-%) were irradiated with UV-lamp type HPM-15 (2000 W, Philips) at room temperature in the presence of air. Two different irradiation times were used, 6 and 12 h respectively.

SCANNING MICROSCOPY

SEM micro-photographs were made with a Jeol JSM-820 scanning microscope. For the study of gel morphology, water or saliva were removed by low-temperature lyophilization procedures [15].

RESULTS AND DISCUSSION

Alginates are a family of polysaccharides with $\beta(1-4)$ linked L-guluronic (G) and D-mannuronic (M) acid units, as depicted in Figure 33.2 [16,17]. The M/G ratio is frequently used to characterize alginate composition and can be determined with FTIR spectroscopy [18–20], ^1H and ^{13}C NMR spectroscopy [21–23] and many other different methods. The M/G ratio of the sodium alginate (NaA) used in this research was estimated to be 1.6 by determining the ratio of absorbencies of the bands at 1320 and 1290 cm^{-1} [18]. However, neither the distribution of lengths of the different types of GG, GM, and MM blocks nor their average

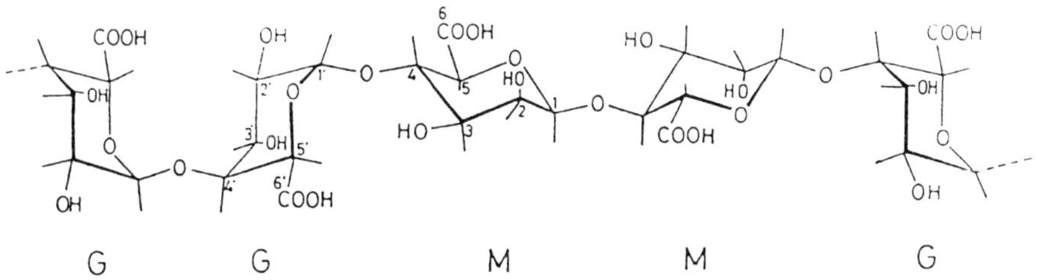

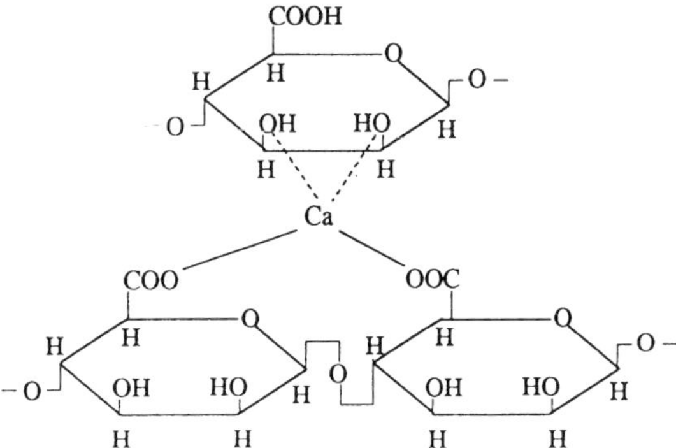

Figure 33.2 Chemical structure of sodium alginate, showing β (1–4) linked L-guluronic (G) and D-mannuronic (M) acid units [20]

length is known. Sodium alginates are, depending on the molecular weight of macromolecular chains, soluble in water up to 3–10 wt-%. The Na^+ content in NaA was 4.6%, close to 5%, which indicated that all potential sites for Na^+ were filled [20]. Water solutions of NaA form *in situ* insoluble ionotropic hydrogels with divalent cations such as Ca^{++} [24–32]. However, Ca^{++} ions bind preferentially with G compared to M acids [29]. Sol–gel transition is characterized by a complex formation in which the carboxyl groups in both guluronate (G) and mannuronate (M) residues are co-ordinated with OH groups with binding to metal ions [24,26,28,32] (Figure 33.3). It has also been suggested by several authors that the ring oxygen atoms are involved in the binding sites [26–29]. Ionotropic hydrogel complexes differ from the classical types of gel in which the long chain molecules are held together by Van der Waals forces. The characteristic features of these hydrogels are: (1) the alginate component forms a three-dimensional structure with Ca^{++} ions, (2) they contain substantial quantities of water and (3) they are soft materials. The CaAH complexes probably have a structure, defined as a homologous structure, which is characterized by a scale-independent correlation over a relatively large range distance [33,34].

Figure 33.3 Chemical structure of calcium alginate hydrogel (CaAH) complex [24]

CaAH may be chemically destabilized in many different ways, i.e. by calcium chelators such as phosphate, citrate, ethylenediaminetetraacetic acid (EDTA), ethylene glycol-*bis*(β-aminoethyl ether)N,N,N',N''-tetraacetic acid (EGTA), or by competition with non-gelling cations [35].

The ion exchange process between NaA and Ca salts in a water solution is very rapid. More than half the ion conversion between Na^+ and Ca^{++} occurs within the first 30 s, and the rate of ion conversion is faster for NaA of lower molecular weights [20]. The calcium alginate hydrogels (CaAH) prepared in artificial saliva (AS) instead of water solutions differ slightly since the saliva consists of many different salts such as $KHPO_4$, $NaHPO_4$, $KHCO_3$, $NaCl$, $MgCl_2$ and $CaCl_2$ [14], which may participate in alginate gelation. The IR spectra in Figure 33.4 show that the chemical structure of CaAH, obtained from NaA and different Ca salts such as $CaCl_2$, $Ca(NO_3)_2$, $Ca(H_2PO_4)_2$ and Ca-gluconate are similar; however, the number of Ca^{++} ions bound differ (Table 33.1). The detailed assignments of IR spectra of alginates are given elsewhere [19,20]. There is an exchange of approximately 2 Na^+ by 1 Ca^{++}, which is the expected ion exchange between monovalent and divalent cations. Anions can probably be co-ordinated to CaAH. This can have an effect on the deswelling/swelling kinetic curves of CaAH prepared in water as seen in Figure 33.5 or in saliva solutions in Figure 33.6 and on the morphology of dried gels as seen in Figure 33.7.

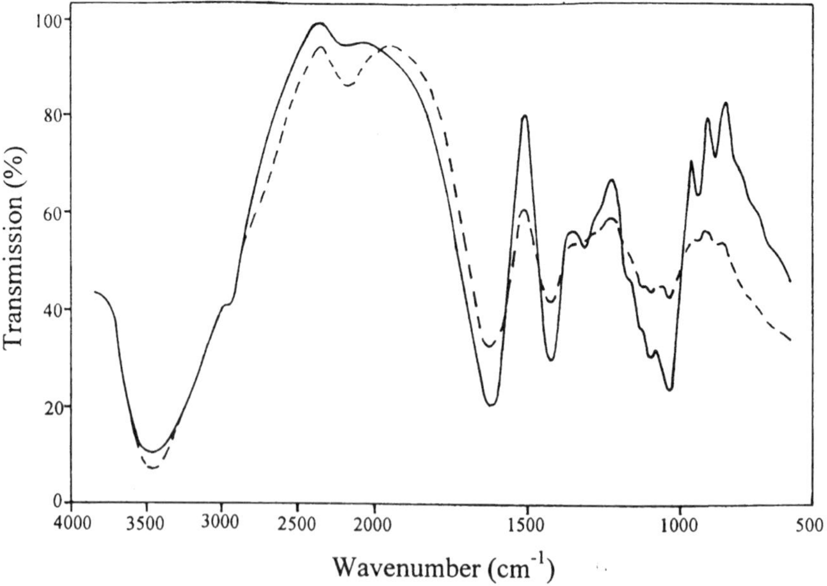

Figure 33.4 IR spectra of: (——) NaA and (– – –) CaAH. The latter IR spectrum is the same for all the Ca salts studied in this paper

Table 33.1 Cation content (%) in NaA and CaAH, measured by atomic absorption spectroscopy

Sample	Na$^+$ (%)	Ca^{++} (%)
Pure NaA	4.60	
NaA 6 h UV irradiated	3.60	
NaA 12 h UV irradiated	2.82	
CaAH from: Ca (NO$_3$)$_2$		4.45
CaCl$_2$		4.39
Ca (H$_2$PO$_4$)$_2$		4.13
Ca-gluconate		3.93

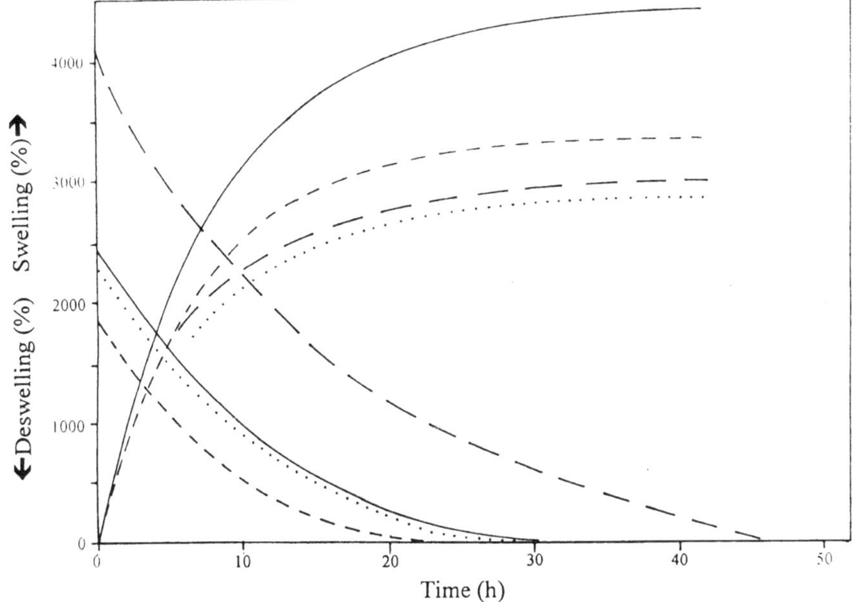

Figure 33.5 Deswelling/swelling curves of CaAH in water obtained from: (———) Ca(NO$_3$)$_2$; (— —) Ca-gluconate; (----) CaCl$_2$ and (····) Ca(H$_2$PO$_4$)$_2$. Gelation took place in water

Kinetics of swelling/deswelling *in vitro* show that CaAH are able to absorb very high contents of water (Figure 33.5) or saliva (Figure 33.6) even up to 4000% and steady-state conditions are reached after 30 h. The absorption of water of the CaAH formed in situ within the rigid dentinal tubules results in a 'swelling pressure' that is higher than that observed for the native hydrogel [2]. The physiological consequences of this higher internal pressure is not elucidated but the SEM images show that the native gel network is partly intact and penetrating the rather dense dehydrated CaAH network (Figure 33.9(b)). The high content

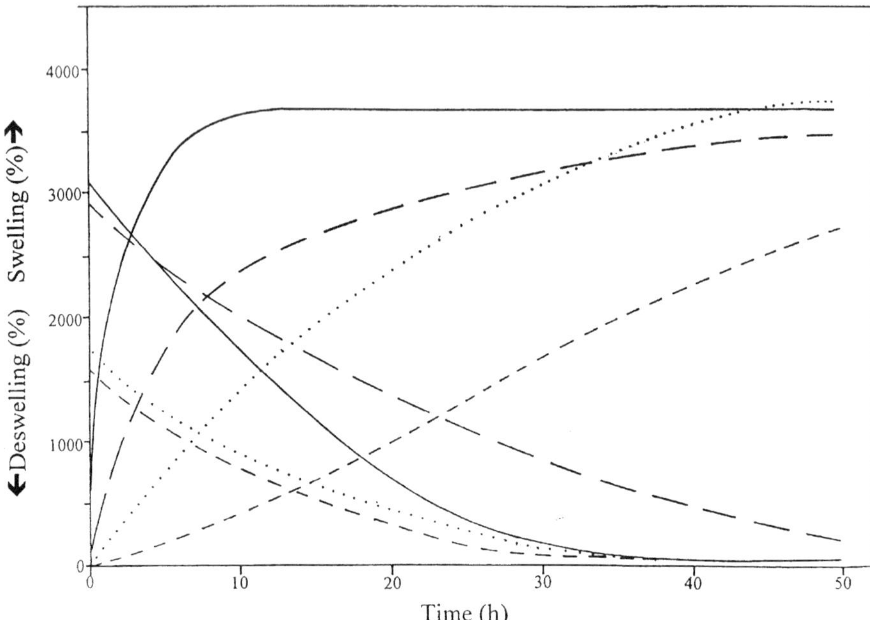

Figure 33.6 Deswelling/swelling curves of CaAH in saliva obtained from: (———)
$Ca(NO_3)_2$; (— —) Ca-gluconate; (----) $CaCl_2$ and (····) $Ca(H_2PO_4)_2$. Gelation
took place in artificial saliva

of water in swollen CaAH, however, still allows for a free diffusion of ions and
molecules similar to that observed for the native hydrogel network with a pore
radius of about 30 nm [2]. The difference in swelling/deswelling behaviour of
the CaAH *in vivo*, due to different ion concentrations in beverages, foodstuffs and
liquid therapeutic agents used by dentists, can have an impact on the blocking
effect of the hydrogel. This may be of importance for the physiology of the
tissues *in vivo*. There is a significant difference in the gel morphology of CaAH
prepared with $Ca(NO_3)_2$, $CaCl_2$ and $Ca(H_2PO_4)_2$ compared to Ca-gluconate as
seen in Figures 33.7(c, d). The preparation method also has an effect on the gel
morphology. In the method where the calcium ions were allowed to diffuse into
an NaA solution from a calcium salt solution, fibrous structures were formed
(Figure 33.8). This could have implications on how to perform application of
hydrogels in the dental clinic.

In the present study, we have favoured the method in which 3 wt-% water solution of calcium ions first, and after that NaA were allowed to diffuse into the tooth
microchannels and infiltrate the native, fibrous, protein hydrogel. The NaA macromolecules are, during the diffusion into microchannels, entangled with the native
protein molecules, whereas the small Ca^{++} ion diffused freely. The rapid ion
exchange process between Na^+ and Ca^{++} ions causes NaA macromolecules to

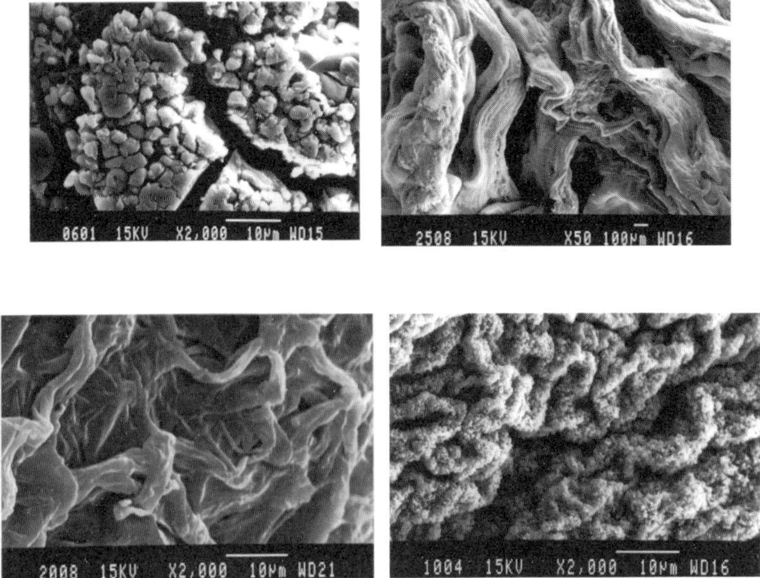

Figure 33.7 SEM microphotographs of dried: (a) NaA, and CaAH obtained from: (b) $Ca(NO_3)_2$; (c) $Ca(H_2PO_4)_2$ (gelation took place in water) and (d) Ca-gluconate (gelation took place in saliva)

Figure 33.8 SEM microphotographs of CaAH fibre structures formed by introducing NaA through capillaries into (a) $CaCl_2$ and (b) Ca-gluconate water solution respectively

be drawn into the microchannels. SEM microphotographs in Figure 33.9 illustrate that CaAH (prepared from $CaCl_2$ or $Ca(NO_3)_2$ and NaA solutions) completely blocked the lumen of the microchannels. The diffusion of the Ca-gluconate into microchannels filled with native hydrogel networks can be hindered by the size of the gluconate anion residue. In this case the blocking of microchannels was weak. The penetration of NaA into native hydrogel in the microchannels depends

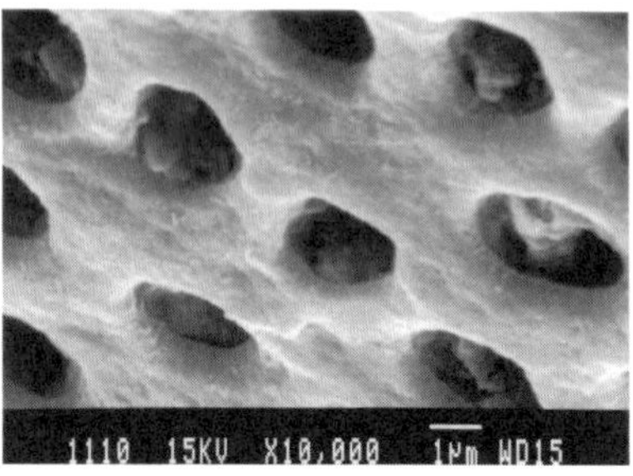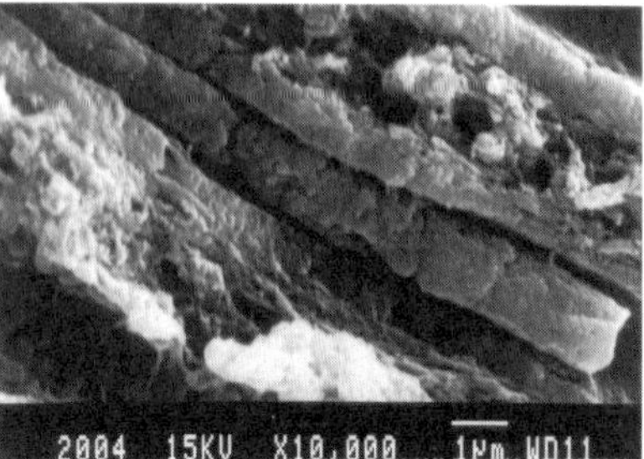

Figure 33.9 SEM microphotographs of dentine tubules blocked by CaAH, prepared by the method in which $Ca(NO_3)_2$ was first diffused into the microchannels and next followed by NaA diffusion: (a) fractured perpendicular to their length and (b) fractured parallel to their length

on the molecular weight of NaA and the stiffness parameters for the three types of GG, MG and MM blocks. The GG block is characterized by a strong binding of Ca^{++} ions [26–29].

The diffusional properties of NaA, which are critical parameters in tightening of tooth microchannels, depend on the conformation of the alginate molecule and its molecular weight (MW). Incomplete knowledge of the monomer sequence in our NaA sample has prevented an experimental determination of the role of

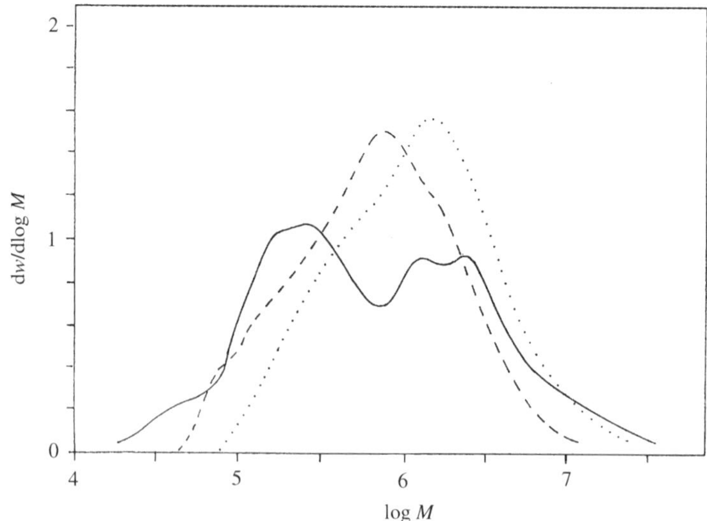

Figure 33.10 Molecular weight distribution curves from gel permeation chromatography (GPC) of: pure NaA (——), NaA UV-irradiated for 6 h (----), and for 12 h (····)

conformation in the diffusion of NaA into the microchannel hydrogel. Macromolecules of NaA with a lower MW should easily diffuse into and through the native hydrogel network. We made an attempt to decrease the MW of our NaA by UV irradiation in water solution for 6 h and 12 h. We obtained, however, unexpected results since the gel permeation chromatography (GPC) showed that during UV irradiation of NaA its MW increased, instead of decreasing (Figure 33.10). With UV irradiation a partial decarboxylation of the NaA occurred and the content of Na^+ in NaA decreased (Table 33.1). The intensities of the COO^- stretching asymmetric band at 1621 cm^{-1} and the COO^- stretching symmetric band at 1413 cm^{-1} also decreased. The free radicals formed from NaA terminate each other to give branched structures of alginate. The NaA, UV-irradiated 6 h, formed with Ca^{++} a very weak, soft hydrogel, with morphology differing from non-irradiated CaAH (Figure 33.11). SEM microphotographs, presented in Figure 33.12, showed that the CaAH prepared from $CaCl_2$ and NaH, which was UV-irradiated for 6 h, may also block the lumen of the microchannels; however, the morphology of these blockings differ from those seen in Figures 33.9a and b. NaA UV-irradiated for 12 h did not cause any gelation in Ca^{++} solutions and is not effective for tightening microchannels in teeth.

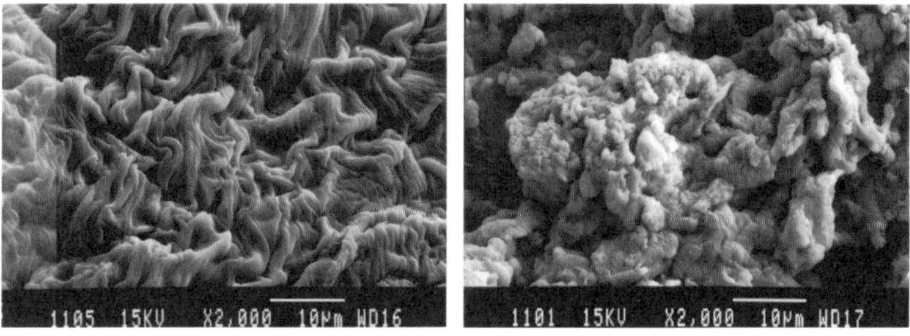

Figure 33.11 SEM microphotographs of dried hydrogel formed by $CaCl_2$ and: (a) obtained from non-irradiated NaA and (b) from NaA UV-irradiated for 6 h

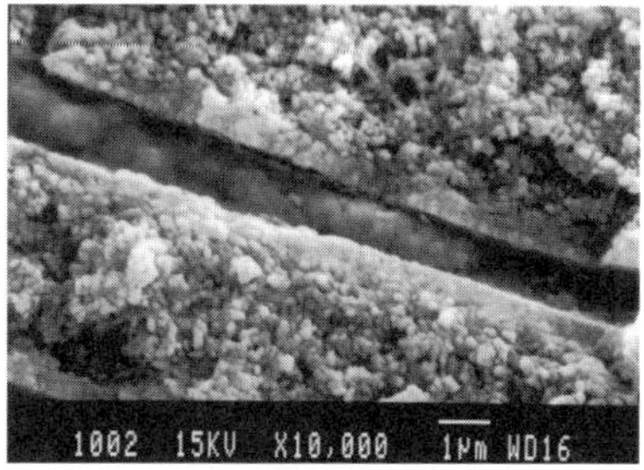

Figure 33.12 SEM microphotograph of dentine tubule blocked by CaAH, prepared from $CaCl_2$ and NaA UV-irradiated for 6 h

It is also expected that diffusional properties of NaA into microchannels vary according to gelling conditions such as NaA viscosity, concentration, temperature and pH because alginate molecules, being polyelectrolytes, change their conformation depending on the solution environment. The viscosity of NaA decreases with increasing pH above 8.0, or with pH values below 6.0 [36]. The pH of NaA (3 wt-% in water, used in our experiments) was 7.8. The decrease in viscosity by pH adjustment corresponds to a decrease in the hydrodynamic volume of the alginate molecules. Smaller conformation of alginate corresponds to an increase in the volume fraction of the polymer segments in the hydrodynamic volume of the alginate, as well as a decrease in intermolecular entanglement. However, the decrease of pH below 6 causes dissolution of hydroxyapatite in the walls of the microchannels in teeth and should be avoided for prophylactic reasons.

We have also found that poly(acrylic acid)(PAA) did not form mixed hydrogels with NaA and calcium salts. After reaction, PAA remains in the solution while CaAH precipitates as a hydrogel.

CONCLUSIONS

Alginate polyelectrolyte ionotropic hydrogels prepared in situ from sodium alginate and calcium salts can be used to effectively block microchannels in teeth. Alginate hydrogels are biocompatible, nontoxic and nonallergenic substances that are acceptable for direct clinical applications. Careful selection of alginates, calcium salts and gelling conditions are essential for successful application of CaAH for tightening of microchannels in teeth. Systematic research on this subject is still necessary in order to investigate the relationship between the diffusion of NaA and its molecular weight; gel morphology and the mechanical strength of CaAH.

ACKNOWLEDGMENTS

We gratefully acknowledge the support of the Swedish Institute who generously provided a post-doctoral stipendium for Nie Jun and A. Wrzyszczynski. This research was sponsored by grant from the Swedish Medical Research Council project Nr. K97-24X-11650-02A.

REFERENCES

1. L.Å. Lindén, and Ö. Källskog, *Swed. Dent J.*, **16**, 56 (1992).
2. L.Å. Lindén, Ö. Källskog, and M. Wolgast, *Arch. Oral. Biol.*, **40**, 991 (1995).
3. L.Å. Lindén, *Odont. Rev.*, **19**, 1 (1968).
4. M. Brännström, *J. Dent. Res.*, **43**, 619 (1961).
5. M. Brännström, G. Johnson and L.Å. Lindén, *Odont. Rev.*, **20**, 15 (1969).
6. L.Å. Lindén, *Odont. Rev.*, **19**, 18 (1968).

7. M.L. Ahlquist, J.P. Coffey, O.G. Franzon and D.H. Pashley. *Eur. J. Neurosci*, Suppl: **176** (1988).

8. F.C. Driessens, H.M. Theuns, J.M.P. Borggreven and J.W.E. van Dijk, *Caries Res.*, **20**, 103 (1986).

9. H.M. Theuns, J.W.E. van Dijk, F.C.M. Driessens and A. Groenveld. *Arch. Oral Biol.*, **30**, 37 (1985).

10. L.Å. Lindén and J.F. Rabek, *J. Appl. Polym. Sci.*, **50**, 1331 (1993).

11. H Kaczmarek, L.Å. Lindén and J.F. Rabek, *J. Appl. Polym. Sci.*, **60**, 2321 (1996).

12. Xin Qu, A. Wrzyszczynski, K.J. Pielichowski, J. Pielichowski, E. Adamczak, S. Morge, L.Å. Lindén and J.F Rabek, *J. Macromol. Sci. Pure Appl. Chem.*, A**34**, 707 (1997).

13. L.Å. Lindén, J.F. Rabek, E. Adamczak, S. Morge, H. Kaczmarek and A. Wrzyszczynski. *Macromol. Symp.*, **93**, 337 (1995).

14. A. Odén, E. Wiatr-Adamczak and S. Olsson. *Tandläkartidningen,*, **83**, 634 (1991).

15. J.F. Rabek. *Experimental Methods in Polymer Chemistry*, Wiley, Chichester 1980, p. 227.

16. G. Skjåk-Braek, *Biochem. Soc. Trans.*, **20**, 27 (1992).

17. B.T. Stokke, O. Smidsrød and D.A. Brant. *Carbohydr. Polym.*, **22**, 57 (1993).

18. M.P. Fillipov and R. Kohn, *Chem. Zvesti*, **28**, 817 (1974).

19. B. Dupuy, A. Arien and A. Perrot Minnot, *Art. Cells Blood Subs. Immob. Biotech.*, **22**, 71 (1994).

20. C. Sartori, D.S. Finch, B. Ralph and K. Gilding, *Polymer*, **38**, 43 (1977).

21. H. Grasdalen, B. Larsen and O. Smidsrød. *Carbohydr. Res.*, **68**, 23 (1979).

22. H. Grasdalen *Carbohydr. Res.*, **118**, 255 (1983).

23. Z.Y. Wang, Q.Z. Zhang, M. Konno and S. Saito. *Biopolym.*, **33**, 703 (1993).

24. R. Schweiger. *J. Org Chem.*, **27**, 1786 (1962).

25. A. Haug and O. Smidsrød, *Acta Chem. Scand.*, **24**, 843 (1970).

26. S.J. Angyal, *Pure Appl. Chem.*, **35**, 131 (1973).

27. T.A. Bryce, A.A. McKinnon, E.R. Morris, D.A. Rees and D. Thom, *Faraday Disc. Chem. Soc.*, **57**, 221 (1974).

28. O. Smidsrød, A. Haug and S.G. Whittington *Acta Chem. Scand.*, **26**, 2563 (1972).

29. O. Smidsrød, *Faraday Disc. Chem. Soc.*, **57**, 263 (1974).

30. Z.Y. Wang, J.W. White, M. Konno, S. Saito and T. Nozawa, *Biopolym.*, **35**, 227 (1995).

31. T. Kozawa, K Yamagawa and A. Ohakawa, *J. Chem. Eng. Japan*, **27**, 833 (1994).

32. E.R. Morris, D.A. Rees and D.J. Thom. *J. Chem Soc, Chem Commun.*, 245 (1973).

33. M. Muthukumar, *Phys. Rev. Lett.*, **50**, 839 (1983).

34. M. Muthukumar, *J. Chem. Phys.*, **83**, 3161 (1983).

35. S. Hertzberg, E. Moen, C. Volgelsang and K. Ostgaard, *Appl. Microbiol. Biotechnol.*, **43**, 10 (1995).

36. K. Yamagawa, T. Kozawa and A. Ohakawa, *J. Chem. Eng. Japan*, **28**, 462 (1995).

34

Surface Modification of Polymer Hydrogel and its Effect on the Wall-Shear Stress in Water

MOHAMMAD HAMIDUL ISLAM, HIROHISA MORIKAWA,
NANANE WATANABE, MASASHI WATANABE and
TOSHIHIRO HIRAI
Faculty of Textile Science and Technology, Shinshu University Tokida 3-15-1,
Ueda Shi 386-8567, Japan

Wiley Polymer Networks Group Review Series Vol. 2. Edited by B.T. Stokke and A. Elgsaeter
© 1999 John Wiley & Sons Ltd

ABSTRACT

We investigated the interaction between the flow and the surface of the hydrogel tube of poly(vinyl alcohol)(PVA) which was found to possess visco-elastic behavior similar to a blood vessel. A transparent PVA hydrogel tube prepared from the solution of dimethyl sulfoxide (DMSO) and water was adequate for the observation of the flow profile near the gel surface in micron scale. The hydrogel was also good for the chemical modification of its surface with poly(acrylic acid)(PAA). Monodisperse polystyrene particle (diameter = 7.21 µm) was used as a tracer of fluid flow and the flow was observed by a CCD camera under a microscope. The flow velocity was measured by particle-tracking (PT) and particle–image deformation (PID) method. Wall-shear stress (WSS) was estimated from the velocity gradient near the wall. As a result, it was found that the WSS depends on the surface structure of the hydrogel. The WSS was smaller in case of a modified hydrogel tube than an intact PVA hydrogel tube in dissociated condition.

INTRODUCTION

Diseases underlying damage of the endothelial cell of blood vessels have been the subject of reviews in biomedical science for a long time. Much work carried out from the standpoint of fluid dynamics has been making a major contribution to understanding basic problems. Concerning the occurrence and development of atherosclerosis there are two major theories: a low-shear-stress theory [1] and a high-shear-stress theory [2]. It has been reported that the shape of an endothelial cell is distorted in the direction of its action by the wall-shear stress [3,4].

In order to measure *in vivo* velocity profiles in animal arteries, the hot film anemometer [5] and the ultrasound pulsed Doppler velocity meter [6] have been used. Both of these methods lack sufficient resolution close to the wall for accurate measurement of the veolocity gradient or the shear stress. The estimation of the viscosity and velocity of a concentrated suspension, such as blood, in the near-wall region presents major difficulties. The velocity of the fluid near the wall cannot be inferred directly from extrapolation data. Therefore, an accurate estimation of the fluid velocity very near to the surface is inevitable.

Many researchers have been using an artificial model tube made of rubber as a substitute of the blood vessel to measure the flow and the shear stress [7,8]. It is not claimed, however, that the flows in the model tubes adequately simulates blood flow because the mechanical properties of the model tubes are quite different from those of arteries.

Hydrophilic gels have attracted considerable attention as candidates for good bio-compatible materials, because the interface between the water-swollen gel and blood or tissues may have very low free energy leading to very low adverse interaction of the gel surface with an aqueous biological environment [9–11]. A large number of investigations have been undertaken to develop biomedical materials which do not elicit thrombus formation when exposed to the blood flow. Recently many studies have shown that PVA is antithrombotic and one of the most promising materials for the inner surface of small vascular prostheses [12–14].

PVA hydrogel tubes could be prepared to have similar mechanical properties to those of an artery by fixing various strain in the structure of the hydrogel tube such as elongation and twisting [15,16]. These properties of the gel tube were based on the excellent shape memory of the PVA gel [17].

In the present study, the velocity profile and wall shear stress of the laminar flow with varying liquid pH in intact and chemically modified PVA hydrogel tubes have been experimentally studied. A transparent PVA hydrogel tube prepared from the solution of dimethyl sulfoxide (DMSO) and water is adequate for the observation of the flow profile near the gel surface in micron scale. The hydrogel is also good for the chemical modification of its surface with poly(acrylic acid) (PAA). Monodisperse polystyrene particle (diameter $= 7.21$ μm) was used as a tracer of fluid flow and the flow was observed by a CCD camera under a microscope. The flow velocity was measured by particle-tracking (PT) and particle-image deformation (PID) method. Wall-shear stress (WSS) was estimated from the velocity gradient near the wall. By these methods we could measure the flow near the surface in micron scale. The results show that the WSS on modified PVA hydrogel surface was affected by pH and ionic strength.

EXPERIMENTAL

MATERIALS

The PVA used, purchased from Kuraray Co., had a degree of polymerization of 1700. PVA was used after complete saponification in our laboratory. PAA (MW $=$ 450 000) was purchased from Aldrich Chemical Co., and used as obtained. Styrene was supplied by Wako Pure Chemical Industries Ltd and was purified by distillation in the nitrogen atmosphere to remove inhibitor before use.

PREPARATION OF MONODISPERSE PARTICLE

For the flow visualization experiment various particles are available, but poly-stryene is particularly useful in water, as it density ($1.05-1.07$ g/cm^3) is almost the same as that of water. To keep the similarity of the red blood cells, we prepared 7.21 μm diameter monodisperse polystyrene particles by suspension and seeded polymerization.

PREPARATION AND MODIFICATION OF PVA HYDROGEL TUBE

Homogeneous PVA solution with concentration of 10 wt% was prepared by heating the mixture of PVA powder, water, and DMSO (water/DMSO $= 20/80$ by weight) for 6 h in a water bath at 90 °C. The solution was cast in a cell of polymethyl methacrylate (PMMA). To prepare the inner hole in the hydrogel tube, we used a PMMA rod and a nylon filament. The polymer solution was changed into hydrogel by a repetitive cooling method. The cooling cycle consisted of cooling at $-22\,^{\circ}$C for 23 h and standing at ambient temperature for 1 h. After completing six cooling cycles, the PMMA rod was withdrawn from the gel. The gel tube was immersed into water to remove the DMSO in the gel. The hydrogel tube was transparent and easy for observation of the inner surface. The inner surface of the PVA hydrogel tube was chemically modified with PAA in a water bath at $30\,^{\circ}$C for 96 h in the presence of hydrochloric acid. The hydrogel tube was filled with the PAA, and the PAA was attached on the PVA hydrogel surface through ester linkage. The modified hydrogel tube was rinsed thoroughly in flowing water to remove unreacted PAA. The inner hole dimension of the tube with square cross-section was 1.85 mm and 100 mm long.

WORKING FLUID

The working fluids were distilled water and different pH buffer with 0.005% neutrally buoyant 7.21 μm diameter monodisperse polystyrene particles as a tracer of fluid flow. Throughout these experiments, the ionic strength of all buffers was 0.35. The fluid temperature was kept at $22\,^{\circ}$C.

FLOW SYSTEM

The schematic of experimental apparatus used for this work is shown in Figure 34.1. PVA hydrogel tube was placed inside a sample cell. The sample

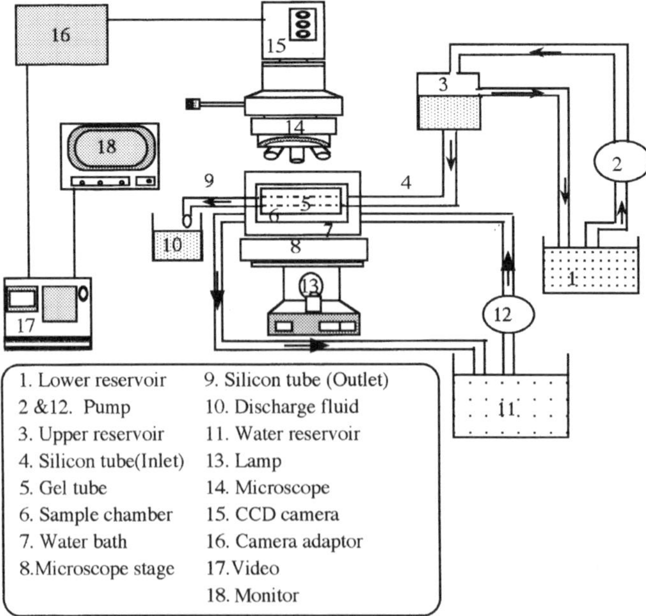

Figure 34.1 Scheme of the flow experimental apparatus

cell was placed in a thermostatic water bath made of PMMA. To get the sharp microscopic image, the hydrogel tube was irradiated by light from below. The mean volumetric flow rate was determined using the measured fluid density and the mass of fluid collected over a timed interval. The flow patterns were observed with a CCD camera (Model Dxc-108, Sony) under a microscope (OLYMPUS BH2-UMA). The CCD camera equipped with a high speed electronic shutter (up to $10\,000$ s^{-1}) enables us to obtain a clear picture of moving particles. The picture was directly shown on the monitor screen and the image was recorded by a video recorder (Hi-band, Beta hi-fi, Sony). The particle flow rates were estimated by analyzing the video images of the flowing particles using a personal computer.

DETERMINATION OF PARTICLE VELOCITY

To measure individual particle velocity in PVA hydrogel tubes, the PT and PID methods have been used. In the PT method the fluid velocity was inferred from the motion of tracking particles. The tracking particles were considered exactly to follow the motion of the fluid. Only images which were reasonably sharp and clearly belonged to the same particle were used. The data was obtained from an optical section around the median plane of the hydrogel tube. Two consecutive images were overlapped. The time differences of the two consecutive images were recorded. Particle displacement for the time interval was measured from the overlapped image (see Figure 34.2). The velocity of the individual particle was estimated from the particle displacement divided by the time interval of the two consecutive images. In the PID method the deformed particle image was captured by the CCD camera by controlling the shutter speed. The velocity was estimated directly from the deformed longitudinal particle-image length multiplying by shutter speed (see Figure 34.3). PID is applicable for the case of rapidly flowing system, and can provide a flow profile with vector image.

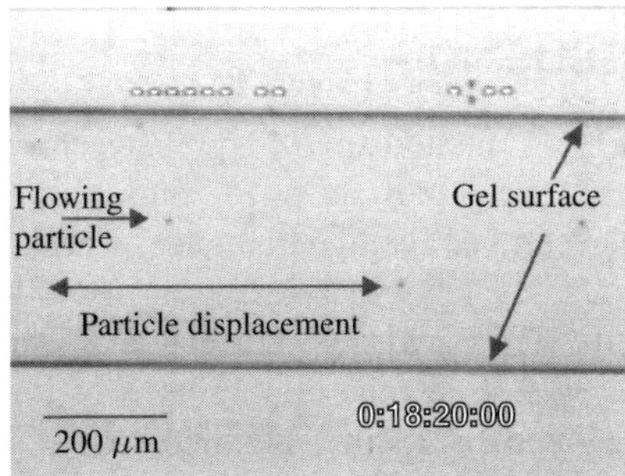

Figure 34.2 An example of velocity measurement in a PVA hydrogel tube by PT method. Overlapped image of two consecutive images

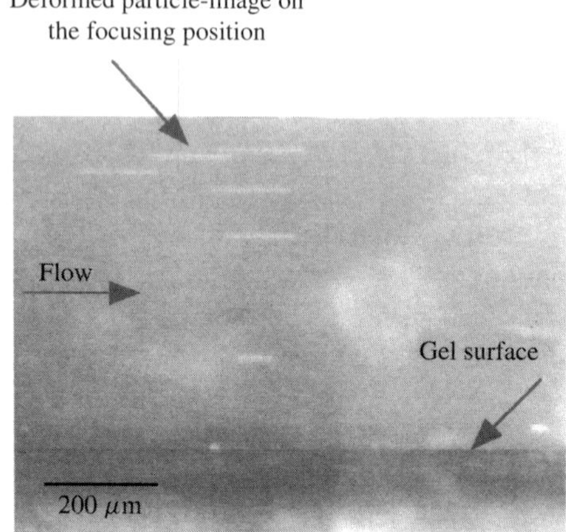

Figure 34.3 An example of velocity measurement in a PVT hydrogel tube by PID method

DATA ANALYSIS

The Reynolds number (Re) for a tube with square cross-section and WSS (τ) are defined as follows:

$$\mathrm{Re} = \frac{WU}{\nu} \tag{34.1}$$

$$\tau = \mu \left(\frac{du}{dx}\right)_0 \tag{34.2}$$

where U is mean velocity (Q/A), A is cross-sectional area, W is the length of a side of the square cross-section, ν is kinematic viscosity, μ is the fluid viscosity, and $(du/dx)_0$ is a velocity gradient on the gel surface.

RESULTS AND DISCUSSION

A SEM micrograph of monodisperse polystyrene particles is shown in Figure 34.4. An average particle size and uniformity ratio (U.R. = weight average diameter/number average diameter) were obtained a 7.21 µm and 1.004, respectively.

The PAA layer on the PVA hydrogel surface was detected by dyeing with methylene blue. Figure 34.5 shows the photographs of intact and modified PVA hydrogels, which were dyed by methylene blue. After dyeing both gels were immersed in water and found that the modified PVA gel stained the dye and showed blue color, while the intact PVA did not stain the dye. Methylene blue is a cation, and can bind electrostatically to anionic species. Esterified (modified)

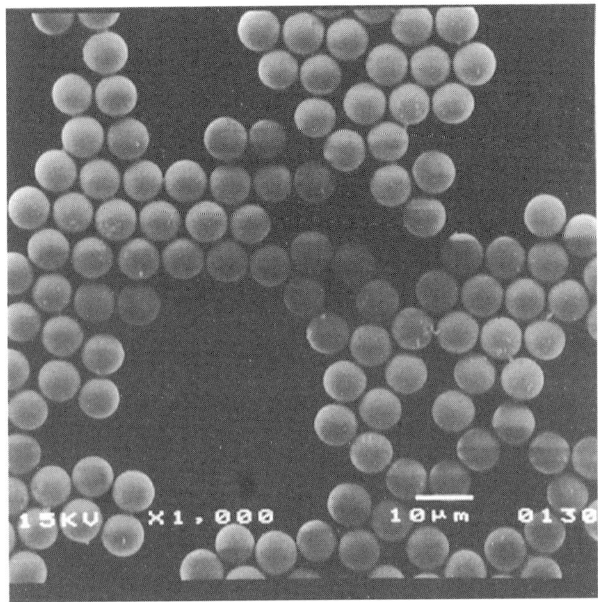

Figure 34.4 SEM micrograph of polystyrene particles prepared from 2.76 μm seed particle by seeded polymerization. Average particle size and U.R. are 7.21 μm and 1.004, respectively

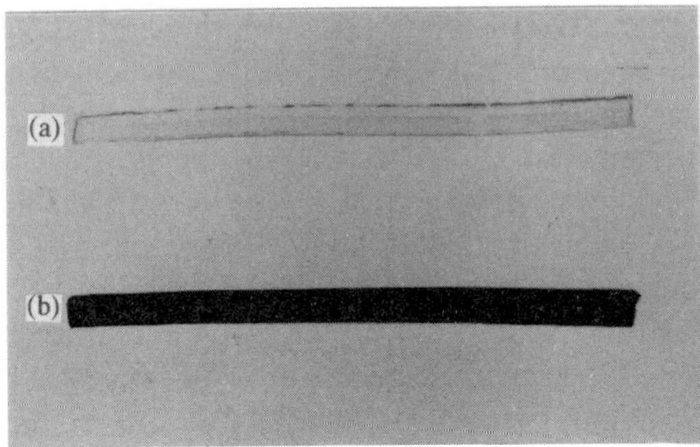

Figure 34.5 Photographs of dyed hydrogel after 5 days immersed in water: (a) intact PVA hydrogel; (b) modified PVA hydrogel

gel has a large amount of residual carboxylic acid, which can easily be bound ionically with methylene blue. On the contrary, the intact PVA hydrogel does not have such activity, due to the absence of ionic groups in it. It has been considered that PAA is difficult to attach to PVA by ester linkage in an aqueous solution. In our previous paper, we clarified that it is possible to chemically crosslink PVA and PAA in a hydrogel [18].

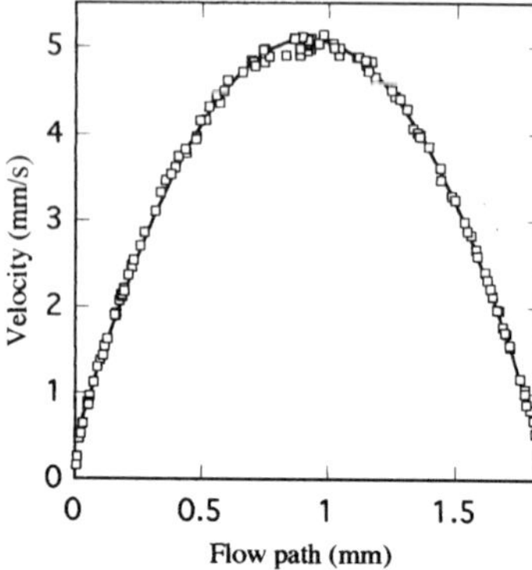

Figure 34.6 Laminar flow profile in a PVA hydrogel tube estimated from PT method in steady flow (Re is 5.23). The error in the flow rate is less than 1%

The PT and PID methods that we employed to estimate the velocity profile in hydrogel tubes provided accurate measurements of velocity profiles. In Figure 34.6 an example of a velocity profile in a modified PVA hydrogel tube at Reynolds Number 5.23, which was estimated by the PT method, is shown. An open square in the graph indicates the velocity of one particle. The velocity profiles were curve-fitted to a second-order polynomial and the WSS was obtained by calculating the velocity slope at the wall.

The swelling/deswelling behavior in modified PVA hydrogel tube at flow condition was observed. The modified surfaces swell at pH 5.6 and 8 and deswell at pH 3. It should be noted that the degrees of dissociation (DD) of PAA at pH 3, 5.6, and 8 are about 5, 95, and 99 percent, respectively. Tentative models of pH-induced swelling/deswelling behavior of the PAA layer depends on pH change in flow conditions are shwon in Figure 34.7. With increasing velocity, the dissociated PAA layers aligned in the flow direction.

Figure 34.8 shows the Reynolds Number dependence of WSS in the intact and the modified PVA hydrogel tubes conveying water and different pH buffer. The WSS of the modified PVA hydrogel tube was lower than that of the intact PVA hydrogel tube. The Figure also shows that the WSS in modified hydrogel tube was lower in water than in different pH buffer. It is not necessary to consider the effect of the ionic charge of the particle, because the polystyrene is non-ionic. In the modified hydrogel in water and pH 5.6, the DD of PAA are almost the same because pH of water is about 5.6, but the WSS was higher in pH 5.6 buffer than in water. It is due to the effect of ionic strength of pH buffer on the anionic gel surface layer. At pH 5.6 and 8 the WSS was almost the same due to the same ionic strength of the pH buffer (0.35) and the same DD of the PAA layer. We also found that the WSS in modified hydrogel was higher at pH 3, because the PAA layers stay almost associated condition at pH 3 and the surface became tight and cause an increase of WSS. The reason for the low WSS in the modified

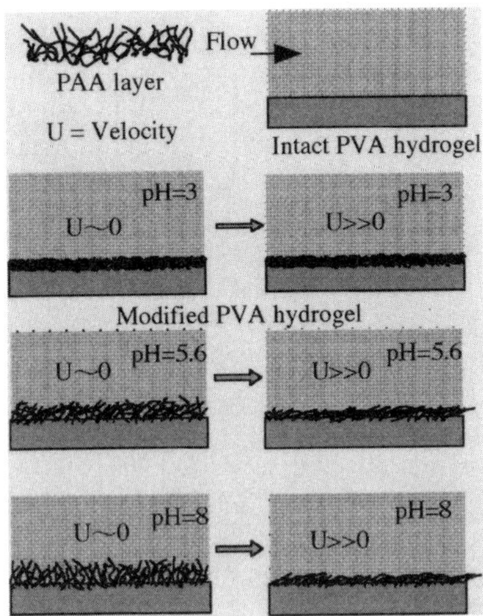

Figure 34.7 Tentative models of pH-induced swelling/deswelling behaviours of the PAA layer on the PVA hydrogel tube, and the deformation under flow condition

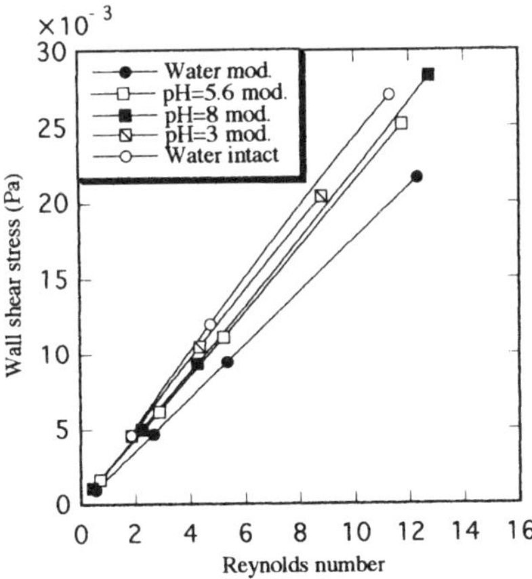

Figure 34.8 Reynolds number dependence of wall shear stress on intact and modified PVA hydrogel tube at different pH (Mod. for modified hydrogel and intact for intact hydrogel)

PVA hydrogel tube is that the diffusibility of the hydrogel surface increased at dissociated condition which is higher than that in intact PVA hydrogel surface and the segment density of the interface is too low to interact with flowing liquid. Therefore, the grafted PAA layers decrease the WSS in the modified PVA hydrogel tube at dissociated condition.

Recently many investigations on the effects of a flow on cultured endothelial cells have focused on the effects of a steady laminar shear stress [19,20]. These studies show that shear stress produced by flowing fluid has been found to cause changes in the structure and function of endothelial cells. In the modified hydrogel, the WSS was changed by varying the pH of working fluid. We can say that at the dissociated condition the hydrogel surface becomes diffused and grafted PAA layer is elongated and is aligned in the flow direction and changes the WSS.

CONCLUSIONS

An elastic and transparent hydrogel tube of PVA was prepared for simulating micro blood vessel. The observation and the measurement of a flow in the hydrogel tube in micron scale accuracy were shown to be possible by the PT and PID methods. The measured velocity profiles were curve-fitted well with a second-order polynomial and the error was less than 1%. WSS was lower for the modified PVA hydrogel tube than for the intact PVA hydrogel. We suggest that the results will provide important information on understanding biological blood vessel system.

ACKNOWLEDGMENT

This work was supported by Grant-in-Aid for COE Research (10CE2003) by the Ministry of Education, Science, Sports and Culture of Japan.

REFERENCES

1. D.L. Fry, *Circulation Res.*, **22**, 165 (1968).
2. C.G. Caro, J.M. FitzGerald, and R.C. Schroter, *Proc. R. Soc. Lond.*, **B177**, 109 (1971).
3. R. Nerman, M.J. Levesque, and J.F. Cornhill, *ASME J. Biomech. Eng.*, **103**, 172 (1981).
4. C.F. Dewey, Jr S.R. Bussolari, Jr. M.A. Gimborne and P.F. Davies, *ASME J. Biomech. Eng.*, **103**, 177 (1981).
5. R.M. Nerem, J.A. Rumbeger, D.R. Gross, R.L. Hamlin and G.L. Geiger, *Circulation Res.*, **34**, 193 (1974).
6. M.B. Histand, C.W. Miller and F.D. McLeod, *Cardiovasc. Res.*, **7**, 703 (1973).
7. K. Rhee and J.M. Tarbell, *J. Biomechanics*, **27** (3), 329 (1994).
8. E.Y. Kwack, L.H. Back, X.M. Runa and A. Chaux, *ASME J. Biomech. Eng.*, **118**, 165 (1996).
9. O. Wichterle and D. Lim, *Nature*, **185**, 117 (1960).
10. J.D. Andrade, *J. Assoc. Advan. Med. Inst.*, **7**, 110 (1973).
11. B.D. Ratner and A.S. Hoffman and J.D. Andrade (ed.), *Am. Chem. Soc. Symp. Ser.*, **31**, 1 (1976).
12. Y. Ikada, H. Iwata, F. Horii, T. Matsunaga, M. Suzuki, W. Taki, S. Yamagata, Y. Yonekawa, and H. Handa, *J. Biomed. Mater. Res.*, **15**, 697 (1981).

13. H. Miyake, H. Handa, W. Taki, Y. Naruo, S. Yamagata, Y. Ikada, H. Iwata and M. Suzuki, *Microsurgery*, **5**, 144 (1984).
14. A.J. Aleymma and C.P. Sharma, *Biomimetic Polymers* Plenum Press, New York, 1990, p.191.
15. T. Hirai, H. Nemoto, H. Morikawa, T. Sakurai and S. Hayashi, *Biorheology*, **29** (1), 150 (1992).
16. H. Morikawa, T. Hirai, J.C. Debes, M. Sakurai, and M. Nakazawa, *Biorheology*, **29** (1), 151 (1992).
17. T. Hirai, H. Maruyama, T. Suzuki and S. Hayashi, *J. Appl. Polym. Sci.*, **45**, 1849 (1992).
18. T. Hirai, T. Okinaka, M. Hirai and S. Hayashi, *Angew. Makromol. Chem.*, **240**, 213 (1996).
19. C.F. Dewey, S.R. Bussolari, Jr, M.A. Gimbrone and P.F Davies, *ASME J. Biomech. Eng.*, **103**, 177 (1981).
20. C.F. Dewey, Jr., *ASME J. Eng.*, **106**, 31 (1984).

PART 6
Polymer Networks in Restricted Geometries

35

Ultrathin Cross-linked Polyorganosiloxane Networks at Interfaces between Fluids: Structure, Properties and Preparative Prospectives

HEINZ REHAGE, BARBARA ACHENBACH and
MARTIN HUSMANN
Universität-Gesamthochschule Essen, Institut für Umweltanalytik,
Universitätsstraße 5–7, D-45141 Essen

ABSTRACT

The tensio-active properties of octadecyltrichlorosilane and octadecyltrimethoxysilane can be used to synthesize ultrathin cross-linked membranes at the surface of water or at the interface between oil and water. At the planar surface we investigated the two-dimensional

Wiley Polymer Networks Group Review Series Vol. 2. Edited by B.T. Stokke and A. Elgsaeter
© 1999 John Wiley & Sons Ltd

rheological behavior by dynamic time-, frequency- and strain-sweep experiments. In addition to these measurements, we have also investigated the molecular structure of these networks by means of Brewster-angle microscopy. The interfacial polymerization is not restricted to flat surfaces, and the corresponding films can be used to stabilize different types of emulsions. Oil droplets which are surrounded by cross-linked membranes, form simple systems, which exhibit interesting mechanical properties of biological cells. In addition to the interface shear rheology, we have developed a new experimental technique that consists in measuring the deformation of a capsule in a spinning drop apparatus. From these experiments it is possible to determine the Young modulus and the Poisson ratio. It turns out that the measured results are in excellent agreement with the existence of area incompressible membranes.

INTRODUCTION

One of the most characteristic properties of amphiphilic compounds is their tendency to form ultrathin stable films at different types of interfaces. These phenomena might occur at the surface between air and water or in the contact layer between two immiscible solvents. Because of their ability to stabilize interfaces, surfactants are often used as emulsifiers, dispersing agents, or foam forming additives. Many natural or biological surface-active compounds, such as proteins or phospholipids, exhibit similar properties. Phospholipids tend to form bilayer structures, which are basic compounds of the plasma membrane of living cells. Although the lipid molecules are known to form stable films, they are still in a liquid analogous state. In living organisms, however, this situation is often more complicated because the bimolecular layers of phospholipid molecules are in close contact with two-dimensional or three-dimensional networks. It is easy to show that these structures enhance the stability of the fluid-like membranes. The functions and importance of biologically relevant two-dimensional networks were thoroughly reviewed in two articles by W. Burchard [1,2]. A typical example, showing the stabilizing influence of two-dimensional structures, is observed in the membrane of erythrocytes. A red blood cell consists of a lipid bilayer that is coupled at well-defined anchor points to a two-dimensional spectrin skeleton. As a consequence, the red blood cell is easily shearable, but strongly resistant to any increase of the local surface area. It was also observed, that the stability of red blood cells depends on the rubber-elastic properties of the spectrin network. Some mutants of mice, which are not able to build up such protein networks, have globular red blood cells instead of biconcave shapes. In addition, these blood cells are so fragile, that they are destroyed even at small mechanical forces. It is hence of foremost interest to understand the stabilizing effects of these networks in living cells.

As biological structures are rather complicated, we have synthesized well-defined model networks, which give new insights into some basic features of these confined systems. In analogy to the spectrin network of red blood cells, we use molecules, which are partially fixed in a planar arrangement. Cross-linked membranes have been prepared with different techniques at fluid interfaces.

These structures can be divided into two separate categories, depending on the nature and strength of bonding forces. There are two extreme cases which are often observed in colloidal systems: transient networks and permanent structures. When the cross-linking process is a result of chemical reactions, the structures are always of the permanent type. In former investigations, we have used the radical polymerization of surface-active methacrylate diesters, in order to synthesize ultrathin cross-linked networks at the interface between oil and water [3]. It turns out that the samples prepared by these techniques exhibit rubber-elastic features and the stabilized films are well characterized in respect to their rheological properties. We have carefully analyzed the two-dimensional sol-gel transition, the kinetics of gelation and the rheological properties of the cross-linked films. The scaling behavior was in excellent agreement with percolation theories. We also observed a power-law relaxation behavior in the vicinity of the sol–gel transition, which coincides with experimental results first published by Winter and co workers [4–7]. It is also interesting to note, that the pore size in these two-dimensional networks could be obtained from electron spin resonance measurements [3]. In a recent publication, we have also investigated the elongational rheological properties of these membranes [8]. For these measurements, we have developed a new experimental technique that consists in measuring the deformation of a microcapsule in the spinning drop apparatus. A theoretical analysis of the deformation process allowed to calculate an interface apparent elastic Young modulus. From a comparison of the surface shear modulus with the Young modulus, it was then possible to evaluate the Poisson ratio, which were of the order of zero. These results point to the existence of area incompressible membranes [9].

Besides these chemically cross-linked membranes, networks can also be formed by physical contacts. A typical surfactant system capable of forming transient cross-linked structures is Span 65 (trioctadecanyl ester of sorbic acid). These surface-active molecules tends to build up fluctuating supermolecular structures, which are stabilized by hydrogen bonds. The fundamental physical properties show a marked dependence on the duration of the measurement; a type of behavior well described by the expression 'transient' or 'temporary' network. A wide variety of other types of bonding, lying between these two extreme cases of chemical or physical network forming mechanisms, are known. In a recent publication, we observed a general mechanism to build up coagulated cross-linked network structures [10]. These systems can be formed by combining positively or negatively charged surfactants like bis-(2-ethylhexyl)sulfosuccinate (AOT) with multivalent counter-ions such as Al^{3+} or Ce^{4+}. The membranes, thus prepared, exhibit striking rubber-elastic properties. It was also possible to visualize these structures using Brewster-angle microscopy. Typical results of these investigations and a comparison to hydrogen bonded network systems are summarized in Figure 35.1.

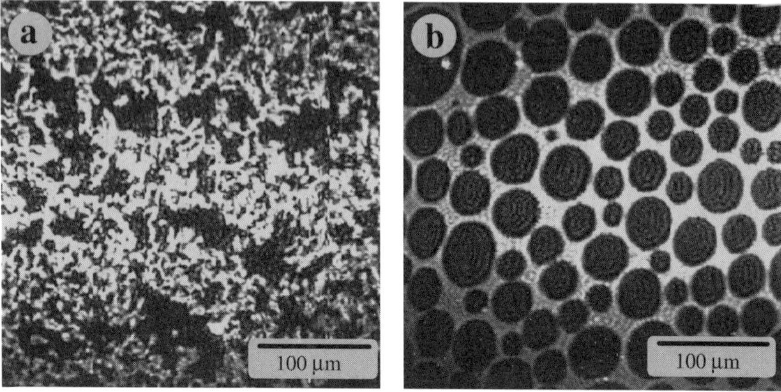

Figure 35.1 Brewster-angle microscopy images of two viscoelastic films: (a) coagulated cross-linked network structure (AOT, $c = 5 \times 10^{-7}$ mol/l; $Al_2(SO_4)_3$, $c = 1 \times 10^{-3}$ mol/l; pH = 5) (b) hydrogen bonded network system (span 65, $\Gamma = 3$ molecules nm^{-2})

It is easy to see that the coagulated network structures consist of disc-like aggregates, having typical sizes of about 10 µm. These particles are in close contact forming network clusters of fractal geometry. The membranes, formed by hydrogen bonds (Span 65) seem to have a different structure which corresponds to a two-dimensional foam. In the vicinity of the sol–gel transition, these networks have many pores, but with increasing concentration the uncovered areas become much smaller. The ultrathin films, stabilized by Coulomb interactions, are in qualitative agreement with calculated clusters of percolation theories [11].

In this article we shall, however, focus on chemically cross-linked systems, which can easily be synthesized from organosilane compounds. Due to the fast cross-linking reactions one obtains stable films which can be used for controlled release processes or for the synthesis of emulsions, foams and microcapsules. These two-dimensional networks open up a new field of interesting technical applications in micro and nano technology.

EXPERIMENTAL PART

Octadecyltrimethoxysilane (ODTMOS, 90%) and octadecyltrichlorosilane (ODTClS, 95%) were purchased from Aldrich, N-cetyl-N,N,N-trimethylammoniumbromide (CTAB, >99%) from Fluka. All monomers and surfactants were used without further purification. Water was obtained from a pure water system (Seralpur PRO 90 CN).

The elongational properties of microcapsules were measured by means of a commercial spinning-drop tensiometer (Krüss SITE 04), inside which, instead of a drop, we introduced a capsule. The capsule consisted of an oil droplet enclosed by an ultrathin cross-linked membrane obtained by polycondensation reactions of organosilanes. Its deformation, caused by the action of centrifugal forces, was measured at increasing rotation rates. The maximum operating speed of the glass tube was limited to 10 000 r.p.m. Above this value vibrations became important and leaking occurred from sealings.

Before starting an experiment, the tube was cleaned and filled with the aqueous solution. Special caution was necessary to avoid air bubbles and dust particles. At a low rotational speed (of the order of 1000 r.p.m.) a dodecane oil droplet containing a specific amount of monomer was injected with a syringe. This droplet, which was held stationary by the rotation of the tube, was still spherical. Under the given conditions, the surface gelation took place in a few minutes and a stable microcapsule was thus formed. The typical polymerization time was of the order of 1 h, but before increasing the rotational speed we waited for about 2 h. After that period all cross-linking reactions were terminated and the ultrathin membrane was stable. At experimental conditions where surface gelation was terminated, an initially spherical capsule can be deformed by increasing the rotation of the glass tube. The shape and size of the capsule was recorded by means of a video camera. The optical system was calibrated with steel wires of known thickness, in order to compensate for optical distortions due to the lens effects of the glass tube and to the refractive indices of the solvents. The capsule diameter and length could then be measured using image analysis.

A theoretical analysis of the mechanics of an initially spherical elastic shell subjected to centrifugal forces allows deduction of the value of the surface elastic coefficients from the experimental measure of capsule deformation. Details of this analysis are summarized in Ref. [8]. The capsule is assumed to be a sphere of radius a in its rest state. It is filled with an incompressible liquid with a density ρ_i and is enclosed by an infinitely thin elastic membrane devoid of bending resistance. The capsule is placed inside a cylindrical tube of radius $b(b > a)$, rotating around its axis with steady angular velocity ω. The tube is filled with another incompressible liquid with a density $\rho_e(\rho_e > \rho_i)$. Buoyancy effects due to gravity are assumed to be much smaller than centrifugal forces, so that the capsule is centered on the tube axis and the system is fully axisymmetric. This condition holds for angular frequencies >3000 revolutions per minute.

The shear rheological properties of flat membranes were evaluated in a Rheometrics fluid spectrometer (RFS II), with a modified shear system [12]. The measuring cell consisted of a quartz dish (diameter 83.6 mm) and a thin biconical titanium plate (angle 2°, diameter 60 mm) which could be placed exactly at the interface between oil and water. The dish was first filled with the aqueous phase. The titanium plate was then positioned at the water surface and a solution of monomers was added. We measured the torque required to hold the plate stationary as the cylindrical dish was rotated with a sinusoidal angular frequency ω. In such experiments, the two-dimensional storage modulus μ' and the loss modulus μ'' can be evaluated from the amplitude and phase angle of the stress and deformation signals.

Optical measurements were performed using a Brewster-angle microscope (BAM 2) designed by Nanofilm Technology GmbH, Göttingen (Germany). The basic principles of BAM experiments are extensively described in Refs. [13–15]. If a p-polarized light beam is incident at Brewster angle to the surface of water (53.1°), no light is reflected. A video camera, arranged in the direction of the reflected light beam, will now observe darkness. In the presence of surface-active compounds, however, the refractive index of the water surface is slightly changed. As a consequence, the video camera will now obtain some light and the image of the network structure can be analyzed. One obtains, hence, a large optical contrast between the pure water surface (black regions) and those parts covered with surfactant molecules (white regions). It is interesting to note that the lateral resolution of the BAM is related to the wave lengths of the incident laser beam (690 nm), but this technique allows to investigate monomolecular films with a thickness of only 1 nm.

EXPERIMENTAL RESULTS

NETWORKS AT THE SURFACE OF WATER

It is well known that octadecyltrimethoxysilane tends to form polymer films when dissolved in an easily evaporable solvent and spread onto the water surface. This occurs by polycondensation reactions which are observed under acid- or alkaline-catalyzed conditions [16]. An extensive description of the film formation in acid- and base-catalyzed regimes was recently published by Sjöblom *et al.* [17]. In the following chapters we shall only focus on the two-dimensional gelation at low pH-values.

Octadecyltrichlorosilane tends to be even more reactive than the corresponding trimethoxysilane compound. This monomer was therefore preferentially used to synthesize ultrathin network structures at the interface between oil (dodecane) and water.

In general the reaction can be divided into hydrolysis and subsequent condensation. The initial hydrolysis of a reactive organosilane leads to the formation of a silanol.

$$\equiv SiX + H_2O \longrightarrow \equiv SiOH + XH \qquad X = OCH_3, Cl$$

In the condensation step the produced silanol reacts either with another silanol

$$2 \equiv SiOH \longrightarrow \equiv Si-O-Si\equiv + H_2O$$

or with an organosilane to generate a siloxane.

$$\geq SiX + \geq SiOH \longrightarrow \geq Si-O-Si\leq + HX$$

The formation of very stable siloxane groups accounts for the fact that the polycondensation occurs at room temperature.

The molecular structure of polysiloxane films was often discussed in literature. From extensive investigations of low-angle X-ray reflectivity and ellipsometry Wassermann *et al.* [18] calculated a film thickness of about 2 nm for complete monolayers. From these measurements, it was also possible to determine the area occupied by each alkylsiloxane group at the surface on silicon–silicon dioxide substrates, which was of the order of 0.21 ± 0.03 nm^2 for each R–Si group. From these experiments Ulman [19] supposed that the alkyl chains are nearly perpendicular aligned in respect to the interface (orientation angle $>75°$). This leads to a possible cyclic trimer where the chains are connected at the axial position as shown in Figure 35.2.

Compression isotherms of films at the surface of water showed that the corresponding molecular area was 0.23–0.25 nm^2 in the regime of condensed monolayers [20]. The experimental results of Ulman lead to a simple model concerning the molecular structure of two-dimensional polysiloxane networks. We can expect that the ultrathin membranes are composed of trimeric clusters, which are cross-linked. A schematic drawing of these membranes is demonstrated in Figure 35.3.

The kinetics of surface gelation of these films can be analyzed by measuring the two-dimensional storage modulus μ' as a function of the reaction time (time-sweep experiment). In this test, a sinusoidal deformation with small amplitude is applied to the sample at a constant angular velocity ω. As the two-dimensional shear modulus does not depend on the frequency, the data describe the evolution

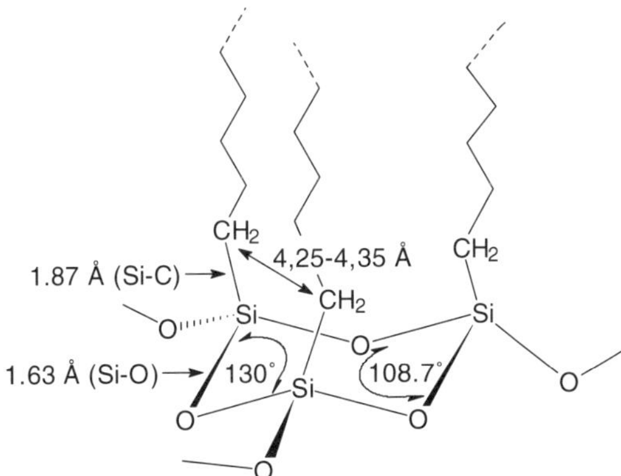

Figure 35.2 Cyclic structure of a siloxane trimer as the basic part of the polyorganosiloxane networks [19]

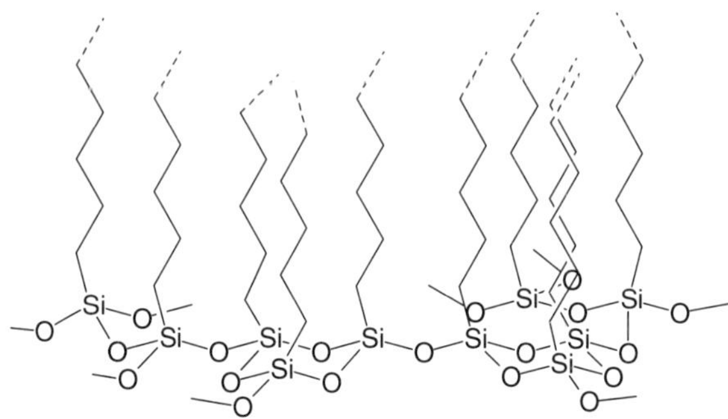

Figure 35.3   Amorphous film structure

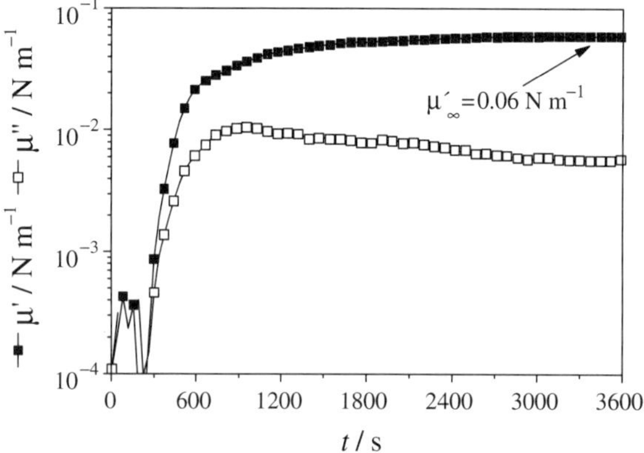

Figure 35.4   The two-dimensional storage modulus μ' and loss modulus μ'' of the ultrathin membrane as a function of the reaction time (surface concentration Γ_m of ODTMOS = 3.9 molecules nm^{-2}, $T = 23\,°C$, $\omega = 6.28$ rad s^{-1}, $\gamma = 0.5\%$). The polymerization was induced by the addition of H_2SO_4 (20 mmol/l, pH 1.6) to the water phase

of cross-linking points as a function of time. A typical curve showing such data is given in Figure 35.4.

In this experiment, the monomers were first dissolved in tert.-butylmethylether and afterwards spread on the surface of water (pH = 1.6). The evaporation of the organic solvent takes about 5 minutes. It is evident that the polymerization proceeds with time constants of the order of about 1 h. As the diffusion of reactive surfactant molecules to the surface occurs within some milliseconds, the characteristic gelation time must be the result of a cooperative process originated by the formation of junction points. It is interesting to note that the cross-linked network structure exhibits rubber–elastic properties. Figure 35.5 gives some insight into the dynamic features of these membranes.

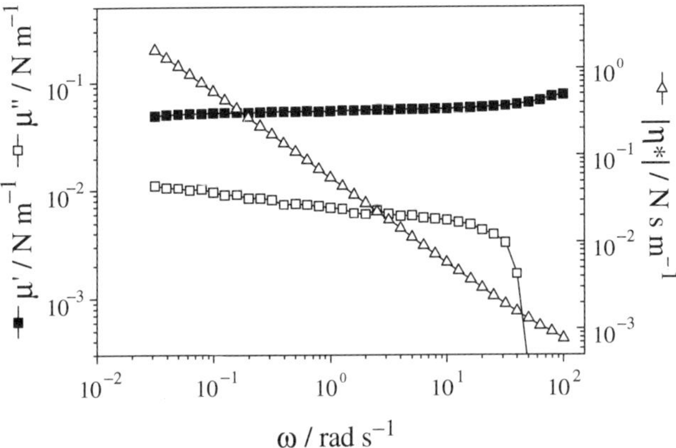

Figure 35.5 The two-dimensional storage modulus μ', two-dimensional loss modulus μ'' and the complex viscosity $|\eta^*|$ as a function of the angular frequency ω for an ultrathin cross-linked polysiloxane film at the surface of water ($\Gamma_m = 3.9$ molecules nm^{-2}, $T = 23\,°C$, $\gamma = 0.5\%$, 20 mmol/l H_2SO_4)

The two-dimensional storage modulus μ' describes the elastic properties of the sample and the two-dimensional loss modulus μ'' is proportional to the energy dissipated as heat. The nearly constant plateau value μ_∞' points a negligible influence of any relaxation processes. Such curves are typical for rubber–elastic materials. This phenomenon can be traced back to the existence of permanent cross-linking points, which remain stable as a function of time. It is interesting to note that the number of elastically effective chains Γ_{eff}, which can be calculated from equation (35.1), where k represents Boltzmann's constant and T the temperature, is of the order of 15 molecules nm^{-2}:

$$\mu_\infty' = \Gamma_{\text{eff}}\,kT \tag{35.1}$$

This value is a factor of 3.8 larger than the number of monomers, fixed in the network structure. As each molecule is connected to three other monomers the total number of elastically effective chains should be a factor of three larger than the number of monomers. The higher value obtained from experiment could also be ascribed to additional physical linkages, such as entanglements or crystal formations.

It is interesting to note that a certain minimum surface concentration of reactive monomers is required for the formation of a two-dimensional network structure. This is clearly shown in Figure 35.6, where the number of elastically effective chains was calculated as a function of the surface concentration.

The threshold interface concentration Γ_c, where gelation occurs, is equal to 2.5 molecules nm^{-2}. Above this limiting value, an infinitely large cluster is formed spanning the whole sample. As the polymerization proceeds on a

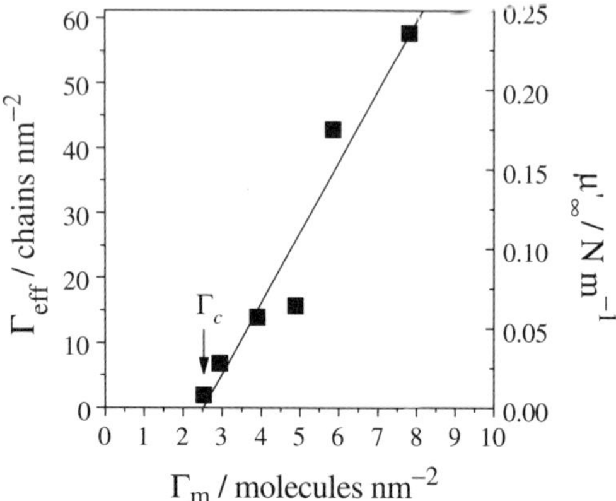

Figure 35.6 The number of elastically effective chains Γ_{eff} and plateau modulus μ'_∞ as a function of the surface concentration Γ_m of ODTMOS for ultrathin cross-linked polysiloxane films at the surface of water. Linear fit: $\Gamma_{eff} = \mu'_\infty k^{-1} T^{-1} = a\Gamma_m + b$ ($a = 10.8$, $b = -27.3$ molecules nm^{-2}, $R = 0.9681$, $\Gamma_c = -b/a = 2.52$ molecules nm^{-2})

hexagonal lattice, the gel point should be equal to the critical conversion point $p_c = 0.653$ [11]. This means that about 65% of all bonds must have reacted in order to build up a continuous two-dimensional network structure. For a square lattice, the gel point is already given at $p_c = 0.5$. These limiting values were derived from percolation theories [11]. Assuming that each molecule forms exactly three contacts, we can easily calculate that a dense film, where each lattice point is occupied by just one molecule, corresponds to a molecular area of 0.258 nm^2. This value is in good agreement with the previous described values of $0.23-0.25$ nm^2 by Lindén *et al.* [20].

In analogy to three-dimensional systems, the elastic modulus remains constant up to quite large deformations. This is clearly shown in the strain-sweep experiment demonstrated in Figure 35.7.

It is interesting to note that the modulus remains nearly constant up to deformations of about 4%. This is similar to three-dimensional rubber-elastic systems, where breakdown of the structure generally starts to occur at deformations between 10% and 500%. In terms of molecular models, this phenomenon can be explained by the high flexibility of the organosiloxane chains. The relative low value of 4% points to the presence of relatively short elastically effective chains. A typical example of the network structure, formed by octadecyltrimethoxysilane monomers at the surface of water, is visualized in Figure 35.8.

Image (a) shows a membrane in the sol-state that already contains some finite clusters. These aggregates resemble honeycombs, in which the units have no close contact to each other. This structure could be interpreted as a skeleton of

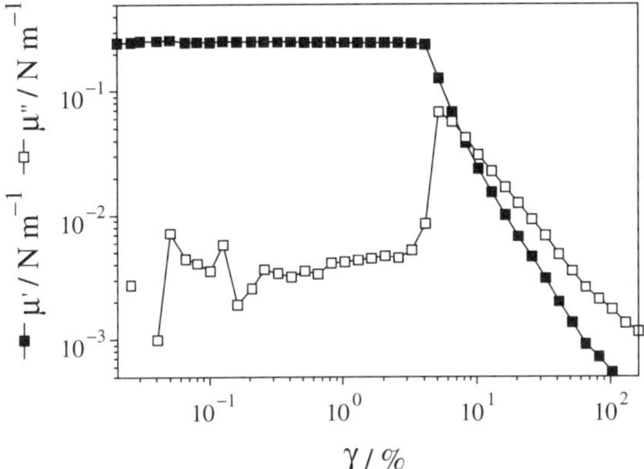

Figure 35.7 The two-dimensional storage modulus μ' and loss modulus μ'' as a function of deformation γ for an ultrathin cross-linked polysiloxane film at the surface of water ($\Gamma_m = 7.8$ molecules nm^{-2}, $T = 23\,^\circ$C, $\omega = 6.28$ rad s^{-1}, 20 mmol/l H$_2$SO$_4$)

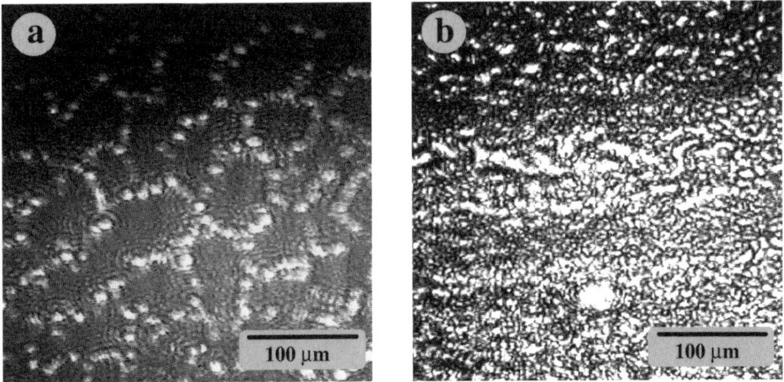

Figure 35.8 Brewster-angle microscopy images of network structures formed by octadecyltrimethoxysilane monomers at the surface of water: (a) film in sol-state, $\Gamma_m = 2$ molecules nm^{-2} (b) film in gel-state, $\Gamma_m = 4$ molecules nm^{-2}

a two-dimensional foam with many defects. It is possible that this structure is induced by evaporation of the solvent.

In the gel state (b) the uncovered areas are much smaller. The film consists of roughly spherical particles that are in close contact forming a dense two-dimensional network. The geometry appears to be of fractal nature. Furthermore, a kind of crumpling or buckling of the flat surface is indicated by optically homogeneous stripes that may be due to a superstructure.

NETWORKS AT THE INTERFACE BETWEEN OIL AND WATER

In order to increase the reaction rates it was helpful to use very small amounts of an additional surfactant. Even at low concentrations of CTAB in the water phase the rates of polymerization increased to a maximum. By using this system it was possible to synthesize soft films at low concentrations in less than 2 h, as shown in Figure 35.9.

As the molecular structure of the cross-linked polysiloxane films is identical to the one prepared from the corresponding trimethoxy compounds, all physical properties should be similar. This behavior is demonstrated in Figure 35.10, where the dynamic features of a typical film is plotted as a function of the angular frequency.

Comparison of these experimental data with Figure 35.5 reveals that both structures have common properties. Although the monomer concentrations are different the evolution of loss and storage modulus is comparable. We can therefore assume that nearly the same type of network structure is formed at the interface between oil and water as at the surface of water.

The interfacial polymerization is not restricted to flat surfaces, and we can use the same technique in order to synthesize microcapsules. These particles were formed by injection of the oil phase into the rotating tube of the spinning-drop tensiometer. Typical microcapsules observed at different rotational speeds are shown in Figure 35.11.

For small values of ω (less than 3000 r.p.m.), the capsule is slightly off center. The capsule radius is about 0.5 mm. At low values of the angular frequency, gravity is important, and the corresponding mechanical model, predicting the

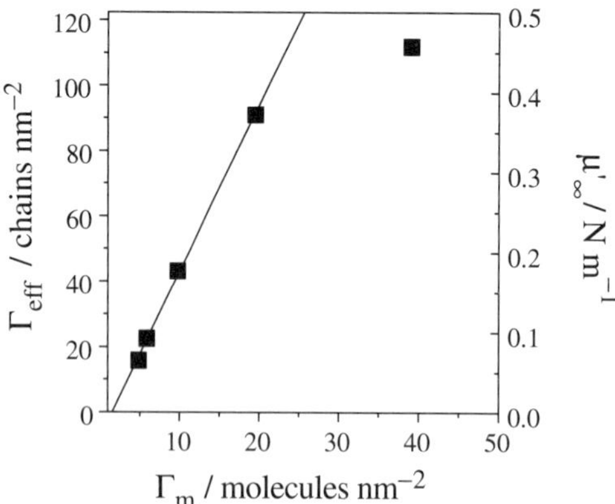

Figure 35.9 The number of elastically effective chains Γ_{eff} and plateau modulus μ_∞' as a function of the surface concentration Γ_m of ODTClS for ultrathin cross-linked polysiloxane films at the interface between dodecane and water (5 μmol l^{-1} CTAB in the water phase)

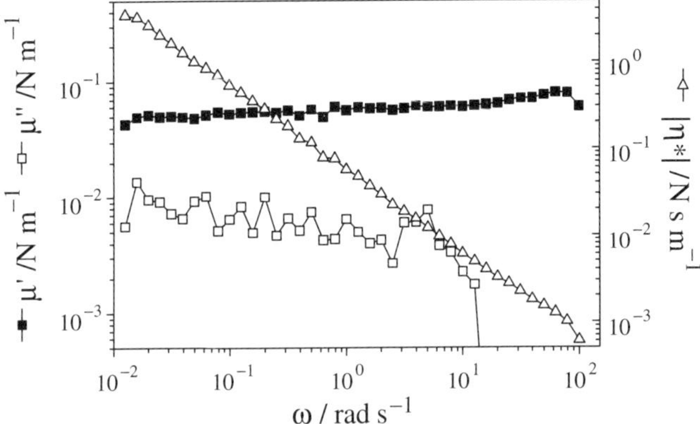

Figure 35.10 The two-dimensional storage modulus μ', two-dimensional loss modulus μ'' and the complex viscosity $|\eta^*|$ as a function of the angular frequency ω for an ultrathin cross-linked polysiloxane film at the interface between dodecane and water ($\Gamma_m = 4.9$ molecules nm^{-2} ODTCIS for 5 ml dodecane, $T = 23\,°$C, $\gamma = 0.2\%$, 5 μmol l^{-1} CTAB in the water phase)

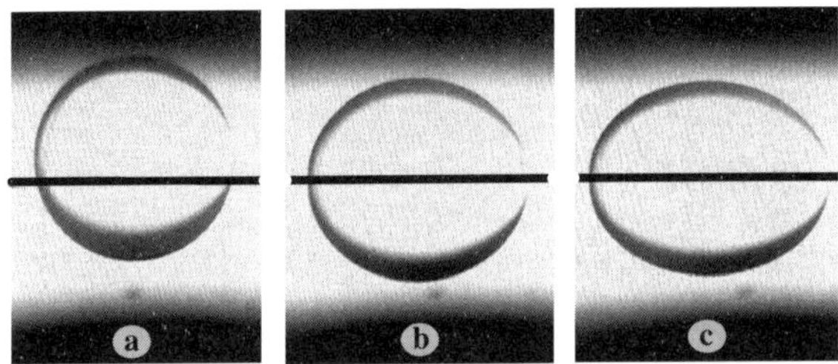

Figure 35.11 Microcapsules observed at different speeds of the rotating tube: (a) 1000 r.p.m.; (b) 5000 r.p.m. (c) 10000 r.p.m. The horizontal line represents the axis of symmetry of the tube

particle deformation, becomes very complicated. Thus, the results presented here were all obtained for large enough rotational speeds, for which buoyancy effects may be ignored and the simple model proposed in Ref. [8] may be used. According to the theory of D. Barthès-Biesel, the capsule deformation can be described by the simple equation

$$D = \frac{l-b}{l+b} = \Delta\rho\omega^2 a^3 \cdot \frac{5+v_s}{16 \cdot E_s} \tag{35.2}$$

Here, l denotes the longest and b the shortest axis of the deformed capsule. The difference of density between dodecane inside the capsule and the continuous water phase is $\Delta\rho = -0.252$ g/cm^3. ω is the angular frequency of the rotating tube and a is the radius of the capsule in the quiescent state. The parameter E_s describes the Young modulus of the ultrathin membrane and v_s is the surface Poisson ratio. As the surface shear modulus μ_s and the Young modulus are related by

$$\mu_s = E_s/2 \cdot (1 + v_s) \tag{35.3}$$

we can replace E_s in equation (35.2) and one finally obtains

$$D = \Delta\rho \cdot \omega^2 \cdot a^3 \cdot \frac{5 + v_s}{32 \cdot \mu_s \cdot (1 + v_s)} \tag{35.4}$$

It is worth while to mention that this simple formula holds only for small deviations from the spherical shape. Non-linear effects, which are often observed at elevated deformations, are not included in this equation. Typical results, where the deformations of the capsules are plotted as a function of $-\Delta\rho\omega^2 a^3$, are summarized in Figure 35.12.

The deformation of the spinning capsule is the same linear function of $-\Delta\rho\omega^2 a^3$, for all three diameters tested. This is in agreement with the model equation (35.4). Assuming that μ_∞' is equal to μ_s, it is thus possible to determine the Poisson number from the slope of the linear correlation, shown in Figure 35.9. In combination with equation (35.3) one also obtains Young's modulus.

Relevant data of these experiments are summarized in Figure 35.13.

It is easy to recognize that the Poisson ratio attains negative or positive values for all three interfaces. This ratio measures the strain in the transverse direction which results from a longitudinal extension. A positive Poisson ratio is usually found for three-dimensional materials. This corresponds to a material that shrinks

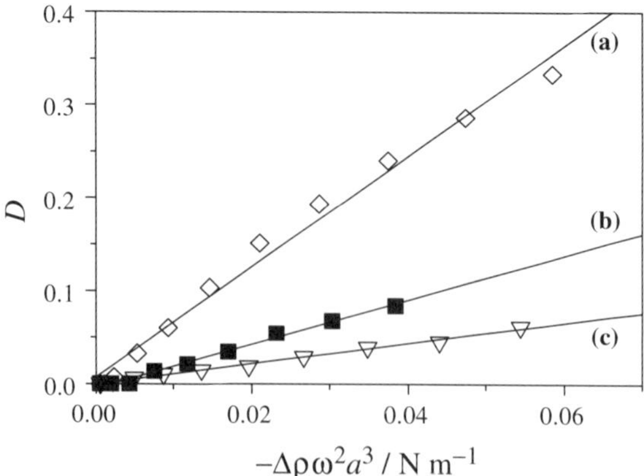

Figure 35.12 The deformation D as a function of $-\Delta\rho\omega^2 a^3$ for three different capsules: (a) $\Gamma_m = 3.6$ molecules nm^{-2}, $a = 0.595$ mm; (b) $\Gamma_m = 4.5$ molecules nm^{-2}, $a = 0.555$ mm; (c) $\Gamma_m = 9.4$ molecules nm^{-2}, $a = 0.585$ mm

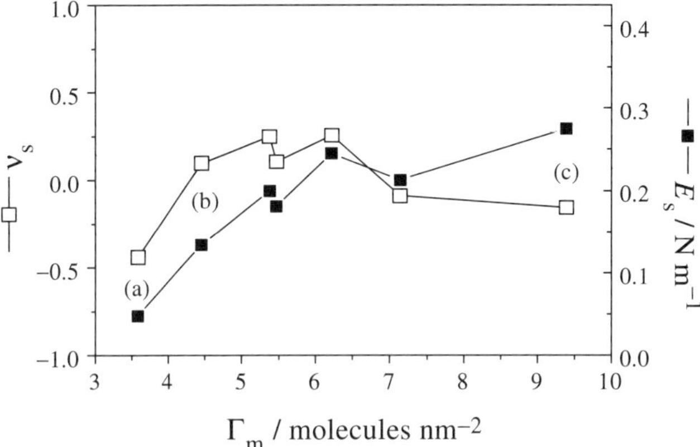

Figure 35.13 The Poisson ratio ν_s and Young's modulus E_s as a function of interface concentration of the monomer Γ_m for different capsules

transversally when it is stretched longitudinally. For two-dimensional membranes the behaviour is different and it seems that negative values of ν_s are also possible. From mean field calculations and Monte Carlo simulations, Boal *et al.* [9] calculated ν_s for different types of tethered networks. For self-avoiding and phantom networks, the Poisson ratio varies between -1 and $+1$. However, for large deformations ν_s was found to be always positive, but at intermediate deformations, negative values were obtained. These results can be understood by imaging a crumpled sheet of paper, which stretches in the direction perpendicular to that of stress. Presently, it is not clear whether the ultrathin cross-linked membranes can really be compared to a phantom or self-avoiding network structure. The values of ν_s that were obtained are at least in general agreement with these theoretical predictions.

CONCLUSIONS

The self-association process of different types of reactive monomers can be used to synthesize ultrathin networks at the surface of water or at the interface between oil and water. Up to now, this phenomenon was observed in chemically and physically cross-linked samples. It is interesting to note that all structures we have investigated so far exhibit similar properties. The basic network structure consists of dense aggregates having close contact and they tend to form infinitely large clusters of fractal geometry. These networks have many pores, but with increasing monomer concentration the uncovered areas become much smaller. The two-dimensional network structure is in qualitative agreement with calculated clusters of percolation theories, although a quantitative comparison has not yet been performed [11]. This work is still in progress. It is worth while to mention that all ultrathin networks exhibit striking elastic properties. They can, hence, be used to form stable micro- or nanocapsules. The cross-linked films on the surfaces of the droplets have an appreciable stabilizing effect on the spherical shape. Such emulsions represent a new kind of encapsulation. As in the case of living cells, the deformation of the droplets depends on the elasticity of the membrane. In a flow field one observes quite marked deformations, orientation processes and bursting similar to those occurring in red blood cell dispersions. The encapsulated oil droplets have a very simple structure, and therefore represent a suitable model system for studying these phenomena. Artificial cells of this kind are also suitable for the controlled release of active substances stored in either the oil or the water phase. The network of the surrounding membrane acts as a semipermeable barrier, allowing only a limited flow though the boundary layer. Thus it is possible, for example, to release drugs in a controlled manner over long periods. It is also possible to break the capsules on applying mechanical forces. Because of these numerous areas of application, there is now worldwide interest in synthesizing new types of polymer films, and the microscopic and macroscopic properties of these substances are the subject of extensive studies ranging from fundamental problems to technological products.

ACKNOWLEDGMENT

Financial support of this work by a grant of the Deutsche Forschungsgemeinschaft (DFG Re 681-4/4) and the Fonds der Chemischen Industrie are gratefully acknowledged.

REFERENCES

1. W. Burchard, *Chem. Unserer Zeit*, **23**, 7 (1989).
2. W. Burchard, *Chem. Unserer Zeit*, **23**, 69 (1989).
3. A. Burger, H. Leonhard, H. Rehage, R. Wagner and M. Schwoerer, *Macromol. Chem. Phys.*, **196**, 1 (1995).
4. H.H. Winter, *Progr. Colloid Polymer Sci.*, **75**, 104 (1987).
5. H.H. Winter, *Polym. Eng. Sci.*, **27**, 1698 (1987).
6. F. Chambon and H.H. Winter, *Polym. Bull.*, **13**, 499 (1985).
7. H.H. Winter and F. Chambon, *J. Rheol.*, **30**, 367 (1986).
8. G. Pieper, H. Rehage and D. Barthès-Biesel, *J. Colloid Interface Sci.*, **202**, 293 (1998).
9. D.H. Boal, U. Seifert and J.C. Shillock, *Phys. Rev. E*, **48**, 4274 (1993).
10. H. Rehage, B. Achenbach and A. Kaplan, *Ber. Bunsenges. Phys. Chem.*, **101**, 1683 (1997).
11. D. Stauffer and A. Aharony, *Percolationstheorie*, VCH, Weinheim, 1995.
12. S.G. Oh, J.C. Slattery, *J. Colloid Interface Sci.*, **67** (1978).
13. D. Hönig and D. Möbius, *J. Phys. Chem.*, **95**, 4590 (1991).
14. D. Hönig, G.A. Overbeck and D. Möbius, *Adv. Mater.*, **4**, 419 (1991).
15. S. Hénon and J. Meunier, *J. Chem. Phys.*, **98**, 9148 (1992).

16. C.J. Brinker and G.W. Scherer, Sol-Gel Science, *The Physics and Chemistry of Sol-Gel Processing*, Academic Press, San Diego, CA (1991).
17. J. Sjöblom, G. Stakkestad and H. Ebelhoft, *Langmuir*, **11**, 2652 (1994).
18. S.R. Wassermann, G.M. Whiteslides, M. Tidswell, B.M. Ocko, P.S. Pershan and J.D. Axe, *J. Am. Chem. Soc.*, **111**, 5852 (1989).
19. A. Ulman, *Adv. Mater.*, **2**, 573 (1990).
20. M. Lindén, J.P. Slotte, J.B. Rosenholm, *Langmuir*, **12**, 4449 (1996).

36

Preparation and Application of Crosslinked Monodisperse Particles

ARVID BERGE[1], RUTH SCHMID[2], ODDVAR AUNE[2], PER STENSTAD[2] and LARS KILAAS[2]

[1]Department of Industrial Chemistry Norwegian University of Science and Technology NTNU, N-7034 Trondheim, Norway
[2]SINTEF Applied Chemistry N-7034 Trondheim, Norway

ABSTRACT

A brief survey is given of the method of activated swelling of polymer particles and its basic thermodynamic principles. Then are described several different types of monodisperse particles including magnetizable particles and their applications within chromatography and biomedicine. Finally a recently developed new method for producing monodisperse polymer particles by step growth polymerization is presented.

Wiley Polymer Networks Group Review Series Vol. 2. Edited by B.T. Stokke and A. Elgsaeter
© 1999 John Wiley & Sons Ltd

INTRODUCTION

The paper is given in honour of the memory of the late Professor John Ugelstad. In 1977 Ugelstad invented the so-called activated swelling method, an unique method whereby it is possible to increase the swelling capacity of polymer particles towards low molecular weight compounds several orders of magnitude. This was a breakthrough for the preparation of polymer particles and especially so for the preparation of large monosized particles of diameters from about 1 to more than 100 µm. The activated swelling method provides unparalleled freedom of choice as regards the type of vinyl monomers to be used, the degree of crosslinking and the option of producing macroporous structures.

PREPARATION OF MONODISPERSE PARTICLES BY CHAIN ADDITION POLYMERIZATION APPLYING ACTIVATED SWELLING

The main point in the preparation of large monodisperse polymer particles is that one starts with a dispersion, normally in water, of small monodisperse polymer particles which are activated by incorporating a substantial amount of a relatively low molecular weight highly water insoluble compound (Y) prior to swelling with a vinyl monomer (Z) and subsequent polymerization. The activated particles can absorb 10 to more than 1000 times the amount of (Z) compounds than can pure polymer particles.

The basic thermodynamic principles behind the swelling have been discussed in terms of the original Flory–Huggins theory for the free energy of mixing [1–5]. Although this theory has several deficiencies it turned out that it suffices to give a good, qualitative and in most cases also an acceptable semiquantitative picture of the often large effect of the different parameters describing the system.

In the case we intend to prepare large monodisperse polymer particles by swelling activated monodisperse seed particles we have a three phase system. The composition of the phases in the most simple cases is:

Phase (a): Spherical seed particles or droplets containing polymer (P) and/or substance (Y) and (Z) compounds being monomers and other slightly water soluble compounds.

Phase (b): (Z) components which may be present in bulk form or as droplets.

Phase (c): Dispersion medium and a small amount of dissolved (Z).

The coexistence of any component (i) in the three phases in contact with each other requires that at equilibrium the following condition must be fulfilled:

$$\overline{\Delta G_{ia}} = \overline{\Delta G_{ib}} = \overline{\Delta G_{ic}} \tag{36.1}$$

i.e. the partial molar free energy of a component $\overline{\Delta G_i}$ must be the same in all three phases.

 463

When water is the dispersion medium it has been assumed that only (Z) compounds may be transported through the continuous phase, while both polymer (P) and compound (Y) are completely insoluble.

If pure Z_i in a plane phase is used as a reference state, the partial molar free energy of mixing of Z_i in a phase consisting of n components may be written:

$$\Delta \overline{G}_i / RT = \ln \phi_i + \sum_{\substack{j=1 \\ j \neq i}}^{n} (1 - m_{ij}) \phi_j + \sum_{\substack{j=1 \\ j \neq i}}^{n} \chi_{ij} \phi_j^2$$

$$+ \sum_{\substack{j=1 \\ j \neq i}}^{n-1} \sum_{\substack{k=j+1 \\ k \neq i}}^{n} \phi_j \phi_k (\chi_{ij} + \chi_{ik} - \chi_{jk} m_{ij}) + 2 \gamma \overline{V}_i / rRT \qquad (36.2)$$

where ϕ denotes volume fractions, $\overline{V}_i$ is the partial molar volume, $m_{ij} = \overline{V}_i / \overline{V}_j$, χ_{ij} is the interaction parameter per molecule of compound Z_i with compound Z_j, γ is the interfacial tension, and r is the radius of the particle.

A thorough discussion of the application of equation (2) and the appropriate modifications has been given by Ugelstad *et al.* [3–5].

Adding monomer (Z) in bulk form to a dispersion of activated seed particles containing polymer (P) and oligomer (Y), the monomer will diffuse through the continuous phase and swell the particles until an equilibrium situation is reached. In this case we have a three component system and equation (2) leads to:

$$\ln \phi_Z + (1 - m_{ZY}) \phi_Y + \phi_P + \phi_Y^2 \chi_{ZY} + \phi_P{}^2 \chi_{ZP} + \phi_Y \phi_P (\chi_{ZY} + \chi_{ZP} - \chi_{YP} m_{ZY})$$

$$+ 2 \gamma_a \overline{V}_{Z_a} / r_a RT = 0 \qquad (36.3)$$

In case the monomer (Z) is added in the form of droplets the right hand side of the equation should be replaced by $2 \gamma_b \overline{V}_{Z_b} / r_b RT$ instead of zero.

The ϕ values are volume fractions in the (a) phase at equilibrium swelling, m_{ZY} is the ratio of the partial molar volumes of the (Z) compound and the (Y) compound; χ_{ZP}, χ_{ZY} and χ_{YP} are interaction parameters of Z with P, of Z with Y and of Y with P respectively; r_a and r_b are radii of the (a) phase and the droplets of pure (Z) (b-phase), at equilibrium swelling; $\overline{V}_{Z_a}$ and $\overline{V}_{Z_b}$ are the partial molar volumes of (Z) in the particles and the droplets respectively; and γ_a and γ_b are interfacial tensions.

Ugelstad *et al.* have worked out several different graphical figures to demonstrate the effect of various parameters on the swelling capacity of particles. Examples of such are shown in Figures 36.1 and Figure 36.2 where equation (36.3) is used to calculate the curves. V_Z, V_Y and V_P are the volumes of (Z), (Y) and (P) respectively at equilibrium swelling with (Z) present as plane bulk phase.

In practice the total swelling capacity is not necessarily utilized. Within wide limits the method of activated swelling allows preparation of monodisperse particles of predetermined size controlled by the swelling ratio chosen. All ingredients

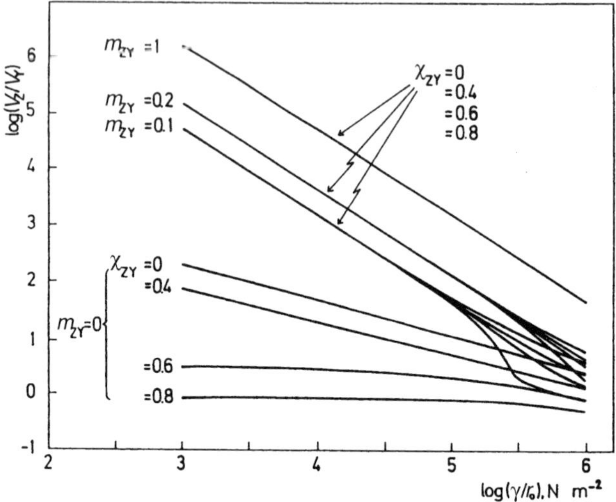

Figure 36.1 Swelling capacity of droplets of water insoluble compound (Y) with substance (Z) versus γ/r_o for various values of m_{ZY} and χ_{ZY}, where r_o is the initial radius of Y-droplets

$$\overline{V}_Z = 10^{-4}\ m^3/mol, \quad T = 323\ K, \quad r_b = \infty$$

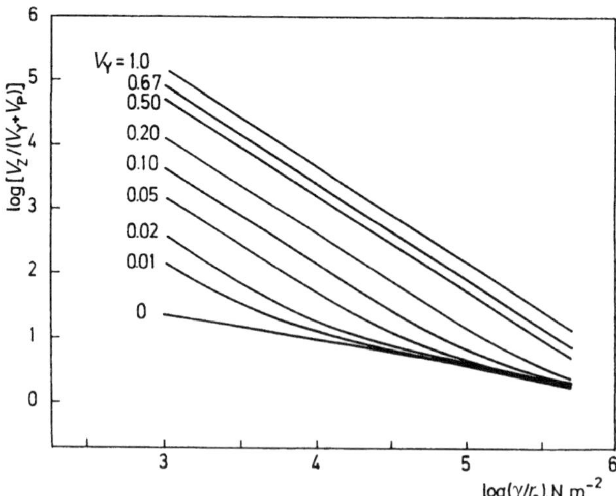

Figure 36.2 Swelling capacity of polymer (P)-oligomer (Y) particles versus γ/r_o for different values of V_Y, where r_o is the radius of the polymer-oligomer particles prior to swelling with (Z).

$$V_Y + V_P = 1, \quad m_{ZY} = 0.2, \quad \chi_{ZY} = \chi_{ZP} = 0.5, \quad \chi_{YP} = 0,$$
$$\overline{V}_Z = 10^{-4}\ m^3/mol, \quad T = 323\ K, \quad r_b = \infty$$

including initiator are introduced into the activated, highly swellable seed particles before polymerization is started by increasing the temperature. The method is robust and reproducible, and carried out properly a coefficient of variation in particle size less than 1% may be achieved.

The method of activated swelling has proved to be especially suitable for the preparation of crosslinked particles, both compact ones and particles with a porous macroreticular structure. Thus, the method has been successfully applied to produce macroporous polymer particles from a number of different vinyl monomers. In the latter case the swelling is carried out applying inert diluents in addition to monomers including crosslinking agents. The pore volume, pore size and pore size distribution may be varied by adjusting the composition and the relative amounts of inert diluents and the polymer system. The monodisperse technology based on the activated swelling method ensures a high degree of batch to batch reproducibility and an equal bead-to-bead morphology.

During the 1990s an increasing number of publications have appeared from several international research groups applying the activated swelling method for preparation of monodisperse, macroporous particles [6,7]. The complete method is sometimes called a shape template method.

The characterization of porous particles with respect to specific surface area, total pore volume, pore size distribution and particle morphology has been carried out by the BET-method, mercury porosimetry, scanning electron microscopy (SEM) and transmission electron microscopy (TEM) of ultrathin sections of the particles.

Examples of the visualization of the particle surface morphology by SEM pictures are shown in Figure 36.3 for two samples of 15 μm macroporous particles of poly(styrene-co-divinylbenzene). These particles have different pore structures brought about by varying the chemical composition of the pore forming agent. Characterizations of the particle surface by AFM, atomic force microscopy, may reveal even more details than SEM pictures. A comparative example of the characterization of 30 μm macroporous particles is shown in the next figures. In Figure 36.4 the particle surface has been examined with a Nanoscope III (AFM) microscope, using the tapping mode procedure. Figure 36.5 shows a SEM picture of the same particles. The higher magnification achieved with AFM gives more details than SEM.

MONODISPERSE PARTICLES IN CHROMATOGRAPHY

One of the most important and largest field of application of monodisperse polymer particles today is as separation media or column packing materials in liquid chromatography. The flow properties are ideal with monodisperse packings, and it is possible to achieve high resolution without sacrificing capacity or speed [8,9].

Monodisperse packings were first brought to the market by Pharmacia in 1982 in connection with the fast protein liquid chromatography system (FPLC) for

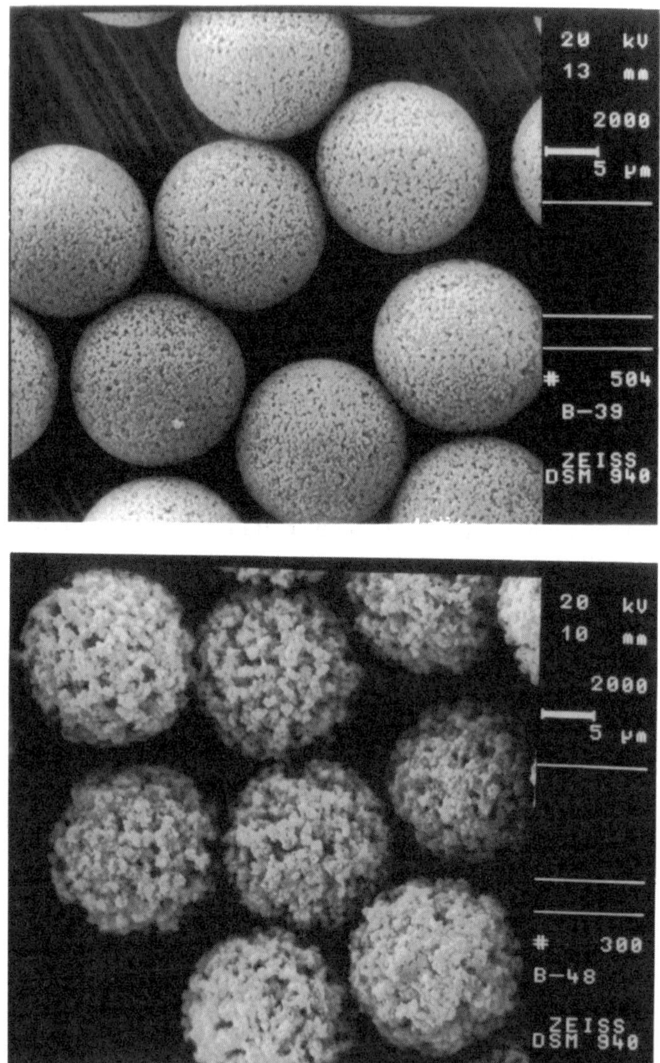

Figure 36.3 SEM picture of two samples of 15 µm macroporous particles with different pore structure

analysis and separation of biological substances. The basis was 10 µm monodisperse macroporous particles of poly(styrene-co-divinylbenzene) prepared by the activated swelling method. The particles were surface modified by a multistep modification to give three types of hydrophilic ion exchange systems, the so-called MonoBeads™. The MonoBeads soon became famous all over the world, and their applications are described in an immense number of scientific papers.

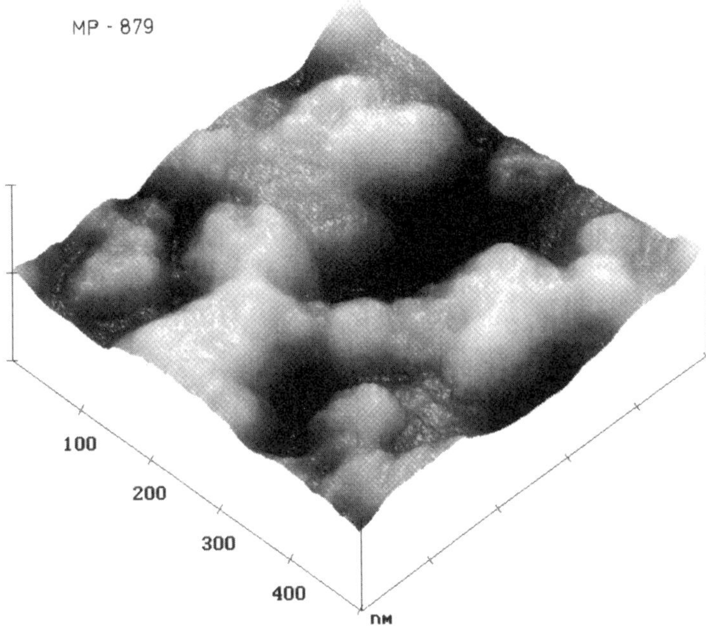

Figure 36.4 AFM picture of details on the surface of 30 μm macroporous particle

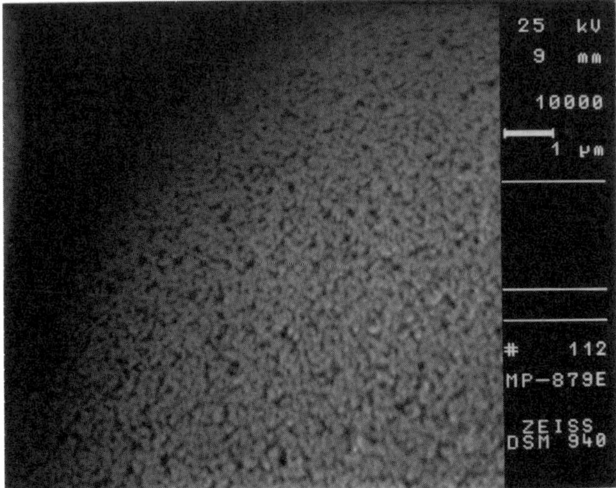

Figure 36.5 SEM picture of 30 μm macroporous particle

In 1993 Pharmacia Biotech (now Amersham Pharmacia Biotech) launched a new range of particle supports named Source^{TM} to give fast, high-resolution preparative separations. The first Source matrices consist of hydrophilized 15 μm monodisperse, macroporous poly(styrene-co-divinylbenzene) particles in which the pore size distribution is balanced to give high capacities for peptides, proteins and oligonucleotides. Pore structure is consistent both bead-to-bead and batch to batch. Now, Source media have also been launched as 30 μm monodisperse particles. They are ideal for the intermediate stages of industrial purification processes where high productivity and maintained performance at large scale are important.

We have to some extent studied the applicability of our monodisperse, macroporous particles in size exclusion chromatography [10,11]. Several interesting results have been achieved both in aqueous and non-aqueous systems. Columns with 5, 10 or 20 μm particles of poly(styrene-co-divinylbenzene) used in inverse SEC calibration with polystyrene standards in THF showed unusually high plate counts. With 5 μm particles the height equivalent to a theoretical plate was 0.006 mm, almost as small as the particle diameter.

Different types of hydrophilized 15 μm particles were tested in aqueous size exclusion chromatography. For particles with 70% of the pore volume confined to pores with radii larger than 2000 Å, it was estimated that the molecular weight range of polysaccharides of the pullulan type accessible to SEC was extended up to $\sim 10^8$. The corresponding value for an extended, rod like, triple-helical polysaccharide (scleroglucan) was $10^{7.}$ [12].

MAGNETIC MONODISPERSE POLYMER PARTICLES

In 1982 Ugelstad and coworkers succeeded in developing new processes for the introduction of magnetic oxides into polymer particles [13]. Rendering magnetic properties to the particles, making it possible to manipulate and collect particles in a magnetic field, was indeed a key to open up many new applications within the biomedical field [5,14,15].

The most common types of particles, which are now used in cell separation, microbiology and molecular biology, are prepared from monodisperse porous polymer particles. The magnetic material is formed inside the pores of the particles, which have been derivatized so that they carry oxidizing groups covalently bound to the surface of the pores. Treatment of these particles with ferrous salts (Fe^{2+}) under alkaline conditions causes magnetic iron oxides to precipitate as fine grains evenly distributed through the pore volume of the particle.

Figure 36.6 shows a TEM of ultrathin sections of magnetic particles confirming that the magnetic iron oxides are evenly distributed. According to analysis by Mössbauer spectroscopy and X-ray powder diffraction, the iron is mainly present as γ-Fe_2O_3, maghemite which is almost as magnetic as magnetite. The saturation magnetizations of maghemite and magnetite, at $20\,°C$ are 76 and 92 e.m.u. respectively.

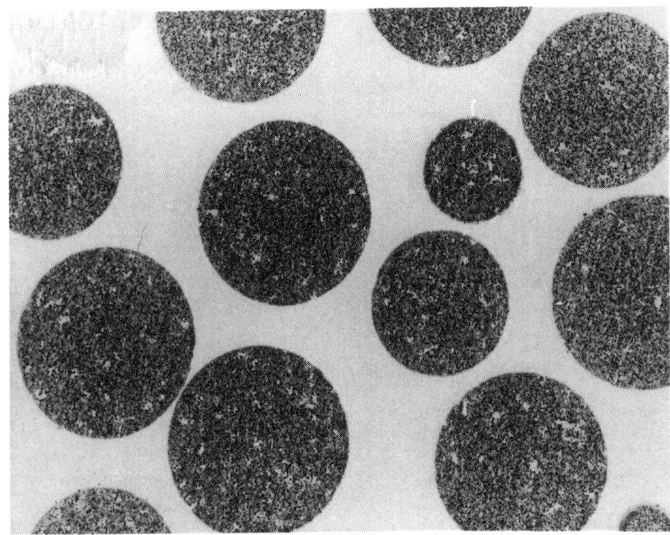

Figure 36.6 Transmission electron micrograph of ultrathin sections of magnetic particles

By the in situ precipitation of magnetic oxides it is possible to prepare particles having high content of iron, up to about 35% iron by weight. The procedure ensures that the particles become superparamagnetic, i.e. they are only magnetic in the presence of a magnetic field, and have no remanent magnetization after the magnetic field has been removed. Figure 36.7 shows the magnetization curve of typical monodisperse magnetic particles. Increasing and decreasing magnetic field shows no hysteresis, accordingly the particles have neither retentivity nor coercivity.

The superparamagnetic property of magnetic polymer particles is very important from a practical point of view as it means that suspended particles may repeatedly be collected by a magnet, and immediately redispersed once the magnetic field is removed.

After the introduction of the magnetic oxides into the particles most particles are treated with polymerizable substances that fill up the pores to give a smooth surface. This treatment also introduces functional surface groups that may be used to bind affinity ligands such as antibodies or other binding proteins, lectins and nucleic acids.

Monosized magnetic polymer particles are commercially available under the trade name Dynabeads (Dynal A/S Oslo, Norway). A SEM picture of Dynabeads M-450 illustrating the excellent monodispersity is shown in Figure 36.8.

The Dynabeads have got a large number of applications within biochemistry, biology as well as analytical and clinical medicine. Important fields of application are: cell separation, both analytical and clinical, tissue typing; organelles

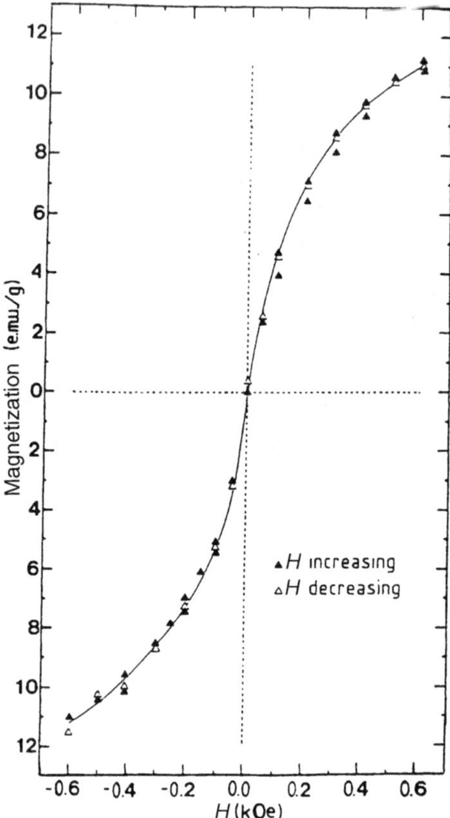

Figure 36.7   Magnetization curve for magnetic particles. Fe content 20 wt%

and viruses separation; DNA-synthesis, -sequencing, -probes, -hybridization; immunoassays, immunoprecipitation; HIV-diagnosis, -research; microbiology.

The use of Dynabeads has been cited in more than 2000 scientific papers since they first were launched to the market in 1986.

PREPARATION OF MONODISPERSE POLYMER PARTICLES BY STEP-GROWTH POLYMERIZATION

Quite recently we have developed a new method which allows the preparation of monodisperse particles by step-growth polymerization processes [16]. The particles can be prepared from different monomers and in controlled sizes up to more than 1000 μm. Suitable monomer systems are: Resorcinol- formaldehyde, urea-formaldehyde, glycidol, epoxy compounds, organosiloxanes. Also in this case one starts with monodisperse seed particles acting as catalytic reaction centres or

Figure 36.8 SEM picture of Dynabeads M-450, 4.5 μm magnetic particles

templates for the polymerization. The catalytic activity is achieved by functional acid or base groups covalently bound to the seed particle matrix. To exemplify, suitable seed particles may consist of slightly crosslinked polyvinyl polymer carrying sulphonic acid or quaternary ammonium hydroxy groups. It is essential that the functional groups attached to the particle matrix bring about a high degree of swelling of the seed when dispersed in water or other ionizing liquids.

As an example of an acid catalysed preparation one may proceed as follows: One prepares in a first step monodisperse seed particles of polystyrene crosslinked with a small amount of divinylbenzene. Preferably the activated swelling method is used for this preparation. The seed particle is then sulphonated in concentrated sulphuric acid. Preparing a seed with a divinylbenzene content of 0.1% the sulphonated particles may swell more than 200 times by volume in water. The swollen particles are almost large 'water' droplets in a continuous water phase, and they are extremely difficult to observe in the light microscope unless phase contrast optics are used. If for example an aqueous solution of resorcinol and formaldehyde is mixed with sulphonated seed particles,

Figure 36.9 Light microscope picture of dry 610 µm monosized particles prepared by acid catalysed polymerization of resorcinol-formaldehyde

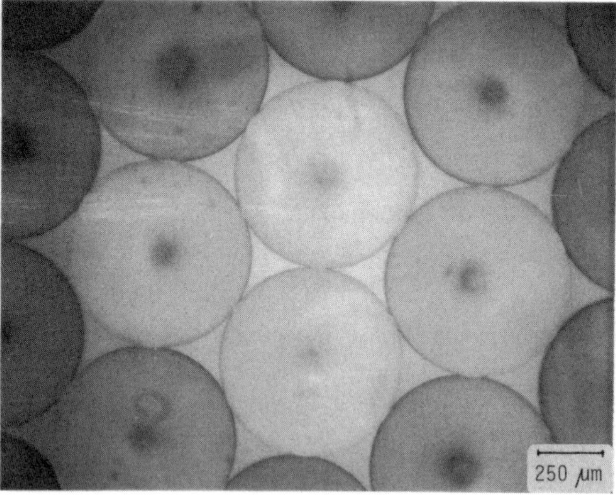

Figure 36.10 Light microscope picture of 680 µm monosized particles in water prepared by acid catalysed polymerization of glycidol

and temperature increased, the reaction between resorcinol and formaldehyde, which is acid catalysed, will take place almost exclusively within the swollen seed particles. During the reaction the monomers are continuously transported into the seed particles where they polymerize. In general the final polymer particles become almost as large as the highly swollen seed, and thereby it is possible to prepare particles where the starting seed constitutes less than 1%

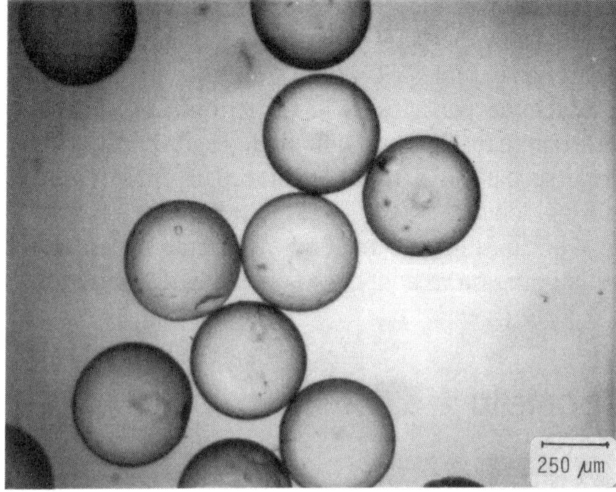

Figure 36.11 Light microscope picture of 410 µm monosized particles in water prepared by acid catalysed polymerization of (3-glycidyloxypropyl)trimethoxysilane

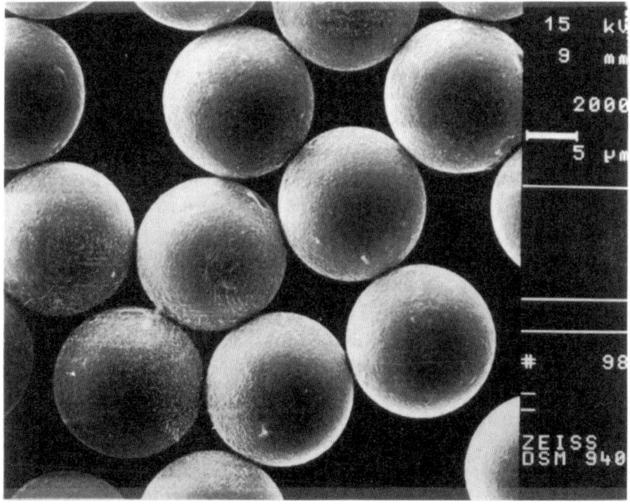

Figure 36.12 SEM picture of 13 µm monosized carbon particles

of the final particle. Figures 36.9–36.11 give examples of different types of monodisperse particles prepared by acid catalysed reactions in highly swollen seed particles applying the monomer systems: resorcinol–formaldehyde, glycidol and (3-glycidyloxypropyl)trimethoxy silane. Particles made from glycidol and organosilane show a high degree of swelling in water increasing their diameters about 150 and 35% respectively compared to the dry state.

As a further aspect of the invention it has been found that especially resorcinol–formaldehyde particles are very suitable for the preparation of monodisperse carbon particles. By carrying out pyrolysis of monodisperse resorcinol–formaldehyde particles in an inert atmosphere at 1000 °C spherical monodisperse carbon particles are obtained. Figure 36.12 shows an SEM of 13 μm monodisperse carbon particles achieved by pyrolysing 20 μm particles of resorcinol–formaldehyde.

Development work on this new method of particle preparation by step-growth polymerizations is going on in cooperation between Dyno Particles A/S, Norway and SINTEF.

ACKNOWLEDGMENT

We greatly acknowledge the invitation to lecture at the P.N. Conference and obtaining the opportunity to publish this paper. We have especially stressed the activated swelling method, an important contribution for polymer chemistry and for the development of modern particle technology. Without this method our research would have been far more tedious.

REFERENCES

1. J. Ugelstad, P.C. Mørk, K.H. Kaggerud, T. Ellingsen and A. Berge, *Adv. Colloid Interface Sci.*, **13**, 101 (1980).
2. J. Ugelstad, P.C. Mørk, A. Berge, T. Ellingsen and A.A. Khan, in *Emulsion Polymerization*, (I. Piirma, ed.), Academic Press, New York, 1982, p. 383.
3. J. Ugelstad, P.C. Mørk, H.R. Mfutakamba, E. Soleimany, I. Nordhuus, R. Schmid, A. Berge, T. Ellingsen, O. Aune, K. Nustad, in *Sci. and Technol. of Polymer Colloids*, NATO ASI, Series **67**, (G.W. Poehlein, R.H. Ottewill and J.W. Goodwin, eds.), Martin Nijhoff, Boston, 1983, p. 51.
4. J. Ugelstad, P.C. Mørk, I. Nordhuus, H. Mfutakamba and E. Soleimany, *Macromol. Chem. Suppl.*, **10/11**, 215 (1985).
5. J. Ugelstad, A. Berge, T. Ellingsen, R. Schmid, T.-N. Nilsen, P.C. Mørk, P. Stenstad, E. Hornes and Ø. Olsvik, *Prog. Polym. Sci.*, **17**, 87 (1992).
6. V. Smigol and F. Svec, , *J. Appl. Polym. Sci.*, **46**, 1439 (1992).
7. Y.-C. Liang, F. Svec and J.M.J. Fréchet, *J. Polym. Sci., Part A, Polym. Chem.*, **33**, 2639 (1995).
8. J. Ugelstad, L. Söderberg, A. Berge and J. Bergström, *Nature*, **303**, 95 (1983).
9. J. Ugelstad, R. Schmid, A. Berge, O. Aune, J. Bjørgum, L. Kilaas, P. Stenstad, A. Skjeltorp, T. Lindmo and L. Korsnes, *The Pol. Mat. Encycl.* (J.C. Salamone, ed.), CRC Press, Boca Raton, 1996, p. 4501.
10. L.I. Kulin, P. Flodin, T. Ellingsen and J. Ugelstad, *J. Chromatogr.*, **514**, 1 (1990).
11. T. Ellingsen, O. Aune, J. Ugelstad and S. Hagen, *J. Chromatogr.*, **535**, 147 (1990).
12. B.E. Christensen, M.H. Myhr, O. Aune, S. Hagen, A. Berge and J. Ugelstad, *Carbohydr. Polym.*, **29**, 217 (1996).
13. J. Ugelstad, T. Ellingsen, A. Berge and B. Helgée, *PCT Int. Appl.* WO 83/03920 (1983).

14. J. Ugelstad, P. Stenstad, L. Kilaas, W.S. Prestvik. R. Herje, A. Berge and E. Hornes, *Blood Purif.*, **11**, 349 (1993).
15. W.S. Prestvik, A. Berge, P.C. Mørk, P. Stenstad and J. Ugelstad, in *Scientific and Clinical Applications of Magnetic Carriers* (Häfeli *et al.*, eds.) Plenum Press, New York, 1997, p. 11.
16. A. Berge, T.-N. Nilsen, J. Bjørgum and J. Ugelstad, U.S. Pat.no. 5, 677, 373 (1993).

37

Polymer Solutions Confined in a Rigid Network

C. KLOSTER[1,2], C. BICA[2], C. LARTIGUE[1], C. ROCHAS[1], D. SAMIOS[2] and E. GEISSLER[1]

[1]Laboratoire de Spectrométrie Physique, CNRS UMR 5588, B.P. 87, 38402 St Martin d'Hères Cedex, France

[2]Laboratorio de Instrumentação e Dinâmica molecular, Instituto de Quimica, Universidade Federal do Rio Grande do Sul, Porto Alegre, Rio Grande do Sul, Brazil

ABSTRACT

Dynamic light scattering measurements are made in solutions of low molecular weight dextran ($M_w = 70\,000$), in the free state and inside agarose hydrogels of varying concentrations. Light scattered by the rigid agarose matrix heterodynes that form the mobile polymer, giving measurements both of the diffusion coefficient D_c and of the Rayleigh ratio R_θ of the dextran. For dextran concentrations c below the overlap concentration c^*, D_c decreases with increasing concentration of the gel c_g. The product $D_c R_\theta$, which depends only on hydrodynamic factors, is independent of c_g. The main effect of the static gel matrix is to reduce the osmotic pressure of the free polymer.

Wiley Polymer Networks Group Review Series Vol. 2. Edited by B.T. Stokke and A. Elgsaeter
© 1999 John Wiley & Sons Ltd

INTRODUCTION

Recently, several investigations have been reported on the behaviour of free polymer chains inside swollen networks. Small angle neutron scattering shows that free chains in a swollen network adopt a collapsed configuration that is smaller than the unperturbed size of the molecule [1,2], owing to the reduction in available configuration space in the random network [3]. Measurements of quasi elastic light scattering have also been made in systems where the host matrix is iso-refractive with the solvent [4]; in this condition the movement of the guest polymer alone is detected. Measurements have also been reported on the dynamics of free chains trapped inside biopolymer hydrogels [5,6]. Generally, the diffusion of the guest polymer is found to be slower than in free solution. To understand the thermodynamics of the system, however, it is essential to measure the intensity of the dynamically scattered light.

Here we report quasi-elastic light scattering investigations of the motion of flexible polymer molecules (dextran) inside a rigid agarose gel [7]. Unlike gels made of flexible chains, the movements in a rigid-rod network are extremely limited, and the overwhelming majority of the light scattered by these systems is elastic. In principle, therefore, if such gels host a solution containing free polymer molecules, any quasi-elastic component in the scattered light is almost entirely attributable to the movement of the free chains in the network. The diffusion coefficient and the osmotic susceptibility of the polymer solution inside the matrix can therefore be measured. In the situations described here, however, as the static scattering can be more than three orders of magnitude more intense than the fluctuating component of the guest solution, the analysis of the correlation spectra is of great importance.

HETERODYNING TECHNIQUE [8–12]

At the temperature of measurement (25 °C), the agarose matrix is practically rigid, generating a static speckle pattern whose intensity $I_s(q)$ and phase depend on the particular speckle observed. The total electric field at the detector is then the sum of the static field and the field $E_f(t)$ from the mobile polymers in the network. The resulting intensity fluctuations yield the field correlation function, $g(t)$, and also the fluctuating intensity, $I_f(t) = |E_f(t)|^2$. For a mobile polymer obeying Fick's equation, the correlation function $g(t)$ is

$$g(t) = \exp(-D_c q^2 t) \tag{37.1}$$

where D_c is the diffusion coefficient and $q (= (4\pi n/\lambda)\sin(\theta/2))$ is the transfer wave vector. Equation (37.1) involves time fluctuations of the concentration only; permanent spatial fluctuations in the system contribute to the static scattering, but not to the dynamics. The correlation spectrum of the intensity $I(t)$ is then

$$\frac{1}{t_{\mathrm{E}}} \int I(t)I(t + \tau)\mathrm{d}t = I_{\mathrm{s}}^2 + 2I_{\mathrm{s}}\langle I_{\mathrm{f}}\rangle + \langle I_{\mathrm{f}}\rangle^2[1 + \beta g^2(\tau)] + 2I_{\mathrm{s}}\langle I_{\mathrm{f}}\rangle\beta g(\tau) \quad (37.2)$$

where the angular brackets are averages over the experimental accumulation time t_{E}, and $\beta(\leqslant 1)$ is the optical coherence factor. Normalizing (37.2) by the square of the average intensity (computer baseline),

$$\langle I\rangle^2 = \langle I_{\mathrm{s}} + I_{\mathrm{f}}\rangle^2 = \langle I_{\mathrm{s}}\rangle^2 + 2\langle I_{\mathrm{s}}\rangle\langle I_{\mathrm{f}}\rangle + \langle I_{\mathrm{f}}\rangle^2 \quad (37.3)$$

yields the total intensity correlation function

$$G(\tau) = \frac{\int I(t)I(t + \tau)\mathrm{d}t}{\langle I\rangle^2 t_{\mathrm{E}}}.$$

As with most gels, however, the speckle pattern evolves slowly with time, and I_{s} is not a true constant. This consideration has stimulated investigations into the notion of ensemble averaging and non-ergodicity [11]. In the present case, the dissolved polymer molecules are free to overlap and are therefore fully ergodic. Any time dependence of I_{s} is assumed to be sufficiently slow for it to be distinct from the fast fluctuations in I_{f} due to the motion of the free polymer. The time average of I_{s}^2, i.e. $\langle I_{\mathrm{s}}^2\rangle$, determines the effective constant baseline ('far point') to which the correlation spectrum decays, i.e.

$$\text{far point} = \frac{1}{t_{\mathrm{E}}} \int I(t)I(t + \tau_\infty)\mathrm{d}t = \langle I_{\mathrm{s}}^2\rangle + 2\langle I_{\mathrm{s}}\rangle\langle I_{\mathrm{f}}\rangle + \langle I_{\mathrm{f}}\rangle^2 \quad (37.4)$$

where τ_∞ is the extended channel delay. Here we are concerned only with the quantities $\langle I_{\mathrm{f}}\rangle$ and $g(t)$. To obtain these, the far point is subtracted from equation (37.2), and the result normalized by the computer baseline. This yields, for the dynamic component of the intensity correlation function,

$$H(\tau) = G(\tau) - \frac{\langle I_{\mathrm{s}}^2\rangle + 2\langle I_{\mathrm{s}}\rangle\langle I_{\mathrm{f}}\rangle + \langle I_{\mathrm{f}}\rangle^2}{\langle I\rangle^2}$$

$$= \beta[2X(1 - X)g(\tau) + X^2 g^2(\tau)] \quad (37.5)$$

where $X = \langle I_{\mathrm{f}}\rangle/\langle I_{\mathrm{f}} + I_{\mathrm{s}}\rangle$, and $H(\tau)$ tends to zero as τ tends to ∞. β is found by extrapolating $G(\tau)$ to $\tau = 0$ for a dilute suspension of latex beads (for which $X = 1$). In the present optical arrangement, $\beta = 0.96$. Hence, for the heterodyne case, equation (37.5) is soluble [11] for X and for $g(\tau)$. The required intensity is thus

$$I_{\mathrm{f}} = X\langle I\rangle \quad (37.6)$$

where $\langle I\rangle = \langle I_{\mathrm{f}} + I_{\mathrm{s}}\rangle$ is the total intensity averaged over the experimental observation time t_{E}. Spatial averaging of the speckle pattern is not necessary to determine I_f. For the absolute value of the fluctuating intensity, R_θ, however, the transmission Tr of the sample must be measured, as well as the intensity $I_{stand} = R_v I_0$ from a standard sample (toluene), where R_v is the Rayleigh ratio of the standard and I_0 is the intensity of the incident beam. Thus, finally,

$$R_\theta = \frac{R_v X \langle I \rangle \sin \theta}{I_{stand} Tr} \tag{37.7}$$

SAMPLE PREPARATION

Dextran, of molecular weight $M_w = 70\,000$, was used as supplied by Sigma. The agarose was kindly provided by R. Armisen (Hispanagar, Spain). Its molecular weight, determined by viscometry [13], is $M_w = 1.2 \times 10^5$. The sulphate content specified by the manufacturer is 0.1%, and the methyl content, found by ^{1}H NMR, was 0.6%. Dextran solutions were investigated in the concentration range $2 \text{ g l}^{-1} \leqslant c \leqslant 100 \text{ g l}^{-1}$; the agarose concentration of the host gel was in the range $0 \leqslant c_g \leqslant 40 \text{ g l}^{-1}$.

Samples were prepared by mixing the appropriate weights of agarose and dextran in de-ionized water, heating the mixture to $100\,^\circ$C, and stirring until complete dissolution. The mixtures were then transferred to cylindrical glass tubes of 10 mm diameter and sealed. The samples were melted again at $100\,^\circ$C and allowed to cool to room temperature; gelation of the agarose occurred as the temperature fell below $45\,^\circ$C. It was found that for agarose concentrations greater than the overlap concentration c^*, the gels became strongly turbid, indicating local phase separation. At higher concentrations still, the samples display macroscopic phase separation with a sparse component of increased turbidity; these samples were not investigated by light scattering.

Dynamic light scattering measurements were made with a Spectra Physics SP162 laser working at 488 nm, and a Malvern Instruments 7032 correlator. All measurements were made at $25\,^\circ$C. Static light scattering measurements were performed with the same instrument. Small angle neutron scattering (SANS) measurements of the unfilled gels were made on the D11 instrument at the ILL, Grenoble.

RESULTS AND DISCUSSION

To characterize the pore sizes prevailing in the gel structure, Figure 37.1 shows the scattering response, obtained by static light scattering and by SANS, of two unfilled agarose–water gels of different concentrations [14]. These measurements were made using toluene and protonated water as standards respectively, and the light scattering data are normalized to SANS units by multiplying by the corresponding ratio of contrast factors. For $q > 10^5 \text{ cm}^{-1}$ these data fall on a master curve that is independent of concentration. The observed straight-line

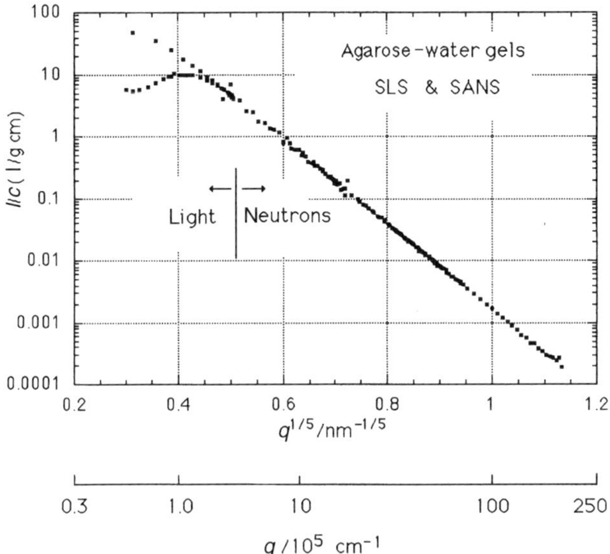

Figure 37.1 Scattering intensity $I(q)/c$ of agarose–water gels in the absence of dextran, from small angle neutron and static light scattering. For $q < 10^5$ cm^{-1} the data depend on the concentration and thermal history of the sample; upper points: $c = 10$ g/l; lower points: 30 g/l. The intensities are expressed in terms of SANS measurements

behaviour in this representation corresponds to a stretched exponential expression of the form

$$I(q) = A \exp\{-(qR)^n\}$$

in which $n = 1/5$ and the value of R is of the order of 100 μm. Moreover, the maximum in Figure 37.1 at $q \approx 10^5$ cm^{-1} indicates that for the 30 g/l sample there is a maximum pore size of approximately 6000 Å. For the 10 g/l sample, a maximum in intensity appears to occur at an even lower value of q. We conclude that the gel structure is characterized by a continuous distribution of pore sizes with an upper cut-off in the region of several thousand angstroms. For the dynamic light scattering, the polymer solutions and gels were measured at angles 60°, 90° and 150°. The measurements were made as a function of dextran concentration c between the dilute and the semi-dilute region, for various concentrations c_g of the agarose gel. The overlap concentration c^* in aqueous solution for the $M_w = 70\,000$ sample is known from intrinsic viscosity measurements [13] to be $c^* = 35$ g/l.

For the pure dextran solutions, the correlation spectra were found to be purely homodyne, the intercept of the reduced total intensity correlation function $G(\tau) - 1$ at $\tau = 0$ being close to the measured value of β. In contrast, for the spectra obtained from the dextran solutions in the agarose gels, the value of $G(0) - 1$

is some two orders of magnitude smaller, due to strong heterodyning by the light scattered from the agarose gel. The resulting values of X lie in the range 10^{-2}–10^{-3}.

As found for dextran in gellan gels [6], the motion of the polymer inside the gel becomes slower as the gel concentration increases. Also, with increasing gel concentration c_g, a second much slower motion develops, whose amplitude becomes increasingly large. In the present case, the spectra could in general be decomposed quite well into a simple sum of two exponentials. The relaxation rates are proportional to q^2, indicating that the motion is diffusive. The diffusion coefficient of the fast process is denoted D_c. The dependence of D_c upon the dextran concentration c is shown in Figure 37.2, for agarose concentrations lying between 0 and 40 g/l. In the free solution, D_c increases monotonically, and remains continuous at c^*. With increasing c_g, however, D_c for the dilute dextran solutions decreases significantly, but tends to the free solution value as c tends to c^*.

The total intensity of the dynamically scattered intensity R_θ (averaged over the three measured angles) was measured as a function of c for different gel compositions. R_θ increases practically linearly with dextran concentration up to c^*. Above this concentration the free solution exhibits a plateau, as expected in the vicinity of the maximum in semi-dilute solutions. In the gels, however, the intensity increases strongly in the semi-dilute region.

For the dextran sample used here, the radius of gyration R_G, measured by static light scattering, was found to be 9.6 nm. This result is consistent with previous measurements made on this system [15]. For the present observations therefore, $qR_G \ll 1$, and, as far as the individual dextran coils are concerned, the experimental condition is close to the thermodynamic limit $q = 0$. In this

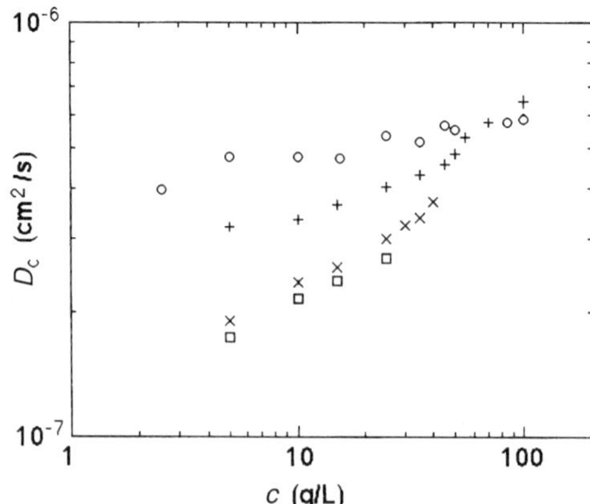

Figure 37.2 Diffusion coefficient D_c as a function of dextran concentration c in agarose gels of various concentrations c_g; O: $c_g = 0$ (free solution); +: $c_g = 5$ g/l; ×: $c_g = 30$ g/l; □: $c_g = 40$ g/l

approximation, the scattered intensity is given by

$$R_\theta = Kc\frac{kT}{\partial\Pi/\partial c} \tag{37.8}$$

where Π is the osmotic pressure of the solution and K is the light scattering contrast factor. Also, the diffusion coefficient can be expressed as

$$D_c = \frac{\partial\Pi/\partial c}{f} \tag{37.9}$$

where f can be construed as the friction coefficient of the dextran molecule. From equations (37.8) and (37.9) it follows that the product

$$D_c \times R_\theta = KkTc/f \tag{37.10}$$

contains no thermodynamic information, but reflects only hydrodynamics.

Figure 37.3 shows the variation of $D_c \times R_\theta$ as a function of c in different gel environments. Below c^*, the points fall on a master curve. This implies that the friction coefficient f of the dextran molecule is invariant whether it is in free solution or in the gel. The observed decrease in the diffusion rate of dextran in the gel with respect to the free solution is therefore due to a decrease in the osmotic modulus and is unrelated to the friction of the surrounding matrix. As the gel matrix is purely static, however, its configuration entropy is zero; its influence is therefore confined to reducing the space available to the dextran molecules. This situation is a corollary of the reduction in radius of gyration of polymers trapped in random networks, where the cross-links reduce the configuration entropy of the guest molecules [1–3].

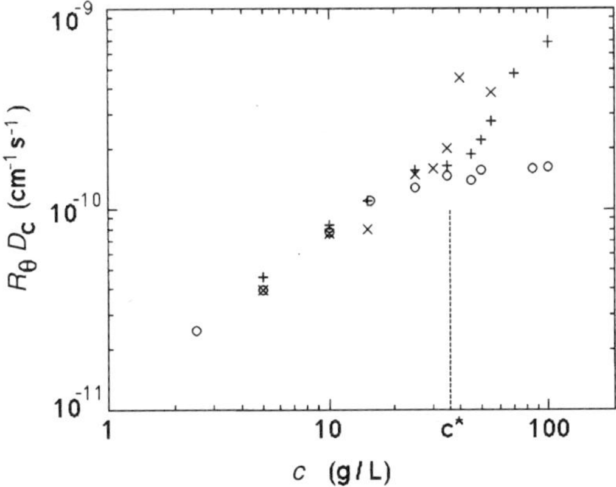

Figure 37.3 Product $R_\theta \times D_c$ as a function of dextran concentration c with various agarose gel concentrations c_g. O: $c_g = 0$ (free solution); +: $c_g = 5$ g/l; ×: $c_g = 30$ g/l

CONCLUSIONS

Low molecular weight dextran ($M_w = 70\,000$), when dissolved inside agarose hydrogels, exhibits single phase behaviour at concentrations $c \leqslant c^*$, where c^* is the overlap concentration of the dextran. Above c^*, phase separation occurs; this behaviour contrasts with that of the free solution, which is continuously soluble in water. For dextran concentrations less than or equal to the c^*, the diffusion coefficient D_c is reduced and the Rayleigh ratio R_θ is enhanced compared to the free solution. In this concentration range, however, the quantity $D_c \times R_\theta$, which depends only on hydrodynamic factors, is found to be independent of the agarose concentration. It is concluded that the rigid gel matrix reduces the osmotic pressure of the dextran solution by reducing the configuration space available to the dextran.

ACKNOWLEDGMENTS

Carmen Kloster thanks the CAPES Foundation of the Ministry of Education of Brazil for financial support. We are also grateful to the Institut Laue Langevin for beam time on D11.

REFERENCES

1. F. Horkay, H.B. Stanley, E. Geissler and S.M. King, *Macromolecules*, **28**, 678 (1995).
2. R.M. Briber, X. Liu and B.J. Bauer, *Science*, **268**, 395 (1995).
3. A. Baumgärtner and M. Muthukumar, *J. Chem. Phys.*, **87**, 3082 (1987).
4. C.-S. Kuo, R. Bansil and C. Koñák, *Macromolecules*, **28**, 768 (1995).
5. P.Y. Key and S.B. Sellen, *J. Polym. Sci. Polym. Phys. Ed.*, **20**, 659 (1982).
6. P.M. Burne and D.B. Sellen, *Biopolymers*, **34**, 371 (1994).
7. C. Kloster, C. Bica, C. Lartigue, C. Rochas, D. Samios and E. Geissler, *Macromolecules*, **31**, 7712 (1998).
8. B.J. Berne and R.Pecora, *Dynamic Light Scattering*, Wiley, New York, 1976.
9. E. Geissler and A.M. Hecht, *J. Chem. Phys.*, **65**, 103 (1976).
10. D.B. Sellen, *J. Polym. Sci. Part B: Polym. Phys.*, **25**, 699 (1987).
11. P.N. Pusey and W. van Megen, *Physica A*, **157**, 705 (1989).
12. E. Geissler, in *Dynamic Light Scattering* (ed. W. Brown), Clarendon Press, Oxford, 1993, Ch 11, and references therein.
13. C. Rochas, A. Domard and M. Rinaudo, *Eur. Polym. J.*, **16**, 135 (1980).
14. C. Rochas, A.M. Hecht, and E. Geissler, *Die Makromolekulare Chemie, Makromolecular Symposia*, **138**, 157 (1999).
15. P. Roger, Thesis, University of Nantes, France, 1993.

38

High Elongation of Deswollen Polysiloxane Networks

KENJI URAYAMA, KEISUKE YOKOYAMA and
SHINZO KOHJIYA
Institute for Chemical Research, Kyoto University, Uji, Kyoto-fu 611, Japan

ABSTRACT

Mechanical properties and cold crystallization behavior of deswollen polydimethylsiloxane (PDMS) networks are investigated as a function of polymer concentration in preparation. Deswollen PDMS networks are prepared by removing solvent from PDMS crosslinked in solution. A deswollen PDMS network prepare from 10% solution shows a remarkable extensibility over 3000% elongation. The high extensibility originates from two features in the network topology: (1) smaller number of trapped entanglements by crosslinking in

low polymer concentration; (2) smaller end-to-end distance of network chains in unde
formed state by a large decrease of gel volume in deswelling. Polymer concentration
dependence of heat of fusion of deswollen networks shows anomaly in low concentration
region. It is considered that this anomaly results from a more compact conformation of
network chains in deswollen networks (supercoil) relative to random-coil.

INTRODUCTION

Deswollen dry polymer networks, made by removing solvent completely from
networks prepared in solution, have been expected to be interesting polymer
network systems whose topological structure is very different from that of dry
polymer networks prepared in the bulk state [1–3]. This difference arises from the
large volume change of the gels in deswelling. For example, in case of a polymer
network prepared from a solution of $\phi = 0.1$ (ϕ is polymer volume fraction), the
gel volume in the fully deswollen state is only 1/10 of that in the preparation state.
Despite a large volume change, densities of network systems are almost constant
before and after deswelling, because the systems are in a liquid (rubbery) state
through the deswelling process, and generally, the difference in densities of the
solvent and polymer is small [4]. As a result, network chains in the deswollen
state are theoretically required to have a more compact conformation relative to
those in the preparation state. The conformation of network chains specific to
deswollen networks has often been called *supercoil* [1,2] or *double-folded* [3].
There have been some studies [5–9] on structure and physical properties of
deswollen networks. The details of supercoil structure and its specific physical
characteristics have not fully been elucidated, though a small angle neutron scattering (SANS) study reported that the gyration radius of the network chains in
deswollen polystyrene networks is smaller than that in the unperturbed state [9].
Recently, we have investigated the mechanical properties of deswollen polydimethylsiloxane (PDMS) networks, and shown that deswollen PDMS network
prepared from a solution of low polymer concentration reveals remarkable high
extensibility over 3000% elongation, and that the strain dependence of stress is
much weaker than that in the preparation state [10–13]. We have also indicated
that the preparation concentration dependence of Young's modulus of deswollen
networks is not explained by classical rubber elasticity theory treating the network
chain as random-coil [11]. These results occur because supercoil has a more
densely packed conformation than random-coil. In this paper, we review our
results for the elastic properties of deswollen PDMS networks, and also show
recent experimental results of the (cold) crystallization behavior of deswollen
PDMS networks.

EXPERIMENTAL

Three kinds of vinyl-terminated PDMS different in average molecular size
and molecular size distribution were used as precursor chain: $M_\mathrm{n} = 2.8 \times 10^4$,

$M_w/M_n = 1.7$ (PDMS-A); $M_n = 9.9 \times 10^4$, $M_w/M_n = 1.2$ (PDMS-B); $M_n = 8.4 \times 10^4$, $M_w/M_n = 1.3$ (PDMS-C), where M_n and M_w are the number- and weight-average molecular weight, respectively. Tetrakisdimethylsiloxysilane was employed as crosslinker. End-linking reaction was carried out by hydrosilylation using $H_2PtCl_6 \cdot 6H_2O$ as catalyst. Distilled toluene was used as solvent.

Polymer volume fractions at preparation were varied from 0.1 to 1 (without solvent). The gels were immersed for about 1 week in toluene to remove unreacted materials. Deswelling was done by drying the swollen gels in air, or adding methanol (poor solvent) to the swelling agent stepwise, and finally drying the samples completely. Fractions of unreacted materials were estimated from sample weights in the dry state, and (effective) polymer volume fraction in preparation state (ϕ_o) was re-calculated by subtracting the fraction of unreacted materials.

Young's modulus and stress–strain relations of (PDMS) networks in the preparation and deswollen state were investigated by uniaxial elongation at constant crosshead speed. It was experimentally checked that the crosshead speed was so slow that the effects of stress relaxation on stress–strain relations were negligible.

Cold crystallization behavior of deswollen PDMS networks prepared with PDMS-C was investigated by means of a differential scanning calorimeter (DSC) (Rigaku DSC8230B). Deswelling of all samples for DSC measurements was gradually made using mixtures of toluene and methanol with a series of compositions as swelling agent. In DSC measurements, the samples were cooled at ca. $10\,^{\circ}C/$ min to $-150\,^{\circ}C$, and then heated at $5\,^{\circ}C/$ min.

RESULTS AND DISCUSSION

HIGH EXTENSIBILITY OF DESWOLLEN PDMS NETWORKS

Figure 38.1 shows stress (σ_e) versus elongation (λ) curve for three kinds of dry PDMS networks. The stress σ_e is nominal stress defined by force per cross-section in the undeformed state, and σ_e in the figure is normalized by the initial Young's modulus (E) of each sample. It is found that deswollen networks prepared at $\phi_o \approx 0.1$ show remarkable extensibility (i.e. elongation at break λ_{max} is well over 10, and especially, λ_{max} of the deswollen network prepared from PDMS-B with larger molecular size and narrower molecular size distribution exceeds 30. On the other hand, λ_{max} of the dry network prepared in the bulk state ($\phi_o = 1$) remains at about 2. The high extensibility of the deswollen networks originates from the following two topological features of the network structure: (1) number of trapped entanglements in the networks, which act as additional chemical crosslinks against external forces, is remarkably decreased by crosslinking is solution with low polymer concentration; (2) end-to-end distance of network chains in the undeformed state is considerably reduced by the large volume decrease in deswelling. The number of segments between neighboring trapped entanglements in a network prepared at ϕ_o is larger than that in a bulk-cured network

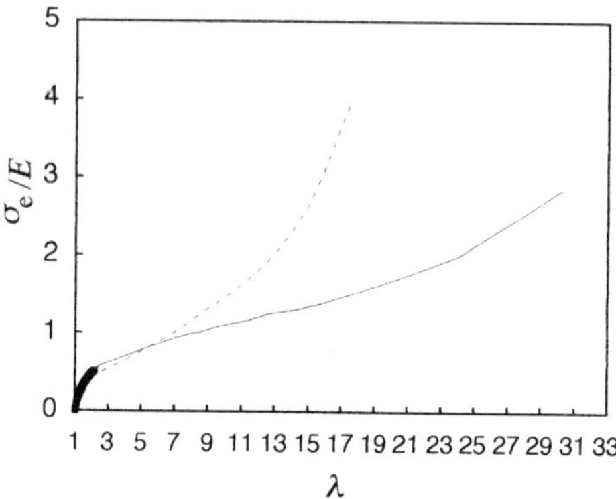

Figure 38.1 Reduced nominal stress (σ_e) versus elongation (λ) curves for dry PDMS networks. E is the initial Young's modulus of each sample. Dashed and solid lines represent the results for deswollen PDMS networks of $\phi_o = 0.1$ with PDMS-A and PDMS-B, respectively. The bold solid line indicates the results for a dry PDMS network prepared in the bulk state with PDMS-A

by a factor of ϕ_o^{-1} [14,15], and the end-to-end distance of network chains in the deswollen state is reduced by a factor of $\phi_o^{1/3}$ relative to that in the preparation state under assumption of affine displacement of crosslinks. As a result, these topological features increase λ_{max} of the deswollen network relative to that of the corresponding bulk-cured network by a factor of $\phi_o^{-5/6}$ [10–12].

$$\lambda_{max}(\phi_o) \approx \lambda_{max}(1)\phi_o^{-5/6} \tag{38.1}$$

where $\lambda_{max}(1)$ is λ_{max} of the network prepared in the bulk state. The possibility of high extensibility of deswollen networks based on this strategy was first theoretically introduced by Obukhov *et al.* [3]. We can expect that λ_{max} of deswollen network of $\phi_o = 0.1$ reaches 40 using $\lambda_{max}(1) = 4 \sim 6$ for conventional elastomers. The approach by equation (38.1) is too primitive to evaluate λ_{max} quantitatively, because generally, fracture phenomena are mainly governed by defects in the sample [16]. However, equation (38.1) is still instructive for quantitatively understanding the origin of the high extensibility of the deswollen network .

DEPENDENCE OF ELASTIC MODULUS OF DESWOLLEN PDMS NETWORKS ON ϕ_O

According to the classical theory of rubber elasticity [17], the elastic modulus of the polymer network with polymer volume fraction ϕ, which is prepared at ϕ_o, $E(\phi, \phi_o)$, is given by

$$E(\phi, \phi_o) = E_d\phi^{2/3}\phi_o^{1/3} \tag{38.2}$$

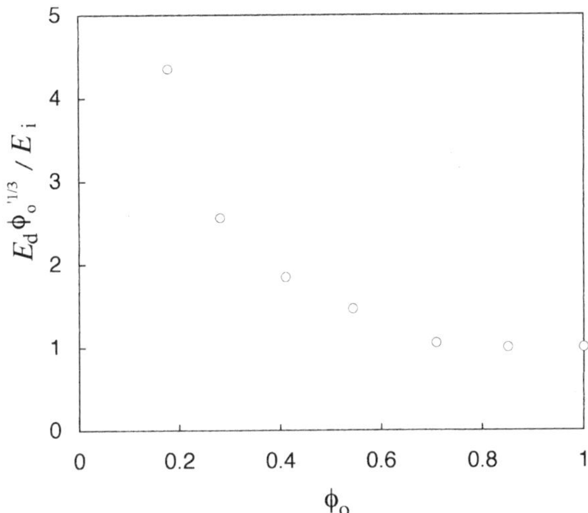

Figure 38.2 Plots of $E_d\phi_o^{1/3}/E_i$ against ϕ_o for PDMS networks prepared with PDMS-A. E_d and E_i are the initial Young's modulus in the fully deswollen and preparation state, respectively. ϕ_o is the polymer volume fraction at preparation

where E_d is the elastic modulus in the fully deswollen state, i.e. $E(1, \phi_o)$. Equation (38.1) assumes a Gaussian chain (random-coil) for conformations of network chains through the deswelling process. The relation between the elastic modulus in the preparation state ($E_i = E(\phi_o, \phi_o)$) and E_d is written as $E_i = E_d\phi_o^{1/3}$. Figure 38.2 shows the plots of $E_d\phi_o^{1/3}/E_i$ against ϕ_o. The classical theory predicts that $E_d\phi_o^{1/3}/E_i$ is unity regardless of ϕ_o. On the other hand, the experimental values of $E_d\phi_o^{1/3}/E_i$ deviate from unity in the low ϕ_o region, though they are close to unity in the high ϕ_o region. This result suggests that the conformation of network chains in deswollen networks of low ϕ_o is not described by a Gaussian chain, while that in deswollen networks of high ϕ_o is approximated by a Gaussian chain. It should be noted that as a network is prepared at lower ϕ_o, the volume change in the deswelling becomes larger. The deviations from unity in the low ϕ_o region are considered to be the result of formation of supercoil, which is not approximated by a Gaussian chain.

The PDMS networks in this study were prepared in good solvent. A closer discussion considering the excluded volume effect for the network chains in the preparation state was made in Ref. [11]. The excluded volume effect does not significantly influence the results in Figure 38.2.

EFFECTS OF DESWELLING ON STRESS–ELONGATION RELATIONS

Figure 38.3 indicates comparison of stress (σ_e)–elongation (λ) relations before and after deswelling for two networks of $\phi_o = 0.41$ and 0.18, respectively. The

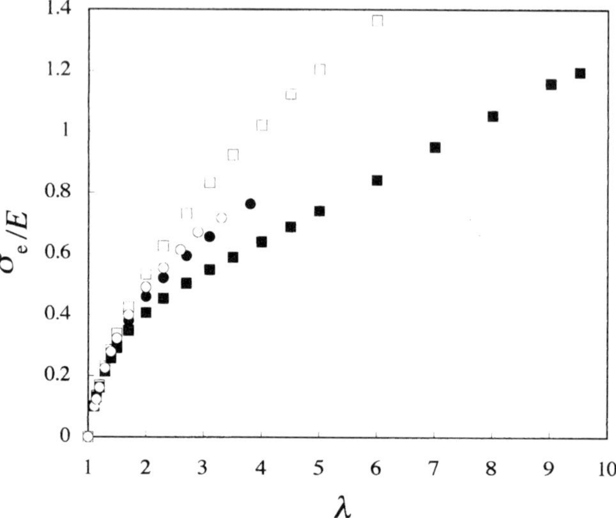

Figure 38.3   Reduced nominal stress (σ_e) versus elongation (λ) curves for PDMS networks of $\phi_o = 0.18$ (circle) and $\phi_o = 0.41$ (rectangular) with PDMS-A. Open and filled symbols indicates the results for preparation and fully deswollen state, respectively, E is the initial Young's modulus for each sample

nominal stress σ_e is reduced by the initial Young's modulus of each sample. It is seen that the stress–elongation relations for the network of high $\phi_o(\phi_o = 0.41)$ are almost the same before and after deswelling, while the λ dependence of σ_e in the deswollen state is much weaker than that in the preparation state for the network of low $\phi_o(\phi_o = 0.18)$. The λ dependence of σ_e for all the networks of $\phi_o > 0.41$ is not significantly influenced by deswelling, although the data are not shown here. These results are qualitatively interpreted in the same way as in the case of the ϕ_o dependence of the elastic modulus. As ϕ_o decreases, the difference in the conformation of network chains before and after deswelling becomes larger due to the larger volume decrease. The difference in the conformation of network chains is recognized by the difference in the stress–elongation relations.

EVALUATION OF SUPERCOIL STRUCTURE BY SCALING ANALYSIS OF STRESS-ELONGATION RELATIONS

According to the scaling concept of Pincus blob [2,18,19] the λ dependence of σ_e for a single (very long) polymer chain under strong stretching is given as follows:

$$\sigma_e \sim \lambda^{1/(D-1)} \tag{38.3}$$

where D is the fractal dimension of the polymer chain. We have experimentally confirmed the validity of equation (38.3) using a polymer network composed

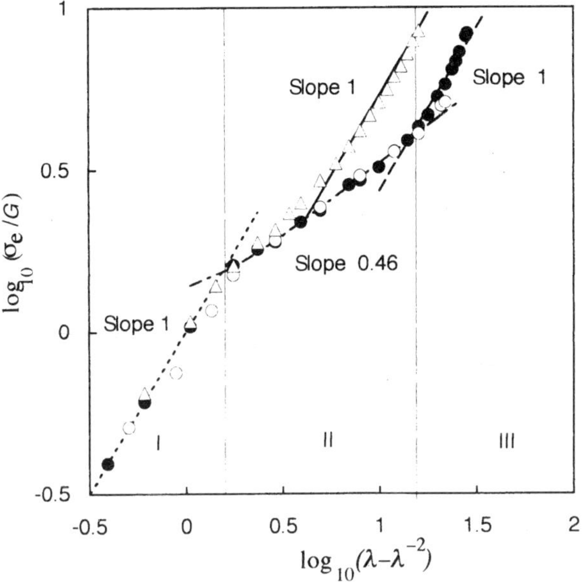

Figure 38.4 Double-logarithmic plots of reduced nominal stress (σ_e) versus $(\lambda - \lambda^{-2})$ for PDMS network in preparation state with $\phi_o = 0.25$ (triangle) and fully deswollen PDMS networks with $\phi_o = 0.15$ (open circle) and $\phi_o = 0.10$ (filled circle). The networks were prepared with PDMS-B. The shear modulus G for each sample was evaluated using $E = 3G$

of random-coil chain $(D = 2)$, and evaluated D for supercoil by analyzing the $\sigma_e - \lambda$ relations of deswollen PDMS networks on the basis of equation (38.3) [13]. Figure 38.4 shows double logarithmic plots of reduced stress vs. $(\lambda - \lambda^{-2})$ for a PDMS network in the preparation state of $\phi_o = 0.25$, and deswollen PDMS networks of $\phi_o = 0.15$ and 0.10. The correction term $-\lambda^{-2}$ originates from the lateral compression effect in elongation, and this term must be included in treating polymer chains with a finite length. However, this correction term can be safely neglected in the strong stretching region $\lambda > 2.5$, i.e., $\lambda - \lambda^{-2} \approx \lambda$.

The PDMS network in the preparation state $(\phi_o = 0.25)$ is a network sample suitable for examining the validity of equation (38.3). Conformation of the network chains is naturally considered as random-coil with $D = 2$. The network prepared in low ϕ_o has a sufficiently high extensibility in which equation (38.3) is applicable, and the network has a small amount of trapped entanglement whose effects on the λ dependence of σ_e are not considered in equation (38.3). PDMS gels in the preparation state of $\phi_o < 0.25$ were so soft that we could not do the elongational measurements. It is found in the figure that the λ dependence of σ_e in the large deformation region $6 < \lambda < 16$ is described by $\sigma_e \sim (\lambda - \lambda^{-2})^{1.0} \approx \lambda^{1.0}$. This dependence agrees well with the prediction of equation (38.3) for random-coil ($D = 2$). Although the network was prepared in good solvent, the excluded volume effect should be fully screened due to the high degree of chain overlapping ($\phi_o/\phi^* \approx 15$ where ϕ^* is the critical polymer volume fraction for the overlapping of precursor chains used here).

It can been seen in Figure 38.4 that the deswollen networks for $\phi_o = 0.15$ and 0.10 have almost the same σ_e–λ relation, suggesting that this λ dependence of σ_e is typical of a deswollen network composed of supercoiled chains. The λ dependence of σ_e can be divided into three regions: $\sigma_e \sim (\lambda - \lambda^{-2})(1 < \lambda < 1.6)$ (Region I); $\sigma_e \sim \lambda^{0.46}(1.6 < \lambda < 16)$ (Region II); $\sigma_e \sim \lambda^{1.0}(16 < \lambda < 31)$ (Region III). Region I is the small deformation region obeying the prediction of the classical theory of rubber elasticity. We consider Region II and III as the pull-out process of supercoil and the elongation process after complete pull-out of supercoil, respectively. The λ dependence of σ_e in Region II is close to the prediction of equation (38.3) for $D = 3$, $\sigma_e \sim \lambda^{0.5}$. This result suggests that supercoil is a more compact conformation relative to random-coil. A polymer chain in the globule state and a polymer chain in an array of obstacles with excluded volume effect are known as examples of densely packed conformation with $D = 3$ [20]. The λ dependence of σ_e in Region III indicates that the elastic response after pull-out of supercoil becomes random-coil type. This type of crossover in λ dependence of σ_e agrees well with the prediction by Obukhov *et al.* [3], though they predicted $D = 4$ for supercoil and the different value of λ for crossover point (λ_c). Their predicted value of λ_c corresponds to $\lambda_c \approx 7.5$ in our system [13]. The discrepancies in D for supercoil and λ_c between their theoretical prediction and our experiments may be due to that length of precursor chain in our system ($M/M_e \approx 12$ where M_e is the molecular weight between neighboring entanglements) is not so long as that in their theoretical model. They treated deswollen networks composed of extremely long precursor chains [3].

COLD CRYSTALLIZATION BEHAVIOR OF DESWOLLEN PDMS NETWORKS

Low-temperature behavior of deswollen PDMS networks is expected as a specific characteristic to elucidate the difference between supercoil and random-coil. Figure 38.5 shows a DSC thermogram in the heating process for a deswollen PDMS network of $\phi_o = 0.44$. Only a single crystalline melting peak was observed at ca. $-38\,^{\circ}$C. The glass transition temperature T_g of amorphous PDMS was not detected. These features were common to all the samples in this study. Clarson *et al.* [21] investigated the low temperature behavior of bulk-cured PDMS networks using precursor chains with a series of molecular size. They reported that the network samples have only a single crystalline melting peak in DSC thermograms, while precursor PDMS show double endothermic peaks originating from melting–recrystallization. It was considered that junction points

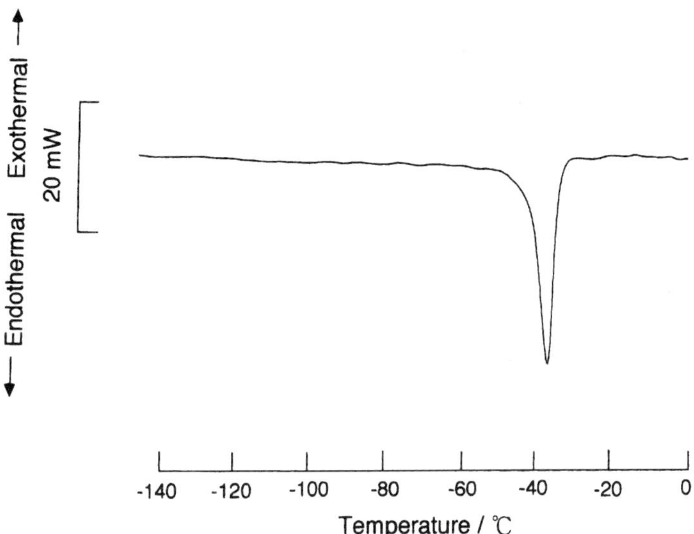

Figure 38.5 A DSC thermogram of a deswollen PDMS network of $\phi_0 = 0.44$ prepared with PDMS-C

inhibited melting–recrystallization. Our results for the deswollen PDMS networks are qualitatively not much different from their results for the dry PDMS networks prepared in the bulk state. It was also reported that PDMS is highly crystallizable in low temperature, and that T_g, which is expected around $-120\,°C$, is not detected unless the cooling rate is quite fast (more than $40\,°C\,min^{-1}$) [21–23]. The cooling rate in our study ($10\,°C\,min^{-1}$) must be too slow to detect T_g.

Figure 38.6 indicates ϕ_0 dependence of enthalpies of fusion (ΔH_f) for deswollen PDMS networks. The values of ΔH_f were determined on the basis of the area of crystalline melting peak. Location of the crystalline melting peaks was almost constant (ca. $-38\,°C$) within experimental error regardless of ϕ_0. It can be seen that ΔH_f increases with decreasing ϕ_0 in the region $\phi_0 > 0.25$, but in $\phi_0 < 0.25$ ΔH_f seems to decrease (or saturate) against the decrease of ϕ_0, although the data points are scattered. The amount of trapped entanglement in deswollen networks decreases with the decrease in ϕ_0. The trapped entanglement should act as an obstacle for crystallization in a similar way to chemical crosslinks. Accordingly, the increase of ΔH_f against the decrease of ϕ_0 in $\phi_0 > 0.25$ is simply explained by the reduction in the amount of trapped entanglement in deswollen networks.

The ϕ_0 dependence of ΔH_f in $\phi_0 < 0.25$ implies some depression effect of supercoil on ΔH_f. There seem to be two physically different interpretations of this result. One is to consider that a crystal formed from supercoil has a smaller heat of fusion than one from random-coil due to a specific conformation of supercoil.

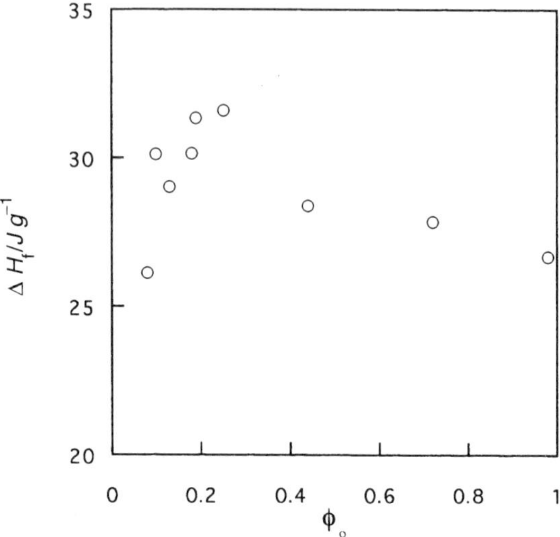

Figure 38.6 Plots of heat of fusion (ΔH_f) against ϕ_o for deswollen PDMS networks prepared with PDMS-C

This means that supercoil and random-coil differ in enthalpy difference ΔH_f^o for transformation of chain conformation from ordered (crystalline) state to disordered (amorphous) state. The other is to regard the depression of ΔH_f as a reduction in the degree of crystallinity by inhibition of densely packed conformation of supercoil on crystal formation. In this case, it is implicitly assumed that ΔH_f^o for supercoil and random-coil are the same. Both the standpoints expect some depression in crystalline melting temperature T_m in the low ϕ_o region, but significant change of T_m against ϕ_o was not experimentally observed. We can not conclude at the moment which of them is a major origin for ϕ_o dependence of ΔH_f in $\phi_o < 0.25$.

As ϕ_o decreases, the degree of volume decrease in deswelling (depression effect of supercoil on ΔH_f) increases, and simultaneously, the amount of trapped entanglement (number of obstacles for crystal formation) decreases. The depression effect of supercoil on ΔH_f must become dominant in $\phi_o < 0.25$. These two opposite effects on ΔH_f compete in the low ϕ_o region, which complicates interpretation of the results in the low ϕ_o region.

Glass transition temperature T_g for deswollen networks of low ϕ_o may be different from that for a dry network prepared in the bulk state, because T_g is expected to reflect densely packed conformation of supercoil. DSC measurements under faster cooling rate to detect T_g is now in progress. This experiment will contribute to the problem for origin of the depression of ΔH_f in the low ϕ_o region. SANS experiments for labelled network chains are also necessary to elucidate the more details of supercoil conformation. This subject is a future subject in our study.

ACKNOWLEDGMENTS

This work was partly supported by a Grant-in-Aid from the Ministry of Education, Science, and Culture of Japan (No. 09750990 and 09875245), and by grants from the Asahi Grass Foundation.

REFERENCES

1. W.W. Graessley, *Adv. Polym. Sci.*, **16**, 1 (1974).
2. P.-G. de Gennes, *Scaling Concepts in Polymer Physics*, Cornell University Press, Ithaca, New York, 1979.
3. S.P. Obukhov, M. Rubinstein and R.H. Colby, *Macromolecules*, **27**, 3191 (1994).
4. We treat here polymer networks whose constituent polymers have glass transition temperatures much lower than room temperature. Furthermore, the systems are supposed to have no strong inter-and intramolecular forces such as hydrogen bonding.
5. V.A. Haug, G. Meyerhoff, *Makromol. Chem.*, **53**, 91 (1962).
6. R.M. Johnson, J.E. Mark, *Macromolecules*, **5**, 41 (1972).
7. C.S.M. Ong, R.S. Stein, *J. Polym. Sci. Polym. Phys. Ed.*, **12**, 1599 (1974).
8. V.G. Vasiliev, L.Z. Rogovina, G.L. Slonimsky, *Polymer*, **26**, 1667 (1985).
9. J. Bastide, in *Physics of Finely Divided Matter*, Springer, Berlin, 1985.
10. K. Urayama, S. Kohjiya, Y. Ikeda, *J. Soc. Rubber Ind., Jpn.*, **68**, 814 (1995).
11. K. Urayama, S. Kohjiya, *Polymer*, **38**. 955 (1997).
12. S. Kohjiya, K. Urayama, Y. Ikeda, *Kautsch. Gum. Kunst.*, **50**, 870 (1997).
13. K. Urayama, S. Kohjiya, *Eur. Phys. J. B.*, **2**, 79 (1998).
14. K. Urayama, S. Kohjiya, *J. Chem. Phys.*, **104**, 3352 (1996).
15. K. Urayama, T. Kawamura, S. Kohjiya, *J. Chem. Phys.*, **105**, 4833 (1996).
16. L.E. Nielsen, *Mechanical Properties of Polymers and Composites*, Marcel Dekker, New York, 1975.
17. L.R.G. Treloar, *The Physics of Rubber Elasticity*, (3rd edn.), Clarendon Press, Oxford, 1975.
18. P. Pincus, *Macromolecules*, **9**, 386 (1976).
19. S.J. Daoudi, *J. Phys. France*, **38**, 1301 (1977).
20. A.Y. Grosberg, S.K. Nechaev, *Adv. Polym. Sci.*, **106** , 1 (1993).
21. S.J. Clarson, J.E. Mark, K. Dodgson, *Polym. Commun.*, **29**, 208 (1988).
22. S.J. Clarson, K. Dodgson, J.A. Semlyen, *Polymer*, **26**, 930 (1985).
23. B. Wang, S. Krause, *Macromolecules*, **20**, 2201 (1987).

Index